Lecture Notes in Physics

W0245932

Springer-Verlag Berlin Heidelberg GmbH

The Editorial Policy for Proceedings

The series Lecture Notes in Physics reports new developments in physical research and teaching – quickly, informally, and at a high level. The proceedings to be considered for publication in this series should be limited to only a few areas of research, and these should be closely related to each other. The contributions should be of a high standard and should avoid lengthy redraftings of papers already published or about to be published elsewhere. As a whole, the proceedings should aim for a balanced presentation of the theme of the conference including a description of the techniques used and enough motivation for a broad readership. It should not be assumed that the published proceedings must reflect the conference in its entirety. (A listing or abstracts of papers presented at the meeting but not included in the proceedings could be added as an appendix.)

When applying for publication in the series Lecture Notes in Physics the volume's editor(s) should submit sufficient material to enable the series editors and their referees to make a fairly accurate evaluation (e.g. a complete list of speakers and titles of papers to be presented and abstracts). If, based on this information, the proceedings are (tentatively) accepted, the volume's editor(s), whose name(s) will appear on the title pages, should select the papers suitable for publication and have them refereed (as for a journal) when appropriate. As a rule discussions will not be accepted. The series editors and Springer-Verlag will normally not interfere with the detailed editing except in fairly obvious cases or on technical matters.

Final acceptance is expressed by the series editor in charge, in consultation with Springer-Verlag only after receiving the complete manuscript. It might help to send a copy of the authors' manuscripts in advance to the editor in charge to discuss possible revisions with him. As a general rule, the series editor will confirm his tentative acceptance if the final manuscript corresponds to the original concept discussed, if the quality of the contribution meets the requirements of the series, and if the final size of the manuscript does not greatly exceed the number of pages originally agreed upon. The manuscript should be forwarded to Springer-Verlag shortly after the meeting. In cases of extreme delay (more than six months after the conference) the series editors will check once more the timeliness of the papers. Therefore, the volume's editor(s) should establish strict deadlines, or collect the articles during the conference and have them revised on the spot. If a delay is unavoidable, one should encourage the authors to update their contributions if appropriate. The editors of proceedings are strongly advised to inform contributors about these points at an early stage.

The final manuscript should contain a table of contents and an informative introduction accessible also to readers not particularly familiar with the topic of the conference. The contributions should be in English. The volume's editor(s) should check the contributions for the correct use of language. At Springer-Verlag only the prefaces will be checked by a copy-editor for language and style. Grave linguistic or technical shortcomings may lead to the rejection of contributions by the series editors. A conference report should not exceed a total of 500 pages. Keeping the size within this bound should be achieved by a stricter selection of articles and not by imposing an upper limit to the length of the individual papers. Editors receive jointly 30 complimentary copies of their book. They are entitled to purchase further copies of their book at a reduced rate. As a rule no reprints of individual contributions can be supplied. No royalty is paid on Lecture Notes in Physics volumes. Commitment to publish is made by letter of interest rather than by signing a formal contract. Springer-Verlag secures the copyright for each volume.

The Production Process

The books are hardbound, and the publisher will select quality paper appropriate to the needs of the author(s). Publication time is about ten weeks. More than twenty years of experience guarantee authors the best possible service. To reach the goal of rapid publication at a low price the technique of photographic reproduction from a camera-ready manuscript was chosen. This process shifts the main responsibility for the technical quality considerably from the publisher to the authors. We therefore urge all authors and editors of proceedings to observe very carefully the essentials for the preparation of camera-ready manuscripts, which we will supply on request. This applies especially to the quality of figures and halftones submitted for publication. In addition, it might be useful to look at some of the volumes already published. As a special service, we offer free of charge LaTeX and TeX macro packages to format the text according to Springer-Verlag's quality requirements. We strongly recommend that you make use of this offer, since the result will be a book of considerably improved technical quality. To avoid mistakes and time-consuming correspondence during the production period the conference editors should request special instructions from the publisher well before the beginning of the conference. Manuscripts not meeting the technical standard of the series will have to be returned for improvement.

For further information please contact Springer-Verlag, Physics Editorial Department II, Tiergartenstrasse 17, D-69121 Heidelberg, Germany

Henrik Aratyn   Tom D. Imbo
Wai-Yee Keung   Uday Sukhatme (Eds.)

Supersymmetry and Integrable Models

Proceedings of a Workshop
Held at Chicago, IL, USA, 12–14 June 1997

 Springer

Editors

Henrik Aratyn
Tom D. Imbo
Wai-Yee Keung
Uday Sukhatme
Department of Physics
The University of Chicago
845 West Taylor Street
Chicago, IL 60607-7059, USA

Cataloging-in-Publication Data applied for.

Die Deutsche Bibliothek - CIP-Einheitsaufnahme

Supersymmetry and integrable models : proceedings of a workshop
held at Chicago, IL, USA, 12 - 14 June 1997 / Henrik Aratyn ...
(ed.).

(Lecture notes in physics ; 502)
ISBN 978-3-662-14188-5   ISBN 978-3-540-69679-7 (eBook)
DOI 10.1007/978-3-540-69679-7

ISSN 0075-8450
ISBN 978-3-662-14188-5

Typesetting: Camera-ready by the authors/editors
Cover design: *design & production* GmbH, Heidelberg
SPIN: 10644042   55/3144-543210 - Printed on acid-free paper

Preface

This volume contains the proceedings of the workshop entitled "Supersymmetry and Integrable Models" held during June 12–14, 1997 at the University of Illinois at Chicago (UIC). This was the second workshop in a series on current problems in fundamental physics organized by the High Energy Group of the Physics Department at UIC. In these workshops we bring together a number of active researchers from areas of current interest. The topics alternate every year between formal theory and phenomenology/experiment.

This year's workshop highlighted the relationship between supersymmetric quantum mechanics and certain integrable systems. The main theme of the workshop has been interpreted rather broadly. The contributions could be grouped into three central (and overlapping) categories: integrable models, the supersymmetric generalizations of integrable models, and related developments in supersymmetric quantum mechanics.

The group of contributions devoted to recent progress in integrable models includes talks by Aratyn, Awata, Carroll, Dickey, Harnad, Kasman, Mulase, Sánchez Guillén, van de Leur, Zachos and Zimerman. The topics mentioned in these talks include: hierarchies of nonlinear evolution equations (such as the KdV, KP and constrained KP hierarchies), solitons, matrix and membrane models, tau-function and vertex methods, isomonodromic deformations, symmetries of integrable models and more.

Another group of papers deals with a formulation of the supersymmetric integrable models. The authors contributing to this group (Das, Gallot, Legare, Mañas, Toppan) use a variety of algebraic and r-matrix methods to formulate and classify the supersymmetric hierarchies and find the corresponding soliton solutions. The presence of many approaches to the problem and lack of one dominating classification scheme remains an open challenge to researchers in this new field.

Supersymmetric quantum mechanics and its recent applications were discussed in talks by Beacom, Comtet, Dunne, Gangopadhyaya, Panigraghi, Russell and Sukhatme. Among the wide-ranging applications were connections with statistical physics, semiclassical approximations, periodic potentials and deterministic chaos.

Both physical and mathematical perspectives were represented and the workshop provided a forum for the interaction between mathematicians and physicists working on related subjects. The talks summarized the current state of the art and presented new results.

The amount of literature on integrable models is overwhelming, and scattered in many journals. It was our aim to produce the proceedings in such a way that it would allow researchers and graduate students to get acquainted rapidly with this constantly growing field. For this reason, we asked authors to include in their papers pedagogical reviews of recent developments in their subject. A majority of papers also present new results which have not appeared previously in any professional journal. As a result of these efforts, we feel that these proceedings go beyond being a compilation of talks, but in fact provide a survey of the rapidly changing field described from different and complementary points of view. We hope that the publication of this volume will contribute to the progress of these constantly developing areas.

Finally, we thank all the speakers, participants and members of the Physics Department at UIC whose efforts made this workshop so successful and enjoyable. Special thanks go to Ms. Ilga Dukats, Ms. Sharon Mistele and Ms. Roslyn Pitts for their help with providing participants with accomodations and other necessities. The editors wish to thank the US Department of Energy (DOE) for partial financial support.

Chicago, November 1997

Henrik Aratyn
Tom Imbo
W.-Y. Keung
Uday Sukhatme

Contents

Part III: Supersymmetric Quantum Mechanics

List of Participants

Aratyn, Henrik   `aratyn@uic.edu`
Department of Physics, University of Illinois at Chicago, 845 W. Taylor St., Chicago, IL 60607-7059, USA

Awata, Hidetoshi   `awata@rainbow.uchicago.edu`
Enrico Fermi Institute and James Frank Institute, University of Chicago, 5640 S. Ellis, Chicago, IL 60637, USA

Beacom, John F.   `beacom@nucth.physics.wisc.edu`
Department of Physics, University of Wisconsin, Madison, WI 53706, USA

Bergvelt, Maarten J.   `bergv@math.uiuc.edu`
University of Illinois at Champaign-Urbana Department of Mathematics, 1409 W. Green St., Urbana, IL 61801, USA

Carroll, Robert W.   `rcarroll@math.uiuc.edu`
University of Illinois at Champaign-Urbana Department of Mathematics, 1409 W. Green St., Urbana, IL 61801, USA

Comtet, Alain   `comtet@ipncls.in2p3.fr`
Division de Physique theorique,IPN,Batiment 100, Universite Paris Sud, 91400 Orsay, France

Curtright, Thomas L.   `curtright@phyvax.ir.miami.edu`
University of Miami, P.O. Box 248046, Coral Gables, FL 33124, USA

Das, Ashok   `das@urhep.pas.rochester.edu`
University of Rochester, Physics Department, Rochester, NY 14627, USA

Dickey, Leonid   `ldickey@ou.edu`
University of Oklahoma, Department of Mathematics, Norman, OK 73019, USA

Dunne, Gerald   `dunne@hep.phys.uconn.edu`
University of Connecticut, Physics Department, U-46, Storrs, CT 06269, USA

Fendley, Paul   `fendley@mmm.lanl.gov`
Los Alamos National Laboratory, Theoretical Division T-8, MS B285, Los Alamos, NM 87544, USA

Gallot, Laurent   `lgallot@enslapp.ens-lyon.fr`
ENSLAPP-Lyon group, 46 #allee d'Italie Lyon, 69007, France

Gangopadhyaya, Asim   agangop@luc.edu
    Department of Physics, Layola University, Chicago, IL 60626, USA

Ghosh, Sasanka   sasanka@iitg.ernet.in
    Physics Department, Indian Institute of Technology, Guwahati, Pan-
    bazar, Guwahati - 781 001, India

Imbo, Tom   imbo@uic.edu
    Department of Physics, University of Illinois at Chicago, 845 W. Taylor
    St., Chicago, IL 60607-7059, USA

Harnad, John   harnad@crm.umontreal.ca
    Centre de recherches mathematiques, Universite de Montreal, C.P. 6128,
    succursale Centre-Ville, Montréal, Québec, H3C 3J7, Canada

Kasman, Alex   kasman@crm.umontreal.ca
    Centre de recherches mathematiques, Universite de Montreal, C.P. 6128,
    succursale Centre-Ville, Montréal, Québec, H3C 3J7, Canada

Keung, W.-Y.   keung@uic.edu
    Department of Physics, University of Illinois at Chicago, 845 W. Taylor
    St., Chicago, IL 60607-7059, USA

Kilmurray, Donough   dkilm@math.uiuc.edu
    University of Illinois at Urbana-Champaign, 273 Altgeld Hall, 1409 West
    Green St. Urbana, IL 61801, USA

Legaré, Martin   mlegare@math.ualberta.ca
    Department of Mathematical Sciences, University of Alberta, Edmonton,
    Alberta, T6G 2G1, Canada

Mañas, Manuel   manuel@dromos.fis.ucm.es
    Departamento de Física Teórica, Universidad Complutense,
    E28040-Madrid, Spain

Mulase, Motohico   mulase@math.ucdavis.edu
    Department of Mathematics, UC Davis, Davis, CA 95616, USA

Pagnamenta, Antonio   pag@uic.edu
    Department of Physics, University of Illinois at Chicago, 845 W. Taylor
    St., Chicago, IL 60607-7059, USA

Panigrahi, Prasanta K   panisp@uohyd.ernet.in
    School of Physics, University of Hyderabad, Hyderabad, 500 046 AP,
    India

Rasinariu, Constantin   costel@uic.edu
    Department of Physics, University of Illinois at Chicago, 845 W. Taylor
    St., Chicago, IL 60607-7059, USA

Rosu, Haret   rosu@ifug.ugto.mx
    Instituto de Física de la Universidad de Guanajuato, Apdo Postal E-143,
    León, Gto, México

Russell, Neil   nerussel@indiana.edu
    Indiana University, Physics Department, Indiana University, Blooming-
    ton, IN 47405, USA

Sánchez Guillén, Joaquin   joaquin@gaes.usc.es
    University of Santiago, Department of Particle Physics, Facultade de
    Fisica 15706 Santiago de Compostela, Spain
Sukhatme, Uday   sukhatme@uic.edu
    Department of Physics, University of Illinois at Chicago, 845 W. Taylor
    St., Chicago, IL 60607-7059, USA
Takebe, Takashi   takebe@math.berkeley.edu
    Department of Mathematics, University of California at Berkeley, Berke-
    ley, CA 94720, USA
Toppan, Francesco   sptoppa@ipcs.shizuoka.ac.jp
    Shizuoka University, Department of Physics, Ohya 836, Shizuoka, Japan
van de Leur, Johan W.   vdleur@math.utwente.nl
    Faculty of Applied Mathematics, University of Twente, P.O. Box 217,
    7500 AE Enschede, The Netherlands
Zachos, C.K.   zachos@hep.anl.gov
    Argonne National Laboratory, HEP 362, Argonne, IL 60439-4815, USA
Zimerman, A.H.   zimerman@axp.ift.unesp.br
    Instituto de Física Teórica - IFT/UNESP, Rua Pamplona 145, 01405-900,
    São Paulo - SP, Brazil

Part I

Integrable Models

Integrable Models

Constrained KP Hierarchy as a Ratio of Differential Operators

Henrik Aratyn

Department of Physics, University of Illinois at Chicago, 845 W. Taylor St., Chicago, IL 60607-7059
e-mail: aratyn@uic.edu

Abstract. In this article, we prove an equivalence between two different approaches to the constrained KP (cKP) hierarchy. One is based on the reduction process from the complete KP hierarchy involving the eigenfunctions of the original KP Lax operator. The other represents the cKP Lax operators as a ratio of differential operators. The elementary proof of equivalence requires only some basic notions of the ordinary differential operator calculus. Relation to the squared eigenfunction potential is briefly discussed.

1 The Standard Eigenfunction Construction of the Constrained KP Hierarchy

We begin with the Sato theory of the KP hierarchy. Let $\{t_j\}$ denote a set of independent variables with $t_1 \equiv x$. The formulation of the KP hierarchy is based on the Lax equations

$$\frac{\partial Q}{\partial t_k} = \left[(Q^k)_+ , Q \right] \quad ; \quad k = 1, 2, \ldots \tag{1}$$

describing isospectral deformations of the pseudo-differential operator:

$$Q = D + \sum_{i=0}^{\infty} u_i(t_1, t_2, \ldots) D^{-i-1} \tag{2}$$

Here u_n are the functions of $\{t_j\}$, and $(Q^n)_+$ is the truncation to the differential part of Q^n. Moreover we denote by D the differential operator $\partial/\partial x$ which acts, according to the Leibniz rule, as an operator on functions to the right of it. The nonlinear evolution equation for the single KP function u_0 (obtained by process of elimination of higher coefficients) will involve $2 + 1$ time flows from (2). Correspondingly, the KP soliton equations are $2 + 1$-dimensional soliton equations.

The constrained KP hierarchy is most conveniently obtained from the KP hierarchy by a process of reduction which involves the so-called eigenfunctions of the KP Lax operator. The eigenfunctions, which we are about to introduce, appear in the constraint relations and introduce a functional dependence between initially infinitely many coefficients of the KP Lax operator.

Correspondingly, only finite number of coefficients appear in the evolution equations of the reduced hierarchy and equations are $1 + 1$-dimensional.

Function Φ (Ψ) is called <u>eigenfunction</u> (respectively <u>adjoint eigenfunction</u>) of the Lax operator Q, satisfying Sato's flow equation (1), if its flows are given by expression:

$$\frac{\partial \Phi}{\partial t_k} = (Q^n)_+ \Phi \quad ; \quad \frac{\partial \Psi}{\partial t_k} = -(Q^n)^*_+ \Psi \tag{3}$$

for the infinite many times t_k. Note that the operator Q^* is obtained from Q via process of conjugation, defined by simple rules $D^* = -D$ and $(AB)^* = B^* A^*$. We now impose the constraint on the KP hierarchy by requiring that the purely pseudo-differential part $(Q^r)_- \equiv Q^r - (Q^r)_+$ of the Lax operator Q^r satisfies a condition:

$$(Q^r)_- = \sum_{a=1}^{m} \Phi_a D^{-1} \Psi_a \tag{4}$$

for r being a fixed positive integer. The constraint (4) generalizes the condition $(Q^r)_- = 0$ which defines the r-th Gelfand-Dickey hierarchy in terms of the purely differential operators Q^r.

It follows that the KP flows defined in (1) are invariant under reduction given by the definition (4). To see it, we first observe that the time evolution of the pseudo-differential operator $(Q^r)_- = \sum_{a=1}^{m} \Phi_a D^{-1} \Psi_a$ is given by:

$$\frac{\partial}{\partial t_k} \sum_{a=1}^{m} \Phi_a D^{-1} \Psi_a = \left[(Q^r)_+ , \sum_{a=1}^{m} \Phi_a D^{-1} \Psi_a \right]_- \tag{5}$$

The proof is a consequence of the following technical identity

$$[K , f D^{-1} g]_- = K(f) D^{-1} g - f D^{-1} K^*(g) \tag{6}$$

which holds for a purely differential operator K and arbitrary functions f, g. We also used relations (3) and $[(Q^r)_+ , \mathcal{L}]_- = [(Q^r)_+ , \mathcal{L}_-]_-$. We now see that if we define the Lax operator $\mathcal{L}$ as $\mathcal{L} \equiv Q^r$ then $\mathcal{L}_-$ automatically satisfies the KP flow equations: $\partial \mathcal{L}_- / \partial t_k = [(\mathcal{L}^{k/r})_+ , \mathcal{L}]_-$.

Furthermore, since

$$\left[\left(\mathcal{L}^{(k/r)} \right)_+ , \mathcal{L} \right] = - \left[\sum_{a=1}^{m} \Phi_a D^{-1} \Psi_a , \mathcal{L} \right]$$

we find that the order of the differential operator

$$\left[\left(\mathcal{L}^{(k/r)} \right)_+ , \mathcal{L} \right]_+ = - \left[\sum_{a=1}^{m} \Phi_a D^{-1} \Psi_a , \mathcal{L}_+ \right]_+$$

must be smaller than r. Hence the manifold of the Lax operators defined in (4) is invariant under the KP hierarchy flows (1).

An alternative construction (Dickey 1995, Krichever 1995) introduces the Lax operator of the constrained KP (cKP) hierarchy as a ratio

$$L_{r,m} \equiv \frac{L_{m+r}}{L_m} \qquad 1 \leq r,\, 0 \leq m \tag{7}$$

of two purely differential operators of orders $m + r$ and m, respectively. In the next sections we will establish a complete equivalence of this formulation to the one based on eigenfunctions. We will only need few basic formulae related to the ordinary differential operators. They will be introduced in the following section.

2   Preliminaries on Differential Operators

Let L_m be a differential operator:

$$L_m = D^m + u_{m-1}D^{m-1} + \ldots + u_1 D^1 + u_0 \tag{8}$$

of order m. The differential operator of order m is called *normalized* if its leading term is D^m.

Let $\{\phi_1, \ldots, \phi_m\}$ be a basis of $\mathrm{Ker}\, L_m$ so that each ϕ_i is a solution of $L_m(\phi_i) = 0$ for $i = 1, \ldots, m$. For the given basis $\{\phi_1, \ldots, \phi_m\}$ we factorize L_m in the first order factors:

$$L_m = (D + v_m)(D + v_{m-1}) \cdots (D + v_1) = T_m \cdots T_1 \tag{9}$$

$$T_j \equiv V_j D V_j^{-1} \; ; \quad V_j \equiv e^{-\int v_j} \; ; \quad v_j = \partial_x \ln \frac{W_{j-1}[\phi_1, \ldots, \phi_{j-1}]}{W_j[\phi_1, \ldots, \phi_j]} \tag{10}$$

where

$$W_k[\phi_1, \ldots, \phi_k] = \det(\mathcal{W}) \; ; \quad (\mathcal{W})_{1 \leq i,j \leq k} = \partial_x^{i-1}\phi_j \tag{11}$$

denotes the Wronskian determinant of $\{\phi_1, \ldots, \phi_j\}$. The functions V_j's and ϕ_j's are related to each other through

$$V_i = \frac{W_i[\phi_1, \ldots, \phi_i]}{W_{i-1}[\phi_1, \ldots, \phi_{i-1}]} \; ; \quad i = 1, \ldots, m \tag{12}$$

or, alternatively, in terms of operators T_j's acting on the functions ϕ_i as

$$V_i = T_{i-1} \cdots T_1(\phi_i) \quad . \tag{13}$$

Connection between (12) and (13) follows from the following composition formula for Wronskians (Ince, Chap. 5, 1926),(Crum 1955):

$$T_k T_{k-1} \cdots T_1(f) = \frac{W_k(f)}{W_k} \tag{14}$$

where

$$T_j = \frac{W_j}{W_{j-1}} D \frac{W_{j-1}}{W_j} = \left(D + \left(\ln \frac{W_{j-1}}{W_j} \right)' \right) \quad ; \quad W_0 = 1 \quad (15)$$

$$W_k \equiv W_k[\phi_1, \ldots, \phi_k] \quad , \quad W_{k-1}(f) \equiv W_k[\phi_1, \ldots, \phi_{k-1}, f] \quad (16)$$

Define now an adjoint operator L_m^* as

$$L_m^* \equiv T_1^* \cdots T_m^* = (-1)^m V_1^{-1} D V_1 \cdots V_m^{-1} D V_m \quad (17)$$

It is easy to find the basis for $\mathrm{Ker}\, L_m^*$:

$$\mathrm{Ker}\, L_m^* = \{\psi_1, \ldots, \psi_m\} \quad \text{such that} \quad V_i^{-1} = T_{i+1}^* \cdots T_m^*(\psi_i) \quad (18)$$

Using the identity

$$\partial \left(\frac{W_{k-1}(f)}{W_k} \right) = \frac{W_k(f) W_{k-1}}{W_k^2} \quad (19)$$

written in notation of (16) one finds that the relations (18) give rise to the following Wronskian expressions for ψ_i's:

$$\psi_i = (-1)^{m-i} \frac{W_{m-1}[\phi_1, \ldots, \widehat{\phi_i}, \ldots \phi_m]}{W_m[\phi_1, \ldots, \phi_m]} \quad ; \quad i = 1, \ldots, m \quad (20)$$

where caret above ϕ_i means that this symbol is to be omitted.

Relation (20) expresses the fact that $(\psi_1, \ldots, \psi_m)^T$ is the last column in the inverse $\mathcal{W}^{-1}$ of the Wronskian matrix $\mathcal{W}$ of $(\phi_1, \ldots, \phi_m)$. It appears that $\{\psi_1, \ldots, \psi_m\}$ can be understood as the dual basis of $\{\phi_1, \ldots, \phi_m\}$ with respect to a canonical pairing defined in terms of the bilinear concomitant (Ince, Chap. 5, 1926, Wilson 1990, Wilson 1992). The latter is given by

$$\langle \phi | \psi \rangle_{L_m} \equiv \sum_{i=1}^{m} \sum_{j=0}^{i-1} (-1)^j \phi^{(i-j-1)} (u_i \psi)^{(j)} \quad (21)$$

with $u_m = 1$. Alternatively, we can write

$$\langle \phi | \psi \rangle_{L_m} = \mathrm{Res}\left(D^{-1} \psi L_m \phi D^{-1} \right) \quad (22)$$

as can easily be verified directly from the definition (21) using the basic properties of the residue Res, which here denotes projection on the coefficient of D^{-1}.

The fundamental property of the bilinear concomitant:

$$\partial \langle \phi | \psi \rangle_{L_m} = L_m(\phi)\, \psi - \phi\, L_m^*(\psi) \quad (23)$$

ensures that $\langle \cdot | \cdot \rangle_{L_m}$ indeed defines a canonical pairing $\mathrm{Ker}\, L_m \times \mathrm{Ker}\, L_m^* \to \mathbb{C}$. As discussed in (Wilson 1990, Wilson 1992) this pairing is nonsingular. Furthermore, in this setting the bases $\{\phi_1, \ldots, \phi_m\}$ and $\{\psi_1, \ldots, \psi_m\}$ related through (20) appear dual to each other in the following sense

$$\langle \phi_i | \psi_j \rangle_{L_m} = \delta_{ij} \quad ; \quad i, j = 1, \ldots, m \tag{24}$$

In view of duality between $\{\phi_1, \ldots, \phi_m\}$ and $\{\psi_1, \ldots, \psi_m\}$ it is not surprising that the relation (20) can be inverted providing an expression for the basis elements ϕ_i in terms of ψ_i:

$$\phi_i = (-1)^{m-i} \frac{W_{m-1}[\psi_m, \ldots, \widehat{\psi_i}, \ldots \psi_1]}{W_m[\psi_m, \ldots, \psi_1]} \quad ; \quad i = 1, \ldots, m \tag{25}$$

The basis $\{\phi_1, \ldots, \phi_m\}$ provides an alternative expression of L_m

$$L_m = \frac{1}{W_m[\phi_1, \ldots, \phi_m]} \begin{vmatrix} \phi_1 & \cdots & \phi_m & 1 \\ \vdots & \cdots & \vdots & \vdots \\ \phi_1^{(m)} & \cdots & \phi_m^{(m)} & D^m \end{vmatrix} \tag{26}$$

Similarly, we have

$$L_m^* = \frac{1}{W_m[\psi_1, \ldots, \psi_m]} \begin{vmatrix} \psi_1 & \cdots & \psi_m & 1 \\ \vdots & \cdots & \vdots & \vdots \\ \psi_1^{(m)} & \cdots & \psi_m^{(m)} & D^m \end{vmatrix} \tag{27}$$

Consider now the pseudo-differential operator $L_m^{-1} = T_1^{-1} \cdots T_m^{-1}$. It turns out that the dual bases $\{\phi_1, \ldots, \phi_m\}$ and $\{\psi_1, \ldots, \psi_m\}$ described above provide a canonical set of coordinates in which L_m^{-1} takes the simple form (Oevel & Strampp 1996) :

$$L_m^{-1} = \sum_{i=1}^m \phi_i D^{-1} \psi_i \tag{28}$$

This statement can be proven by induction. Let us assume that indeed the right hand side of (28) is equal to $T_1^{-1} \cdots T_{m-1}^{-1}$. Applying T_m^{-1} from the left on both sides of equation (28) we get

$$L_m^{-1} = \sum_{i=1}^{m-1} \phi_i D^{-1} V_m^{-1} \int^x (\psi_i V_m) + \sum_{i=1}^{m-1} \phi_i \int^x (\psi_i V_m) D^{-1} V_m^{-1} \tag{29}$$

It is clear that the functions $\phi_i, i = 1, .., m-1$ are also in $\mathrm{Ker}\, L_m$. The remaining element of the kernel is given by $\phi_m = \sum_{i=1}^{m-1} \phi_i \int^x (\psi_i V_m)$ since due to the induction assumption $\sum_{i=1}^{m-1} \phi_i \int^x (\psi_i V_m) = T_1^{-1} \cdots T_{m-1}^{-1}(V_m)$, which in turn is equal to ϕ_m according to (13). The result follows now from the fact that $V_m^{-1} = \psi_m$ and $V_m^{-1} \int^x (\psi_i V_m)$ are all in $\mathrm{Ker}\, L_m^*$ as seen from (18).

3   Construction of the Pseudo-differential Operators

Define two pseudo-differential operators:

$$\mathcal{L}_1 \equiv L_m^{-1} L_{m+r} = \sum_{i=1}^{m} \phi_i D^{-1} L_{m+r}^*(\psi_i) + (\mathcal{L}_1)_+ \tag{30}$$

$$\mathcal{L}_{m+1} \equiv L_{m+r} L_m^{-1} = \sum_{i=1}^{m} L_{m+r}(\phi_i) D^{-1} \psi_i + (\mathcal{L}_{m+1})_+ \tag{31}$$

as ratios of two normalized differential operators L_m, L_{m+r} of order m and $r + m$, respectively. In the above equations $(\mathcal{L})_+$ denotes the differential part of the pseudo-differential operator $\mathcal{L}$. In obtaining the right hand sides of (30) and (31) we used (28). We will assume that L_m and L_{m+r} are co-prime, i.e., their kernels do not intersect. This ensures that none of the terms in the negative part of $\mathcal{L}_{m+1}$ in (31) are zero.

It has been shown (Krichever 1995, Dickey 1995) that the restriction of the KP hierarchy to the pseudo-differential operators of the form given in (31) is equivalent to imposing the following flows on the differential operators:

$$\partial_n L_m = \left(L_{m+r} L_m^{-1} \right)_+^{n/r} L_m - L_m \left(L_m^{-1} L_{m+r} \right)_+^{n/r} \tag{32}$$

$$\partial_n L_{m+r} = \left(L_{m+r} L_m^{-1} \right)_+^{n/r} L_{m+r} - L_{m+r} \left(L_m^{-1} L_{m+r} \right)_+^{n/r} \tag{33}$$

Indeed, it follows easily from (32) and (33) that

$$\partial_n \mathcal{L}_1 = \partial_n \left(L_m^{-1} L_{m+r} \right) = \left[(\mathcal{L}_1)_+^{n/r} , \mathcal{L}_1 \right] \tag{34}$$

$$\partial_n \mathcal{L}_{m+1} = \partial_n \left(L_{m+r} L_m^{-1} \right) = \left[(\mathcal{L}_{m+1})_+^{n/r} , \mathcal{L}_{m+1} \right] \tag{35}$$

On the other hand, one can also show (Krichever 1995) that (34) or (35) define uniquely $\partial_n L_m$ and $\partial_n L_{m+r}$ of orders $(m - 1)$ and $(m + r - 1)$, respectively. Equivalence follows now by observing that the right hand sides of equations (32) and (33) have the right orders.

Lemma 3.1 *Equations (32) and (33) imply that $\phi_i \in \mathrm{Ker}\, L_m$ and $\psi_i \in \mathrm{Ker}\, L_m^*$ are "up to a gauge rotation" (adjoint) eigenfunctions satisfying:*

$$\partial_n \phi_i = (\mathcal{L}_1)_+^{n/r} (\phi_i) \qquad i = 1, \ldots, m \tag{36}$$

$$\partial_n \psi_i = - \left(\mathcal{L}_{m+1}^* \right)_+^{n/r} (\psi_i) \qquad i = 1, \ldots, m \tag{37}$$

Proof. From $\partial_n L_m(\phi_i) = 0$ and (32) we find that $(\mathcal{L}_1)_+^{n/r} (\phi_i) - \partial_n \phi_i \in \mathrm{Ker}\, L_m$. Hence we can write

$$(\mathcal{L}_1)_+^{n/r}(\phi_i) - \partial_n \phi_i = -\sum_{j=1}^{m} \phi_j \, c_{ji}^{(n)}(\tilde{t}) \qquad (38)$$

where $\tilde{t} = (t_2, t_3, \ldots)$. We now proceed using variation of the argument given, in a slightly different setting, in (Aratyn, Nissimov, Pacheva 1997a). Define $(\Delta_n)_{jk} \equiv \partial_n \delta_{jk} - c_{kj}^{(n)}$ so that we can compactly rewrite (38) as $(\Delta_n)_{jk} \phi_k = (\mathcal{L}_1)_+^{n/r}(\phi_j)$. The Zakharov-Shabat equations for $\mathcal{L}_1$, which follow from (34), ensure the zero curvature equation $([\Delta_n, \Delta_l])_{ik} \phi_k = 0$. Thus the "connection" $c_{ij}^{(n)}$ is a pure gauge and can be written in a form

$$c_{ij}^{(n)}(\tilde{t}) = (c^{-1})_{ik}(\tilde{t}) \, \partial_n \, c_{kj}(\tilde{t}) \qquad ; \quad n \geq 2 \qquad (39)$$

Define accordingly

$$\bar{\phi}_j \equiv \phi_k \, (c^{-1})_{kj} \qquad (40)$$

It is easy to verify that $\bar{\phi}_j$ is an eigenfunction of $\mathcal{L}_1$, i.e.

$$\partial_n \bar{\phi}_j = (\Delta_n \phi)_k \, (c^{-1})_{kj} = (\mathcal{L}_1)_+^{n/r}(\bar{\phi}_j) \qquad (41)$$

Similarly, from $\partial_n L_m^*(\psi_i) = 0$ we arrive at

$$(\mathcal{L}_{m+1}^*)_+^{n/r}(\psi_i) + \partial_n \psi_i = \sum_{j=1}^{m} \bar{c}_{ij}^{(n)}(\tilde{t}) \, \psi_j \qquad (42)$$

We will establish now a relation between coefficients $c_{ij}^{(n)}$ and $\bar{c}_{ij}^{(n)}$. We find

$$(\partial_n \mathcal{L}_{m+1})_- = \sum_{i=1}^{m} (\mathcal{L}_{m+1})_+^{n/r} (L_{m+r}(\phi_i)) \, D^{-1} \psi_i \qquad (43)$$

$$- \sum_{i=1}^{m} L_{m+r}(\phi_i) D^{-1} (\mathcal{L}_{m+1}^*)_+^{n/r}(\psi_i)$$

as follows from projecting (35) on the negative part and using the technical identity (6). Equations (38) and (42) agree with (43) provided

$$\sum_{i,j=1}^{m} \left(c_{ij}^{(n)} + \bar{c}_{ij}^{(n)} \right) L_{m+r}(\phi_i) D^{-1} \psi_j = 0 \qquad (44)$$

Define a differential operator of $m - 1$ order

$$K[\phi] \equiv \sum_{s=1}^{m} \sum_{l=0}^{s-1} u_l D^l(\phi)^{(s-l-1)} \qquad (45)$$

such that $K^*[\phi](\psi) = \langle \phi | \psi \rangle_{L_m}$. From (44) and (24) we find

$$\left(\sum_{i,j=1}^{m} \left(c_{ij}^{(n)} + \bar{c}_{ij}^{(n)} \right) L_{m+r}(\phi_i) D^{-1} \psi_j \, K[\phi_k] \right)_{-} = 0 \qquad (46)$$

or

$$\sum_{i=1}^{m} \left(c_{ik}^{(n)} + \bar{c}_{ik}^{(n)} \right) L_{m+r}(\phi_i) = 0 \qquad ; \quad k = 1, \dots, m \qquad (47)$$

Since L_{m+r} and L_m are co-prime and $\{\phi_i\}$ linearly independent we find from (47) that $c_{ij}^{(n)} = -\bar{c}_{ij}^{(n)}$ for all $i, j = 1, \dots, m$. Accordingly, $(\Delta_n^*)_{jk} \, \psi_k = - \left(\mathcal{L}_{m+1}^* \right)_+^{n/r} (\psi_j)$, with $(\Delta_n^*)_{jk} \equiv \partial_n \delta_{jk} + c_{jk}^{(n)}$. Define, next

$$\bar{\psi}_j \equiv c_{jk} \, \psi_k \qquad (48)$$

It follows that

$$\partial_n \bar{\psi}_j = c_{jk} \left(\Delta_n^* \psi \right)_k = - \left(\mathcal{L}_{m+1}^* \right)_+^{n/r} (\bar{\psi}_j) \qquad (49)$$

Hence we succeeded to find a mutually inverse gauge rotations taking $\phi_i \in \text{Ker} \, L_m$ and $\psi_i \in \text{Ker} \, L_m^*$ into (adjoint) eigenfunctions of $\mathcal{L}_1$ and $\mathcal{L}_{m+1}$, respectively. $\square$
Since $\sum_{i=1}^{m} \phi_i D^{-1} \psi_i = \sum_{i=1}^{m} \bar{\phi}_i D^{-1} \bar{\psi}_i$ we arrive at the following main result:

Corollary 3.1 *The Lax operators $\mathcal{L}_1, \mathcal{L}_{m+1}$ can be rewritten as*

$$\mathcal{L}_a = \sum_{i=1}^{m} \Phi_i^{(a)} D^{-1} \Psi_i^{(a)} + (\mathcal{L}_a)_+ \qquad a = 1, m+1 \qquad (50)$$

with $\Phi_i^{(a)}$ and $\Psi_i^{(a)}$ being the (adjoint) eigenfunctions of $\mathcal{L}_a$.

4   Darboux-Bäcklund Linkage Between $\mathcal{L}_1$ and $\mathcal{L}_{m+1}$

Let us now analyze the consequences of being able to write the negative part of $(\mathcal{L}_1)$ as $\sum_{i=1}^{m} \Phi_i D^{-1} \Psi_i$ with $\Phi_i = \phi_i$ and $\Psi_i = L_{m+r}^*(\psi_i)$ being the (adjoint) eigenfunctions of $\mathcal{L}_1$, i.e.

$$\partial_n \Phi_i = (\mathcal{L}_1)_+^{n/r} (\Phi_i) \quad ; \quad \partial_n \Psi_i = - (\mathcal{L}_1^*)_+^{n/r} (\Psi_i) \qquad i = 1, \dots, m \qquad (51)$$

We will define a chain of pseudo-differential operators:

$$\mathcal{L}_1 \to \mathcal{L}_2 = T_1 \mathcal{L}_1 T_1^{-1} \to \cdots \to \mathcal{L}_j = T_{j-1} \mathcal{L}_{j-1} T_{j-1}^{-1} \qquad (52)$$

constructed by successive Darboux-Bäcklund (DB) transformations generated by factors T_j, $j = , \dots, m$ from (10). The chain (52) stops for $j = m+1$ at

$$\mathcal{L}_{m+1} = \mathcal{L}_2 \mathcal{L}_1^{-1} = T_m \cdots T_1 \, \mathcal{L}_1 T_1^{-1} \cdots T_m^{-1} \qquad (53)$$

We need few basic facts about the DB transformations in the setting of the constrained KP hierarchy (Aratyn, Nissimov, Pacheva 1997a). Under an arbitrary DB transformation the Lax operator $L = L_+ + \sum_{i=1}^{m} \Phi_i D^{-1} \Psi_i$ of "degree" m (meaning a number of non-zero terms $\Phi_i D^{-1} \Psi_i$) transformes into $\tilde{L} = (\chi D \chi^{-1}) L (\chi D^{-1} \chi^{-1})$, where χ is an eigenfunction of L. The transformed Lax operator $\tilde{L}$ reads:

$$\tilde{L} = \chi D \chi^{-1} \left(L_+ + \sum_{i=1}^{m} \Phi_i D^{-1} \Psi_i \right) \chi D^{-1} \chi^{-1} \equiv \tilde{L}_+ + \tilde{L}_- \tag{54}$$

$$\tilde{L}_+ = L_+ + \chi \left(\partial_x \left(\chi^{-1} L_+ \chi \right)_{\geq 1} D^{-1} \right) \chi^{-1} \tag{55}$$

$$\tilde{L}_- = \tilde{\Phi}_0 D^{-1} \tilde{\Psi}_0 + \sum_{i=1}^{m} \tilde{\Phi}_i D^{-1} \tilde{\Psi}_i \tag{56}$$

$$\tilde{\Phi}_0 = \chi \left[\partial_x \left(\chi^{-1} L_+(\chi) \right) + \sum_{i=1}^{m} \left(\partial_x \left(\chi^{-1} \Phi_i \right) \partial_x^{-1} \left(\Psi_i \chi \right) + \Phi_i \Psi_i \right) \right]$$

$$\equiv \left(\chi D \chi^{-1} L \right) (\chi) \tag{57}$$

$$\tilde{\Psi}_0 = \chi^{-1} \quad , \quad \tilde{\Phi}_i = \chi \partial_x \left(\chi^{-1} \Phi_i \right) \quad , \quad \tilde{\Psi}_i = -\chi^{-1} \partial_x^{-1} \left(\Psi_i \chi \right) \tag{58}$$

In case when χ coincides with one of the original eigenfunctions of L, e.g., $\chi = \Phi_{i_0}$, it follows that $\tilde{\Phi}_{i_0} = 0$ and the DB transformation (54) preserves the "degree" of the Lax operator involved. The transformations used in connecting $\mathcal{L}_1$ and $\mathcal{L}_{M+1}$ in (53) do not increase the degrees of the Lax operators they connect. The first DB transformation in (52) is induced by T_1 and removes $\Phi_1 = \phi_1$ due to $T_1(\phi_1) = 0$. The remaining functions $\tilde{\Phi}_i$, $i = 2, \ldots, m$ are different from zero. Note, that $\tilde{\Phi}_2 = T_1(\phi_2) = V_2$ will be removed by the next DB transformation induced by T_2. Generally, after the k-th DB transformation from (52) the "lowest" remaining eigenfunction is V_{k+1} and since $T_{k+1}(V_{k+1}) = 0$ it is being removed by the next transformation.

In view of the above it is natural to inquiry about V_k being eigenfunctions of the Lax operators from (52). Indeed we have a following lemma (for $1 \leq j \leq m$) (Aratyn, Nissimov, Pacheva 1997a) addressing this issue:

Lemma 4.1 V_j is an eigenfunction of the Lax operator $\mathcal{L}_j$,i.e.

$$\frac{\partial V_j}{\partial t_n} = \partial_n V_j = \left(\mathcal{L}_j^{n/r} \right)_+ (V_j) \tag{59}$$

if and only if

$$\partial_n T_j = \left(\mathcal{L}_{j+1}^{n/r} \right)_+ T_j - T_j \left(\mathcal{L}_j^{n/r} \right)_+ \tag{60}$$

Proof. Another way to rewrite (60) is to cast it in the form of the gauge (Toda) system:

$$\left(\mathcal{L}_{j+1}^{n/r}\right)_+ = T_j \left(\mathcal{L}_j^{n/r}\right)_+ T_j^{-1} + (\partial_n T_j) T_j^{-1} \tag{61}$$

Compare now (61) with the following identity valid for an arbitrary pseudo-differential operator A and an arbitrary function χ:

$$\left(\chi D \chi^{-1} A \chi D^{-1} \chi^{-1}\right)_+ = \chi D \chi^{-1} (A)_+ \chi D^{-1} \chi^{-1}$$
$$- \chi \partial_x \left(\chi^{-1}((A)_+ \chi)\right) D^{-1} \chi^{-1} \tag{62}$$

After identifying $A = \mathcal{L}_j^{n/r}$ and $\chi = V_j$ and making use of an identity:

$$(\partial_n T_j) T_j^{-1} = -V_j \partial_x \left(V_j^{-1} \partial_n V_j\right) D^{-1} V_j \tag{63}$$

one finds that the equation (62) becomes identical to (61) provided V_j is an eigenfunction of L_j. $\square$

Lemma 4.2 *Let $\Phi_i = \phi_i$ be eigenfunctions of $\mathcal{L}_1$. Then the relations (59) hold for $V_j = W_j[\phi_1, \ldots, \phi_j]/W_{j-1}[\phi_1, \ldots, \phi_{j-1}]$ from (10) and (12).*

Proof. We prove by induction that the relations (59) hold for V_j from (12), provided we assume relations (51), i.e. that $\Phi_i = \phi_i$ are eigenfunctions of $\mathcal{L}_1$. We start with $j = 1$. Since in this case $V_1 = \phi_1$ due to (13), the result follows immediately. Assume now, that the result holds for $j-1$ and V_{j-1} is an eigenfunctions of $\mathcal{L}_{j-1}$. According to (4.1) the equation (60) holds (with $j - 1$ instead of j). It follows then using (60) that:

$$\partial_n V_j = \partial_n \left(T_{j-1} \cdots T_1(\phi_j)\right) = \left(\mathcal{L}_j^{n/r}\right)_+ (V_j) \tag{64}$$

Hence, both eqs.(59) and (60) are proven for all $j = 1, \ldots, m$. $\square$
As a consequence, we easily find from (60) :

$$\partial_n L_m = \partial_n \left(T_m \cdots T_1\right) = \left(\mathcal{L}_{m+1}^{n/r}\right)_+ L_m - L_m \left(\mathcal{L}_1^{n/r}\right)_+ \tag{65}$$

proving (32). Equation (65) implies that

$$\partial_n L_m^{-1} = \left(\mathcal{L}_1^{n/r}\right)_+ L_m^{-1} - L_m^{-1} \left(\mathcal{L}_{m+1}^{n/r}\right)_+ .$$

From the expression (28) for L_m^{-1}, we are able to reproduce the result (37):

$$\partial_n (\psi_i) = -\left(\mathcal{L}_{m+1}^{n/r}\right)_+^* (\psi_i) \qquad i = 1, \ldots, m \tag{66}$$

Comparing it with (51) for $\Psi_i = L_{m+r}^* (\psi_i)$ we find that

$$\partial_n L_{m+r}^* = L_{m+r}^* \left(\mathcal{L}_{m+1}^{n/r}\right)_+^* - \left(\mathcal{L}_1^{n/r}\right)_+^* L_{m+r}^* \quad \text{on} \quad \text{Ker } L_m^*.$$

Similarly, we conclude that

$$\partial_n L_{m+r} = \left(\mathcal{L}_{m+1}^{n/r}\right)_+ L_{m+r} - L_{m+r}\left(\mathcal{L}_1^{n/r}\right)_+ \quad \text{on} \quad \text{Ker } L_m.$$

Since $\partial_n L_{m+r}$ has the order $\leq m + r - 1$ we see that $\partial_n L_{m+r}$ is defined uniquely on Ker L_m for $r = 1$. To obtain (33) for $r > 1$, one has to make an additional assumption, e.g. about $\mathcal{L}_{m+1}$ belonging to the KP hierarchy.

The next result concerns commutativity of flows defined in (60).

Lemma 4.3 *The flows for the DB chain defined by (60) commute if and only if the flows defined by*

$$\partial_n \mathcal{L}_j = \left[\left(\mathcal{L}_j^{n/r}\right)_+, \mathcal{L}_j\right] \tag{67}$$

commute.

Proof. A proof follows from a simple calculation:

$$(\partial_i \partial_j - \partial_j \partial_i) T_k = \left[\partial_i \left(\mathcal{L}_{k+1}^{j/r}\right)_+ - \partial_j \left(\mathcal{L}_{k+1}^{i/r}\right)_+ + \left[\left(\mathcal{L}_{k+1}^{j/r}\right)_+, \left(\mathcal{L}_{k+1}^{i/r}\right)_+\right]\right] T_k$$

$$+ T_k \left[\partial_j \left(\mathcal{L}_{k+1}^{i/r}\right)_+ - \partial_i \left(\mathcal{L}_{k+1}^{j/r}\right)_+ + \left[\left(\mathcal{L}_{k+1}^{i/r}\right)_+, \left(\mathcal{L}_{k+1}^{j/r}\right)_+\right]\right]$$

□

Let us factorize L_{m+r} as

$$L_{m+r} \equiv (D + \bar{v}_{m+r})(D + \bar{v}_{m+r-1})\cdots(D + \bar{v}_1) = S_{m+r}S_{m+r-1}\cdots S_1$$
$$S_j \equiv U_j^{-1} D U_j \quad ; \quad U_j \equiv e^{\int \bar{v}_j} \tag{68}$$

and construct the DB chain of transformations:

$$\mathcal{L}_{m+2} \equiv S_{m+r}^{-1}\mathcal{L}_{m+1}S_{m+r} \to \cdots \to S_1^{-1}\mathcal{L}_{2m+r}S_1 = \mathcal{L}_{2m+r+1} = \mathcal{L}_1 \tag{69}$$

Let now τ_j be a tau-function for the Lax operator $\mathcal{L}_j$. From the fact that the tau-function transforms under DB transformations in (52) and (69) as $\tau_{j+1} = V_j \tau_j$ for $1 \leq j \leq m$ and $\tau_{j+1} = U_j^{-1}\tau_j$ for $m + 1 \leq j \leq 2m + r$ we conclude

$$v_j = \partial \ln \frac{\tau_{j+1}}{\tau_j}, \quad j = 1, \ldots, m \tag{70}$$

$$\bar{v}_j = -\partial \ln \frac{\tau_{j+1}}{\tau_j}, \quad , \tau_{2m+r+1} \equiv \tau_1, \quad j = m+1, \ldots, 2m+r \tag{71}$$

$$0 = \sum_{j=1}^{m} v_j - \sum_{j=m+1}^{2m+r} \bar{v}_j \tag{72}$$

Recall now relation (10) between v_j and Wronskians of $\{\phi_i\}$. Similarly we have a relation between $\bar{v}_j$ and Wronskians of $\{f_i\}$, $i = 1, \ldots, m+r$ where $f_i \in \text{Ker } L_{m+r}$

$$\bar{v}_j = \partial_x \ln \frac{W_{j-1}[f_1, \ldots, f_{j-1}]}{W_j[f_1, \ldots, f_j]} \qquad j = 1, \ldots, m+r \tag{73}$$

In this parametrization the consistency condition (72) translates into

$$W_m[\phi_1, \ldots, \phi_m] = W_{m+r}[f_1, \ldots, f_{m+r}] \tag{74}$$

Note also that from (70)

$$\tau_{m+1} = W_m[\phi_1, \ldots, \phi_m]\tau_1. \tag{75}$$

5   Connections to Other Formalisms

Consider $\mathrm{Res}\,(\mathcal{L}_{m+1}) = \sum_{i=1}^m \Phi_i \Psi_i$ with $\Phi_i = L_{m+r}(\phi_i)$, $\Psi_i = \psi_i$ and note that according to (23):

$$\sum_{i=1}^m \partial_n \left(\Phi_i \Psi_i\right) = \sum_{i=1}^m \left[\left(\mathcal{L}_{m+1}^{(n/r)}\right)_+ (\Phi_i)\Psi_i - \Phi_i \left(\mathcal{L}_{m+1}^{(n/r)}\right)_+^* (\Psi_i)\right]$$

$$= \sum_{i=1}^m \partial_x \langle \Phi_i | \, \Psi_i \rangle_{B_n} \quad ; \quad B_n \equiv \left(\mathcal{L}_{m+1}^{(n/r)}\right)_+ \tag{76}$$

Recall, that $\mathrm{Res}\,(\mathcal{L}_{m+1}) = \partial^2 \ln \tau_{m+1}/\partial x \partial t_r$. Hence, ignoring the integration constant, we can write

$$\frac{\partial^2 \ln \tau_{m+1}}{\partial t_n t_r} = \sum_{i=1}^m \langle \Phi_i | \, \Psi_i \rangle_{B_n} \tag{77}$$

In (Aratyn, Nissimov, Pacheva 1997b) the constrained KP hierarchy was discussed in terms of the so-called squared eigenfunction potential $S\,(\Phi_i, \Psi_i)$. The function $S\,(\Phi_i, \Psi_i)$, for Φ_i and Ψ_i as given above, is defined in such way that (in the notation of the present paper) (Oevel 1993) :

$$\partial_n S\,(\Phi_i, \Psi_i) = \langle \Phi_i | \, \Psi_i \rangle_{B_n} \tag{78}$$

Hence (77) is implied by

$$\frac{\partial \ln \tau_{m+1}}{\partial t_r} = \sum_{i=1}^m S\,(\Phi_i, \Psi_i) \tag{79}$$

This is the relation which as found, in a slightly different way, in (Aratyn, Nissimov, Pacheva 1997b) characterizes the constrained KP hierarchy.

Let Π and Ω be the dressing operators for $\mathcal{L}_1$ and $\mathcal{L}_{m+1}$, i.e.

$$\mathcal{L}_1 = \Pi D^r \Pi^{-1} \quad ; \quad \mathcal{L}_{m+1} = \Omega D^r \Omega^{-1} \tag{80}$$

Correspondingly, we find

$$L_m = \Omega D^m \Pi^{-1} \quad ; \quad L_{m+r} = \Omega D^r \Pi^{-1} \tag{81}$$

The Baker-Akhiezer eigenfunctions of $\mathcal{L}_1$ and $\mathcal{L}_{m+1}$ are given by

$$\psi_{BA}^{(1)}(t,\lambda) = \Pi e^{\xi(t,\lambda)} \quad ; \quad \psi_{BA}^{(m+1)}(t,\lambda) = \Omega e^{\xi(t,\lambda)} \tag{82}$$

where

$$\xi(t,\lambda) \equiv \sum_{n=1}^{\infty} t_n \lambda^n \quad ; \quad t_1 = x. \tag{83}$$

They satisfy the linear spectral equations : $\mathcal{L}_a \psi_{BA}^{(a)} = \lambda^r \psi_{BA}^{(a)}$ for $a = 1, m+1$. From (82) and (81) we find that the Baker-Akhiezer eigenfunctions are related via

$$L_m \psi_{BA}^{(1)}(t,\lambda) = \lambda^m \psi_{BA}^{(m+1)}(t,\lambda) \quad ; \quad L_{m+r} \psi_{BA}^{(1)}(t,\lambda) = \lambda^{m+r} \psi_{BA}^{(m+1)}(t,\lambda) \tag{84}$$

The above system was introduced in (Helminck & van de Leur 1997a) and (Helminck & van de Leur 1997b) within the framework of the Segal-Wilson formalism.

Using the inverse L_m^{-1} (28) and the squared eigenfunction potential (78) the first of equations in (84) can be rewritten as

$$\psi_{BA}^{(1)}(t,\lambda) = \lambda^m \sum_{i=1}^{m} \phi_i(t) S \left(\psi_{BA}^{(m+1)}(t,\lambda), \psi_i(t) \right) \tag{85}$$

Recall, from (Aratyn, Nissimov, Pacheva 1997b) that the adjoint eigenfunction ψ_i can be reproduced by

$$\psi_i(t) = \int d\lambda S \left(\psi_{BA}^{(m+1)}(t',\lambda), \psi_i(t') \right) \psi_{BA}^{(m+1)*}(t,\lambda) \tag{86}$$

Therefore from (86) and (85) we derive a relation

$$\sum_{i=1}^{m} \phi_i(t') \psi_i(t) = \int d\lambda\, \lambda^{-m}\, \psi_{BA}^{(1)}(t',\lambda)\, \psi_{BA}^{(m+1)*}(t,\lambda) \tag{87}$$

or

$$\sum_{i=1}^{m} \left(\partial_x^k \phi_i(t) \right) \psi_i(t) = \int d\lambda\, \lambda^{-m} \left(\partial_x^k \psi_{BA}^{(1)}(t,\lambda) \right) \psi_{BA}^{(m+1)*}(t,\lambda) \tag{88}$$

Since ψ_i from (20) are elements of the last column of $\mathcal{W}^{-1}$ (where $(\mathcal{W})_{1\leq i,j \leq m} = \partial_x^{i-1} \phi_j$) we have

$$\sum_{i=1}^{m} \phi_i^{(k)}(t) \psi_i(t) = \delta_{k,m-1} \quad ; \quad k = 0,1,\ldots,m-1 \tag{89}$$

Furthermore

$$\sum_{i=1}^{m} \phi_i^{(m)}(t)\psi_i(t) = \sum_{i=1}^{m}(-1)^{m-i}\frac{W_{m-1}[\phi_1,\ldots,\widehat{\phi}_i,\ldots\phi_m]\phi_i^{(m)}}{W_m[\phi_1,\ldots,\phi_m]}$$

$$= \partial_x \ln W_m[\phi_1,\ldots,\phi_m] \tag{90}$$

The defining relations for the Baker-Akhiezer eiegnfunctions in terms of the τ-function are:

$$\psi_{BA}(t,\lambda) = e^{\xi(t,\lambda)}\sum_{n=0}^{\infty}\frac{p_n\left(-[\partial]\right)\tau(t)}{\tau(t)}\lambda^{-n} \tag{91}$$

$$\psi_{BA}^*(t,\lambda) = e^{-\xi(t,\lambda)}\sum_{n=0}^{\infty}\frac{p_n\left([\partial]\right)\tau(t)}{\tau(t)}\lambda^{-n} \tag{92}$$

where $[\partial] \equiv (\partial/\partial t_1, \partial/2\partial t_2, \partial/3\partial t_3, \ldots)$ and the Schur polynomials $p_n(t)$ are defined through: $\exp\left(\sum_{l\geq 1}\lambda^l t_l\right) = \sum_{n=0}^{\infty}\lambda^n p_n(t_1,t_2,\ldots)$. Now, it is easy to see directly from the contour integral that indeed

$$\int d\lambda\lambda^{-m}\partial_x^k\psi_{BA}^{(1)}(t,\lambda)\psi_{BA}^{(m+1)*}(t,\lambda) = \delta_{k,m-1} \quad \text{for} \quad k < m$$

as also follows from (87) and (89). Furthermore, we find

$$\int d\lambda\lambda^{-m}\partial_x^m\psi_{BA}^{(1)}(t,\lambda)\psi_{BA}^{(m+1)*}(t,\lambda) = \partial_x\ln(\tau_{m+1}/\tau_1) \tag{93}$$

comparing with (90) we see that at this level (87) is just reproducing the relation (75) between τ_{m+1} and τ_1. For the higher moments of (87) we will get equations for τ_1 (or τ_{m+1}) which supplement the usual Hirota equations of the KP hierarchy. Acting with L_{m+r} on boths sides of (87) we get

$$\sum_{i=1}^{m}\Phi_i(t')\Psi_i(t) = \int d\lambda\lambda^r\psi_{BA}^{(m+1)}(t',\lambda)\psi_{BA}^{(m+1)*}(t,\lambda) \tag{94}$$

see (Cheng & Zhang 1994). Some comments are in order concerning relation (79) versus (87) (or alternatively (94)). The linear (in the tau function) relation (79) implies the bilinear relations (87) and (94). This single relation characterizes the constrained KP hierarchy in addition to the usual Hirota equation for the complete KP hierarchy. It interesting to observe how the bilinear version of this constraint in the form of (87) involves alone the Wronskians of the elements of the kernel of L_m.

Acknowledgments: I am especially grateful to Johan van de Leur and Gerard Helminck for correspondence on the subject, interest and comments on the proof and also for making available an early version of their paper (Helminck & van de Leur 1997a). I also thank Alex Kasman and Leonid Dickey for discussions during the workshop. This work was supported in part by the U.S. Department of Energy Grant No. DE-FG02-84ER40173.

References

Aratyn H., Nissimov E. and Pacheva S. (1997): *Int. J. Mod. Phys.* **A12** 1265 (hep-th/9607234)

Aratyn H., Nissimov E. and Pacheva S. (1997): "Method of Squared Eigenfunction Potentials in Integrable Hierarchies of KP Type", to appear in *Commun. Math. Phys.* (solv-int/9701017)

Cheng Y. and Zhang Y.J. (1994): *Journ. Math. Phys.* **35** 5869

Crum M.M. (1955): *Quart. J. Math. Oxford* **6** 121; Adler M. and Moser J. (1978): *Commun. Math. Phys.* **61** 1

Dickey L. (1995): *Letters in Math. Phys.* **35** 229 (hep-th/9411005)

Helminck G.F. and van de Leur J.W. (1997a): "An analytic description of the vector constrained KP hierarchy" (solv-int/9706004)

Helminck G.F. and van de Leur J.W. (1997b): "Constrained and Rational Reductions of the KP Hierarchy", in this volume

Ince E.L. (1926): *Ordinary Differential Equations*, London, 1926

Krichever I. (1995): *Physica* **D87** 14; *Functional Analysis and Its Application* **29** (1995) 75

Oevel W. (1993): *Physica* **A195** 533

Oevel W. and Strampp W. (1996): *J. Math. Phys.* **37** 6213;

Wilson G. (1992): *Nonlinearity* **5** (1992) 109

Wilson G. (1990): *On the Adler-Gelfand-Dikii bracket* in *CRM Workshop on Hamiltonian Systems ...*, eds. Harnad J.P. and Marsden J.E., Montreal, CRM (1990)

Infinite Dimensional Symmetries in Massive Integrable Models

Hidetoshi Awata

Enrico Fermi Institute and James Frank Institute of University of Chicago
5640 S. Ellis, Chicago, IL 60637, U.S.A.

Abstract. We review some recent results on the Calogero-Sutherland and Ruijsenaars models and massive integrable models with emphasis upon their algebraic aspects.

1 Introduction

The Virasoro algebra is the most important symmetry of string theory and statistical critical phenomena and has played an essential role in the remarkable progress of them. For the non-critical case, our q-Virasoro algebra [SKAO] is expected to take the place of the Virasoro algebra, and, indeed, it has been found relations with sine-Gordon model [L] and RSOS model [LP].

The Virasoro algebra and its q-analogue appear also in a quantum mechanical system with a long-range interaction and its relativistic analogue, e.g., the Calogero-Sutherland [Su] and Ruijsenaars model [R]. Their excited states, Jack polynomials [St] and Macdonald polynomials [M], are described by the (q-) Virasoro algebras.

On the other hand, when there is an internal symmetry, e.g., XXX spin chain, the affine Lie (Kac-Moody) or Yangian algebra takes the place of the Virasoro algebra; and non-critical case, e.g., XXZ chain, the quantum affine Lie algebra does [DFJMN]. However, because of our little knowledge for the elliptic affine Lie algebra, the most general (integrable) non-critical case, e.g., XYZ chain, still be underdeveloped. We find the q-Virasoro algebra relates with level-one elliptic affine Lie algebra and obtain a free boson realization for it [AKOS1], which could opens further avenue to the exploration of elliptic models.

This talk consists of the following three parts; (i) We start from introducing the relation between $1+1$ dimensional long-range interaction systems and (q-) Virasoro algebras. (ii) Next we define our q-Virasoro algebra and show screening currents of this enjoy an elliptic affine Lie algebra and give their free boson realization. (iii) Finally, we present relations between massive integrable model and above two infinite dimensional algebras, which are hidden symmetries.

2   Hidden symmetries in long-range interaction systems

We start with recapitulating the relation between the Jack polynomials, that are the excited states of the Calogero-Sutherland model, and Virasoro singular vectors.

2.1   Calogero-Sutherland model and Jack polynomial

Jack polynomials $J(z) \equiv J(z_1, z_2, \cdots, z_N)$ are defined as symmetric polynomials in z_i's which are eigenfunctions of [Su][St]

$$H = \sum_{i=1}^{N} D_i^2 + \beta \sum_{i<j} \frac{z_i + z_j}{z_i - z_j} (D_i - D_j), \tag{1}$$

with $D_i = z_i \frac{\partial}{\partial z_i}$. Here β is a coupling constant.

This Hamiltonian H relates with the Calogero-Sutherland model, which is a N–body problem on a unit circle with the following Hamiltonian

$$H_{CS} = \sum_{j=1}^{N} \left(\frac{1}{i} \frac{\partial}{\partial q_j} \right)^2 + \frac{1}{2} \sum_{i<j} \frac{\beta(\beta - 1)}{\sin^2 (q_i - q_j)/2}, \tag{2}$$

by changing the variables of coordinates q_i to $z_j \equiv \exp(iq_j)$ on a complex plane. The vacuum wave-function Δ of H_{CS} is

$$\Delta = \prod_{i<j} \sin^\beta \left(\frac{q_i - q_j}{2} \right) = \prod_{i \neq j} (1 - z_i/z_j)^{\beta/2}, \tag{3}$$

and the excited states are written as $J(z)\Delta(z)$ with the Jack polynomial $J(z)$. Thus the relation between H and H_{CS} is $H = \Delta^{-1} H_{CS} \Delta - \varepsilon_0$ with the vacuum energy ε_0.

The reason of that the Jack polynomial $J(z)$ should be "symmetric" and "polynomial" is as follows;

(i). Because of eq. (3), the statistic of the particle is governed by the coupling β: if β is even (odd) then the particles become bosonic (fermionic). To possess the same statistic as the vacuum, excited states $J(z)$ have to be symmetric functions.

(ii). Because of the periodicity, the power of z_i's should be integers. By multiplying the factor $\prod_{i=1}^{N} z_i^M$ with sufficiently large number M, all powers can become positive.

2.2   Integral formula for Jack polynomial

The Jack polynomial $J(z)$ is known to be labeled by a decreasing set of non-negative integers, $\lambda = (\lambda_1 \geq \lambda_2 \geq \cdots \geq \lambda_N \geq 0)$, which is identified with a Young diagram with $\lambda_i (\geq 1)$ squares in i-th row. When $\beta = 1$, the Jack polynomial reduces to the Schur polynomial.

There exists two types of (integral) transformations which maps the eigenstate into another while changing its energy and the number of particles. First, we introduce the Galilean boost $\mathcal{G}_N$, which uniformly shifts the pseudo-momentum of the pseudo-particles from $\lambda = (\lambda_1, \cdots, \lambda_N)$ to $\lambda + (1^N) = (\lambda_1 + 1, \cdots, \lambda_N + 1)$. It can be realized by multiplying the wave-function by $\prod_j e^{iq_j} = \prod_j z_j$. When it is operated to the eigenstate, the Young diagram is changed by adding a rectangle (1^N) from the left:

$$\mathcal{G}_N \cdot J_\lambda(z_1, \cdots, z_N) = J_{\lambda+(1^N)}(z_1, \cdots, z_N)$$

$$= J_\lambda(z_1, \cdots, z_N) \cdot \prod_{i=1}^{N} z_i. \tag{4}$$

The second integral transformation $\mathcal{N}_{NM}$ changes the number of particles from M to N:

$$\mathcal{N}_{NM} \cdot J_\lambda(t_1, \cdots, t_M) = J_\lambda(z_1, \cdots, z_N)$$

$$\propto \oint \prod_{j=1}^{M} \frac{dt_j}{t_j} \cdot \Pi(z, 1/t) J_\lambda(t) \, \Delta^2(t), \tag{5}$$

with

$$\Pi(x, y) \equiv \prod_{i=1}^{N} \prod_{j=1}^{M} (1 - x_i y_j)^{-\beta}. \tag{6}$$

The integration path is along the unit circle around the origin in the complex plane. This property follows from the orthogonality and the Cauchy formula for the Jack polynomial [St][M].

Then, we can construct arbitrary excited state by applying these two transformations $\mathcal{G}_n$ and $\mathcal{N}_{nm}$ successively to the vacuum, $i.e.$ "1". For the Young diagram $\lambda = (\lambda_1, \cdots, \lambda_N)$, we have;

Theorem. *Integral formula [MY2][AMOS]*

$$J_\lambda(z) = \mathcal{G}_N^{\lambda_N} \mathcal{N}_{N,N-1} \mathcal{G}_{N-1}^{\lambda_{N-1}-\lambda_N} \cdots\cdots\cdots\cdots \mathcal{G}_2^{\lambda_2-\lambda_3} \mathcal{N}_{2,1} \mathcal{G}_1^{\lambda_1-\lambda_2} \cdot 1 \tag{7}$$

2.3 Large N limit

Next let us consider the infinite number of particles case, $i.e.$, $N \to \infty$. Since H is symmetric in z_i's, it can be expressed by the power-sums $p_n \equiv \sum_{i=1}^{N} z_i^n$ and their derivatives $\partial_n \equiv \frac{n}{\beta}\frac{\partial}{\partial p_n}$ as follows[AMOS]:

$$H - (\beta - 1 + \beta N)\sum_i D_i = \beta \sum_{n>0} p_n \mathcal{L}_n,$$

$$\mathcal{L}_n = \beta \sum_{m=1}^{n-1} \partial_m \partial_{n-m} + \beta \sum_{m>0} p_m \partial_{n+m} - (n+1)(\beta-1)\partial_n. \qquad (8)$$

Here we must treat p_n's as formally independent variables, i.e., $\partial_n p_m = \frac{n}{\beta}\delta_{n,m}$ for all $n, m > 0$. As we will see, these relations with p_n (free bosons) or $\mathcal{L}_n$ (non-relativistic Virasoro generators) are the Keys to the algebraic aspects of the model.

Finally, we present some examples of Jack polynomials in power-sum p_n.

$$J_{(1)} = p_1, \qquad (9)$$

$$J_{(2)} = p_2 + \beta p_1^2, \qquad (10)$$

$$J_{(22)} = p_4 + \frac{4\beta}{1-\beta}p_3 p_1 - \frac{1+\beta+\beta^2}{1-\beta}p_2^2 - 2\beta p_2 p_1^2 - \frac{\beta^2}{1-\beta}p_1^4. \qquad (11)$$

2.4 Virasoro singular vectors

Virasoro algebra $\{L_n\}_{n\in\mathbf{Z}}$ is an infinite dimensional Lie algebra defined as

$$[L_n, L_m] = (n-m)L_{n+m} + \frac{c}{12}n(n^2-1)\delta_{n+m,0}. \qquad (12)$$

Here central charge $c \in \mathbf{C}$ is an important parameter. This algebra is realized by simpler algebra called free boson algebra $\{a_n\}_{n\in\mathbf{Z}}$ defined with

$$[a_n, a_m] = n\delta_{n+m,0}, \qquad (13)$$

as follows;

$$L_n = \frac{1}{2}\sum_{m\in\mathbf{Z}} : a_{n-m}a_m : -(n+1)\alpha_0 a_n, \qquad \alpha_0 = \frac{1}{\sqrt{2}}\left(\sqrt{\beta} - \frac{1}{\sqrt{\beta}}\right). \qquad (14)$$

Here $: \ast\ast :$ stands for the normal ordering such that positive mode generator should be right, for example, $: a_3 a_{-3} := a_{-3}a_3$.

Let $\mathcal{V}_h$ be the Verma module over the Virasoro algebra generated by the highest weight state $|h\rangle$, such that

$$L_{n>0}|h\rangle = 0, \qquad L_0|h\rangle = h|h\rangle. \qquad (15)$$

In terms of bosons, $|h\rangle$ is

$$a_{n>0}|h\rangle = 0, \qquad a_0|h\rangle = \alpha|h\rangle, \qquad h = \frac{1}{2}\alpha(\alpha - 2\alpha_0). \tag{16}$$

This module $\mathcal{V}_h$ is irreducible if and only if there is no singular vectors $|\chi\rangle \in \mathcal{V}_h$ of grade $N \in \mathbf{Z}_{>0}$ such that

$$L_{n>0}|\chi\rangle = 0, \qquad L_0|\chi\rangle = (h+N)|\chi\rangle, \quad (N \in \mathbf{Z}_{>0}). \tag{17}$$

In case of such singular vectors exists, to get an irreducible representation $\mathcal{M}_h$, we should factor out the unnecessary modules generated by them. Thus it is important to know when and where such singular vector appears, and it is described by the following theorem *Singular vector of grade $N = rs$ exists if and only if central charge c and highest weight h are*

$$c = 1 - 6\frac{(\beta-1)^2}{\beta}, \qquad h_{rs} - \frac{c-1}{24} = \frac{(r\beta-s)^2}{4\beta}, \tag{18}$$

with $\beta \in \mathbf{C}_{\neq 0}$ and $r, s \in \mathbf{Z}_{>0}$, and if β takes generic value, there is only one. In terms of bosons, the highest weight α_{rs} in eq. (16) is

$$\alpha_{rs} = \frac{1}{\sqrt{2}}\left((r+1)\sqrt{\beta} - (s+1)\frac{1}{\sqrt{\beta}}\right), \tag{19}$$

Here we present some example of singular vectors in bosons;

$$|\chi_{11}\rangle = a_{-1}|\alpha_{11}\rangle, \tag{20}$$

$$|\chi_{12}\rangle = \left(a_{-2} + \sqrt{2\beta}a_{-1}^2\right)|\alpha_{12}\rangle, \tag{21}$$

$$|\chi_{22}\rangle = \left(a_{-4} + \frac{4\sqrt{2\beta}}{1-\beta}a_{-3}a_{-1} - 2\frac{1+\beta+\beta^2}{\sqrt{2\beta}(1-\beta)}a_{-2}^2 \right. \tag{22}$$

$$\left. -4a_{-2}a_{-1}^2 - \frac{2\sqrt{2\beta}}{1-\beta}a_{-1}^4\right)|\alpha_{22}\rangle.$$

2.5   Relation between Jack polynomials and singular vectors

It is easy to see that these singular vectors correspond to Jack polynomials in power-sum in eq. (11) by the following rule,

$$a_{-n} \to \sqrt{\frac{\beta}{2}}p_n, \qquad |\alpha_{rs}\rangle \to 1. \tag{23}$$

This rule holds for the Jack polynomials with rectangular Young diagram and we have

Proposition. *[MY1] The Virasoro singular vector $|\chi_{rs}\rangle$ has one to one correspondence with the Jack symmetric polynomial with the rectangular Young diagram $\{s^r\}$*

$$|\chi_{rs}\rangle \sim J_{\{s^r\}}(x; \beta), \tag{24}$$

by the above rule eq. (23).

Proof is given by comparing the integral formula for the Jack polynomial in power-sum and that for the Virasoro singular vector in bosons.

Notice that we can also prove this proposition by using eq. (8). With the rule eq. (23), $\mathcal{L}_n$ in eq. (8) is nothing but the positive mode generator L_n of Virasoro algebra in eq. (14). Since any singular vectors are annihilated by $\mathcal{L}_n$ part of H and are trivially eigenfunction for the remaining part, we complete the proof. It is very important that Hamiltonian H in power-sum is constructed by positive mode Virasoro generators.

It can be shown that the Jack polynomials for arbitrary Young diagrams are realized as the singular vector of the W_N algebra.

Theorem. *[AMOS] The W_n singular vector has one to one correspondence with the Jack symmetric polynomial with arbitrary Young diagram.*

2.6   Macdonald polynomial

Let us turn to an analysis of q-analogue and define our new algebra. The excited states of trigonometric Ruijsenaars model are called Macdonald symmetric functions $P_\lambda(z)$, which defined as eigenfunctions of

$$H = \sum_{i=1}^{N} \prod_{j \neq i} \frac{tz_i - z_j}{z_i - z_j} \cdot q^{D_i}, , \tag{25}$$

where $D_i \equiv z_i \frac{\partial}{\partial z_i}$ and $t = q^\beta$. Note that q^{D_i} is the q-shift operator such that $q^{D_i} f(z_1, \cdots, z_N) = f(z_1, \cdots, qz_i, \cdots, z_N)$. In the limit of $q \to 1$, this Macdonald polynomial reduces to the Jack polynomial.

The integral formula for the Macdonald polynomial is the same as Jack case with the replacement

$$\Pi(x, y) \equiv \prod_{i=1}^{N} \prod_{j=1}^{M} \prod_{k \geq 0} \frac{1 - tq^k x_i y_j}{1 - q^k x_i y_j}, \qquad \Delta^2(x) \equiv \prod_{i \neq j} \prod_{k \geq 0} \frac{1 - q^k x_i/x_j}{1 - tq^k x_i/x_j}. \tag{26}$$

Then there appear the following natural question.

Problem. Find algebras which singular vectors realize the Macdonald polynomials.

A key property is also the same as Jack case; Hamiltonian should be constructed by positive mode generators. The solution looks like unique and is the following q-deformed Virasoro and W_n algebra.

3   q-Virasoro algebra
and level-one elliptic affine Lie algebra

Next we define new Virasoro-type algebra and elliptic algebra that are hidden symmetries of long-range interacting systems and massive integrable models.

3.1   q-Virasoro algebra

The answer for the last problem is the q-Virasoro algebra defined as follows;

Definition. *[SKAO] Let q, t and $p = q/t$ be complex parameters. The q-Virasoro algebra $\{T_n\}_{n\in\mathbf{Z}}$ is an associative algebra such that*

$$[T_n , T_m] = - \sum_{\ell>0} f_\ell \left(T_{n-\ell}T_{m+\ell} - T_{m-\ell}T_{n+\ell}\right)$$
$$- \frac{(1-q)(1-t^{-1})}{1-p}(p^n - p^{-n})\delta_{m+n,0}, \qquad (27)$$

with

$$f(x) \equiv \sum_{\ell\geq 0} f_\ell z^\ell = \exp\left\{\sum_{n=1}^{\infty} \frac{(1-q^n)(1-t^{-n})}{1+p^n}\frac{x^n}{n}\right\}. \qquad (28)$$

And δ is a delta function $\delta(x) \equiv \sum_{n\in\mathbf{Z}} x^n$ and has a property

$$\delta(x)f(x) = \delta(x)f(1). \qquad (29)$$

In the limit of $q \to 1$, this algebra reduces to the Virasoro algebra with center eq. (25).

In terms of the current $T(z) = \sum_{n\in\mathbf{Z}} T_n z^{-n}$, the above relation is equivalent to

$$f(w/z)T(z)T(w)-T(w)T(z)f(z/w) = -\frac{(1-q)(1-t^{-1})}{1-p}\left[\delta\left(\frac{pw}{z}\right) - \delta\left(\frac{w}{pz}\right)\right],$$
$$(30)$$

The structure of the highest weight module is very similar to the $q = 1$ case. And q-Virasoro generator $T(z)$ and singular vectors are also realized by free bosons.

Proposition. *[SKAO] The q-Virasoro singular vector $|\chi_{rs}\rangle$ has one to one correspondence with the Macdonald symmetric polynomial with the rectangular Young diagram $\{s^r\}$.*

The $q\text{-}\mathcal{W}_n$ algebra can be defined by using q-deformed Miura transformation [FF][AKOS2] and we can show the relation with Macdonald polynomials.

Theorem. *[AKOS2] The $q\text{-}\mathcal{W}_n$ singular vector has one to one correspondence with the Macdonald symmetric polynomial with arbitrary Young diagram.*

3.2   Boson realization for q-Virasoro algebra

Let us introduce the fundamental Heisenberg algebra h_n ($n \in \mathbf{Z}$), Q_h having the commutation relations

$$[h_n, h_m] = \frac{1}{n} \frac{(q^{\frac{n}{2}} - q^{-\frac{n}{2}})(t^{\frac{n}{2}} - t^{-\frac{n}{2}})}{p^{\frac{n}{2}} + p^{-\frac{n}{2}}} \delta_{n+m,0}, \qquad [h_n, Q_h] = \frac{1}{2}\delta_{n,0}. \quad (31)$$

By these, the q-Virasoro current $T(z)$ and the screening current $S_\pm(z)$, which commute with q-Virasoro $[T(z), \oint dt S_\pm(t)] = 0$ and generate the singular vectors, are written as

$$T(z) = \Lambda^+(z) + \Lambda^-(z), \tag{32}$$

$$\Lambda^\pm(z) = \; : \exp\left\{ \pm \sum_{n \neq 0} h_n p^{\pm \frac{n}{2}} z^{-n} \right\} : q^{\pm\sqrt{\beta}h_0} p^{\pm\frac{1}{2}}, \tag{33}$$

$$S_+(z) = \; : \exp\left\{ -\sum_{n \neq 0} \frac{p^{\frac{n}{2}} + p^{-\frac{n}{2}}}{q^{\frac{n}{2}} - q^{-\frac{n}{2}}} h_n z^{-n} \right\} : e^{2\sqrt{\beta}Q_h} z^{2\sqrt{\beta}h_0}, \tag{34}$$

$$S_-(z) = \; : \exp\left\{ \sum_{n \neq 0} \frac{p^{\frac{n}{2}} + p^{-\frac{n}{2}}}{t^{\frac{n}{2}} - t^{-\frac{n}{2}}} h_n z^{-n} \right\} : e^{-\frac{2}{\sqrt{\beta}}Q_h} z^{-\frac{2}{\sqrt{\beta}}h_0}. \tag{35}$$

Note that under the isomorphism σ such that:

$$\sigma \; : \quad q \leftrightarrow 1/t, \qquad \sqrt{\beta} \leftrightarrow -\sqrt{1/\beta}, \tag{36}$$

$\sigma \cdot \Lambda^\pm(z) = \Lambda^\pm(z)$ and $\sigma \cdot S_\pm(z) = S_\mp(z)$.

The free boson realization for $T(z)$ is expressed as the following deformed Miura transformation [FR]

$$: \left(p^D - \Lambda^+(z)\right)\left(p^D - \Lambda^-(z)\right) : = p^{2D} - T(z)p^D + 1, \tag{37}$$

which has been generalized to define the q-deformed $\mathcal{W}$ algebra [FF], [AKOS2]. By using this transformation, Frenkel-Reshetikhin [FR] proposed a generalization of their quasi-classical q-Virasoro algebra to $ABCD$-type cases. An analogy to the Baxter's dressed vacuum form Q defined by $: \left(p^D - \Lambda^-(z)\right) Q(z):= 0$, so $\Lambda^\pm(z) = :Q(zp^{\mp 1}) Q(z)^{-1}:$, seems to be of some interest.

3.3   Vertex operators

The vertex operator defined by

$$V_{2,1}(z) \equiv \; : \exp\left\{ \sum_{n \neq 0} \frac{h_n}{q^{\frac{n}{2}} - q^{-\frac{n}{2}}} z^{-n} \right\} : e^{-\sqrt{\beta}Q_h} z^{-\sqrt{\beta}h_0}, \tag{38}$$

satisfies

$$g\left(\frac{w}{z}p^{\pm\frac{1}{2}}\right)T(z)V_{2,1}(w) - V_{2,1}(w)T(z)g^{-1}\left(\frac{z}{w}p^{\mp\frac{1}{2}}\right) \tag{39}$$

$$= t^{\pm\frac{1}{4}}(t^{\frac{1}{2}} - t^{-\frac{1}{2}})\,\delta\left(t^{\pm\frac{1}{2}}\frac{w}{z}\right)V_{2,1}(q^{\pm\frac{1}{2}}w)p^{\mp\frac{1}{2}},$$

with

$$g(x) = t^{\pm\frac{1}{4}}\exp\left\{\pm\sum_{n>0}\frac{1}{n}\frac{t^{\frac{n}{2}} - t^{-\frac{n}{2}}}{p^{\frac{n}{2}} - p^{-\frac{n}{2}}}x^n\right\}. \tag{40}$$

Note that $V_{2,1}(q^{\pm\frac{1}{2}}w)p^{\mp\frac{1}{2}} =: \Lambda^{\mp}(t^{\mp\frac{1}{2}}w)V_{2,1}(w):$. If we let $V_{1,2}(z) \equiv \sigma\cdot V_{2,1}(z)$ and their fusion as

$$V_{\ell+1,k+1}(z) \equiv :\prod_{i=1}^{\ell} V_{2,1}(q^{\frac{\ell+1-2i}{2\ell}}z)\prod_{j=1}^{k} V_{1,2}(t^{\frac{k+1-2j}{2k}}z): , \tag{41}$$

then they also obey a similar commutation relation as (40), which reduces to the usual defining relation for the Virasoro primary field of the conformal weight $h_{\ell+1,k+1}$, in the limit $q \to 1$ [AKMOS]. The adjoint action of the generator $T(z)$ on this fused vertex operator $V_{\ell+1,k+1}$ may be closely connected with a coproduct of the q-Virasoro algebra.

The fundamental vertex operators $V_{2,1}(z)$ and $V_{1,2}(z)$, that satisfy fermion like anti-commutation relation, are especially important. Because the q-Virasoro generator and screening currents are expressed by them as follows;

$$\Lambda^{+}(zp^{\frac{1}{2}}) = :V_{2,1}^{+}(zq^{-\frac{1}{2}})V_{2,1}^{-}(zq^{\frac{1}{2}}):p^{\frac{1}{2}}, \qquad S^{+}(z) = :V_{2,1}^{-}(zp^{\frac{1}{2}})V_{2,1}^{-}(zp^{-\frac{1}{2}}):, \tag{42}$$

and the relations obtained by σ. Here $V_{\ell+1,k+1}^{\pm}(z) \equiv V_{\ell+1,k+1}^{\pm1}(z)$. Hence, they must play more important role in the q-Virasoro algebra.

3.4   Level-one elliptic affine Lie algebra generated by $S^{\pm}(z)$

The properties of screening currents are quite important in the representation theory of the infinite-dimensional algebra; they govern the irreducibility and the physical states. Moreover they relate with hidden quantum symmetries.

Here, we show that the screening currents generate an elliptic hidden symmetry, which reduces to the (quantum) affine Lie algebra with a special center. Let us introduce a new current $\Psi(z) \equiv :S_{+}(q^{\pm\frac{1}{2}}z)\,S_{-}(t^{\pm\frac{1}{2}}z):$, i.e.,

$$\Psi(z) = \exp\left\{\sum_{n\neq0}\frac{p^n - p^{-n}}{(q^{\frac{n}{2}} - q^{-\frac{n}{2}})(t^{\frac{n}{2}} - t^{-\frac{n}{2}})}h_n z^{-n}\right\}e^{2\alpha Q}z^{2\alpha h_0}, \tag{43}$$

with $\alpha = \sqrt{\beta} - 1/\sqrt{\beta}$, then we have

Proposition 5. *[AKOS1] Screening Currents $S_\pm(z)$ and $\Psi(z)$ generate the following elliptic two-parameter algebra;*

$$f_{00}\left(\frac{w}{z}\right)\Psi(z)\Psi(w) = \Psi(w)\Psi(z)f_{00}\left(\frac{z}{w}\right), \tag{44}$$

$$f_{0\pm}\left(\frac{w}{z}\right)\Psi(z)S_\pm(w) = S_\pm(w)\Psi(z)f_{\pm 0}\left(\frac{z}{w}\right), \tag{45}$$

$$f_{\pm\pm}\left(\frac{w}{z}\right)S_\pm(z)S_\pm(w) = S_\pm(w)S_\pm(z)f_{\pm\pm}\left(\frac{z}{w}\right), \tag{46}$$

$$[S_+(z), S_-(w)] = \frac{1}{(p-1)w}\left[\delta\left(p^{\frac{1}{2}}\frac{w}{z}\right)\Psi(t^{-\frac{1}{2}}w) - \delta\left(p^{-\frac{1}{2}}\frac{w}{z}\right)\Psi(q^{-\frac{1}{2}}w)\right], \tag{47}$$

where $f_{00}(x) = f_{++}(x)f_{+-}^2(xp^{\frac{1}{2}})f_{--}(x)$ and $f_{0\pm}(x) = f_{\pm 0}(x) = f_{\pm\pm}(xq^{\frac{1}{2}})f_{-\pm}(xt^{\frac{1}{2}})$ with

$$f_{+-}(x) = f_{-+}(x) = \exp\left\{-\sum_{n>0}\frac{1}{n}(p^{\frac{n}{2}} + p^{-\frac{n}{2}})x^n\right\}x^{-1}, \tag{48}$$

$$f_{++}(x) = \exp\left\{-\sum_{n>0}\frac{1}{n}\frac{t^{\frac{n}{2}} - t^{-\frac{n}{2}}}{q^{\frac{n}{2}} - q^{-\frac{n}{2}}}(p^{\frac{n}{2}} + p^{-\frac{n}{2}})x^n\right\}x^\beta, \tag{49}$$

and $f_{--}(x) = \omega \cdot f_{++}(x)$.

In the limit of q and t tend to 0 with p and $t^{-\frac{|n|}{2}}h_n$ fixed, the relations (44)–(47) reduce to those of level-one $U_q(\widehat{sl}_2)$. Therefore, the algebra generated by $S^\pm(z)$ and $\Psi(z)$ can be regarded as an elliptic generalization of $U_q(\widehat{sl}_2)$ with level-one. Remark that, in this limit, the algebra with $\Psi_\pm(z) \equiv :S_+(p^{\pm\frac{1}{4}}z)\,S_-(p^{\mp\frac{1}{4}}z):$ are more natural than that with $\Psi(z)$.

In the sense of analytic continuation, these relations are also rewritten by using elliptic theta functions[FF],

$$S_\pm(z)S_\pm(w) = U_\pm\left(\frac{w}{z}\right)S_\pm(w)S_\pm(z), \tag{50}$$

with

$$U_\pm(x) = -x^{1-2\beta}\exp\left\{\sum_{n\neq 0}\frac{1}{n}\frac{q^{\frac{n}{2}}t^{-n} - q^{-\frac{n}{2}}t^n}{q^{\frac{n}{2}} - q^{-\frac{n}{2}}}x^n\right\} = -x^{2(1-\beta)}\frac{\vartheta_1(px; q)}{\vartheta_1(px^{-1}; q)}, \tag{51}$$

and $U_-(x) = \omega \cdot U_+(x)$. Note that $U_\pm(x)$ are quasi-periodic functions, namely for $U_+(x)$, we have

$$U_+(qx) = U_+(x), \qquad U_+(e^{2\pi i}x) = e^{-4\pi i\beta}U_+(x). \tag{52}$$

It should be noted that the screening currents of q-$\mathcal{W}$ algebra and $U_q(\widehat{sl}_N)$ also obey similar elliptic relations.

4   Hidden symmetries in ABF model

In the papers [LP], the explicit formula for the multipoint correlation functions is successfully obtained. We review their method and the relation to the q-Virasoro algebra.

4.1   Commutation relations of vertex operators of ABF model

The q-Virasoro algebra can be applied to the off-critical phenomena, especially to the ABF model in the regime III [ABF]. Let $z \equiv p^v$ and the vertex operators $\Phi_\pm(z)$ be

$$\Phi_+(z) \equiv V_{2,1}(z), \qquad \Phi_-(z') \equiv \oint \frac{dz}{2\pi i z} V_{2,1}(z') S_+(z) z^\beta f(v - v', \pi), \quad (53)$$

where $[v] \equiv p^{\frac{1}{2}((1-\beta)v^2 - v)}(z;q)_\infty (qz^{-1};q)_\infty (q;q)_\infty$ and

$$f(v, w) \equiv \frac{[v + \frac{1}{2} - w]}{[v - \frac{1}{2}]}, \qquad \pi \equiv -\frac{2h_0}{\sqrt{\beta} - \sqrt{1/\beta}}. \quad (54)$$

The integration contour is a closed curve around the origin satisfying $p|z'| < |z| < p^{-1}|z'|$.

Then the vertex operators satisfy the following commutation relation;

Theorem 5.[LP] [1]

$$\Phi_{\ell_3-\ell_2}(z_1)\Phi_{\ell_2-\ell_1}(z_2) = \sum_{\ell_4} W \begin{pmatrix} \ell_3 & \ell_4 \\ \ell_2 & \ell_1 \end{pmatrix} \begin{vmatrix} z_1 \\ z_2 \end{vmatrix} \Phi_{\ell_3-\ell_4}(z_2)\Phi_{\ell_4-\ell_1}(z_1), \quad (55)$$

where $\ell_{i+1} - \ell_i = \pm 1$ with $\ell_5 \equiv \ell_1$. Here $W(\ell|z)$ is the Boltzmann weight of the ABF model such that

$$W \begin{pmatrix} \ell \pm 2 & \ell \pm 1 \\ \ell \pm 1 & \ell \end{pmatrix} z \end{pmatrix} = R(z), \qquad R(z) = z^{\frac{\beta}{2}} \frac{g(z^{-1})}{g(z)},$$

$$W \begin{pmatrix} \ell & \ell \pm 1 \\ \ell \pm 1 & \ell \end{pmatrix} z \end{pmatrix} = R(z) \frac{[\ell \pm v][1]}{[\ell][1 - v]}, \quad g(z) = \frac{(pz; q, p^2)_\infty (pqz; q, p^2)_\infty}{(qz; q, p^2)_\infty (p^2z; q, p^2)_\infty},$$

$$W \begin{pmatrix} \ell & \ell \mp 1 \\ \ell \pm 1 & \ell \end{pmatrix} z \end{pmatrix} = -R(z) \frac{[\ell \pm 1][v]}{[\ell][1 - v]}, \quad (56)$$

with $(z; q, p)_\infty \equiv \prod_{n,m \geq 0}(1 - zq^n p^m)$.

Besides this commutation relation these vertex operators also satisfy the other defining relations, homogeneity and normalization condition [FJMMN], of those of the ABF model in the regime III; $0 < p < z < 1$. Here $p = 0$ and 1 correspond to the zero temperature and the critical point, respectively. Thus these vertex operators of the q-Virasoro algebra can be regarded as those of the ABF model.

---

[1] The notations in [LP] are $x = p^{\frac{1}{2}}$ and $r = 1/(1 - \beta)$.

4.2   Felder resolution and the space of ABF model

Not only the vertex operators themselves but also the Hilbert space on which the physical operators act are able to be identified with that of the ABF model. The Hilbert space of the ABF model can be constructed through a deformed Felder-type BRST resolution.

Let $\mathcal{F}_{r,s}$ be the Fock module generated by the highest weight state $|r, s\rangle$ such that

$$h_{n>0}|r, s\rangle = 0, \qquad h_0|r, s\rangle = -\frac{1}{2}\left(r\sqrt{\beta} - s\sqrt{1/\beta}\right)|r, s\rangle. \tag{57}$$

Suppose $\beta = P_-/P_+$ with coprime integers $P_+ > P_- \in \mathbf{N}$ and let the screening charge $X_+ : \mathcal{F}_{r,s} \to \mathcal{F}_{r-2,s}$ be

$$X_+ = \oint \frac{dz}{2\pi i z} S_+(z) z^\beta f(v, \pi), \tag{58}$$

and define the BRST charges Q_j^+ ($j \in \mathbf{Z}$) as

$$
\begin{aligned}
Q_{2j}^+ &= X_+^r &: \mathcal{F}_{r-2jP_+,s} &\to \mathcal{F}_{-r-2jP_+,s}, \\
Q_{2j+1}^+ &= X_+^{P_+-r} &: \mathcal{F}_{-r-2jP_+,s} &\to \mathcal{F}_{r-2(j+1)P_+,s}.
\end{aligned} \tag{59}
$$

We also define the dual screening charge $X_- : \mathcal{F}_{r,s} \to \mathcal{F}_{r,s-2}$ and the dual BRST charges Q_j^- by the replacement $\sqrt{\beta} \leftrightarrow -\sqrt{1/\beta}$, $q \leftrightarrow 1/t$ and $r \leftrightarrow s$.

Proposition 4. [LP], [JLMP] The screening charges $X_\pm$ commute with each other and with q-Virasoro generators

$$
\begin{aligned}
&[X_+, X_-] = 0, \\
&[T(z), X_\pm^{n_\pm}] = 0, \quad \text{on } \mathcal{F}_{r_+,r_-}, \quad \text{with } n_\pm \equiv r_\pm \bmod P_\pm,
\end{aligned} \tag{60}
$$

and are also nilpotent

$$Q_j^\pm Q_{j-1}^\pm = X_\pm^{P_\pm} = 0, \quad P_\pm > 1. \tag{61}$$

Hence we can construct Felder type BRST complexes, for example, by X_+

$$\xrightarrow{X_+^r} \mathcal{F}_{-r+2P_+,s} \xrightarrow{X_+^{P_+-r}} \mathcal{F}_{r,s} \xrightarrow{X_+^r} \mathcal{F}_{-r,s} \xrightarrow{X_+^{P_+-r}} \mathcal{F}_{r-2P_+,s} \xrightarrow{X_+^r}. \tag{62}$$

From the Kac determinant of the q-Virasoro algebra, the Fock module $\mathcal{F}_{r,s}$ with $r, s \in \mathbf{N}$ is reducible. To obtain an irreducible one $\mathcal{L}_{r,s}$, we have to factor out the submodules by the Felder resolution. In a special case, this irreducible module coincides with the space of the ABF model.

To see this, we have to introduce a grading operator, which plays the role of the corner Hamiltonian in the ABF model,

$$H_c = \sum_{n>0} n^2 \frac{p^{\frac{n}{2}} + p^{-\frac{n}{2}}}{\left(q^{\frac{n}{2}} - q^{-\frac{n}{2}}\right)\left(t^{\frac{n}{2}} - t^{-\frac{n}{2}}\right)} h_{-n} h_n + h_0^2 - \frac{1}{24}. \tag{63}$$

This commutes with screening currents up to a total divergence

$$[H_c, S_\pm(z)]z^{\beta^{\pm 1}} = \frac{\partial}{\partial z}\left(S_\pm(z)z^{\beta^{\pm 1}}\right), \tag{64}$$

and its eigenvalues $\varepsilon_{r,s}$ on the Fock module $\mathcal{F}_{r,s}$ are

$$\varepsilon_{r,s} = h_{r,s} - \frac{c}{24} + n, \qquad n \in \mathbf{Z}_{n\geq 0}, \tag{65}$$

with $h_{r,s}$ in eq. (18). When $P_- = P_+ - 1$, these values coincide with the eigenvalues of the corner Hamiltonian of the ABF model corresponding to the 1-d configurations given by the rule: (i) each height takes an integer value between 1 and $P_+ - 1$, (ii) the allowed values of difference in any neighboring heights are ± 1, (iii) the height at the origin is r, (iv) the asymptotic configuration is $\cdots, s, s+1, s, s+1, \cdots$. However, we should note that the multiplicities of the bosonic Fock space and that of ABF model are different.

Lukyanov and Pugai [LP] showed that, after the Felder-type BRST resolution by the dual screening current $S_-(z)$, the multiplicities in the irreducible Fock module $\mathcal{L}_{r,s}$ coincide those of the ABF model. Therefore, ABF model is completely described by the representation of the q-Virasoro algebra and the multi-point local height probabilities of ABF model [FJMMN] are realized as correlation functions of the vertex operators. For example, the probability that the heights at the same vertical column sites have the values $1 \leq r_1, r_2, \cdots, r_n \leq P_+ - 1$ is proportional to

$$Tr_{\mathcal{L}_{r_1,s}}\left[p^{2H_c}\Phi_{-\sigma_1}(z_1/p)\cdots\Phi_{-\sigma_{n-1}}(z_{n-1}/p)\Phi_{\sigma_{n-1}}(z_{n-1})\cdots\Phi_{\sigma_1}(z_1)\right], \tag{66}$$

where $\sigma_s = r_{s+1} - r_s$.

Thus we can conclude that the q-Virasoro algebra (27) and the level-one elliptic affine Lie algebra (44)–(47) are hidden symmetries of ABF model in the following sence; (i) The vertex operator of ABF model (53) is nothing but that of the q-Virasoro algebra. (ii) The q-Virasoro screening currents (34), (35), which play the important role in the vertex operator and Felder resolution of ABF model, enjoy the level-one elliptic affine Lie algebra.

References

[ABF] G. Andrews, R. Baxter and J. Forrester, "Eight-vertex SOS model and generalized Rogers-Ramanujan identities", J. Stat. Phys. **35**,(1984) 193-266 (1984).

[AKMOS] H. Awata, H. Kubo, Y. Morita, S. Odake and J. Shiraishi, "Vertex Operators of the q-Virasoro Algebra; Defining Relations, Adjoint Actions and Four Point Functions", preprint EFI-96-14, DPSU-96-7, UT-750 (April 1996) q-alg/9604023, Revised July 1996. To appear in Lett. Math. Phys.

[AKOS1] H. Awata, H. Kubo, S. Odake and J. Shiraishi, "Virasoro-type Symmetries in Solvable Models", hep-th/9612233

[AKOS2] H. Awata, H. Kubo, S. Odake and J. Shiraishi, "Quantum W_N Algebras and Macdonald Polynomials", Comm. Math. Phys. **179** (1996) 401-416.

[AMOS] H. Awata, Y. Matsuo, S. Odake and J. Shiraishi, "Collective Field Theory, Calogero-Sutherland Model and Generalized Matrix Models", Phys. Lett. **B347** (1995) 49-55; "Excited States of Calogero-Sutherland Model and Singular Vectors of the W_N Algebra", Nucl. Phys. **B449** (1995) 347-374.

[DFJMN] B. Davies, O. Foda, M. Jimbo, T. Miwa and A. Nakayashiki, "Diagonalizaton of the XXZ Hamiltonian by vertex operators", Comm. Math. Phys. **151** (1993) 89-153.

[FF] B. Feigin and E. Frenkel, "Quantum W-Algebras and Elliptic Algebras", Comm. Math. Phys. **178** (1996) 653-678.

[FJMMN] O. Foda, M. Jimbo, K. Miki, T. Miwa and A. Nakayashiki, "Vertex operators in solvable lattice models", J. Math. Phys. **35** (1994) 13–46.

[FR] E. Frenkel and N. Reshetikhin, "Quantum Affine Algebras and Deformations of the Virasoro and W-Algebra", Comm. Math. Phys. **178** (1996) 237-264.

[JLMP] M. Jimbo, M. Lashkevich, T. Miwa and Y. Pugai, "Lukyanov's Screening Operators for the Deformed Virasoro Algebra", preprint RIMS-1087, July 1996, hep-th/9607177.

[L] S. Lukyanov, "A Note on the Deformed Virasoro Algebra", Phys. Lett. **B367** (1996) 121-125.

[LP] S. Lukyanov and Y. Pugai, "Multi-point Local Height Probabilities in the Integrable RSOS Model", Nucl. Phys. **B473** (1996) 631-658.

[M] I. Macdonald, *Symmetric Functions and Hall Functions* (2nd ed.), Oxford University Press 1995.

[MY1] K. Mimachi and Y. Yamada, *Singlar vectors of the Virasoro algebra in terms of Jack symmetric polynomials*, Comm. Math. Phys. **174** (1995) 447-455

[MY2] K. Mimachi and Y. Yamada, RIMS Kokyuroku **919** (1995) 68-78 (in Japanese).

[R] S.N.M. Ruijsenaars, *Complete Integrability of Relativistic Calogero-Moser Systems and Elliptic Function Identities*, Comm. Math. Phys. **110** (1987) 191-213

[St] R.P. Stanley, "Some Combinatorial Properties of Jack Symmetric Functions", Adv. Math. **77** (1989) 76–115

[Su] B. Sutherland, Phys. Rev. **A4** (1971) 2019–2021; **A5** (1972) 1372–1376

[SKAO] J. Shiraishi, H. Kubo, H. Awata and S. Odake, "A Quantum Deformation of the Virasoro Algebra and the Macdonald Symmetric Functions", Lett. Math. Phys. **38** (1996) 33-51.

On the Whitham Equations and (X, ψ) Duality

Robert Carroll

Mathematics Department, University of Illinois, Urbana, IL 61801
email: rcarroll@symcom.math.uiuc.edu

August, 1997

Abstract. We give a sketch of some material on (X, ψ) duality and enhanced dKdV on a Riemann surface from (Carroll (1997a)) plus a new scheme of calculation for enhanced dKdV which simplifies and explicates the procedure. An example is developed on an elliptic curve.

1 Introduction

Some connections of the (X, ψ) duality theory of Faraggi-Matone (Faraggi and Matone (1997a), Faraggi and Matone (1997b)) to Seiberg-Witten (SW) theory for $N = 2$ supersymmetric Yang-Mills, and to dispersionless Korteweg - deVries (dKdV) theory was developed in (Carroll (1997a)) (cf. also (Carroll (1996a), Carroll (1997b), Carroll (1997c))). We give some background in Sections 1 - 2 and in Section 3 we develop a new scheme of calculation for enhanced dKdV; an example is given on the original SW elliptic curve in Section 4.

2 (X, ψ) Duality

We extract here from (Carroll (1997a)) which is based on the (X, ψ) duality theory of (Faraggi and Matone (1997a), Faraggi and Matone (1997b)) (cf. also (Carroll (1996a), Carroll (1997b)) for some preliminary versions of (Carroll (1997a))). The point of departure is the Schrödinger equation

$$\mathcal{H}\psi_E = -\frac{\hbar^2}{2m}\psi_E'' + V(X)\psi_E = E\psi_E \tag{2.1}$$

where X is the quantum mechanical (QM) space variable with $\psi_E' = \partial\psi_E/\partial X$ and we write $\epsilon = \hbar/\sqrt{2m}$ (E is assumed real). One has a possible origin of this from a Kadomtsev-Petviashvili (KP) situation $L_+^2 \psi = \partial\psi/\partial t_2$ where $L_+^2 = \partial_x^2 - v(x, t_i)$ and e.g. $T_2 = \epsilon t_2$ with $\tau_2 = -i\sqrt{2m}T_2$ so $\partial_{t_2} = \epsilon\partial_{T_2} = -i\hbar\partial_{\tau_2}$ This leads to an approximation $(v(x, t_i) = V(X, T_i) + O(\epsilon))$

$$\epsilon^2\psi_E'' - V(X, T_i)\psi_E \sim \epsilon\frac{\partial\psi_E}{\partial T_2} = -i\hbar\frac{\psi_E}{\partial\tau_2} \tag{2.2}$$

corresponding to the Schrödinger equation (details below) and is primarily related to the Korteweg-deVries (KdV) equation and it's dispersionless form dKdV as indicated later.

For the approximation of potentials one assumes e.g. $v = v(x, t_i) \to v(X/\epsilon, T_i/\epsilon) = V(X, T_i) + O(\epsilon)$. This is standard in dispersionless KP $=$ dKP and certainly realizable by quotients of homogeneous polynomials for example. In fact it is hardly a restriction since given e.g. $F(X) = \sum_0^\infty a_n X^n$ consider $\tilde{f}(x, t_i) = a_0 + \sum_1^\infty (x^n / \prod_2^{n+1} t_i)$. Then

$$\tilde{f}(X/\epsilon, T_i/\epsilon) = a_0 + \sum_1^\infty (X^n / \prod_2^{n+1} T_i)$$

and one can choose the T_i recursively so that $1/T_1 = a_1$, $1/T_1 T_2 = a_2, \cdots$, leading to $F(X) = \tilde{F}(X, T_i)$.

Returning to (2.2), when $\psi_E = exp(S/\epsilon)$ for example, one has $\epsilon \psi'_E = S_X \psi_E$ with $\epsilon^2 \psi''_E = \epsilon S_{XX} \psi_E + (S_X)^2 \psi_E$ so in (2.2) we are neglecting an $O(\epsilon)\psi_E$ term from v, and for $\psi_E = exp(S/\epsilon)$ another $\epsilon S_{XX} \psi_E$ term is normally removed in dispersionless theory. Then for $\mathcal{H}$ independent of τ_2 for example one could assume V is independent of T_2 and write formally in (2.2), $\hat{\psi}_E = exp(E\tau_2/i\hbar) \cdot \psi_E$, with $\mathcal{H}\psi_E = E\psi_E$, which is (2.1). Since in the QM problem one does not however run $\hbar \to 0$ (hence $\epsilon \not\to 0$) one should argue that these $O(\epsilon)$ terms should be retained, and we will develop this approach, which essentially corresponds to WKB (with some background structure). In fact, to establish a connection of dKdV, and eventually of our enhanced dispersionless KdV ($dKdV_\epsilon$), with quantum mechanics and the Schrödinger equation, the passage from $v \to V$ or $V + \epsilon\hat{V}$ is the only "assumption" in the development below and this admits various realizations; the impact here only involves some possible minor restrictions on the class of quantum potentials to which the theory applies. The background mathematics behind V determined by KP or KdV essentially generates some additional structure which allows us to insert X into the theory in a manner commensurate with its role in (Faraggi and Matone (1997a), Faraggi and Matone (1997b)). The formulation of (Faraggi and Matone (1997a), Faraggi and Matone (1997b)) then entails some constraints on the background objects as indicated in the text. We emphasize that inserting S is familiar from WKB; we are introducing in an ad hoc manner additional variables T_i or T_i, λ or k, etc. to spawn a KP or KdV theory. We do not assume or even suggest that this is in any way connected a priori with the physics of the quantum mechanical problem (although of course it conceivably could be since integrability ideas are important in quantum mechanics). This procedure generates a nice Hamilton-Jacobi (HJ) theory which guides one to insert X into the machinery, but the insertion itself is at "ground level" and simply reflects a WKB formulation; neither the underlying KdV or KP dynamics nor the

HJ theory is directly used here. Once X is involved connections to (Faraggi and Matone (1997a), Faraggi and Matone (1997b)) are immediate. Actually the procedure could be reversed as a way of introducing duality ideas into the ϵ-dispersionless theory of (Carroll (1997a), Carroll (1996a), Carroll (1997b)) and this will be related to the duality already studied in Whitham theory (cf. (Carroll and Chang (1996), Carroll (1996b), Itoyama and Morozov (1995), Krichever and Phong (1997), Matone (1995), Nakatsu and Takasaki (1996))), given a finite zone theory on a Riemann surface. Thus start with KdV or KP, go to the Schrödinger equation and $dK dV_\epsilon$ or $dK P_\epsilon$, develop the HJ theory, and then use (Faraggi and Matone (1997a), Faraggi and Matone (1997b)) to create duality. More generally, start from a finite zone KdV situation with associated Whitham dynamics on a Riemann surface and compare dualities; this is the theme of (Carroll (1997a)) and we will sketch this here first with examples, further calculations, and some additional clarification.

We list first a few of the equations from (Faraggi and Matone (1997a), Faraggi and Matone (1997b)), as written in (Carroll (1997a)), without a discussion of philosophy (some of which will be mentioned later). Thus $\mathcal{F}$ is a prepotential and, since E is real, ψ_E and $\bar{\psi}_E = \psi_E^D$ both satisfy (2.1) with $\psi_E^D = \partial \mathcal{F} / \partial \psi_E$. The Wronskian in (2.1) is taken to be $W = \psi' \bar{\psi} - \psi \bar{\psi}' = 2\sqrt{2m}/i\hbar = 2/i\epsilon$ and one has $(\psi = \psi(X)$ and $X = X(\psi)$ with $X_\psi = \partial X / \partial \psi = 1/\psi')$

$$\mathcal{F}' = \psi' \bar{\psi}; \quad \mathcal{F} = \frac{1}{2} \psi \bar{\psi} + \frac{X}{i\epsilon}; \quad \frac{\partial \bar{\psi}}{\partial \psi} = \frac{1}{\psi} \left[\bar{\psi} - \frac{2}{i\epsilon} X_\psi \right] \tag{2.3}$$

(ψ always means ψ_E but we omit the subscript occasionally for brevity). Setting $\phi = \partial \mathcal{F} / \partial(\psi^2) = \bar{\psi}/2\psi$ with $\partial_\psi = 2\psi \partial / \partial(\psi^2)$ and evidently $\partial \phi / \partial \psi = -(\bar{\psi}/2\psi^2) + (1/2\psi)(\partial \bar{\psi}/\partial \psi)$ one has a Legendre transform pair

$$-\frac{X}{i\epsilon} = \psi^2 \frac{\partial \mathcal{F}}{\partial(\psi^2)} - \mathcal{F}; \quad -\mathcal{F} = \phi \frac{1}{i\epsilon} X_\phi - \frac{X}{i\epsilon} \tag{2.4}$$

One obtains also $|\psi|^2 = 2\mathcal{F} - (2X/i\epsilon)$ $(\mathcal{F}_\psi = \bar{\psi})$; $-(1/i\epsilon)X_\phi = \psi^2$; $\mathcal{F}_{\psi\psi} = \partial \bar{\psi}/\partial \psi$. Further from $X_\psi \psi' = 1$ one has $X_{\psi\psi} \psi' + X_\psi^2 \psi'' = 0$ which implies

$$\mathcal{F}_{\psi\psi\psi} = \frac{E - V}{4} (\mathcal{F}_\psi - \psi \partial_\psi^2 \mathcal{F})^3 = \frac{E - V}{4} \left(\frac{2X_\psi}{i\epsilon} \right)^3 \tag{2.5}$$

In fact (2.5) corresponds exactly to the Gelfand-Dickey resolvent equation

$$\epsilon^2 \mathcal{F}''' + 4(E - V) \left(\mathcal{F}' - \frac{1}{i\epsilon} \right) - 2V' \left(\mathcal{F} - \frac{X}{i\epsilon} \right) = 0 \tag{2.6}$$

(since $\Xi = |\psi|^2 = 2\mathcal{F} - (2X/i\epsilon)$ - cf. (Carroll (1997a), Carroll (1991))).

We refer now to (Carroll and Kodama (1995), Carroll (1994a), Carroll

(1994b), Guha and Takasaki (1997), Kodama and Gibbons (1990), Takasaki and Takebe (1992)) for dispersionless KP (= dKP) and consider here $\psi = exp[(1/\epsilon)S(X,T,\lambda)]$. Thus $P = S' = S_X$ with $P^2 = V - E$ and $E = \pm\lambda^2$ real will involve us in a KdV situation as indicated below. Some routine calculation yields (recall $X_\psi = 1/\psi'$ and $\psi' = (P/\epsilon)\psi$)

$$\phi = \frac{1}{2}e^{-(2i/\epsilon)\Im S}; \quad \frac{1}{\epsilon}X_\phi = -ie^{(2/\epsilon)S}; \quad X_\psi = \frac{\epsilon}{P}e^{-S/\epsilon} \qquad (2.7)$$

$$\frac{1}{\epsilon}X_{\psi\psi} = \frac{E-V}{P^3}e^{-S/\epsilon}; \quad \mathcal{F}_\psi = \bar\psi = e^{\bar S/\epsilon};$$

$$\mathcal{F}_{\psi\psi} = e^{-(2i/\epsilon)\Im S} - \frac{2}{iP}e^{-2S/\epsilon} \qquad (2.8)$$

$$|\psi|^2 = e^{(2/\epsilon)\Re S}; \quad \frac{S}{\epsilon} = \frac{1}{2}log|\psi|^2 - \frac{1}{2}log(2\phi);$$

$$\bar P = \bar S_X = P - \frac{2}{i\psi\bar\psi} \qquad (2.9)$$

Summarizing one has

$$\Im\mathcal{F} = -\frac{X}{\epsilon}; \quad \Re\mathcal{F} = \frac{1}{2}|\psi|^2 = \frac{1}{2}e^{\frac{2}{\epsilon}\Re S} = -\frac{1}{2\Im P} \qquad (2.10)$$

In the present situation the variables

$$|\psi|^2 = exp[(2/\epsilon)\Re S]$$

and

$$2\phi = exp[-(2i/\epsilon)\Im S]$$

can play the roles of independent variables (cf. (2.9) and we see that $|\psi|^2\Im P = -1$ while $\psi^2\phi = (1/2)|\psi|^2$. Now note that for $L = \partial + \sum_1^\infty u_i\partial^{-i}$, $L_+^2 = \partial^2 + 2u_1$, and $u_1 = \partial^2 log(\tau)$ where τ is the famous tau function. This implies $v = -2\partial^2 log(\tau)$ here, from which $V = -2F_{XX}$ for $\tau = exp[(1/\epsilon^2)F + O(1/\epsilon)]$ in the dispersionless theory (cf. (Carroll (1997a))). Then writing out the Gelfand-Dickey resolvent equation (2.6) yields

$$\epsilon^2\mathcal{F}''' + \left(\mathcal{F}' - \frac{1}{i\epsilon}\right)(8F'' + 4E) + 4F'''\left(\mathcal{F} - \frac{X}{i\epsilon}\right) = 0 \qquad (2.11)$$

which provides a relation between F and $\mathcal{F}$. We will see below how to embellish all this with a new modification of the dKP and dKdV theory. Thus we state here heuristically

THEOREM 2.1. Under the hypotheses indicated of first order WKB type approximation, the equation (2.11) yields a relation between the prepotential $\mathcal{F}$ of (X,ψ) duality defined via (2.3) and the free energy F of dispersionless theory.

COROLLARY 2.2. In the presence of a background hyperelliptic Riemann surface for a finite zone KdV theory, the dispersionless F corresponds to the Seiberg-Witten (SW) prepotential $F(a, T)$ (see below) and (2.11) determines a relation between $\mathcal{F}$ and $F(a, T)$.

PROOF. The details are in (Carroll (1997a)) and we recall here only a few facts. Dispersionless KP (dKP) is built up from $\epsilon x = X = T_1$, $\epsilon t_i = T_i$ $(i \geq 2)$, and $L = \partial + \sum_1^\infty u_{n+1} \partial^{-n} \to L_\epsilon = \epsilon \partial + \sum_1^\infty u_{n+1}(\epsilon, T)(\epsilon \partial)^{-n}$ where $\partial \sim \partial/\partial X$ now and $u_{n+1}(\epsilon, T) \to U_n(T) + O(\epsilon)$. One sets (recall $L\psi = \lambda\psi$)

$$\psi = exp\left(\frac{1}{\epsilon} S(T, \lambda) + O(1)\right);$$

$$\tau = exp\left(\frac{1}{\epsilon^2} F(T) + O\left(\frac{1}{\epsilon}\right)\right) \tag{2.12}$$

and the KP hierarchy $\partial_n L = [B_n, L]$, $B_n = L^n_+$, and $L\psi = \lambda\psi$ with the vertex operator equation (VOE) defined by the rule

$$\psi = \tau(T - (1/n\lambda^n)) exp\left[\sum_1^\infty T_n \lambda^n\right]/\tau(T)$$

lead to $(\partial_n = \partial/\partial T_n, \; P = S_X)$

$$\lambda = P + \sum_1^\infty U_{n+1} P^{-n}; \; P = \lambda - \sum_1^\infty W_i \lambda^{-1};$$

$$\partial_n S = \mathcal{B}_n(P) \Rightarrow \partial_n P = \hat{\partial}\mathcal{B}_n(P) \tag{2.13}$$

where $\hat{\partial} \sim \partial_X + (\partial P/\partial X)\partial_P$. Further for $\mathcal{B}_n = \sum_0^n b_{nm}\partial^m$ one has $\mathcal{B}_n = \sum_0^n b_{nm} P^m$. We list a few additional formulas which are easily obtained (cf. (Carroll (1991))); thus, writing $\{A, B\} = \partial_P A \partial A - \partial A \partial_P B$ one has

$$\partial_n \lambda = \{\mathcal{B}_n, \lambda\}; \; S = \sum_1^\infty T_n \lambda^n + \sum_1^\infty S_{j+1} \lambda^{-j};$$

$$S_{n+1} = -(\partial_n F/n); \; \partial_m S_{n+1} = (F_{mn}/n);$$

$$\mathcal{B}_n = \lambda^n + \sum_1^\infty \partial_n S_{j+1} \lambda^{-j}; \; \partial S_{n+1} \sim -\frac{\partial \partial_n F}{n} \tag{2.14}$$

Next (cf. (Carroll (1997a), Carroll and Chang (1996), Nakatsu and Takasaki (1996))) for a Riemann surface Σ_g of genus g based on a finite zone KP situation let (A_i, B_i) be a canonical homology basis, dw_j a basis of normalized holomorphic differentials $(\oint_{A_j} dw_i = \delta_{ij})$, $\mathcal{A}(P) = (\int_{P_0}^P dw_k)$ the Abel-Jacobi map $(P_o \neq P_\infty \sim \infty)$, and $\Theta(z) = \Theta[0](z)$ the Riemann theta function. Let λ^{-1} be a local coordinate near ∞ with $\lambda(P_\infty) = \infty$ and take $d\Omega_j = d(\lambda^j + O(\lambda^{-1}))$ to be normalized meromorphic differentials of the second kind ($\oint_{A_j} d\Omega_i = 0$). Other normalizations are also used (e.g. $\Re \oint_{A_i} d\Omega_j = \Re \oint_{B_i} d\Omega_j = 0$) but we will not dwell on this. We set also $\Omega_{jk} = \oint_{B_k} d\Omega_j$. Now let $D = P_1 + \cdots + P_g$ be a nonspecial divisor of degree g and set $z_0 = -K - \mathcal{A}(D)$ where $K \sim (K_j)$ corresponds to Riemann constants. One can now introduce "time" coordinates t_j via a uniquely defined BA function (**BA**)

$$\psi = \exp\left[\sum_1^\infty t_j \left(\int_{P_0}^P d\Omega^j + \Omega^j(P_0)\right) + i\sum_1^g \alpha_j \left(\int_{P_0}^P d\omega_j + \omega_j(P_0)\right)\right] \quad (2.15)$$

$$\times \frac{\Theta\left(\mathcal{A}(P) + \sum_1^\infty (t_j/2\pi i)(\Omega_{jk}) + i\sum_1^g \alpha_j(B_{jk}) + z_0\right)\Theta\left(\mathcal{A}(P_\infty) + z_0\right)}{\Theta\left(\mathcal{A}(P_\infty) + \sum_1^\infty (t_j/2\pi i)(\Omega_{jk}) + i\sum_1^g \alpha_j(B_{jk}) + z_0\right)\Theta\left(\mathcal{A}(P) + z_0\right)}$$

(note $\int^P d\Omega_j \sim \int_{P_0}^P d\Omega_j + \Omega_j(P_0)$) and $\mathcal{A}(P) = (\int_{P_\infty}^P d\omega_j) + \mathcal{A}(P_\infty))$ and explicitly now ($z = \lambda^{-1}$ and $q_{mj} = q_{jm}$)

$$d\Omega_j = d\Omega^j \sim d\left(\lambda^j - \sum_1^\infty \frac{q_{mj}}{m} z^m\right); \quad (2.16)$$

$$d\omega_j \sim d\left(-\sum_1^\infty \sigma_{jm}\frac{z^m}{m}\right); \quad \Omega_{nj} = 2\pi i \sigma_{jn}$$

There is also a general theory of prepotential related to SW theory. Following (Carroll and Chang (1996), Itoyama and Morozov (1995), Nakatsu and Takasaki (1996)) for example this involves ($T_n \sim \epsilon t_n$ as indicated below)

$$dS = \sum_1^g a_j d\omega_j + \sum_1^\infty T_n d\Omega_n; \quad \frac{\partial dS}{\partial a_j} = d\omega_j; \quad \frac{\partial dS}{\partial T_n} = d\Omega_n \quad (2.17)$$

Then one considers "prepotentials" $F(a, T)$ related to dS via

$$\frac{\partial F}{\partial a_j} = \frac{1}{2\pi i}\oint_{B_j} dS; \quad \partial_n F = -Res_\infty z^{-n} dS$$

and one has an expanded Whitham hierarchy with (cf. (Carroll and Chang (1996), Krichever (1994)))

$$2F = \sum_1^g a_j \frac{\partial F}{\partial a_j} + \sum_1^\infty T_n \frac{\partial F}{\partial T_n} \quad (2.18)$$

Writing now, in the notation of (Nakatsu and Takasaki (1996)), $d\omega_j = -\sum_1^\infty \sigma_{jm} z^{m-1} dz$ with $d\Omega_n = [-nz^{-n-1} - \sum_1^\infty q_{mn} z^{m-1}]dz$, and using (2.17), one obtains (B_{jk} is the period matrix)

$$2F = \frac{1}{2\pi i}\sum_{j,k=1}^g B_{jk} a_j a_k + 2\sum_1^g a_j \sum_1^\infty \sigma_{jk} T_k + \sum_{k,l=1}^\infty q_{kl} T_k T_l \quad (2.19)$$

Thus the expression (2.19) comes from the Riemann surface theory, without explicit reference to the BA function, and we consider now (**BA**) and

$$\psi = exp\left(\sum_1^\infty t_i\lambda^i\right) \times \frac{\tau(t - [\lambda^{-1}], \alpha)}{\tau(t, \alpha)} \qquad (2.20)$$

to which ideas of dKP can be applied to introduce the slow variables T_k. This means that we will be able to introduce slow variables in two different ways and the resulting comparisons will show an equivalence of procedures. In practice this will enable one to treat ϵ on the same footing in the Whitham theory and in the dispersionless theory. Thus from (2.15) and (2.20) one obtains an expression for τ of the form $(t_1 = x, \ t_2 = y, \ t_3 = t, \ \cdots)$

$$\tau(t, \alpha) = exp[\hat{F}(\alpha, t)]\Theta\,(\mathcal{A}(P_\infty)+ \qquad (2.21)$$

$$+ \sum_1^\infty (t_j/2\pi i)(\Omega_{jk}) + i\sum_1^g \alpha_j(B_{jk}) + z_0\Bigg)$$

where $k = 1, \cdots, g$ and

$$\hat{F}(\alpha, t) = \frac{1}{2}\sum_{k,l=1}^\infty q_{kl}t_kt_l - \frac{1}{4\pi i}\sum_{j,k=1}^\infty B_{jk}\alpha_j\alpha_k+$$

$$+ i\sum_1^g \alpha_j\sum_1^\infty \sigma_{jk}t_k + \sum_1^\infty d_kt_k \qquad (2.22)$$

(see also (Krichever (1988)) for a similar form - recall here $\mathcal{A}(P) = (\int_{P_0}^P d\omega_j)$ and $P_0 \neq P_\infty$ is required). Putting in the slow variables $T_k = \epsilon t_k$ and $a_k = i\epsilon\alpha_k$ one will find that the quadratic part of $\hat{F}(T/\epsilon, a/i\epsilon)$ in T and a is exactly $F(a, T)/\epsilon^2$ for F in (2.19); here $\tau = exp[(1/\epsilon^2)F + O(1/\epsilon)]$ as in (2.12). Suitable calculations are displayed in (Carroll (1997b)) to establish the relations between F and $\hat{F}$ as indicated. We note in addition that $F_{mn} \sim q_{mn}$. Indeed the BA function is defined from the Riemann surface via $d\Omega_n, \ d\omega_j$, and normalizations. It then produces a unique asymptotic expansion at ∞ which characterizes ψ near ∞ in terms of λ and hence must characterize the $d\Omega_n$ and $d\omega_j$ asymptotically. Moreover the normalizations must be built into these expansions since they were used in determining ψ. Thus we must have $F_{mn} \sim q_{mn}$ as a consequence of the BA function linking the differentials and the asymptotic expansions More explicitly one can identify dS and $d\mathcal{S}$ via uniqueness of the BA function and then derive $\partial_n dS = dB_n = \partial_n d\mathcal{S} = d\Omega_n$. **QED**

The exposition to follow using an expanded $dKdV_\epsilon$ theory based on (Carroll (1996a)) will establish more refined relations. One also sees that the Riemann surface background produces the a_i variables naturally here and we want now to find a definition of $\mathcal{F}$ which is based on dKdV quantities and not on ψ directly. Perhaps this will suggest another way to view duality based on $\mathcal{F}$. A priori there seems to be no conceptual reason why SW theory should have any relation to (X, ψ) duality, except perhaps that the background mathematics and development in (Faraggi and Matone (1997a), Faraggi and Matone (1997b), Matone (1995)) has many features related to SW mathematics. The connection indicated by (2.11) relating $F(a, T)$ and $\mathcal{F}$ seems to be purely formal.

3   Enhanced dKdV Theory

First, following (Carroll (1994a), Guha and Takasaki (1997), Kodama and Gibbons (1990)) there is a Hamilton-Jacobi (HJ) theory for KP with $P = S_X$. In particular one can take (P, X, T_n) (for any fixed $n \geq 2$) as Hamiltonian variables satisfying

$$\dot{P}_n = \frac{dP_n}{dT_n} = \partial \mathcal{B}_n; \quad \dot{X}_n = \frac{dX}{dT_n} = -\partial_P \mathcal{B}_n \tag{3.1}$$

Next from (2.8) one has $\Im \mathcal{F} = -X/\epsilon$ and $|\psi|^2 = exp[(2/\epsilon)\Re S]$. In order to have $|\psi|^2 \leq 1$ as a fundamental variable in the (X, ψ) theory some control over $\Re S$ is needed. One notes that dKdV involves $\lambda = \pm ik = P(1 + \sum_1^\infty U_m P^{-2m})$ with $\lambda^2 = -k^2$ real for k real (cf. (Carroll (1997a)) for details). In order to satisfy $|\psi|^2 \Im P = -1$ we will want $\lambda = -ik$ so that $Q = \Im P$ is negative. The U_m are real so $P = iQ$ corresponds to k real and $(-ik)_+^{2n+1} = \mathcal{B}_{2n+1}$ will be purely imaginary (only "times" t_{2n+1} arise in KdV). Further $\partial_P \mathcal{B}_{2n+1}$ will be real so the equations (3.1) make sense. In addition $\Re S = \Re[\sum_0^\infty T_{2n+1}(-ik)^{2n+1} + \sum_1^\infty S_{j+1}(-ik)^{-j}] = 0$ for k real since $S_{j+1} = -\partial_j F/j$ with $\partial_{2n} F = 0$. Now this would imply $|\psi|^2 = 1$, which is absurd, unless we introduce an expansion $S = \sum \epsilon^j S^j$ (i.e. take into account the $O(1)$ terms in $\psi = exp[(1/\epsilon)S + O(1)]$). Such $O(1)$ terms arise quite naturally from the VOE via $log\psi = (S/\epsilon) + O(1) = log\tau[\epsilon, T_n - (\epsilon/n\lambda^n)] - log\tau + \sum_1^\infty T_n \lambda^n/\epsilon$ with $log\tau = (F/\epsilon^2) + O(1/\epsilon)$ and $S_{n+1}^0 = -(\partial_n F/n)$. Thus e.g.

$$F\left(T_n - \frac{\epsilon}{n\lambda^n}\right) - F(T_n) = \tag{3.2}$$

$$= -\epsilon \sum_1^\infty \left(\frac{\partial_n F}{n\lambda^n}\right) + \frac{\epsilon^2}{2} \sum \left(\frac{F_{mn}}{mn}\right) \lambda^{-m-n} + O(\epsilon^3)$$

This leads to first terms of the form $S = S^0 + \epsilon S^1$ with

$$S^1 = \frac{1}{2} \sum \left(\frac{F_{mn}}{nm}\right) \lambda^{-m-n}; \quad S_X \sim P + \frac{\epsilon}{2} \sum \left(\frac{F_{1mn}}{nm}\right) \lambda^{-m-n} \tag{3.3}$$

Eventually we will want also $F = \sum \epsilon^j F^j$ as well. With this development one finds as a first approximation

$$|\psi|^2 = e^{2\Re S^1} = exp\left[\Re \sum \left(\frac{F^0_{mn}}{mn}\right) \lambda^{-m-n}\right] \qquad (3.4)$$

where only terms $F^0_{2m+1,2n+1}\lambda^{-2(m+n)-2}$ arise and these will be real for k real. Thus S^0 and $P^0 = S^0_X = iQ$ are imaginary while S^1 and $P^1 = S^1_X$ are real. Before embarking on a full expansion we note from (2.10) that in a first order expansion $\Im(P^0 + \epsilon P^1) = Q$ which implies $P = iQ = -i/2\Re\mathcal{F}$ and $X/\epsilon = -\Im\mathcal{F}$. Hence

THEOREM 3.1. A possibly fundamental symplectic form in the (X, ψ) theory arises from

$$\frac{dX}{\epsilon} \wedge dP \sim \frac{i}{2(\Re\mathcal{F})^2}d(\Im\mathcal{F} \wedge d(\Re\mathcal{F}) \qquad (3.5)$$

Now on a Riemann surface for finite zone KP (or KdV) one has formulas $F_{mn} = q_{mn}$ as indicated before and via $dS \sim dS$ one obtains also

$$- F_p = Res\, z^{-p}dS = Res\, z^{-p}dS = -\sum a_j\sigma_{jp} - \sum T_n q_{pn} \qquad (3.6)$$

Further there will be Whitham equations (cf. (Carroll and Chang (1996), Dubrovin (1996), Dubrovin (1992), Flaschka et al (1980), Fucito et al (1992), Krichever and Phong (1997), Krichever (1988), Krichever (1994), Nakatsu and Takasaki (1996)))

$$\frac{\partial d\omega_j}{\partial a_i} = \frac{\partial d\omega_i}{\partial a_j}; \quad \partial_n d\omega_j = \frac{\partial d\Omega_n}{\partial a_j}; \quad \partial_n d\Omega_m = \partial_m d\Omega_n \qquad (3.7)$$

which implies (cf. (2.16))

$$\partial_p q_{mn} = \partial_n q_{mp}; \quad \partial_n \sigma_{jm} = \frac{\partial q_{mn}}{\partial a_j}; \quad \frac{\partial \sigma_{jm}}{\partial a_i} = \frac{\partial \sigma_{im}}{\partial a_j} \qquad (3.8)$$

Thus (recall $\partial_n a_j = 0$) from (3.6) we obtain $F'' = \sum_1^g a_j\sigma'_{j1} + q_{11} + \sum_1^\infty T_n q'_{1n}$ leading to $\sum_1^g a_j(\partial F_{11}/\partial a_j) + \sum_1^\infty T_n\partial_n F_{11} = 0$ so F_{11} is in fact homogeneous of degree zero in (a, T) (cf. (Chang (1997))).

In (Carroll (1997a)) by way of motivation one examines various partial expansions for S and F but we go here to the general case. Thus take a finite zone KdV situation and go to the Schrödinger equation $\epsilon^2\psi'' + 2F''\psi = -E\psi$ directly with $S = \sum \epsilon^j S_j$ and $F = \sum \epsilon^j F_j$. Then write $\tilde{P} = S_X = \sum \epsilon^j P_j$ and for $\psi = exp[(1/\epsilon)S^0 + \sum_1^\infty \epsilon^{j-1}S^j]$ one obtains $\epsilon\psi_X = \tilde{P}\psi$ with $\epsilon^2\psi_{XX} = \epsilon\tilde{P}_X\psi + \tilde{P}^2\psi$ leading to

$$\left(\sum_0^\infty \epsilon^j P_j\right)^2 + \epsilon\sum_0^\infty \epsilon^j P'_j + 2\sum_0^\infty \epsilon^j F''_j = -E \qquad (3.9)$$

Consequently we have equations $P_0^2 + 2F_0'' = -E;\ 2P_0P_1 + P_0' + 2F_1'' = 0;\ P_1^2 + 2P_0P_2 + P_1' + 2F_2'' = 0;\ \cdots$ and in general for $k \geq 1$

$$\epsilon^{2k} \left[P_k^2 + 2\sum_0^{k-1} P_iP_{2k-i} + P_{2k-1}' + 2F_{2k}'' \right] = 0; \qquad (3.10)$$

$$\epsilon^{2k+1} \left[2\sum_0^{k} P_iP_{2k+1-i} + P_{2k}' + 2F_{2k+1}'' \right] = 0$$

Expanding (3.2) now yields

$$F\left(T_n - \frac{\epsilon}{n\lambda^n}\right) - F(T_n) = \qquad (3.11)$$

$$= \sum_{k=0}^{\infty} \epsilon^k \left[-\epsilon \sum_1^{\infty} \left(\frac{F_n^k}{n\lambda^n}\right) + \frac{\epsilon^2}{2} \sum \frac{F_{mn}^k}{mn}\lambda^{-m-n} + O(\epsilon^3) \right]$$

and for

$$S \sim \sum_1^{\infty} T_n\lambda^n + \sum_1^{\infty} \epsilon^k \sum_1^{\infty} S_{j+1}^k \lambda^{-j} \quad \text{with} \quad S^0 \sim \sum_1^{\infty} T_n\lambda^n + \sum_1^{\infty} S_{j+1}^0 \lambda^{-j}$$

$$\frac{1}{\epsilon}\sum_0^{\infty} \epsilon^k \left[-\sum_1^{\infty} \left(\frac{F_n^k}{n}\right)\lambda^{-n} + \frac{\epsilon}{2}\sum \left(\frac{F_{mn}^k}{nm}\right)\lambda^{-m-n} + O(\epsilon^2) \right] =$$

$$= \frac{1}{\epsilon}\sum_0^{\infty} \epsilon^k \sum_1^{\infty} S_{j+1}^k \lambda^{-j} \qquad (3.12)$$

There will be further terms $-(\epsilon^2/3!) \sum (F_{mnp}^k/mnp)\lambda^{-m-n-p} + \cdots$ in [ ] but the extra terms present no problem conceptually. This leads to the equations $S_{j+1}^0 = -F_j^0/j,\ S_2^k = -F_1^k$, and

$$S_{j+1}^k = -\frac{F_j^k}{j} + \frac{1}{2}\sum_1^{j-1} \left(\frac{F_{m,(j-m)}^{k-1}}{m(j-m)}\right) + \cdots +$$

$$+ \frac{(-1)^{k+1}}{(k+1)!} \sum_{\sum m_i = j} \frac{F_{m_1,\cdots,m_{k+1}}^0}{\prod_1^{k+1} m_i} \qquad (3.13)$$

We proceed now as indicated in Remark 6.9 of (Carroll (1997a)). One recalls the derivation of the dispersionless Hirota equations in (Carroll and Kodama (1995), Carroll (1994b)) via the differential Fay identity developed as in (Takasaki and Takebe (1992)). Thus we will expand the formula

$$\frac{1}{\epsilon^2} \sum_{m,n=1}^{\infty} \mu^{-m}\lambda^{-n}\chi_n(-\epsilon\tilde{\partial})\chi_m(-\epsilon\tilde{\partial})\tilde{F} = \qquad (3.14)$$

$$= log \left[1 + \frac{1}{\epsilon} \sum_1^\infty \frac{\mu^{-s} - \lambda^{-s}}{\mu - \lambda} \chi_s(-\epsilon \tilde\partial) \partial_X \tilde{F} \right]$$

where $\chi_n(t)$ is the Schur polynomial ($\sum j k_j = k$)

$$\chi_n(t) = \sum \left[\frac{t_1^{k_1} \, t_2^{k_2}}{k_1! \, k_2!} \cdots \right]; \quad \tilde\partial = (\partial_1, \frac{1}{2}\partial_2, \frac{1}{3}\partial_3, \cdots) \qquad (3.15)$$

in powers of ϵ (cf. here (Carroll (1997c), Chang (1997)) and remarks below for more details on this). The result should provide all compatibility relations between the coefficients $F^k_{ab \cdots q}$ at once and we will see how this goes below. Actually one can arrive at many different formulations of dispersionless Hirota equations for F^0 (and of enhanced equations for the F^k) by working with any of the dKdV hierarchy equations, upon stipulating the canonical forms for S^0, P_0 (and S^j, P_j). In using the KdV background via $L^2\psi = \lambda^2\psi$ we are in fact essentially defining consistency relations between the $F^k_{ab \cdots q}$ and thus the relations obtained from (3.10) should at least be a subset of those obtained in (3.14) (although perhaps expressed differently). We should include all possible $F^k_{ab \cdots q}$ of course and here $F^{2n}_{1,2m} = 0$ is entirely natural (with $F^0_{1,2m} = 0$ mandatory since we will preserve the dKdV origins in the F^0 terms). Now in order to maintain S^{2n} imaginary with S^{2n+1} real for real k we could adopt a strategy based on $KP1$. Thus introduce imaginary $T_{2n} = i\hat{T}_{2n}$ and think of $F^{2j+1} = i\hat{F}^{2j+1}$ as imaginary such that F^{2j+1}_{2n+1} is imaginary and F^{2j+1}_{2n} is real. This can be achieved e.g. via power series $F^{2j+1} \sim i \sum a_\alpha T_1^{\alpha_1} T_2^{2\alpha_2} \cdots$ with a_α real. In this context $V \sim \tilde{V} = -2 \sum \epsilon^j F^j_{11}$ could be complex however and this would interfere with the introduction of (X, ψ) duality; we will see that the consistency calculations indicate $F^{2j+1} = 0$ but we include them at times for completeness (cf. (Carroll (1997c), Chang (1997)) for more on this). Now look at the compatibility conditions arising from (3.10) (we assume first $F^{2j+1} = 0$ along with $F^{2n}_{1,2m} = 0$). The first terms are ($P_{2k} = i\hat{P}_{2k}$)

$$-Q^2 + 2F^0_{XX} + E = 0; \quad 2QP_1 + Q' = 0;$$

$$P_1' + P_1^2 - 2Q\hat{P}_2 + 2F^2_{XX} = 0; \qquad (3.16)$$

$$\hat{P}_2' + 2QP_3 + 2P_1\hat{P}_2 = 0; \quad P_3' - \hat{P}_2^2 - 2Q\hat{P}_4 + 2P_1P_3 + 2F^4_{XX} = 0; \quad \cdots$$

and look at the dKdV situation ($\epsilon \to 0$) in (3.14). Writing $\dot{P} \sim \dot{P}_0 = dP_0/d\lambda$ one obtains (cf. (Carroll and Kodama (1995)))

$$log \, \dot{P} = log \left(1 + \sum_1^\infty F_{1j} \lambda^{-j-1} \right) = \sum_2^\infty \lambda^{-j} \sum_{m+n=j} \frac{F_{mn}}{mn} \qquad (3.17)$$

whereas from (3.13) ($P_1 = \partial S^1$)

$$S^1_{j+1} = \frac{1}{2} \sum_1^{j-1} \frac{F^0_{m(j-m)}}{m(j-m)}; \quad P^1 = \frac{1}{2} \sum_2^\infty \lambda^{-j} \sum_{m+n=j} \frac{F^0_{1mn}}{mn} \qquad (3.18)$$

Now the first question arises about compatibility in $2QP_1 + Q' = 0$ or equivalently $2P_0 P_1 + P_0' = 0$ from (3.16). One can write ($\lambda \sim -ik$)

$$iQ = P_0 = \lambda \left[1 - \frac{2F_{11}^0}{\lambda^2} \right]^{1/2} = \tag{3.19}$$

$$= \lambda - \frac{F_{11}^0}{\lambda} - \sum_2^\infty \frac{(2m-3)(2m-5)\cdots 1 (F_{11}^0)^m}{m! \lambda^{2m-1}}$$

Now $P_1 = -P_0'/2P_0$ and $P_0 = iQ = \lambda[1 - (2F_{11}^0/\lambda^2)]^{1/2}$ which implies that $P_0' = -(F_{111}^0/P_0)$ and hence $P_1 = F_{111}^0/2P_0^2$. From (3.17) and the definition of P_1 in (3.18) we see that $(1/2)\partial \log \dot{P}_0 = \dot{P}_0'/2\dot{P}_0^2 = P_1$ and hence the compatibility of the (3.18) definition of P_1 with $2P_0 P_1 + P_0' = 0$ requires

$$\dot{P}_0'/2\dot{P}_0 = F_{111}^0/2P_0^2 . \tag{3.20}$$

To see that this is true we write from $P_0' = -F_{111}^0/P_0$ the equation $\dot{P}_0' = (F_{111}^0/P_0^2)\dot{P}_0$ so in fact (3.20) holds. Actually we have accomplished significantly more, namely we seem to obtain the dispersionless equations of (Carroll and Kodama (1995), Carroll (1994b)) by using the coefficient formulas for P_0 and P_1 in terms of F_{mn}^0 and the relation $2P_0 P_1 + P_0' = 0$. This is then apparently an example of a derivation of the dispersionless Hirota equations starting from a hierarchy equation and stipulating the canonical form of S and P in terms of $F \sim F^0$. In the present situation we can produce such equations for a whole hierarchy F^j at once, using $\tilde{F} = \sum \epsilon^j F^j$ and the machinery of (Carroll and Kodama (1995), Carroll (1994b), Takasaki and Takebe (1992)) involving the differential Fay identity (with suitable insertion of ϵ). One can continue this process and the next stage from (3.16) (to be confirmed via (3.14)) involves the creation of P_2 via $2P_0 P_2 + P_1^2 + P_1' + 2F_{11}^2 = 0$ and the compatibility equation $P_2' + 2(P_0 P_3 + P_1 P_2) = 0$ involving P_3. We must have $F_{11}^3 = 0$ here (which seems to imply in fact that $F^3 = 0$ - cf. (Carroll (1997c), Chang (1997))) since $exp[(2/\epsilon)\Re \tilde{S}]\Im \tilde{P} = -1$ implies that

$$2 \sum_0^\infty \epsilon^{2j} P_{2j+1} + [(\sum_0^\infty \epsilon^{2j} P_{2j}')/(\sum_0^\infty \epsilon^{2j} P_{2j})] = 0 \tag{3.21}$$

leading to $2P_0 P_1 + P_0' = 0$, $2(P_0 P_3 + P_1 P_2) + P_2' = 0, \cdots$. Thus our odd compatibility equations from (3.16) are exactly the conditions required for (X, ψ) duality (note that (3.21) yields the second formula in (3.10) with $F_{11}^{2j+1} = 0$), and see Section 5 for still another point of view.

Consider now (3.14) and let us try to write this in a form analogous to (3.17). Thus recall from (Carroll and Kodama (1995)) that one used

$$\sum_1^\infty \frac{F_{1n}^0}{n} \lambda^{-n} = \lambda - P_0(\lambda) \Rightarrow \tag{3.22}$$

$$1 - \sum_1^\infty \frac{\mu^{-n} - \lambda^{-n}}{\mu - \lambda} \frac{F_{1n}^0}{n} = \frac{P_0(\mu) - P_0(\lambda)}{\mu - \lambda} \to \dot{P}_0(\lambda)$$

as $\mu \to \lambda$. In the present situation we recall also that $S = S^0 + \sum_1^\infty \epsilon^j S^j$, $S^0 = \sum T_n \lambda^n + \sum_1^\infty S_{j+1}^0 \lambda^{-j}$, and $S^k = \sum_1^\infty S_{j+1}^k \lambda^{-j}$. The differential Fay identity arises from the Hirota bilinear identity, using the vertex operator equation (VOE) (2.20). Let us write now

$$\lambda - \Xi(\lambda, \epsilon) = -\frac{1}{\epsilon} \sum_1^\infty \chi_s(-\epsilon \tilde{\partial}) \lambda^{-s} \tilde{F}_1 \tag{3.23}$$

so that the right side of (3.14) involves

$$\mu - \lambda + \frac{1}{\epsilon} \sum_1^\infty \left(\mu^{-s} - \lambda^{-s} \right) \chi_s(-\epsilon \tilde{\partial}) \tilde{F}_1 = \Xi(\mu, \epsilon) - \Xi(\lambda, \epsilon) \tag{3.24}$$

Then (3.14) becomes for $\mu \to \lambda$

$$\frac{1}{\epsilon^2} \sum_1^\infty \lambda^{-m-n} \chi_n(-\epsilon \tilde{\partial}) \chi_m(-\epsilon \tilde{\partial}) \tilde{F} = log\, \dot{\Xi}(\lambda, \epsilon) \tag{3.25}$$

where

$$\dot{\Xi}(\lambda, \epsilon) = 1 + \frac{1}{\epsilon} \sum_1^\infty (-s) \lambda^{-s-1} \chi_s(-\epsilon \tilde{\partial}) \tilde{F}_1 = \dot{P}_0 + \tag{3.26}$$

$$+\frac{1}{\epsilon} \sum_1^\infty (-s) \lambda^{-s-1} \left(\left[\chi_s(-\epsilon \tilde{\partial}) + \frac{\epsilon \partial_s}{s} \right] F_1^0 + \chi_s(-\epsilon \tilde{\partial}) \sum_1^\infty \epsilon^{2j} F_1^{2j} \right)$$

In any event from (3.25) we have

$$\frac{1}{\epsilon^2} \sum_1^\infty \lambda^{-m-n} \chi_m(-\epsilon \tilde{\partial}) \chi_n(-\epsilon \tilde{\partial}) \tilde{F}_1 = \frac{\dot{\Xi}'}{\dot{\Xi}} \Rightarrow \tag{3.27}$$

$$\left(\frac{1}{\epsilon^2} \sum_1^\infty \lambda^{-m-n} \chi_m \chi_n \tilde{F}_1 \right) \times \left(1 - \frac{1}{\epsilon} \sum_1^\infty s \lambda^{-s-1} \chi_s \tilde{F}_1 \right) =$$

$$= -\frac{1}{\epsilon} \sum_1^\infty s \lambda^{-s-1} \chi_s \tilde{F}_{11}$$

(cf. (Carroll and Kodama (1995), Carroll (1994b)) for more on Schur polynomials), and repeat (3.9) in the form $\epsilon \tilde{P}' + \tilde{P}^2 + 2\tilde{F}_{11} = -E = \lambda^2$ with $\tilde{P} = \sum_0^\infty \epsilon^j P_j$, so that

$$\epsilon \partial \partial_\lambda \tilde{P} + 2\tilde{P} \partial_\lambda \tilde{P} = 2\lambda \Rightarrow P_0 \dot{P}_0 = 2\lambda;$$

$$\dot{P}'_{j-1} + 2\sum_0^j P_i \dot{P}_{j-i} = 0 \qquad (3.28)$$

We suggest now that the $F^j_{ab\cdots q}$ equations obtained from (3.27) and (3.28) are equivalent. More explicitly we want to show that Ξ of (3.23) corresponds to $\tilde{P}$ and secondly that (3.27) is equivalent to (3.9) - (3.10) and (3.28). We will prove most of this and refer to (Chang (1997)) for a more complete discussion. First one notes the fit at the first stages since $\Theta = (1/\epsilon^2)\sum \lambda^{-m-n}\chi_m\chi_n\tilde{F}_1$ begins with $\sum_1^\infty \lambda^{-m-n}(F^0_{1mn}/mn) \sim 2P_1$ and $1 - (1/\epsilon)\sum s\lambda^{-s-1}\chi_s\tilde{F}_1$ begins with $1 + \sum_1^\infty \lambda^{-s-1}F^0_{1s} \sim \dot{P}_0$. Hence the first term in (3.27) in an ϵ expansion is $2P_1\dot{P}_0 = \dot{P}'_0$ (known to be true) while $\dot{P}'_0 + 2(P_0\dot{P}_1 + P_1\dot{P}_0) = 0$ corresponds then to $4P_1\dot{P}_0 + 2P_0\dot{P}_1 = 0$ with $P_0 = \lambda[1 - (2F^0_{11}/\lambda^2)]^{1/2}$ and $P_1 = F^0_{111}/2P_0^2$. Thus $\dot{P}_1 = -2P_1\dot{P}_0/P_0 = -F^0_{111}\dot{P}_0/P_0^3$ which agrees with $\dot{P}_1 = \partial_\lambda(F^0_{111}/2P_0^2)$. Let us examine the next terms now. We will assume $F^{2j+1} = 0$ with $F^{2j}_{2n} = 0$ for convenience (and eventually by necessity). First let us recast the desiderata in simpler terms.

In this direction, since $exp(\sum_1^\infty t_n\lambda^n) = \sum_0^\infty \chi_j(t)\lambda^j$ with $\chi_0 = 1$, it follows that

$$exp(\sum_1^\infty 2t_n\lambda^n) = \sum_0^\infty \chi_\ell(2t)\lambda^\ell = \sum_0^\infty \lambda^\ell \sum_0^\ell \chi_m(t)\chi_{\ell-m}(t) \Rightarrow \qquad (3.29)$$

$$\Rightarrow \chi_\ell(2t) = \sum_0^\ell \chi_m(t)\chi_{\ell-m}(t) = 2\chi_\ell(t) + \sum_1^{\ell-1} \chi_m(t)\chi_{\ell-m}(t)$$

Hence in (3.14) we can write the left side as

$$\frac{1}{\epsilon^2}\sum_1^\infty \lambda^{-m-n}\chi_n\chi_m\tilde{F} = \frac{1}{\epsilon^2}\sum_2^\infty \lambda^{-s}\sum_1^{s-1}\chi_n\chi_{s-n}\tilde{F} = \qquad (3.30)$$

$$= \frac{1}{\epsilon^2}\sum_2^\infty \lambda^{-s}[\chi_s(-2\epsilon\tilde{\partial}) - 2\chi_s(-\epsilon\tilde{\partial})]\tilde{F}$$

Then (3.27) becomes

$$\left[\frac{1}{\epsilon^2}\sum_2^\infty \lambda^{-s}\left(\chi_s(-2\epsilon\tilde{\partial}) - 2\chi_s(-\epsilon\tilde{\partial})\right)\tilde{F}_1\right] \times \qquad (3.31)$$

$$\times \left[1 - \frac{1}{\epsilon}\sum_1^\infty s\lambda^{-s}\chi_s(-\epsilon\tilde{\partial})\tilde{F}_1\right] = -\frac{1}{\epsilon}\sum_1^\infty s\lambda^{-s-1}\chi_s(-\epsilon\tilde{\partial})\tilde{F}_{11}$$

There is still a lot of calculation so consider an argument based on (Carroll and Kodama (1995)) and (3.24) - (3.25) due to J.H. Chang. First to prove

$\tilde{P} = \Xi$ observe that via $exp[-\sum_1^\infty (\epsilon \partial_n / n\lambda^n)]\tilde{F}(T) = \tilde{F}(T_n - (\epsilon/n\lambda^n))$ one has

$$S = \sum_1^\infty T_n \lambda^n + \frac{1}{\epsilon} \left[\tilde{F}\left(T_n - \left(\frac{\epsilon}{n\lambda^n}\right)\right) - \tilde{F}(T_n) \right] = \tag{3.32}$$

$$= \sum_1^\infty T_n \lambda^n + \frac{1}{\epsilon} \sum_1^\infty \lambda^{-j} \chi_j(-\epsilon\tilde{\partial})\tilde{F}$$

leading to

$$\tilde{P} = \lambda + \frac{1}{\epsilon} \sum_1^\infty \lambda^{-j} \chi_j(-\epsilon\tilde{\partial})\tilde{F}_1 \tag{3.33}$$

which is exactly Ξ from (3.23). In passing Chang also notes that from (3.25) one can extend a formula of (Carroll and Kodama (1995)) to obtain

$$\dot{\Xi} = \sum_0^\infty \chi_\ell(\tilde{Z}_1 = 0, \tilde{Z}_2, \cdots, \tilde{Z}_\ell)\lambda^{-\ell}; \tag{3.34}$$

$$\tilde{Z}_j = \sum_{m+n=j} \chi_n(-\epsilon\tilde{\partial})\chi_m(-\epsilon\tilde{\partial})\frac{\tilde{F}}{\epsilon^2}$$

This corresponds to (3.27) in an integrated form and we then ask if this gives the same equations as (3.9) - (3.10) and (3.28). For this we recall that the differential Fay identity is equivalent to the Hirota bilinear identity in characterizing the KP (or KdV) hierarchy equations, so equations arising from $L^2\psi = \lambda^2\psi$ must be compatible with equations generated in (3.28). Whether all such equations arise via (3.9) - (3.10), (3.28) is actually immaterial for the connection to (X, ψ) duality, etc. but in any case numerous explicit calculations also show agreement at the early stages and provide expressions for the F_{mn}^2 in terms of F_{11}^2, F_{11}^0, and derivatives theoreof. Finally observe that (3.34) provides coefficient formulas via

$$\sum_0^\infty \chi_\ell(\tilde{Z}_1 = 0, \tilde{Z}_2, \cdots, \tilde{Z}_\ell)\lambda^{-\ell} = \dot{\Xi} = 1 - \frac{1}{\epsilon} \sum_2^\infty (\ell - 1)\lambda^{-\ell}\chi_{\ell-1}\tilde{F}_1$$

$$\Rightarrow -\frac{1}{\epsilon}(\ell - 1)\chi_{\ell-1}(-\epsilon\tilde{\partial})\tilde{F}_1 = \chi_\ell(\tilde{Z}_1 = 0, \tilde{Z}_2, \cdots, \tilde{Z}_\ell) \tag{3.35}$$

for $\ell \geq 2$, and this can be broken down into ϵ series to produce coefficient formulas. Summarizing one can state (cf. also Section 5)

THEOREM 3.2. The procedure indicated is formally consistent and allows determination of the P_n and F^{2n} from Q (and the F_{11}^{2n}). The ensuing compatibility equations for the $F_{ab\cdots q}^{2n}$ extend the dispersionless Hirota equations of (Carroll and Kodama (1995), Carroll (1994b)) and are included in those derived from the differential Fay identity via (3.27), (3.31), or (3.35). The adaption to (X, ψ) duality via $|\psi|^2 \Im \tilde{P} = -1$ (and hence to the third order differential equation (2.11) and to the derivations of enhanced dKdV in (Carroll (1997a), Carroll (1996a))) is automatic in this context (cf. Section 5) and corresponds to the conditions $F_{11}^{2j+1} = 0$ (which seems to also imply $F^{2j+1} = 0$).

4   An Example

Let us take as an example the famous elliptic curve for $SU(2)$ susy Yang-Mills (YM) introduced by Seiberg-Witten (SW) in (Seiberg and Witten (1994)). We utilize here also the relation of this to the Gurevich-Pitaevskij (GP) solutions developed in (Gurevich and Pitaevskij (1973)) (cf. further e.g. (Gorsky (1996), Gorsky et al (1997), Itoyama and Morozov (1995), Klemm (1997), Lerche (1996), Marshakov (1996))). Thus one puts branch points at $(-1, 1, u)$, with cuts $(-1, 1)$ and (u, ∞) (u real), and considers the hyperelliptic Riemann surface determined via $y^2 = (z^2 - 1)(z - u)$ (∞ will also be a branch point). Here a normalization $\Lambda \to 1$ has already been made from the original form, say $y^2 = (z^2 - \Lambda^4)(z - u)$; the scaling parameter Λ will be important when discussing renormalization theory in connection with the Whitham equations but we do not deal with this here (cf. (Carroll (1997c))). In any event the SW differential is now ($\lambda_1 = dz/y$, $\lambda_2 = zdz/y$)

$$\lambda_{SW} = \frac{1}{\pi\sqrt{2}} \frac{\sqrt{z - u}}{\sqrt{z^2 - 1}} dz = \frac{1}{\pi\sqrt{2}}[\lambda_2 - u\lambda_1] \tag{4.1}$$

with cycles A around $(-1, 1)$ and B around $(1, u)$ leading to

$$a = \oint_A \lambda = -2i\psi_1(u) = \sqrt{2}(u + 1)^{1/2} F(-\frac{1}{2}, \frac{1}{2}, 1, \frac{2}{u + 1}); \tag{4.2}$$

$$a^D = \oint_B \lambda = i\psi_2 = i\frac{u - 1}{2} F(\frac{1}{2}, \frac{1}{2}, 2, \frac{1 - u}{2})$$

where $(-\partial_z^2 + W(z))\psi = 0$ with $W = -(1/4)[1/(z^2 - 1)]$. Here λ_1 is a holomorphic differential. Now in some sense $\lambda_{SW} \sim dS$ in (2.17) with suitable fixing of time variables. Let us examine this via the GP solution of (Gurevich and Pitaevskij (1973)) which is an elliptic one gap solution to KdV, namely ($\mathcal{P} \sim$ Weierstrass function)

$$\tilde{u}(t_1, t_3, \cdots | u) = \frac{\partial^2}{\partial t_1^2} log\tau(t_1, t_3, \cdots | u) =$$

$$u_0 \mathcal{P}(k_1 t_1 + k_3 t_3 + \cdots + \Phi_o | \omega, \omega') + \frac{u}{3} \tag{4.3}$$

Here one recalls that for KdV there are differentials Ω_n, $n > 0$ as in Section 2 such that

$$d\Omega_{2j+1}(z) = \frac{P_{j+g}(z)}{y(z)} dz; \quad y^2 \sim (z^2 - 1)(z - u) \tag{4.4}$$

In particular one writes $(dp \sim d\Omega_1, \; dE \sim d\Omega_3)$

$$dp = \frac{z - \alpha(u)}{y(z)} dz; \quad dE = \frac{z^2 - \frac{1}{2}uz - \beta(u)}{y(z)} dz \qquad (4.5)$$

The normalization conditions $\oint_A d\Omega_i = 0$ yield $\alpha(u)$ and $\beta(u)$ immediately. Associated with this situation we have the classical Whitham theory giving

$$\frac{\partial d\Omega_i(z)}{\partial T_j} = \frac{\partial d\Omega_j(z)}{\partial T_i}; \quad d\Omega_i(z) = \frac{\partial dS(z)}{\partial T_i} \qquad (4.6)$$

where dS is some action term which traditionally was thought of in the form $dS = \sum T_i d\Omega_i$. Further taking coordinates u_α as the branch points of the corresponding hyperelliptic (here elliptic) surface one has the hydrodynamic type equations

$$\frac{\partial u_\alpha}{\partial T_i} = v_{ij}^{\alpha\beta}(u) \frac{\partial u_\beta}{\partial T_j}; \quad v_{ij}^{\alpha\beta} = \delta^{\alpha\beta} \left. \frac{d\Omega_i(z)}{d\Omega_j(z)} \right|_{z=u_\alpha} \qquad (4.7)$$

Now what happens is that after one switches on the Whitham dynamics the periods of the differential dS become the periods of the "modulated" function in (4.3). To be more precise it is shown in (Gorsky (1996)) that $(z \sim z^{-1}$ in Section 2) $dS = g(z|T_i, u)\lambda$ in the form

$$dS(z) = \left(T_1 + T_3(z + \frac{1}{2}u) + O(T_5) + \cdots \right) \times \frac{z - u}{y(z)} dz \qquad (4.8)$$

where λ is the SW differential $(z - u)dz/y(z)$. The demonstration is sort of ad hoc and goes as follows. Computing from (4.8) one gets

$$\frac{\partial dS(z)}{\partial T_1} = \left(z - u - (\frac{1}{2}T_1 + \frac{3}{4}uT_3) \frac{\partial u}{\partial T_1} \right) \frac{dz}{y(z)}; \qquad (4.9)$$

$$\frac{\partial dS(z)}{\partial T_3} = \left(z^2 - \frac{1}{2}uz - \frac{1}{2}u^2 - (\frac{1}{2}T_1 + \frac{3}{4}uT_3) \frac{\partial u}{\partial T_3} \right) \frac{dz}{y(z)}$$

and comparing with (4.5) gives

$$(\frac{1}{2}T_1 + \frac{3}{4}uT_3) \frac{\partial u}{\partial T_1} = \alpha(u) - u;$$

$$(\frac{1}{2}T_1 + \frac{3}{4}uT_3) \frac{\partial u}{\partial T_3} = \beta(u) - \frac{1}{2}u^2 \qquad (4.10)$$

Hence the construction gives a solution to the general Whitham equation of the form $\partial u/\partial T_3 = v_{31}(u)(\partial u/\partial T_1)$ with

$$v_{31} = \frac{\beta(u) - \frac{1}{2}u^2}{\alpha(u) - u} = \left. \frac{d\Omega_3(z)}{d\Omega_1(z)} \right|_{z=u} \qquad (4.11)$$

which is what it should be from the general Whitham theory (cf. (4.7)). It follows that for $T_3 = T_5 = \cdots = 0$

$$a = \frac{1}{T_1} \oint_A dS(z); \quad a_D = \frac{1}{T_1} \oint_B dS(z) \tag{4.12}$$

and also

$$\frac{\partial}{\partial T_i} \oint_A dS = \oint_A d\Omega_i = 0; \quad \frac{\partial}{\partial T_i} \oint_B dS = \oint_B d\Omega_i = k_i \tag{4.13}$$

where the k_i are the frequencies in the original KdV solution (4.3). We note that in (4.12) $(1/T_1)dS|_{T_3=0} = [(z-u)/y(z)]dz = \lambda_{SW}$ is fine but one does not have the form $dS = ad\omega + \sum T_n d\Omega_n$ as in (Itoyama and Morozov (1995)) or Section 2, where $cd\omega = dz/y(z) = dv$ is the canonical holomorphic differential with $\oint_A d\omega = 1$ and $\oint_B d\omega = \tau$ (note $c = c(u)$). It is at this point that one appreciates the subtlety of the argument in (Itoyama and Morozov (1995)) expressing dS as $ad\omega + \sum T_n d\Omega_n$. We can in fact write our present dS in the form of a "canonical" dS as follows. A little calculation gives, for $\gamma = (1/2)T_1 + \frac{3}{4}uT_3$ and $dz/y = cd\omega$, $\oint_A d\omega = 1$,

$$dS = g\lambda = T_1 dp + T_3 dE + c\gamma[T_1 \partial_1 u + T_3 \partial_3 u]d\omega =$$

$$= T_1 dp + T_3 dE + \hat{a} d\omega \tag{4.14}$$

This would then imply that $\oint_A dS = \hat{a} = c[(1/2)T_1 + \frac{3}{4}uT_3][T_1 \partial_1 u + T_3 v_{31} \partial_1 u]$ which should also equal $a = \oint_A \lambda_{SW}$. Evidently for $T_3 = 0$ one has $\hat{a} = (1/2)cT_1^2 \partial_1 u$ while from $(1/T_1)dS \sim dp + (1/2)cT_1 \partial_1 u d\omega$ there results then via (4.12) $a = (1/2)cT_1 \partial_1 u$. This suggests that a correct version of (4.12) should involve setting $T_1 = 1$ or else one should use a calculation based on $a = \oint_A dS$. We recall from (Itoyama and Morozov (1995)) that either explicit calculations or expressions as $a_i(h)$ with $h_k = h_k(T)$ suggest that a_i depends on T; in fact this is an illusion since the condition $\partial a_j / \partial T_i = 0$ is built into the constructions of (Itoyama and Morozov (1995)) or Section 2. Here we should check this from (4.14) (for $T_3 \neq 0$). Hence this suggests that the device of (Gorsky et al (1995)) was a little too ad hoc and one should instead use $dS = g\lambda + \mu(T, u)d\omega$ where μ is chosen so that $a = \oint_A dS = \oint_A dS$ or $\hat{a} + \mu = a$; this would involve conditions $\partial_i(\hat{a} + \mu) = 0$, which could in principle be realized.

In order to introduce a prepotential one compares to (Itoyama and Morozov (1995)) where $dS_{min} \sim \lambda_{SW}$ as in (4.1) with

$$\partial dS_{min}/\partial u = -(1/2\pi\sqrt{2})(dz/y) = -(1/2\pi\sqrt{2})c(u)d\omega .$$

Note also that $\oint_A \sim 2\int_{-1}^1$ and $\oint_B \sim 2\int_1^u$. Then $F_{red}(a)$ is defined as $F(a, T_n)$ for $T_n = 0$ when $n > 1$ or $n < -1$. Note here that a Toda theory with times T_0, $T_{\pm n}$ $(n \geq 1)$ is used in (Itoyama and Morozov (1995)) with two points

$P_\pm \sim \infty$; the approach of (Gorsky et al (1995)) sketched above uses a KP (or KdV) format with T_n ($N \geq 1$) and we saw that $T_n = 0$ for $n > 1$ with $T_1 = 1$ could be used in describing λ_{SW}. In (Itoyama and Morozov (1995)) this leads to $F_{red} = (1/2)aa^D - (iu/\pi)$ so F_{red} is not homogeneous whereas $2F = \sum T_n d\Omega_n + \sum a_i d\omega_i = \sum T_n \partial_n F + \sum a_i \partial F / \partial a_i$ is homogeneous of degree two (cf. also (Carroll and Chang (1996))). This is also developed in (Eguchi and Yang (1995)) for example where for susy YM coupled to massless hypermultiplets ($\sim T_0 = 0$) a basic formula is

$$aa^D - 2F_{red} = -T_1 \partial_1 F_{red} = 8\pi i b_1 u \qquad (4.15)$$

Thus $F_{red} = (1/2)aa^D - 4\pi i b_1 u \Rightarrow (i/\pi) = 4\pi i b_1$ or $b_1 = (1/4\pi^2)$ (cf. also (Matone (1995))). Note in (4.12) etc. in a KP format we could think of F_{red} in the same way since for $T_1 = 1$ and $T_n = 0$ ($n > 1$) (4.5) is $dS = \lambda_{SW} = dS_{min}$. Now the definition of a^D as $\partial F_{red}/\partial a$ implies

$$a^D = \frac{1}{2}a^D + \frac{1}{2}a\frac{\partial a^D}{\partial a} + \frac{1}{i\pi}\frac{\partial u}{\partial a};$$

$$\frac{\partial a}{\partial u} = -\frac{1}{\pi\sqrt{2}}\int_{-1}^{1}\frac{dz}{y} = -\frac{c(u)}{2\pi\sqrt{2}} \qquad (4.16)$$

Then e.g.

$$(1/2)a^D = (1/2)a(\partial a^D/\partial a) + (1/i\pi)(\partial u/\partial a)$$

and

$$\partial_a a^D = \partial_a \oint_B dS_{min} = \oint_B d\omega = \tau(u) = (4\pi i/g^2) + (\theta/2\pi)$$

where the important objects g, θ are functions of a or u. There are many other formulas and perspectives here but we don't need them at the moment.

Next note from (Alvarez-Gaumé and Hassan (1997)) that (4.2) can be rewritten for $k^2 = (2/1 + u)$ and $\hat{k}^2 = 1 - k^2$ with $(1 - u)/2 = -\hat{k}^2/k^2$ as (do not confuse k here with $\lambda = \pm ik$)

$$a(u) = \frac{4}{\pi k}E(k) = \frac{4}{\pi k}F\left(-\frac{1}{2}, \frac{1}{2}, 1, k^2\right);$$

$$a^D(u) = \frac{4}{i\pi}E\left(\frac{i\hat{k}}{k}\right) + \frac{4i}{\pi k^2}K\left(\frac{i\hat{k}}{k}\right) \qquad (4.17)$$

where $K(k) = (\pi/2)F(1/2, 1/2, 1, k^2)$ and the identity

$$c(1 - z)F(a, b, c, z) - cF(a, b - 1, c, z) + (c - a)F(a, b, c + 1, z) = 0$$

has been used. Further $F(a, b, c, z) = (1 - z)^{-a}F(a, c - b, c, z/(z - 1)) = (1 - z)^{-b}F(b, c - z, c, z/(z - 1))$ so one can write $K(i\hat{k}/k) = kK(\hat{k}) = k\hat{K}(k)$ and $E(i\hat{k}/k) = k^{-1}E(\hat{k}) = k^{-1}\hat{E}(k)$ leading to

$$a(u) = \frac{4}{\pi k} E(k); \quad a^D(u) = \frac{4}{\pi i} \frac{\hat{E}(k) - \hat{K}(k)}{k} \qquad (4.18)$$

and after some calculation

$$\tau = \partial^2 F/\partial a^2 = \partial a^D/\partial a = (da^D/dk)/(da/dk) = i\hat{K}/K .$$

To continue with expressions for a and u we go to (Dorey et al (1996), Matone (1995)) and recall

$$F(a) = F(a,0) = \frac{i}{2\pi} a^2 \log a^2 + \sum_0^\infty F_k a^{2-4k}; \qquad (4.19)$$

$$u = G(a) = \sum_0^\infty G_k a^{2-4k}; \quad G_0 = \frac{1}{2}; \quad G_k = 2\pi i k F_k$$

Note that any T_n dependence is removed here or distorted if $u = u(T_n)$ is assumed (see below).

Now we want to determine the nature of the dependence of Q on a in order to give explicit formulas in Theorem 3.2 for example. One has $V = -2F_{11}$ where $F^0 \sim F$ and F is given e.g. by (2.19) with dependence on a indicated via (3.8) for example. In particular, writing $B_{ij} \sim B \sim \tau$ here one has $(d\omega = d\omega_1 = -\sum_1^\infty \sigma_{1m} z^{m-1} dz$ where z as in Section 2 corresponds to z^{-1} in (4.8) etc.)

$$2F = \frac{1}{2\pi i} a^2 \tau + 2a \sum_1^\infty \sigma_{1k} T_k + \sum F_{k\ell} T_k T_\ell;$$

$$\partial_n \sigma_{1k} = \partial_a q_{kn} = \partial_a F_{kn} \qquad (4.20)$$

Further by construction as in (Itoyama and Morozov (1995)) $\partial_n a = 0$ but $\tau = \tau(u)$ and u as a branch point modulus depends on the T_n. Thus e.g. $\partial_X \tau \neq 0$ in general and further in general $\partial_X \sigma_{1k} = \partial_a F_{1k} \neq 0$ as well, since the σ_{1k} depend on u. Recall also here from (4.7)

$$\partial_i u = v_{ij} \partial_j u; \quad v_{ij} = \frac{d\Omega_i}{d\Omega_j}(z)\Big|_{z=u} \qquad (4.21)$$

Now explicitly $Q = k[1 + (2F_{11}/k^2)]^{1/2}$ which can be expanded as a power series in the $k^{\pm n}$ with coefficients which are powers of F_{11}. Thus in particular

$$\frac{\partial Q}{\partial a} = \frac{k}{2}\left[1 + \frac{2F_{11}}{k^2}\right]^{-1/2} \frac{2}{k^2} \partial_a F_{11} = \frac{\partial_a F_{11}}{Q} \qquad (4.22)$$

Consequently

THEOREM 4.1. Following (Carroll (1997a)) we have $\tilde{\mathcal{F}} = (X/i\epsilon) -$

$(1/2\Im\tilde{P})$ from (2.10) which we write as $\tilde{\mathcal{F}} = (X/i\epsilon) + \sum_0^\infty \epsilon^j \tilde{\mathcal{F}}_j$ so $\partial\tilde{\mathcal{F}}/\partial a = \partial\Re\tilde{\mathcal{F}}/\partial a$ leading to

$$\frac{\partial\mathcal{F}_0}{\partial a} = \frac{\partial Q/\partial a}{2Q^2} = \frac{\partial_a F_{11}}{2Q^3} \qquad (4.23)$$

and from (4.20)

$$\partial_a F_{11} = \partial_1 \sigma_{11} = \partial_X \sigma_{11} \qquad (4.24)$$

where for $z \sim \zeta \to \infty$ one has $d\omega = \sum_1^\infty \sigma_{1m}\zeta^{-m-1} = \sigma_{11}\zeta^{-1} + \cdots$.

5   General Remarks

There are several important themes connected with topics in this paper which need further development. One compelling direction seems to be the establishment of further connections to the Olavo theory of (Olavo (1995)) which derives quantum mechanics (QM) via the density matrix and statistical mechanics using an infinitesimal Wigner-Moyal transformation (cf. (Carroll (1997a), Carroll (1996a), Carroll (1997c)) for some preliminary remarks on this). Next is the determination of the roles of integrability in string theory and D-brane dynamics. In this direction one knows that e.g. SW theory with its Riemann surfaces and integrability connections has a stringy geometrical background based on Calabi-Yau (CY) manifolds (cf. (Alvarez-Gaumé and Hassan (1997), Donagi (1997), Donagi and Witten (1996), Gorsky et al (1995), Gorsky (1996), Gorsky et al (1997), Itoyama and Morozov (1995), Klemm (1997), Lerche (1996), Makeenko (1997a), Marshakov (1996), Marshakov et al (1997), Witten (1997))) but the structures of integrable models have not played as fundamental a role as in the early works on 2-D toy models in quantum gravity, matrix models, topological field theories (TFT), etc. Since D-brane dynamics seems to be a rather labyrinthine zoo at the moment it would be nice to have some coherent mathematical structures (such as integrability) for description and understanding of strings, M(matrix) theory, etc. (see here (Gorsky (1996), Gorsky et al (1997), Makeenko (1997a), Marshakov et al (1997), Witten (1997)) for some intelligible forays). Finally "everyone knows" that the Whitham equations are connected to renormalization theory but a good perspective on this seems to be missing (cf. (Gorsky et al (1995), Gorsky (1996), Gorsky et al (1997), Marshakov (1996))). Let us make a few remarks on the latter theme following (Faraggi and Matone (1997a)) (cf. also (Bonelli and Matone (1996))). Thus in (X, ψ) duality one considers $\hbar$ or $\epsilon = \hbar/\sqrt{2m}$ as a scale parameter for a statistical theory and looks at the quantity $G(\tau) = \sqrt{2mE}X/\hbar = \sqrt{E}X/\epsilon$ where $\tau = \partial^2\mathcal{F}/\partial(\psi^2) = \bar{\psi}/2\psi$ (note $\tau \sim \partial^2 F/\partial a^2$ in Section 4). Following (Bonelli and Matone (1996)) one writes

$$\beta(\tau) = \hbar\partial_\hbar\tau = \epsilon\partial_\epsilon\tau \qquad (5.1)$$

corresponding to a nonperturbative beta function in renormalization theory. Note the analogy here between X/ϵ and u/π (cf. (4.15)) or better, adjusting dimensional factors and scaling, $X\sqrt{2mE} \sim u$. Then from (5.1) one has

$$\hbar \partial_\hbar G(\tau) = \hbar \left(\frac{\partial \tau}{\partial \hbar} \right) \partial_\tau G = \beta \partial_\tau G = -G \tag{5.2}$$

Consequently $log\, G = - \int_{\tau_0}^{\tau} \beta^{-1}(y) dy$ which implies

$$G = G_0 e^{- \int_{\tau_0}^{\tau} \beta^{-1}(y) dy} \equiv X = \left(\frac{\hbar}{\hbar_0} \right) X_0 e^{- \int_{\tau_0}^{\tau} \beta^{-1}(y) dy} \tag{5.3}$$

This can be interpreted by saying that the space coordinate has an anomalous dimension determined by the beta function, and one notes that the Heisenberg uncertainty principle depends on the scale via

$$\Delta X \Delta p \geq \hbar = \hbar_0 + corrections \tag{5.4}$$

After the fact one could insert formulas for X and τ based on the (enhanced) dispersionless theory but this does not seem too productive. We remark also that the general form for a prepotential $\mathcal{F}$ in situations involving masses and scaling is (cf. (D'Hoker et al (1996)))

$$2\mathcal{F} = (\Lambda \partial_\Lambda + \sum m_i \partial_{m_i} + \sum a_j \partial_{a_j}) \mathcal{F} \tag{5.5}$$

to which can also be added $\sum T_n \partial_n \mathcal{F}$.

Let us call attention here also (following (Messiah (1995))) to a standard WKB approach for $\psi'' + (2m/\hbar^2)(E - V)\psi = 0$ (this is (2.1). One puts $\psi = exp(iW/\hbar)$ with $W = s + (\hbar/i)log\, A$ (where s and A are even functions of $\hbar$) to obtain

$$(s')^2 - 2m(E - V) = \hbar^2 \frac{A''}{A}; \quad 2A's' + As'' = 0 \tag{5.6}$$

From the second equation $A = c(s')^{-1/2}$ and the first equation becomes

$$(s')^2 = 2m(E - V) + \frac{\hbar^2}{2} \left[\frac{3}{2} \left(\frac{s''}{s'} \right)^2 - \frac{s'''}{s'} \right] \tag{5.7}$$

This equation is rigorously equivalent to the Schrödinger equation and the last term (Schwartzian derivative) is said to be the quantum correction in (Floyd (1997)) (this is of course D. Bohm's quantum potential). Note $exp[(i/\hbar)(s + (\hbar/i)log\, A)] = Aexp(is/\hbar)$ so one is extracting an amplitude factor A. In our notation we have $\psi = exp[\tilde{S}/\epsilon]$ with $\tilde{S} = \sum_0^\infty \epsilon^{2j+1} S^{2j+1} + \sum_0^\infty \epsilon^{2j} S^{2j}$ where S^{2j+1} is real and S^{2j} is imaginary (for k real, $\lambda = -ik$) so that $\Re \tilde{S} = \tilde{S}_{odd}$ and $i\Im \tilde{S} = \tilde{S}_{even}$. Consequently from $\tilde{S}/\epsilon \sim iW/\hbar$ we have $(1/\epsilon)(\Re \tilde{S} + i\Im \tilde{S}) \sim (is/\hbar) + log\, A$ so $s \sim \sqrt{2m} \Im \tilde{S}$ and $log\, A \sim (1/\epsilon)\tilde{S}_{odd}$. Writing $\tilde{P} = \tilde{S}' = \Re \tilde{S}' + i\Im \tilde{S}'$ we have $s' \sim \sqrt{2m}\Im \tilde{P}$ (or $is' \sim \sqrt{2m}\tilde{P}_{even}$) and $A'/A \sim (1/\epsilon)\tilde{P}_{odd}$. Hence $A = c(\Im \tilde{P})^{-1/2}$ is automatic or $|\psi|^2 = A^2 = c^2/\Im \tilde{P}$ in agreement with $|\psi|^2 \Im \tilde{P} = -1$ for $c = i$ (the i arises from $\Im \tilde{P} < 0$ - cf. (Carroll (1997a), Carroll (1996a))). This indicates that the condition $|\psi|^2 \Im \tilde{P} = -1$ is better viewed as a standard consequence of using the WKB formulation (hence no constraint) rather that as an adaption to (X, ψ) duality. The quantum correction term in (5.7), expressed by the Schwartzian $-(\hbar^2/2)\mathbf{SCH}(s')$ is interesting in itself since the Schwartzian appears in many contexts involving KdV and/or the Schrödinger equation (cf. (Matone (1995))). Here we can write this as ($s' \sim \sqrt{2m}\Im\tilde{P}$)

$$\mathbf{SCH}(s') = \frac{\Im\tilde{P}''}{\Im\tilde{P}} - \frac{3}{2}\left(\frac{\Im\tilde{P}'}{\Im\tilde{P}}\right)^2 \tag{5.8}$$

One is reminded of the third order equation (2.6) but (5.7) seems conceptually different (for more on this see (Carroll (1997c))). Actually from (Floyd (1997)) s' satisfies (5.7) and $1/s'$ satisfies the GD resolvant equation leading to (2.6).

References

Alvarez-Gaumé L. and Hassan S. (1997): hep-th/9701069

Bonelli G. and Matone M. (1996): *Phys. Rev. Lett.* **76** 4107-4110

Carroll R. (1997a): hep-th/9705229, *Nucl. Phys.* **B502** to appear

Carroll R. (1996a): hep-th/9607219, hep-th/9610216, and hep-th/9702138

Carroll R. and Chang J. (1996): solv-int/9612010, *Applicable Anal.* **64** (1997) 343-378

Carroll R. (1997b): Integrability and the Schrödinger equation, Proc. Conf. Gen. Analytic Functions, Graz, Austria, Jan. 6-10, 1997, to appear

Carroll R. (1996b): solv-int/9606005, Proc. Second World Congress Nonlin. Analysts, Athens, 1996, North-Holland, to appear

Carroll R. and Kodama Y. (1995): *Jour. Phys.* **A28** 6373-6387

Carroll R. (1994a): *Jour. Nonlin. Sci.* **4** 519-544; *Teor. Mat. Fizika* **99** (1994) 220-225

Carroll R. (1994b): Proc. NEEDS'94, World Scientific, 1995, pp. 24-33

Carroll R. (1993): *Applicable Anal.* **49** (1993) 1-31; **56** (1995) 147-164

Carroll R. (1991): Topics in soliton theory, North-Holland

Carroll R. (1992): Proc. First World Congress Nonlinear Analysts, Tampa, 1992, deGruyter, 1996, pp. 241-252

Carroll R. (1997c): Proc. ISAAC Conf., Univ. Delaware, June 1997, Kluwer, to appear

Chang J.H. (1997): Thesis, Univ. of Illinois, In preparation

D'Hoker E. Krichever I. and Phong D. (1996): hep-th/9610156

Donagi R. (1997): alg-geom/9705010

Donagi R. and Witten E. (1996): *Nucl. Phys.* **B460** 299-334

Dorey N. Khoze V. and Mattis M. (1996): hep-th/9606199

Dubrovin B. (1996): Lect. Notes Math. 1620, Springer, 1996, pp. 120-348

Dubrovin B. (1992): *Nucl. Phys.* **B379** 627-689

Eguchi T. and Yang S. (1995): hep-th/9510183

Faraggi A. and Matone M. (1997a): *Phys. Rev. Lett.* **78** 163-166

Faraggi A. and Matone M. (1997b): hep-th/9705108

Flaschka H. Forest M. and McLaughlin D. (1980): *Comm. Pure Appl. Math.* **33** 739-784

Floyd E. (1997): quant-ph/9707051

Fucito F. Gamba A. Martellini M. and Ragnisco O. (1992): *Inter. Jour. Mod. Phys.* B 6 2123-2147

Gorsky A. Krichever I. Marshakov A. Mironov A. and Morozov A. (1995): *Phys. Lett.* **B355** 466-474

Gorsky A. (1996): hep-th/9612238

Gorsky A. Gukov S. and Mironov A. (1997): hep-th/9707120

Guha P. and Takasaki K. (1997): solv-int/9705013

Gurevich A. and Pitaevskij L. (1973): *JETP* **65** 590-604

Itoyama H. and Morozov A. (1995): hep-th/9511126, 9512161, and 9601168

Klemm A. (1997): hep-th/9705131

Kodama Y. and Gibbons J. (1990): Workshop on nonlinear and turbulent processes in physics, World Scientific, 1990, pp. 166-180

Krichever I. and Phong D. (1997): *Journ. Diff. Geometry* **45** 349-389 (hep-th/9604191)

Krichever I. (1988): *Funkts. Anal. Prilozh.* **22** 200-213

Krichever I. (1994): *Comm. Pure Appl. Math.* **47** 437-475

Lerche W. (1996): hep-th/9611190

Makeenko Y. (1997a): hep-th/9704075

Marshakov A. (1996): hep-th/9602005, 9702083

Marshakov A. Martellini M. and Morozov A. (1997): hep-th/9706050

Matone M. (1995): *Phys. Lett.* **B357** 342-348

Messiah A. (1995): *Quantum mechanics* Vol. 1, North-Holland

Nakatsu T. and Takasaki K. (1996): *Mod. Phys. Lett.* **A11** 157-168

Olavo L. (1995): quant-ph/9503020, 9503021, 9503022, 9503024, 9503025, 9509012, 9509013, 9511028, 9511039, 9601002, 9607002, 9607003, 9609003, 9609023, 9703006, 9704004

Seiberg N. and Witten E. (1994): *Nucl. Phys.* **B426** 19-52; **B431** 484-550

Takasaki K. and Takebe T. (1992): *Inter. Jour. Mod. Phys.* **A Supp.** pp. 889-922; *Rev. Math. Phys.* **7** (1995) 743-808

Witten E. (1997): hep-th/9703166

Poisson Brackets for Densities of Functionals

Leonid A. Dickey

University of Oklahoma, Norman OK 73019, USA

Abstract. In the theory of integrable systems and in other field theories one usually deals with Poisson brackets between functionals. The latter are integrals of densities. Densities are defined up to divergence (boundary) terms. A question arises, is it possible to define a reasonable Poisson bracket for densities themselves? A general theory was suggested by Barnich, Fulp, Lada, Markl and Stasheff which has led them to the notion of a strong homotopy Lie group, (sh Lie). We are giving a few concrete examples.

1. Recall the Poisson bracket for the KdV equation (the Gardner-Zakharov-Faddeev bracket). There is one independent variable x and one field $u(x)$. All smooth functions of u and derivatives of u of arbitrary orders, $f(u, u', u'', ...)$ form a differential algebra $\mathcal{A}$. "Functionals" are integrals $\int \mathbf{f}$ where $\mathbf{f}$ is a differential form $\mathbf{f} = f dx$. The Poisson bracket between two functionals is

$$\{\int \mathbf{f}, \int \mathbf{g}\} = \int X'Y dx$$

where $X = \delta f/\delta u = \sum(-\partial)^i \partial f/\partial u^{(i)}$, $\partial = d/dx$ and $Y = \delta g/\delta u$.

A more general situation is when there are several variables $x_1, ..., x_n$, one or several fields $u_k(x_1, ..., x_n)$ (for simplicity of writing we will consider one u) and smooth functions $f, g, ...$ of these fields and their derivatives $u_{(\alpha)} = u_{\alpha_1,...,\alpha_n} = \partial^n u/\partial x_1^{\alpha_1}...\partial x_n^{\alpha_n}$. Differential forms $\Omega_k = \{f dx_{i_1}...dx_{i_k}\}$ make a complex

$$0 \rightarrow C \rightarrow \Omega_0 \xrightarrow{\mathrm{d}} \cdots \xrightarrow{\mathrm{d}} \Omega_n \tag{1}$$

known to be exact.

Elements of Ω_n will be denoted as $\mathbf{f} = f dx_1...dx_n$ and called "densities". Suppose Poisson bracket between integrals of densities be defined having the form $\{\int \mathbf{f}, \int \mathbf{g}\} = \int \{\mathbf{f}, \mathbf{g}\}$ where $\{\mathbf{f}, \mathbf{g}\}$ is some expression involving partial derivatives of f and g and their derivatives with respect to x_i, multiplied by $dx_1...dx_n$. For example, in the KdV case this is $X'Y dx$. The Poisson bracket is supposed to be bi-linear, skew-symmetric, and to have the Jacobi property:

$$\{\{\int \mathbf{f}, \int \mathbf{g}\}, \int \mathbf{h}\} + \text{cyclic} = 0 \ .$$

There is a consistency condition: if $\mathbf{f} = dQ$ where $Q \in \Omega_{n-1}$ then

$$\{dQ, \mathbf{g}\} = dR \tag{2}$$

where $R \in \Omega_{n-1}$.

The question is: can the expression $\{\mathbf{f}, \mathbf{g}\}$ be considered as a Poisson bracket of densities $\mathbf{f}$ and $\mathbf{g}$? In other words: can the density of the Poisson bracket be considered as a Poisson bracket of densities? There is very much freedom of choice here. Every density is defined up to a coboundary. It can be shown that changing the definition of $\{\mathbf{f}, \mathbf{g}\}$ by a coboundary, one can always make it bi-linear and skew-symmetric. Moreover, the consistency condition (2) can be made strong:

$$\{dQ, \mathbf{g}\} = 0 . \tag{3}$$

However, the Jacobi identity cannot be forced. The only thing we can be sure is that an integral of the expression $\{\{\mathbf{f}, \mathbf{g}\}, \mathbf{h}\}$+cyclic vanishes. This implies that

$$\{\{\mathbf{f}, \mathbf{g}\}, \mathbf{h}\} + \text{cyclic} = d\Psi_1(\mathbf{f}, \mathbf{g}, \mathbf{h}) \tag{4}$$

where $\Psi_1(\mathbf{f}, \mathbf{g}, \mathbf{h}) \in \Omega_{n-1}$ is also defined up to a coboundary. It is convenient to rename the Poisson bracket as $\{\mathbf{f}, \mathbf{g}\} = \Psi_0(\mathbf{f}, \mathbf{g})$. Using the freedom of choice, Ψ_1 can be made a tri-linear, skew symmetric form with a property

$$\Psi_1(dQ, \mathbf{g}, \mathbf{h}) = 0 . \tag{5}$$

We have mappings:

$$\Psi_0 = \{ \, , \, \} : \Omega_n \otimes \Omega_n \to \Omega_n \ ,$$

$$\Psi_1 : \Omega_n \otimes \Omega_n \otimes \Omega_n \to \Omega_{n-1} \ .$$

Is there a "Jacobi identity of the second order", involving both, Ψ_0 and Ψ_1 such that the left-hand side of it is equal some $d\Psi_2$, where $\Psi_2 : \otimes^3 \Omega_n \to \Omega_{n-2}$? The answer is positive. This theory was given in works by Barnich, Fulps, Lada, Markl and Stasheff, see Barnich et al. (1997). There is, indeed, a sequence of multi-linear, skew symmetric mappings

$$\Psi_k : \otimes^{k+2} \Omega_n \to \Omega_{n-k}$$

and some "higher Jacobi" between them. The strong consistency condition can be imposed: Ψ_k vanishes if one of its arguments is a coboundary. The sequence must truncate on the nth step. The sequence of mappings and higher Jacobi identities is called a sh Lie algebra (sh stands for strong homotopy).

Our goal is to provide this theory with a few examples, and explain the theory on these examples.

2. The first example is the Poisson bracket we started with: the Gardner-Zakharov-Faddeev bracket. Changing a density by divergence terms, one can write the same bracket in various forms, e.g.,

$$\left\{ \int \mathbf{f}, \int \mathbf{g} \right\} = \int X'Y \, \mathrm{d}x = -\int XY' \, \mathrm{d}x = (1/2) \int (X'Y - XY') \, \mathrm{d}x$$

$$= -\int \xi_f g \, \mathrm{d}x = \int \xi_g f \, \mathrm{d}x = (1/2) \int (\xi_g f - \xi_f g) \, \mathrm{d}x \ .$$

Here we used the following notations: $\xi_f = \partial_{X'}$ where ∂_a is a derivation

$$\partial_a = \sum a^{(i)} \partial / \partial u^{(i)}, \text{ where } a \in \mathcal{A} .$$

The derivation ξ_f is called the vector field corresponding to a Hamiltonian f.

Some of these densities are "better", the other are "worse". We chose the only density which satisfies both the exact skew-symmetry and exact consistency (3) conditions, this is

$$\{ \int \mathbf{f}, \int \mathbf{g} \} = (1/2) \int (X'Y - XY') \mathrm{d}x .$$

One of the quasi-Jacobi identities we have already: this is Eq. (4).

Proposition 1. The second, and the last, Jacobi identity is

$$\sum_{i<j,k<l} (-1)^{i+j} \Psi_1(\{\mathbf{f}_i, \mathbf{f}_j\}, \mathbf{f}_k, \mathbf{f}_l) = 0 \qquad (6)$$

where sequences (i, j, k, l) are permutations of $(1, 2, 3, 4)$.

Proof. Take the differential d of the left-hand side of (6) and use the definition (4) of Ψ_1 two times:

$$\sum_{i<j,k<l} (-1)^{i+j} [\{\{\{\mathbf{f}_i, \mathbf{f}_j\}, \mathbf{f}_k\}, \mathbf{f}_l\}$$

$$-\{\{\{\mathbf{f}_i, \mathbf{f}_j\}, \mathbf{f}_l\}, \mathbf{f}_k\}] = \sum_{i<j<k} (-1)^l \{\mathrm{d}\Psi_1(\mathbf{f}_i, \mathbf{f}_j, \mathbf{f}_k), \mathbf{f}_l\}$$

which vanishes by virtue of (3). The terms of type $\{\{\mathbf{f}_k, \mathbf{f}_l\}, \{\mathbf{f}_i, \mathbf{f}_j\}\}$ appearing in the process of the first transformation cancel. Thus, the left-hand side of (6) is a constant. The form Ψ_1 is defined up to a constant. One can always take it as an element of Ω_0 without a constant ($f(u, u', ...)$ is without a constant if $f(0, 0, ...) = 0$). Then Eq. (6) holds. $\square$

We do not know what is Ψ_1 except for the fact that it exists. Now we are going to express it in terms of physically significant quantity: the Poincaré invariant. Let $g(u, u', ...)$ be a smooth function (a "Lagrangian"). Take its variation. Integrating by parts, one obtains $\delta g = Y \delta u + \partial \Omega_g^1(\delta u)$ where $Y = \delta g / \delta u$ and $\Omega_g^1(\delta u)$ is 1-form in $\delta u, \delta u', ...$; in canonical variables this is $\sum p_i \delta q_i$, the Poincaré invariant related to the Lagrangian g. Computing this form on a vector ξ_f, one gets

$$\xi_f g = X'Y + \partial \Omega_g^1(X') .$$

Now, let $\Phi(\mathbf{f}, \mathbf{g}) = \Omega_g^1(X') - \Omega_f^1(Y')$. The statement is:

Proposition 2. The form Ψ_1 is the Lie-Cartan derivative of Φ:

$$\Psi_1(\mathbf{f}, \mathbf{g}, \mathbf{h}) = D\Phi(\mathbf{f}, \mathbf{g}, \mathbf{h}) = \xi_f \Phi(\mathbf{g}, \mathbf{h}) - \Phi(\{\mathbf{f}, \mathbf{g}\}, \mathbf{h}) + \text{cyclic} .$$

We skip a proof here.

A similar theory can be given for the Poisson bracket generated by a matrix first-order differential operator $I\partial + U$ where U is a matrix.

3. Now we go to the case of two independent variables, x and y. This will be a Poisson bracket borrowed from the theory of two-dimensional flow of inviscid incompressible fluid (see, e.g., Olver (1993)). The main field variable is here $\omega(x,y)$, the vorticity (it can depend also on t, this dependence is not interesting to us now). The space Ω_0 consists on smooth functions $f(\omega, ..., \omega_\alpha, ...)$ where (α) is a multi-index, $\omega_{(\alpha)} = \partial^{\alpha_1+\alpha_2}/\partial x^{\alpha_1}\partial y^{\alpha_2}$. If Ω_1 and Ω_2 are the spaces of 1- and 2-forms, then we have the complex:

$$0 \to C \to \Omega_0 \xrightarrow{d} \Omega_1 \xrightarrow{d} \Omega_2 \qquad (7)$$

which is exact. Let a and b belong to Ω_0. Then $(a,b) = a_x b_y - a_y b_x$ is the Jacobian. It satisfies the Jacobi identity

$$((a,b),c) + \text{cyclic} = 0 \ .$$

Vector fields are derivations $\partial_a = \sum a_{(\alpha)}\partial/\partial\omega_{(\alpha)}$ which can act in all Ω_i and commute with d.

Vector fields of a special type: $\partial_{(\omega,a)}$ form a Lie subalgebra since

$$[\partial_{(\omega,a)}, \partial_{(\omega,b)}] = \partial_{(\omega,c)}, \quad \text{where } c = (a,b) + \partial_{(\omega,a)}b - \partial_{(\omega,b)}a \ .$$

If $a = \delta f/\delta\omega$, we denote $\partial_{(\omega,a)} = \xi_f$.

Now, a 2-form is introduced in the space of all vector fields $\partial_{(\omega,a)}$:

$$\Omega^2(\partial_{(\omega,a)}, \partial_{(\omega,b)}) = \omega \cdot (a,b)dxdy \in \Omega_2.$$

It is easy to show by a direct calculation that this form is closed with respect to the Lie-Cartan derivative D:

$$D\Omega^2(\partial_{(\omega,a)}, \partial_{(\omega,b)}, \partial_{(\omega,c)})$$

$$\equiv \partial_{(\omega,a)}\Omega^2(\partial_{(\omega,b)}, \partial_{(\omega,c)}) - \Omega^2([\partial_{(\omega,a)}, \partial_{(\omega,b)}], \partial_{(\omega,c)}) + \text{cyclic} = 0 \ .$$

We define now Poisson bracket in Ω_2: let $\mathbf{f} = fdxdy$, and $\mathbf{g} = gdxdy$, then

$$\{\mathbf{f}, \mathbf{g}\} = \Omega^2(\xi_f, \xi_g) \ .$$

There are two well-known corollaries of the closedness of the form Ω^2:

$$\xi_{\{\mathbf{f},\mathbf{g}\}} = [\xi_f, \xi_g] \qquad (8)$$

and

$$\int [\{\{\mathbf{f}, \mathbf{g}\}, \mathbf{h}\} + \text{cyclic}] = 0$$

which imply the existence of a form $\Psi_1 : \otimes^3\Omega_2 \to \Omega_1$ such that

$$\{\{\mathbf{f},\mathbf{g}\},\mathbf{h}\} + \text{cyclic} = d\Psi_1(\mathbf{f},\mathbf{g},\mathbf{h}) \ . \tag{9}$$

It can be shown that the freedom of option makes it possible to make Ψ_1 tri-linear, skew symmetric, and satisfying the strict condition (5).

Proposition 3. There is a four-linear, skew symmetric mapping Ψ_2 : $\otimes^4 \Omega_2 \to \Omega_0$ such that

$$\sum_{i<j,k<l} (-1)^{i+j}\Psi_1(\{\mathbf{f}_i,\mathbf{f}_j\},\mathbf{f}_k,\mathbf{f}_l) = d\Psi_2(\mathbf{f}_1,\mathbf{f}_2,\mathbf{f}_3,\mathbf{f}_4) \ ; \tag{10}$$

the mapping can be chosen to satisfy the strict condition

$$\Psi_2(dQ,\mathbf{f}_2,\mathbf{f}_3,\mathbf{f}_4) = 0 \ . \tag{11}$$

Proof. In the same way as before, we prove that the operator d annihilates the left-hand side of the equality. However, this does not mean that the left-hand side is zero, it only means that this is a coboundary, $d\Psi_2$. We skip the detail of the choice of Ψ. □

Proposition 4. The third (and the last) Jacobi identity is

$$\sum_{i<j,k<l<m} (-1)^{i+j}\Psi_2(\{\mathbf{f}_i,\mathbf{f}_j\},\mathbf{f}_k,\mathbf{f}_l,\mathbf{f}_m) = 0 \ . \tag{12}$$

Proof. Let us apply the operator d to the left-hand side of the equality. Using Eq. (10) twice, we get

$$\sum_{i<j<k,l<m} (-1)^{l+m}\Psi_1(d\Psi_1(\mathbf{f}_i,\mathbf{f}_j,\mathbf{f}_k),\mathbf{f}_l,\mathbf{f}_m)$$

which vanishes by virtue of (5). This means that the left-hand side of (12) is a constant; if Ψ_2 does not contain constants which always can be assumed, then the left-hand side is zero. □

Markl's observation. There still remains some freedom of choice of Ψ_1 after all the above requirements are satisfied. One can choose Ψ_1 such that $\Psi_2 = 0$ and

$$\sum_{i<j,k<l} (-1)^{i+j}\Psi_1(\{\mathbf{f}_i,\mathbf{f}_j\},\mathbf{f}_k,\mathbf{f}_l) = 0 \ .$$

Proof. A usual proof of the exactness of the complex (7) is based on the existence of a contracting homotopy operator s:

$$\Omega_0 \overset{s}{\leftarrow} \Omega_1 \overset{s}{\leftarrow} \Omega_2$$

such that the operator identity $sd + ds = 1$ holds on elements of Ω_1. It is also true on elements of $d\Omega_1 \subset \Omega_2$: $(sd+ds)dq = d(sd)q = d(1-ds)q = dq$.

Now, let

$$\Psi_1(\mathbf{f},\mathbf{g},\mathbf{h}) = s(\{\{\mathbf{f},\mathbf{g}\},\mathbf{h}\} + \text{cyclic}) \ .$$

Let us show that this Ψ_1 satisfies Eq. (9), i.e., this is a legitimate choice of Ψ_1.

$$\mathrm{d}\Psi_1(\mathbf{f},\mathbf{g},\mathbf{h}) = \mathrm{d}s(\{\{\mathbf{f},\mathbf{g}\},\mathbf{h}\} + \text{cyclic})$$

$$= (1 - s\mathrm{d})(\{\{\mathbf{f},\mathbf{g}\},\mathbf{h}\} + \text{cyclic}) = \{\{\mathbf{f},\mathbf{g}\},\mathbf{h}\} + \text{cyclic}$$

as required. Finally,

$$\sum_{i<j,k<l} (-1)^{i+j}\Psi_1(\{\mathbf{f}_i,\mathbf{f}_j\},\mathbf{f}_k,\mathbf{f}_l) = s \sum_{i<j,k<l} (-1)^{i+j}[\{\{\{\mathbf{f}_i,\mathbf{f}_j\},\mathbf{f}_k\},\mathbf{f}_l\}$$

$$-\{\{\{\mathbf{f}_i,\mathbf{f}_j\},\mathbf{f}_l\},\mathbf{f}_k\}] = s \sum_{i<j<k} (-1)^l\{\mathrm{d}\Psi_1(\mathbf{f}_i,\mathbf{f}_j,\mathbf{f}_k),\mathbf{f}_l\} = 0 \ .$$

Soloviev (Soloviev (1993-6)) tried to find out whether it is possible to truncate the sequence on the first step by a different choice of the Poisson bracket, i.e., to obtain a pure Jacobi identity for the Poisson bracket Ψ_0. In some cases, among them KdV, he managed to do that, in a general form he failed.

Skipped proofs and comments will be published in a more detailed paper.

References

Barnich, G., Fulp, R., Lada, T., and Stasheff, J. (1997): The sh Lie structure of Poisson brackets in field theory. To be published

Olver, P.J. (1993): Applications of Lie groups to differential equations. 2nd ed., Springer

Soloviev, V.O. (1993), (1995), (1996): Boundary values as Hamilton variables. I, hep-th/9305133
Boundary values as Hamilton variables. II, q-alg/9501017
Boundary terms and their Hamiltonian dynamics. hep-th/9601107

Hamiltonian Dynamics, Classical R-Matrices and Isomonodromic Deformations

J. Harnad[1,2]

[1] Department of Mathematics and Statistics, Concordia University, 7141 Sherbrooke W., Montréal, Canada H4B 1R6,
[2] Centre de recherches mathématiques, Université de Montréal C. P. 6128-A, Montréal, Canada H3C 3J7

Abstract. The Hamiltonian approach to the theory of dual isomonodromic deformations is developed within the framework of rational classical R–matrix structures on loop algebras. Particular solutions to the isomonodromic deformation equations appearing in the computation of correlation functions in integrable quantum field theory models are constructed through the Riemann-Hilbert problem method. The corresponding τ–functions are shown to be given by the Fredholm determinant of a special class of integral operators.

Keywords. Integrable systems, isomonodoromic deformations, classical R–matrix, loop algebras, Riemann–Hilbert problem, τ–function, Fredholm determinants

1 Monodromy Preserving Hamiltonian Systems

1.1 Isomonodromic Deformation Equations

Monodromy preserving deformations of rational covariant derivative operators of the form:

$$\mathcal{D}_\lambda := \frac{\partial}{\partial \lambda} - \mathcal{N}(\lambda), \tag{1.1}$$

where

$$\mathcal{N}(\lambda) := B + \sum_{i=1}^{n} \frac{N_i}{\lambda - \alpha_i}, \tag{1.2}$$

B is the diagonal $r \times r$ matrix

$$B = \mathrm{diag}(\beta_1, \ldots, \beta_r), \tag{1.3}$$

and the matrices $\{N_i\}_{i=1\cdots n}$ are $r \times r$ matrix functions of the $n+r$ deformation parameters $\{a_i, \beta_a\}$, were studied by Jimbo $et.$ $al.$ in (Jimbo et al (1980)), (Jimbo et al (1981)), (Jimbo and Miwa (1981)). It was shown there that the most general differentable monodromy preserving deformations of such operators are determined by the integrable Pfaffian system:

$$dN_i = -\sum_{\substack{j=1 \\ j\neq i}}^{n}[N_i, N_j]d\log(\alpha_i - \alpha_j) - [N_i, d(\alpha_i B) + \Theta], \qquad (1.4)$$

where Θ is the $r \times r$ matrix with elements

$$\Theta_{ab} = (1 - \delta_{ab})(\sum_{i=1}^{n} N_i)_{ab}d\log(\beta_a - \beta_b). \qquad (1.5)$$

Such operators and their monodromy are of great importance in the theory of quantum integrable systems, since the computation of correlation functions in such systems very often leads to particular solutions to such systems (Its et al. (1990)), (Korepin et al. (1993)). In the following subsection, it will be shown how these equations may be understood as a compatible set of nonautonomous Hamiltonian systems generated by commuting Hamiltonians that are spectral invariants of the matrix $\mathcal{N}(\lambda)$. In subsequent sections, the classical R-matrix approach to such systems will be explained and the computation of certain solutions related to Fredholm determinant calculations via the matrix Riemann–Hilbert problem will be decribed.

1.2   Nonautonomous Hamiltonian Structure

We begin with the Lie Poisson structure on $(\oplus_{i=1}^{n}\mathfrak{gl}(r))^*$, defined by the following Poisson brackets between the various matrix elements

$$\{(N_i)_{ab}, (N_j)_{cd}\} = \delta_{ij}\left(\delta_{bc}(N_i)_{ad} - \delta_{ad}(N_i)_{cb}\right). \qquad (1.6)$$

The system (1.4) can be expressed in multi–Hamiltonian form by introducing the following Hamiltonian 1–form on the parameter space

$$\theta := \sum_{i=1}^{n} H_i d\alpha_i + \sum_{a=1}^{r} K_a d\beta_a \qquad (1.7)$$

where

$$H_i := \operatorname{tr}(BN_i) + \sum_{\substack{j=1 \\ j\neq i}}^{n} \frac{\operatorname{tr}(N_i N_j)}{\alpha_i - \alpha_j}, \qquad i = 1,\ldots,n \qquad (1.8)$$

$$K_a := \sum_{i=1}^{n} \alpha_i (N_i)_{aa} + \sum_{\substack{b=1 \\ b\neq a}}^{r} \frac{(\sum_{i=1}^{n} N_i)_{ab}(\sum_{j=1}^{n} N_j)_{ba}}{\beta_a - \beta_b}, \qquad a = 1,\ldots,r (1.9)$$

Equations (1.4) may then equivalently be written in multi–Hamiltonian form as

$$dN_i = \{N_i, \theta\}. \qquad (1.10)$$

Involutiveness of the H_i's and K_a's, which will be explained in the following section, then implies that the differential form θ on the parameter space is in fact closed

$$d\theta = 0, \tag{1.11}$$

which allows one to introduce the τ–function (cf. Jimbo et al (1980), Jimbo et al (1981)) by the formula

$$\theta = d\log\tau. \tag{1.12}$$

In the next section, we show how the above Hamiltonian structure may very naturally be viewed as the restriction of the rational R–matrix structure on a loop algebra to a finite dimensional Poisson submanifold. From this the commutativity of the Hamiltonians (H_i, K_a) defined above follows.

2 Loop Algebra Moment Maps, Spectral Invariants and Isomonodromic Deformation Equations

The following discussion is based on the approach developed in Harnad (1994).

2.1 Dual Moment Maps and Split R-matrix Structure

We introduce an auxiliary symplectic vector space (M, ω), which will be referred to as the *generalized Moser space*, consisting of pairs (F, G) of complex $N \times r$ matrices whose elements are viewed as canonically conjugate variables. Thus, the symplectic form ω is just

$$\omega := \operatorname{tr}(dF \wedge dG^T). \tag{2.1}$$

We use the following notation to denote the loop algebra of $r \times r$ matrices depending on a loop parameter λ, viewed as a point on a circle S^1 in the complex λ–plane, and its splitting into negative a positive Fourier components

$$\widetilde{\mathfrak{gl}}(r) = \widetilde{\mathfrak{gl}}(r)_+ + \widetilde{\mathfrak{gl}}(r)_- \sim \widetilde{\mathfrak{gl}}(r)^* \tag{2.2}$$

$$\widetilde{\mathfrak{gl}}(r)_\pm^* \sim \widetilde{\mathfrak{gl}}(r)_\mp \tag{2.3}$$

The identification with the dual space indicated in eq. (2.2) is defined through the Ad–invariant scalar product

$$< X_1, X_2 > = \oint_{S^1} \operatorname{tr}\left(X_1(\lambda)X_2(\lambda)\right) d\lambda. \tag{2.4}$$

The rational R–matrix structure is obtained by just redefining the Lie algebra structure in such a way that the new algebra splits into a Lie algebraic direct sum of the positive and negative frequency parts, with a change of sign in the Lie product for the second summand

$$\widetilde{\mathfrak{gl}}_R(r) := \widetilde{\mathfrak{gl}}(r)_+ \ominus \widetilde{\mathfrak{gl}}(r)_- \sim \widetilde{\mathfrak{gl}}_R(r)^*. \tag{2.5}$$

The rational R–matrix structure is then just the corresponding Lie Poisson structure on $\widetilde{\mathfrak{gl}}_R(r)^*$. Expressed in terms of the individual matrix elements this gives

$$[\mathcal{N}_{ab}(\lambda), \mathcal{N}_{cd}(\mu)] = \frac{\delta_{ad}(\mathcal{N}_{cb}(\lambda) - \mathcal{N}_{cb}(\mu)) - (ad \leftrightarrow cb)}{\lambda - \mu}. \tag{2.6}$$

This can be expressed more succintly in the tensorial (St. Petersburg) notation as follows

$$\{\mathcal{N}(\lambda) \otimes \mathcal{N}(\mu)\} = [r(\lambda - \mu),\ \mathcal{N}(\lambda) \otimes \mathbf{I} + \mathbf{I} \otimes \mathcal{N}(\mu)], \tag{2.7}$$

where $\mathcal{N}(\lambda)$ and $\mathcal{N}(\mu)$ are viewed as endomorphisms of $\mathbf{C}^n$ and the rational R–matrix $r(\lambda - \mu)$ is the endomorphism of $\mathbf{C}^n \otimes \mathbf{C}^n$ defined by

$$r(\lambda - \mu) := \frac{P_{12}}{\lambda - \mu} \in \mathrm{End}(\mathbf{C}^n \otimes \mathbf{C}^n), \tag{2.8}$$

where P_{12} denotes the endomorphism that interchanges the first and second factors in $\mathbf{C}^n \otimes \mathbf{C}^n$.

Now let A and B be the diagonal $N \times N$ and $r \times r$ matrices, respectively, defined by

$$A := \mathrm{diag}(\alpha_i) \in \mathfrak{gl}(N), \qquad B := \mathrm{diag}(\beta_a) \in \mathfrak{gl}(r), \tag{2.9}$$

where the eigenvalues $\{\alpha_i\}_{i=1\cdots n}$ have multiplicities $\{k_i\}_{i=1,\dots,n}$, and the eigenvalues $\{\beta_a\}_{a=1\cdots r}$ are multiplicity free. Define the Poisson subspace $\mathbf{g}_B^A \subset \widetilde{\mathfrak{gl}}_R(r)^*$ by

$$\mathbf{g}_B^A := \{\mathcal{N}(\lambda) = B + \sum_{i=1}^n \frac{N_i}{\lambda - \alpha_i},\ N_i \in \mathfrak{gl}(r)\} \sim \sum_{i=1}^n \mathfrak{gl}^*(r). \tag{2.10}$$

Then the following defines a Poisson quotient map of the symplectic space M, such that the image is identified with a Poisson submanifold of $\mathbf{g}_B^A$.

$$\widetilde{J}_B^A : M \longrightarrow \mathbf{g}_B^A$$
$$\widetilde{J}_B^A : (F, G) \longmapsto B + G^T(A - \lambda I_r)^{-1}F =: \mathcal{N}(\lambda) \tag{2.11}$$

$$\mathcal{N}(\lambda) = B + \sum_{i=1}^n \frac{N_i}{\lambda - \alpha_i} \in \widetilde{\mathfrak{gl}}^*(r)_R \tag{2.12}$$

$$N_i := -G_i^T F_i, \qquad F_i, G_i \in M^{k_i \times r}. \tag{2.13}$$

We may now apply the standard classical R–matrix theory to deduce a set of commuting Hamiltonian flows on the space M, generated by the spectral invariant Hamiltonians on $\widetilde{\mathfrak{gl}}(r)^*$, pulled back to M via the above Poisson map, for which the Hamiltonian flows are represented by Lax equations. The resulting Hamiltonian flows are therefore isospectral for the matrix $\mathcal{N}(\lambda)$. We denote by

$$\mathcal{I}_B^A := \mathcal{I}(\widetilde{\mathfrak{gl}}(r)^*)|_{\mathbf{g}_B^A} \tag{2.14}$$

the ring of spectral invariants restricted to $\mathbf{g}_B^A$. Then the classical R–matrix theory in this case tells us that:

(i) $\mathcal{I}_B^A$ is Poisson commutative.
(ii) For $H \in \mathcal{I}_B^A$, Hamilton's equations have the Lax form:

$$\frac{d\mathcal{N}}{dt} = [\mathcal{A}_\sigma^H, \mathcal{N}], \tag{2.15}$$

$$\mathcal{A}_\sigma^H := \sigma dH_+(\mathcal{N}) + (\sigma - 1)dH_-(\mathcal{N}) =: \mathcal{P}_\sigma(dH(\mathcal{N})), \tag{2.16}$$

where the subscripts $\pm$ denote projections to the positive and negative Fourier components and $\sigma \in \mathbf{R}$ is arbitrary.
It follows that the spectral curve defined by the characteristic equation

$$\det(\mathcal{N}(\lambda) - zI_r) = 0, \tag{2.17}$$

where

$$\mathcal{N}(\lambda)B + G^T(A - \lambda I_N)^{-1}, \tag{2.18}$$

is invariant under the resulting Hamiltonian flows.

In order to apply this to the isomonodromic deformation equations considered above, we must adapt these results to the case of nonautonomous Hamiltonian systems, in which the flow parameters are reinterpreted as deformation parameters upon which the spectral invariant Hamiltonians may explicitly depend. This will be done in the next subsection.

2.2 Nonautonomous Systems: Isomonodromic Deformations

Letting the matrices A and B depend explicitly on some deformation parameter t

$$A = A(t), \quad B = B(t), \tag{2.19}$$

the above Lax equations must be modified to take the resulting explicit t–dependence of the matrix $\mathcal{N}(\lambda)$ into account. This gives the nonautonomous system

$$\frac{d\mathcal{N}}{dt} = [\mathcal{A}_\sigma^H, \mathcal{N}] + \frac{\partial \mathcal{N}}{\partial t}. \tag{2.20}$$

Suppose now that, for some σ and H, the following special condition holds:

$$\frac{\partial \mathcal{N}}{\partial t} = \frac{\partial \mathcal{A}_\sigma^H}{\partial \lambda}. \tag{2.21}$$

It follows that Hamilton's equations become *isomonodromic deformation equations*, since eq. (2.20) then takes the form of commutativity conditions

$$[\mathcal{D}_\lambda, \mathcal{D}_t] = 0, \tag{2.22}$$

where

$$\mathcal{D}_\lambda := \frac{\partial}{\partial \lambda} - \mathcal{N}(\lambda) \qquad (2.23)$$

$$\mathcal{D}_t := \frac{\partial}{\partial t} - \mathcal{A}_\sigma^H. \qquad (2.24)$$

These are precisely the necessary conditions for the invariance of the monodromy of the operator $\mathcal{D}_\lambda$ under deformations in the parameter t.

To apply this to the system (1.4), we choose the following set of spectral invariant Hamiltonians $\{H_i \in \mathcal{I}_B^A\}_{i=1,\dots n}$

$$H_i(\mathcal{N}) := \frac{1}{4\pi i} \oint_{\lambda=\alpha_i} \mathrm{tr}((\mathcal{N}(\lambda))^2 d\lambda = \mathrm{tr}(BN_i) + \sum_{\substack{j=1 \\ j \neq i}}^n \frac{\mathrm{tr}(N_i N_j)}{\alpha_i - \alpha_j}. \qquad (2.25)$$

The autonomous form of Hamilton's equations that result are then

$$\frac{\partial \mathcal{N}}{\partial t_i} = -[(dH_i)_-, \mathcal{N}], \qquad (2.26)$$

where

$$(dH_i)_- = \frac{N_i}{\lambda - \alpha_i} \in \widetilde{\mathfrak{gl}}(r)_-. \qquad (2.27)$$

Identifying the various deformation parameters now with the eigenvalues of the matrix A, $\{t_i = \alpha_i\}_{i=1,\dots,n}$, the Lax equations are modified to the following form:

$$\frac{\partial \mathcal{N}}{\partial \alpha_i} = -[(dH_i)_-, \mathcal{N}] - \frac{\partial (dH_i)_-}{\partial \lambda}. \qquad (2.28)$$

These are just the commutativity conditions

$$[\mathcal{D}_\lambda, \mathcal{D}_i] = 0, \quad i = 1, \dots, n, \qquad (2.29)$$

$$\mathcal{D}_\lambda := \frac{\partial}{\partial \lambda} - \mathcal{N}(\lambda) \qquad (2.30)$$

$$\mathcal{D}_i := \frac{\partial}{\partial \alpha_i} + (dH_i)_- = \frac{\partial}{\partial \alpha_i} + \frac{N_i}{\lambda - \alpha_i}. \qquad (2.31)$$

guaranteeing the preservation of the monodromy of the operator $\mathcal{D}_\lambda$ under the deformations generated by varying the α_i's. Evaluating residues at $\{\alpha_i\}_{i=1,\dots n}$ gives

$$\frac{\partial N_j}{\partial \alpha_i} = \frac{[N_j, N_i]}{\alpha_j - \alpha_i}, \quad j \neq i, \quad i, j = 1, \dots, n, \qquad (2.32)$$

$$\frac{\partial N_i}{\partial \alpha_i} = [B + \sum_{\substack{j=1 \\ j \neq i}}^n \frac{N_j}{\alpha_i - \alpha_j}, N_i]. \qquad (2.33)$$

which are just the α_i components of the differential system (1.4).

2.3   Dual Isomonodromic System

To obtain the β_a components of this system, it is convenient to introduce another representation, in terms of a second system of rational covariant derivative operators whose monodromy will also be preserved: the *dual* isomonodromic system. Define another loop algebra $\widetilde{\mathfrak{gl}}(N)$, consisting of $N \times N$ matrices depending similarly on a loop parameter z that lies on a circle S^1 in the complex z–plane, with corresponding splitting into positive and negative Fourier components:

$$\widetilde{\mathfrak{gl}}(N) = \widetilde{\mathfrak{gl}}(N)_+ + \widetilde{\mathfrak{gl}}(N)_- \sim \widetilde{\mathfrak{gl}}(N)^*, \tag{2.34}$$

and corresponding rational R–matrix structure $\widetilde{\mathfrak{gl}}(N)^*_R$ on the dual space. We also define a corresponding Poisson subspace $\mathbf{g}^B_A \subset \widetilde{\mathfrak{gl}}(N)^*_R$ consisting of rational elements $\mathcal{M}(z)$ of the form

$$\mathbf{g}^B_A := \{\mathcal{M}(z) = -A + \sum_{a=1}^{r} \frac{M_a}{z - \beta_a}\} \sim \sum_{a=1}^{r} \mathfrak{gl}^*(N). \tag{2.35}$$

Introduce the "dual" Poisson map from M to $\mathbf{g}^B_A$ as:

$$\widetilde{J}^B_A : M \longrightarrow \mathbf{g}^B_A$$
$$\widetilde{J}^B_A : (F, G) \longmapsto -A - F(B - zI_N)^{-1}G^T := \mathcal{M}(z), \tag{2.36}$$

and denote the dual ring of spectral invariants restricted to $\mathbf{g}^B_A$ as

$$\mathcal{I}^B_A := \mathcal{I}(\widetilde{\mathfrak{gl}}(N)^*)|_{\mathbf{g}^B_A}. \tag{2.37}$$

We then have the remarkable fact that the spectral rings $\mathcal{I}^B_A$ and $\mathcal{I}^A_B$ coincide when pulled back under the respective Poisson maps $\widetilde{J}^B_A$ and $\widetilde{J}^A_B$ to M.

THEOREM 2.1 (DUALITY THEOREM):   *The two spectral invariant rings* $\widetilde{J}^{B*}_A(\mathcal{I}^B_A)$ *and* $\widetilde{J}^{A*}_B(\mathcal{I}^A_B)$ *coincide.*

Proof: This essentially follows from the simple linear algebra identity

$$\det(A - \lambda I_N)\det(B + G^T(A - \lambda I_N)^{-1}F - zI_r)$$
$$= \det(B - zI_r)\det(A + F(B - zI_r)^{-1}G^T - \lambda I_N), \tag{2.38}$$

which implies that the spectral curves of $\mathcal{N}(\lambda)$ and $\mathcal{M}(z)$ are identical.

Now define the set of spectral invariant Hamiltonians $\{K_a \in \mathcal{I}^B_A\}_{a=1,...r}$, similarly to the H_i's, as:

$$K_a := \frac{1}{4\pi i} \oint_{z=\beta_a} \mathrm{tr}(\mathcal{M}(z))^2 dz. \tag{2.39}$$

On $\mathbf{g}^B_A$, these similarly generate the equations:

$$\frac{\partial \mathcal{M}}{\partial \beta_a} = -[(dK_a)_-, \mathcal{M}] - \frac{\partial (dK_a)_-}{\partial z}, \tag{2.40}$$

where

$$(dK_a)_-(z) = \frac{M_a}{z - \beta_a} \in \widetilde{\mathfrak{gl}}(N)_-, \tag{2.41}$$

which imply the invariance of the monodromy of the rational covariant derivative operator

$$\mathcal{D}_z := \frac{\partial}{\partial z} - \mathcal{M}(z) \tag{2.42}$$

under the deformations generated by changes in the parameters $\{\beta_a\}_{a=1,\cdots r}$.

Using Theorem 2.1, we may also pull back the K_a's under the map $\tilde{J}_B^A$, to determine corresponding spectral invariant Hamiltonian functions of $\mathcal{N}(\lambda)$. These again generate isomonodromic deformation equations for the operator $\mathcal{D}_\lambda$, which may be expressed

$$\frac{\partial \mathcal{N}}{\partial \beta_a} = [(dK_a)_+, \mathcal{N}] + \frac{\partial (dK_a)_+}{\partial \lambda}, \tag{2.43}$$

where

$$(dK_a)_+(\lambda) = \lambda E_a + \sum_{\substack{b=1 \\ b \neq a}}^{r} \sum_{i=1}^{n} \frac{E_a N_i E_b + E_b N_i E_a}{\beta_a - \beta_b}, \tag{2.44}$$

and E_a denotes the elementary $r \times r$ matrix with diagonal entry 1 in the aa position and zero elsewhere. Evaluating residues at $\lambda = \alpha_i$ gives

$$\frac{\partial N_i}{\partial \beta_a} = [(dK_a)_+(\alpha_i), N_i], \tag{2.45}$$

which are precisely the β_a components of the isomonodromic system (1.4). Similarly, viewing $\{H_i\}_{i=1,\ldots n}$ as Hamiltonians defined on $\mathbf{g}_A^B$, these generate the *dual* equations, which imply the invariance of the monodromy of the operator $\mathcal{D}_z$:

$$\frac{\partial \mathcal{M}}{\partial \alpha_i} = [(dH_i)_+, \mathcal{M}] + \frac{\partial (dH_i)_+}{\partial z}, \tag{2.46}$$

where

$$(dH_i)_+(z) = -z E_i + \sum_{\substack{j=1 \\ j \neq i}}^{n} \sum_{a=1}^{r} \frac{E_i M_a E_j + E_j M_a E_i}{\alpha_i - \alpha_j}, \tag{2.47}$$

and E_i denotes the elementary $n \times n$ matrix with entry 1 in the ii position. Evaluating residues at $z = \beta_a$ gives these in the form

$$\frac{\partial M_a}{\partial \alpha_i} = [(dH_i)_+(\beta_a), M_a]. \tag{2.48}$$

3   Isomonodromic Deformations and the Riemann –Hilbert Problem

In this last section, we discuss certain specific solutions of the above isomonodromic deformation equations which can be constructed through application
of the Zakharov–Shabat "dressing" method (Novikov et al. (1984)). This class
of solutions is of particular interest from the viewpoint of applications, since
they arise in the calculation of correlation functions for quantum integrable
systems (Korepin et al. (1993)), (Harnad and Its (1997)) and spectral distributions in the theory of random matrices (Tracy and Widom (1994)), (Harnad et al. (1994)). The results quoted in this section are based on joint work
with A. Its; the full details may be found in the joint paper (Harnad and Its
(1997)).

The particular class of solutions to the isomonodromic deformation equations in question may be constructed by applying the *dressing method*, based
on the matrix Riemann–Hilbert problem, suitably adapted to this case. To do
this, we first introduce a *vacuum* solution Ψ_0, which is chosen as the invertible
$r \times r$ matrix function obtained by exponentiating λB

$$\Psi_0(\lambda) := e^{\lambda B}. \tag{3.1}$$

We then introduce a family of loop group elements $H_0(\lambda)$, viewed as $Gl(r)$–
valued functions defined along some oriented, closed curve Γ chosen, in this
case, to pass *through* the points $\{\alpha_i\}$, consecutively, with the latter ordered
by their subscript labels. We also assume in the following that the number
n of such points is even, and write $n = 2m$. (If the number happens to be
odd, we just increase it by adding $\lambda = \infty$ as the last point.) Let $\{\Gamma_j\}_{j=1\cdots m}$
denote the segment of Γ between α_{2j-1} and α_{2j} and let $\theta_j(\lambda)$ denote the
characteristic function, along Γ, of the interval Γ_j. We define $H_0(\lambda)$ as the
piecewise constant element of the form

$$H_0(\lambda) := \mathbf{I}_r + 2\pi i \sum_{j=1}^{n} \mathbf{f}_j \mathbf{g}_j^T \theta_j(\lambda), \tag{3.2}$$

where $\{\mathbf{f}_j, \mathbf{g}_j\}_{j=1\cdots m}$ is any fixed set of $r \times p$ complex, rectangular matrices,
with $p \leq r$, satisfying the null conditions

$$\mathbf{g}_j^T \mathbf{f}_k = 0, \quad \forall j, k. \tag{3.3}$$

The relevant matrix Riemann–Hilbert problem consists of finding a nonsingular $r \times r$ matrix valued function $\chi(\lambda)$ that is analytic on the complement
of Γ, extending to $\lambda = \infty$ off Γ, with asymptotic form

$$\chi(\lambda) \sim \mathbf{I}_r + \mathcal{O}\left(\lambda^{-1}\right) \tag{3.4}$$

for $\lambda \to \infty$, and has cut discontinuities across Γ given by

$$\chi_-(\lambda) = \chi_+(\lambda)H(\lambda), \quad \lambda \in \Gamma, \tag{3.5}$$

where $\chi_+(\lambda)$ and $\chi_-(\lambda)$ are the limiting values of $\chi(\lambda)$ as Γ is approached from the left and the right, respectively, and $H(\lambda)$ is the $r \times r$ invertible matrix valued function along Γ defined by

$$H(\lambda) = \Psi_0(\lambda)H_0(\lambda)\Psi_0^{-1}(\lambda). \tag{3.6}$$

Following the Zakharov–Shabat dressing method, we define the *dressed* wave function as

$$\Psi_\pm(\lambda) := \chi_\pm(\lambda)\Psi_0(\lambda), \tag{3.7}$$

with limiting values $\Psi_\pm$ on either side of the segments of Γ given by

$$\Psi_\pm(\lambda) := \chi_\pm(\lambda)\Psi_0(\lambda). \tag{3.8}$$

We then have the following result, which is quoted here from Harnad and Its (1997),

THEOREM 3.1:   *The wave function $\Psi(\lambda)$ defined by (3.8) satisfies the equations*

$$\frac{\partial\Psi}{\partial\lambda} - \left(B + \sum_{j=1}^{n} \frac{N_j}{\lambda - \alpha_j}\right)\Psi = 0, \tag{3.9}$$

$$\frac{\partial\Psi}{\partial\alpha_j} + \frac{N_j}{\lambda - \alpha_j}\Psi = 0, \tag{3.10}$$

with the N_j's given by

$$N_j := -\mathbf{F}_j\mathbf{G}_j^T, \tag{3.11}$$

where

$$\mathbf{F}_j := \lim_{\lambda \to \alpha_j} \mathbf{F}(\lambda) \quad \mathbf{G}_j := (-1)^j \lim_{\lambda \to \alpha_j} \mathbf{G}(\lambda). \tag{3.12}$$

This implies the commutativity

$$[\mathcal{D}_\lambda, \mathcal{D}_{\alpha_j}] = 0, \quad [\mathcal{D}_{\alpha_i}, \mathcal{D}_{\alpha_j}] = 0, \quad i, j = 1, \ldots, n \tag{3.13}$$

of the operators

$$\mathcal{D}_\lambda := \frac{\partial}{\partial\lambda} - B - \sum_{j=1}^{n} \frac{N_j}{\lambda - \alpha_j} = \frac{\partial}{\partial\lambda} - \mathcal{N}(\lambda) \tag{3.14}$$

$$\mathcal{D}_{\alpha_j} := \frac{\partial}{\partial\alpha_j} + \frac{N_j}{\lambda - \alpha_j}, \tag{3.15}$$

and hence the invariance of the monodromy data of the operator $\mathcal{D}_\lambda$ under changes in the parameters $\{\alpha_j\}$.

Thus, the operators defined in eqs. (3.14)-(3.15) represent a solution to the α_j components of the isomonodromic system (1.4). The following result, also quoted from Harnad and Its (1997), shows that the same construction also provides a solution to the β_a components.

THEOREM 3.2:   *The wave function $\Psi(\lambda)$ also satisfies the equations*

$$\mathcal{D}_{\beta_a}\Psi = 0, \quad a = 1, \ldots, r, \tag{3.16}$$

where the operators $\{\mathcal{D}_{\beta_a}\}_{b=1,\ldots,r}$ are defined by

$$\mathcal{D}_{\beta_a} := \frac{\partial}{\partial \beta_a} - \lambda E_a - \sum_{\substack{b=1 \\ b \neq a}}^{r} \frac{E_a \left(\sum_{j=1}^{n} N_j\right) E_b + E_b \left(\sum_{j=1}^{n} N_j\right) E_a}{\beta_a - \beta_b}, \tag{3.17}$$

with the N_i's given by eqs. (3.11), (3.12). This implies the commutativity conditions

$$[\mathcal{D}_\lambda, \mathcal{D}_{\beta_a}] = 0, \quad [\mathcal{D}_{\beta_a}, \mathcal{D}_{\beta_b}] = 0, \quad a, b = 1, \ldots, r, \tag{3.18}$$

and hence the invariance of the monodromy data of $\mathcal{D}_\lambda$ under the deformations parameterized by $\{\beta_a\}_{a=1,\ldots,r}$.

From the viewpoint of applications to quantum integrable systems (Its et al. (1990), Korepin et al. (1993)) and the spectral theory of random matrices (Tracy and Widom (1994), Harnad et al. (1994)) this construction has particular importance, since the underlying τ–function, as defined in eq. (1.12), is just the Fredholm determinant of an integral operator that may be constructed from the same data, and which gives the correlation functions and spectral distribution generating functions in question. This result is contained in the following theorem, also quoted from Harnad and Its (1997).

THEOREM 3.3:   *Let $\mathbf{K}$ be the $p \times p$ matrix Fredholm integral operator acting on $\mathbf{C}^p$–valued functions $\mathbf{v}(\lambda)$,*

$$\mathbf{K}(\mathbf{v})(\lambda) = \int_\Gamma K(\lambda, \mu)\mathbf{v}(\mu)d\mu, \tag{3.19}$$

defined along the curve Γ, with integral kernel given by

$$K(\lambda, \mu) = \frac{\mathbf{f}^T(\lambda)\mathbf{g}(\mu)}{\lambda - \mu}, \tag{3.20}$$

where $\mathbf{f}, \mathbf{g}$ are the rectangular $r \times p$ matrix valued functions

$$\mathbf{f}(\lambda) := \Psi_0(\lambda) \sum_{j=1}^{m} \mathbf{f}_j \theta_j(\lambda) \tag{3.21}$$

$$\mathbf{g}(\lambda) := \left(\Psi_0^T(\lambda)\right)^{-1} \sum_{j=1}^{m} \mathbf{g}_j \theta_j(\lambda). \tag{3.22}$$

Then the *logarithmic derivative of the Fredholm determinant is given by*

$$d \ln \det(\mathbf{I} - \mathbf{K}) = \omega = \sum_{k=1}^{n} H_k d\alpha_k + \sum_{a=1}^{r} K_a d\beta_a, \qquad (3.23)$$

where *the individual factors may be expressed*

$$H_k = \frac{\partial \ln \det(\mathbf{I} - \mathbf{K})}{\partial \alpha_k} = \mathrm{tr}(BN_k) + \sum_{j=1, j \neq k}^{n} \frac{\mathrm{tr}\, N_j N_k}{\alpha_k - \alpha_j} \qquad (3.24)$$

$$K_a = \frac{\partial \ln \det(\mathbf{I} - \mathbf{K})}{\partial \beta_a}$$

$$= \sum_{j=1}^{n} \alpha_j \left(N_j\right)_{aa} + \sum_{\substack{b=1 \\ b \neq a}}^{r} \frac{\left(\sum_{j=1}^{n} N_j\right)_{ab} \left(\sum_{k=1}^{n} N_k\right)_{ba}}{\beta_a - \beta_b}. \qquad (3.25)$$

Hence, $\det(\mathbf{I} - \mathbf{K})$ *may be identified as the* τ-*function defined in eq. (1.12).*

Finally, it should be mentioned that the dual isomonodromic systems defined in eqs. (2.42)–(2.48) may be derived in exactly the same way, by interchanging the rôles of the matrices A and B when defining the vacuum wave function (3.1). The corresponding curve $\tilde{\Gamma}$ must be chosen, in the complex z-plane, so as to pass through the parameters $\{\beta_a\}$ giving the diagonal elements of the matrix B. The resulting τ-function turns out to just be given by the Fredholm determinant of the *dual* Fredholm integral operator $\tilde{\mathbf{K}}$, which is related to the operator $\mathbf{K}$ appearing in Theorem 3.2 by taking a Fourier–Laplace transform along the curves Γ and $\tilde{\Gamma}$. Full details regarding this result, as well as a number of related results, including generalizations to isomonodromic deformations of operators having higher order pole singularities, may be found in Harnad and Its (1997).

Acknowledgements. This research was supported in part by the Natural Sciences and Engineering Research Council of Canada and the Fonds FCAR du Québec.

References

Jimbo, M., Miwa, T., Môri, Y. and Sato, M. (1980): "Density Matrix of an Impenetrable Bose Gas and the Fifth Painlevé Transcendent", *Physica* **1D**, 80-158.

Jimbo, M., Miwa, T., and Ueno, K. (1981): "Monodromy Preserving Deformation of Linear Ordinary Differential Equations with Rational Coeefficients I.", *Physica* **2D**, 306-352.

Jimbo, M., and Miwa, T. (1981): "Monodromy Preserving Deformation of Linear Ordinary Differential Equations with Rational Coeefficients II, III.", *Physica* **2D**, 407-448; *ibid.*, **4D**, 26-46 (1981).

Harnad, J. (1994): "Dual Isomonodromy Deformations and Moment Maps to Loop Algebras", *Commun. Math. Phys.* **166**, 337–365.

Harnad, J., and Its, A. (1997): "Integrable Fredhom Operators and Dual Isomonodromic Deformations", preprint CRM-2477 (1997), solv-int/9706002.

Harnad, J., Tracy, C., and Widom, H. (1994): "Hamiltonian Structure of Equations Appearing in Random Matrices", in: *NATO ASI Series B*, Vol. **314**, 231–245, *Low Dimensional Topology and Quantum Field Theory* (Plenum, New York, 1994), ed. H. Osborn.

Its, A. R., Izergin, A. G., Korepin, V. E., and Slavnov, N. A. (1990): "Differential Equations for Quantum Correlation Functions," *Int. J. Mod. Phys.* **B4**, 1003–1037 .

Korepin, V. E., Bogolyubov, N .M., and Izergin, A. G. (1993): *Quantum Inverse Scattering Method and Correlation Functions*, Cambridge Monographs on Mathematical Physics, Cambridge .

Novikov, S. P., Zakharov, V. E., Manakov. S. V., and Pitaevski, L. V. (1984): *Soliton Theory: The Inverse Scattering Transform*, Plenum, New York.

Tracy, C. A., and Widom, H. (1994): "Fredholm determinants, differential equations and matrix models" *Commun. Math. Phys.* **163**, 33–72.

Integrable Systems with Singular Rational Spectral Varieties

Alex Kasman[1,2]

[1] Department of Mathematics and Statistics, Concordia University, Montréal,
Québec H4B 1R6 CANADA
[2] Centre de recherches mathématiques, Université de Montréal, Montréal, Québec
H3C 3J7 CANADA

Abstract. A standard technique for investigating "rank one" solutions to the KP equation with singular rational spectral curves constructively determines a solution whose wave function is annihilated by a choice of finitely supported distributions. More recently, this technique has been reinterpreted in the language of Darboux transformations and generalized. The present paper is a non-technical review of some of these generalizations and their implications. In particular, it will focus on the application of these methods to higher rank KP Solutions with singular rational spectral curves and to algebraically integrable quantum systems with rational spectral varieties having non-planar singularities.

1 Introduction

Commutative rings of ordinary (resp. partial) differential operators play a key role in the theory of classical (resp. quantum) integrable system as the "integrals of motion". Being commutative with a natural grading, the ring automatically has a geometric interpretation as an projective curve (resp. variety). In this paper we will consider the "inverse spectral problem" of constructing rings – and thus KP solutions or algebraically integrable quantum systems – having a given singular rational spectral variety.

It is convenient to categorize commutative rings of differential operators by two associated integers. Let $\mathcal{R}$ be a commutative ring of partial differential operators, then we say that the dimension of $\mathcal{R}$ ($n = \dim \mathcal{R}$) is the dimension of the variety $\operatorname{Spec} \mathcal{R}$ and the rank of $\mathcal{R}$ ($r = \operatorname{rank} \mathcal{R}$) is the greatest common divisor of the order of its elements. In general, when one thinks of the algebro-geometric approach to integrable systems, one thinks of the case $n = r = 1$. For example, the well known solutions of the KP equation which can be solved by ϑ-functions of a complex projective curve C [15, 19] are associated to rank one commutative rings whose spectral variety is the curve C.

As implied by the title, this paper will primarily be concerned with the special case in which the spectral variety is rational. Rationality of the spectral variety is equivalent to the existence of an isomorphism between $\mathcal{R}$ and a subring of the polynomial ring $\mathbb{C}[x_1, \dots, x_n]$. Moreover, the spectral variety is non-singular only if $\mathcal{R}$ is isomorphic to the entire polynomial ring. In the next section we will recall a method of Krichever [16] for constructing rank one KP solutions having singular rational spectral curves. The main focus of this paper will then be to discuss the methods and implications of recent generalizations of this technique to higher dimension ($n > 1$) and higher rank ($r > 1$).

New results about Darboux transformations play an important role in these generalizations. Since these transformations are of interest to researchers outside of the field of integrable systems, these results may be of more general interest as well. In particular, just as in the well known methods for constructing quantum systems with equivalent energy spectra [2], the higher dimensional generalization to be discussed here involves the factorization of partial differential operators and the transformation given by exchanging the factors. However, the method to be described here represent a completely different approach having been inspired by the recent work of Veselov and Chalykh [7–8,23–24] and Wilson [25]. As in those papers, we will apply a Darboux transformation to every element of a subring of a polynomial ring of differential operators to produce a commutative ring with a singular rational spectral variety.

1.1   Outline

The remainder of Sect. 1 will be a very brief introduction to the KP hierarchy and the methods of algebraic geometry for integrable systems. For additional information please consult [15, 20, 22] and their references.

Then, in Sect. 2 we will recall a particular technique for constructing explicit rank one solutions to the KP hierarchy with singular rational spectral curves. In this method, one requires as an ansatz that the *wave function* should be annihilated by a chosen set of finitely supported distributions, thereby effectively specifying the singularities of the corresponding spectral curve. The key observation which allows this method to be generalized is that on the level of the associated commutative ring, this technique is a procedure for producing rank one commutative rings of ordinary differential operators by *Darboux transformation* of a ring of constant coefficient operators. The notion of Darboux transformations for the KP hierarchy is reviewed in Sect. 3. Section 4 will then prove two elementary theorems which relate these transformations to the inverse method.

Since Darboux transformations of ordinary differential operators are well understood, it is not difficult now to extend this same technique to the higher rank situation. This was done separately in the papers [4] and [12] and will be briefly outlined in Sect. 5. This provides a constructive means for producing and studying KP solutions of arbitrary rank having singular rational spectral curves. As an example, I provide an explicit KP solution of rank two having the same spectral curve as the KdV 2-soliton.

Finally, Sect. 6 will discuss the higher dimensional generalization of this technique. In particular, a further generalization of the same approach yields commutative rings of partial differential operators with singular rational spectral varieties. However, the higher dimensional situation is much more complicated and is not fully understood at present. For example, unlike the one dimensional case, one is not free to choose an *arbitrary* a set of distributions to annihilate the wave function. One of the main problems, therefore, is to determine which sets of distributions are "good" in the sense that they will lead to a non-trivial commutative ring. The main focus of this section will be the recent result showing that there exist good sets of "distributions" in any dimension with arbitrary algebraic support [6].

1.2   The KP Hierarchy

A formal pseudo-differential operator

$$L = \partial + \sum_{j=1}^{\infty} u_j(t_i)\partial^{-j} \qquad \left(\partial = \frac{\partial}{\partial x}, \ x \equiv t_1 \right) \tag{1}$$

with coefficients depending on $t_1, t_2, \ldots$ is said to be a solution of the KP hierarchy if there is a *wave function*

$$\psi(t_i, z) = \left(1 + O(z^{-1})\right) e^{\sum t_i z^i} \tag{2}$$

satisfying

$$L\psi(x, z) = z\psi \qquad \text{and} \qquad (L^i)_+ \psi = \frac{\partial}{\partial t_i}\psi. \tag{3}$$

The KP hierarchy is a completely integrable system and contains as reductions many non-linear PDEs of physical interest such as the KdV equation and plays a significant role in current theories of quantum gravity.

1.3   The Algebro-Geometric Approach

In 1895, Korteweg and de Vries [18] made history by deriving the solitary travelling wave which we now know as the 1-soliton solution to the KdV equation. The travelling waves of the KdV equation can be easily rederived today by anyone familiar with the algebraic geometry of elliptic curves. Simply presume that the solution has the ansatz form $u(x, t) = u(x+ct)$, integrate the equation twice (multiplying by u' before the second integration) and one finds an equation known to be satisfied by the Weirstrass $\wp$-function of an elliptic curve. Then, in particular, one may degenerate the elliptic curve to a curve with a nodal singularity and the corresponding limiting solution is the solitary wave. This singular rational curve is what would be called the *spectral curve* of the solution. Alternatively, it also arises as the variety corresponding to the ring of all ordinary differential operators commuting with the Schrödinger operator $\partial^2 + u(x, t)$. As I will explain below, many solutions to the KP hierarchy are associated to spectral curves in a similar way.

The algebro-geometric approach to integrable systems similarly takes advantage of our knowledge of geometry to allow us to construct and study these solutions.

Let L be any first order pseudo-differential operator in x and let A_L be the commutative ring of all *ordinary* differential operators which commute with L. If A_L is non-empty then Spec A_L is a complex algebraic curve (i.e. $\dim A_L = 1$). In such a case, as was discovered in the 1970's and '80's, results of algebraic geometry can be used to determine the time evolution of L under the KP hierarchy (i.e., the unique KP solution $L(t_i)$ such that $L(0) = L$).

The main idea behind these constructions is the fact that the spectral curve Spec A_L is an invariant of the KP flow. Moreover, one may also consider the common eigenfunctions of the ring A_L as a torsion free coherent sheaf[1] of rank r (where $r = \text{rank}\, A_L$) over the curve. This sheaf is *not* an invariant of the KP flows. One may thus view the curve and sheaf as a geometric version of "action-angle" variables for finite dimensional integrable systems. Consequently, with enough knowledge of the moduli of rank r sheaves on Spec A_L one can hope to construct the time evolution $L(t_i)$ given by the KP hierarchy. It is therefore common to refer to $L(t_i)$ as a rank r solution.

The best known solutions (e.g. finite gap and soliton) are rank one solutions meaning that they can be integrated using sufficient knowledge of the Jacobian varieties of the spectral curves. Although much work has also been done on higher rank KP solutions, they remain more difficult and obscure. In fact, very few explicit higher rank solutions are known.

1.4   Notation

Throughout this review, I will be working with two copies of $\mathbb{C}^n$. Let $x, z \in \mathbb{C}^n$ denote the vectors $x = (x_1, \ldots, x_n)$ and $z = (z_1, \ldots, z_n)$ of the "spacial" and "spectral" parameters respectively. Furthermore, the differential operators to be discussed below will always be operators only in the variables x or in the variables z but never both simultaneously. Therefore, Let $\partial = (\partial_1, \ldots, \partial_n)$ denote the differential operators $\partial_i = \partial/\partial x_i$ and similiarly $\underline{\partial}$ denotes the differential operators $\underline{\partial}_i = \partial/\partial z_i$. Finally, since differential operators in these rings will occasionally be singled out, note that $\mathbb{C}[\partial]$ is the ring of constant coefficient differential operators and that $\mathbb{C}(x)[\partial]$ will denote the ring of differential operators with rational coefficients in the spacial variables (and similarly for the spectral parameters.)

2   Rational Spectral Curves in Rank One: Conditions on the Wave Function

As I mentioned, most (all?) of the well known solutions are rank one algebro-geometric solutions. The pure soliton and "vanishing" rational solutions are

---

[1] a vector bundle in the non-singular case

examples of rank one solutions with singular rational spectral curves. The former correspond to spectral curves having nodal singularities and the latter to spectral curves having cuspidal singularities. (See Fig. 1.) (In addition, there are "hybrid" solutions having spectral curves with combinations of these two types of singularities.) The soliton solutions are of obvious interest to mathematical physicists. The rational solutions are also of interest due to their connection to Calogero-Moser particle systems [15, 26] and the fact that they are *bispectral* [11, 25].

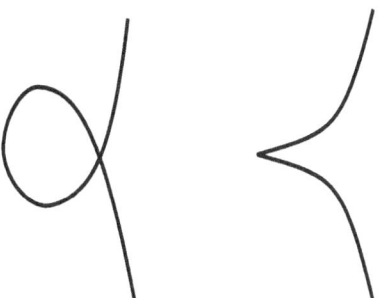

Nodal Singularity   Cuspidal Singularity
(Soliton)   (Rational)

Fig. 1. Two types of singularities

A standard method for constructing either of these type of solutions is to require as an ansatz that the wave function ψ should satisfy some equations involving differentiation and evaluation in the spectral parameter. For example, Krichever [16] constructs rational solutions to the KP equation by requiring ψ to satisfy the equations

$$\left.\frac{\partial}{\partial z}\psi\right|_{z=\lambda_i} = 0 \qquad \lambda_i \in \mathbb{C},\ 1 \le i \le N$$

and Segal and Wilson [22] construct N-soliton solutions of KdV by requiring ψ to satisfy

$$\left.\psi\right|_{z=\lambda_i} + \gamma_i\psi\big|_{z=-\lambda_i} = 0 \qquad \lambda_i \in \mathbb{C},\ \gamma_i \in \mathbb{C}^*,\ 1 \le i \le N. \tag{4}$$

Then, following this ansatz, a solution with this property can be easily determined using little more than basic linear algebra[2].

---

[2] For details see the original papers. Rather than discussing the particulars of these methods here I would instead like to explain *why* such a simple construction is possible.

The most general form of this construction in the case $n = r = 1$ was in [25], which is in fact the inspiration for the further developments to be described below. In this method, one may choose a set of N independent finitely supported distributions[3] in z, $C = \{c_i | 1 \leq i \leq N\}$ and a polynomial $q(z)$ of degree N (which only affects the gauge of the solution and will not be of importance here). Then, using only finite dimensional determinants one may compute the τ-function and wave function of a KP solution whose wave function satisfies

$$c_i(q\psi) = 0 \qquad 1 \leq i \leq N. \tag{5}$$

Of course the previous two examples are special cases of this more general one. In fact, *every* rank one KP solution with a singular rational spectral curve can be obtained in this way.

It is clear (and remarked in both [16] and [25]) that the support of these distributions is directly related to the singularity structure of the spectral curve of the corresponding solution. It is reasonable to think about it in this way: starting with a nonsingular rational curve (a Riemann sphere) with local parameter z away from an identified point, each distribution identifies the points in its support and introduces singularities there to produce the spectral curve. Thus, for instance, (4) would create N nodes by identifying the point $z = \lambda_i$ with $z = -\lambda_i$. Moreover, the coefficients γ_i can be viewed as coordinates on the generalized Jacobian, $(\mathbb{C}^*)^N$, specifying the sheaf.

So, as we see, it is relatively simple to explicitly compute a rank one KP solution with any specified singular rational spectral curve. Our goal at this point is to generalize this technique so that it can be applied to the "more interesting" cases $n > 1$ or $r > 1$. Thus we are interested in answering the question:

Why is it possible to compute these solutions so simply?

Certainly one true answer to this question is merely that it reflects the simplicity of the algebraic geometry of rank one sheaves over singular rational curves. However, this does not lead to an obvious approach to the more general situation. To generalize we need a better understanding of the role played by the "conditions" (5). As we shall see, we can provide a useful answer to the question in the language of *Darboux transformations*.

3   A General Method: Darboux Transformation

Here we suppose that we have a fixed solution L of the KP hierarchy and its corresponding wave function ψ and we wish to construct a new solution through the method of Darboux transformation. In general, Darboux transformation produces a new operator/eigenfunction pair $(\hat{L}, \hat{\psi})$ from a known pair (L, ψ) through conjugation by an ordinary differential operator K:

---

[3] This means that they are linear functionals which differentiate and evaluate functions of the complex parameter z.

$$\hat{L} = KLK^{-1} \qquad \hat{\psi} = K\psi.$$

Of course, given that the original pair (L, ψ) satisfy the equations of the KP hierarchy (3), the new pair will satisfy the same equations only for certain choices of the operator K.

One characterization of such operators K involves distributions in the spectral parameter: Let $\bar{C}$ be a vector space of distributions (not necessarily finitely supported) in the spectral parameter with basis c_i:

$$c_i(\cdot) := \int_{\Gamma_i} \cdot \, k_i(z)\, dz \qquad 1 \le i \le N.$$

(In other words, c_i is the linear functional on functions in the variable z which multiplies by the kernel k_i and integrates around the path Γ_i.) If $c_i(\psi)$ are well defined functions, let K be the unique monic ordinary differential operator with kernel

$$\ker K = \{c(\psi(t_i, z)) \mid c \in \bar{C}\}.$$

This operator K is guaranteed to exist and can be computed explicitly using Wronskian determinants. Then $\hat{L} = KLK^{-1}$ is also a solution of the KP hierarchy (3). (See, for instance, [3].)

4   Some Useful Observations

4.1   Darboux Transformations and Conditions on the Wave Function

The method we discussed for handling rank one solutions with rational spectral curves requires the wave function to be annihilated by distributions in the spectral parameter. On the other hand, the Darboux dressing operator used to construct new solutions is also built out of distributions in the spectral parameter. Thus, both methods have in common the use of these distributions. However, the following simple observation relates the two by generalizing the "conditions on the wave function" approach to arbitrary Darboux transformations:

Theorem 4.1 *Suppose* $(\hat{L} = KLK^{-1}, \hat{\psi})$ *is a Darboux transform of* (L, ψ) *and let* C *be the vector space of all distributions annihilating* $\hat{\psi}$, *then*

$$\ker K = \{c(\psi)|c \in C\}.$$

This provides an answer to the question: *given a KP solution, one may always produce a new solution whose wave function is annihilated by a given set of distributions by Darboux transformation.* In fact, one may reinterpret Wilson's construction as the Darboux transformations of the trivial KP solution $L = \partial$ specified by the choice of finitely supported distributions [4, 13].

The proof of this observation (Thm. 4.1) is quite elementary. Using [3] we know that ker K contains only functions of the form $c(\psi)$. Thus it is sufficient to notice that $c(\hat{\psi}) = Kc(\psi)$ and thus c annihilates $\hat{\psi}$ iff $c(\psi)$ is in the kernel of K.

4.2 Conditions and the New Spectral Curve

Recall that the spectral curve associated to a solution L is given by its centralizer A_L in the ring of ordinary differential operators. According to the Section 3 we can choose a set $C = \{c_i\}$ of distributions and get a new solution $\hat{L}$. It turns out that one can also determine information about the spectral curve of the new solution $\hat{L}$ in terms of the vector space $\bar{C}$ of distributions with basis C. Recall that the composition of the function $q(z)$ with $c(\cdot) = \int_\Gamma \cdot k(z)\, dz$ is

$$c \circ q(\cdot) = \int_\Gamma \cdot k(z) q(z)\, dz.$$

Then one may determine ordinary differential operators commuting with $\hat{L}$ by considering the eigenvalues of the operators in A_L as distributions and composing them with the elements of C.

Theorem 4.2 *Let $Q \in A_L$ satisfy the eigenvalue equation $Q\psi = q(z)\psi$. Then $\hat{Q} = KQK^{-1}$ is an element of $A_{\hat{L}}$ if $\bar{C} \circ q \subset \bar{C}$.*

By conjugating the equation $[Q, L] = 0$ we find that $[\hat{Q}, \hat{L}] = 0$, and so it only remains to check whether $\hat{Q}$ is an ordinary differential operator. Let $c \in C$ and note that since $c \circ q = c'$ for some $c' \in C$ we conclude that

$$KQc(\psi) = Kc(q(z)\psi) = Kc \circ q(\psi) = Kc'(\psi) = 0.$$

We therefore find that every element of ker K is contained in the kernel of the operator KQ. This implies that KQ is in the left ideal generated by K and thus that there exists an ordinary differential operator $\hat{Q}$ satisfying $KQ = \hat{Q}K$.

Although Theorem 4.2 is quite general, it works best in the case that L has a non-singular rational spectral curve and the distributions are finitely supported. In that case, just as in Wilson's rank one construction, one may conclude that the spectral curve is a singular rational curve with singularities introduced at the support of the distributions.

5   The Singular Rational Case: Higher Rank

As we have seen, the theorems of the previous section allow us to reinterpret Wilson's construction for KP solutions with singular rational spectral curves in terms of Darboux transformations. However, this does not limit the application of these ideas to the well known rank one solutions described in Section 2. In fact, after only slight modifications the same ideas lead to useful techniques for dealing with the *higher rank* solutions with singular rational spectral curves. Different approaches to the application of the ideas discussed above to the higher rank case were developed separately in [4] and [12] and used in the papers [5] and [13, 14]. Here I will only briefly discuss the latter approach and mention some of its applications.

So, what is the analogous situation in higher rank and how can we generalize the rank one techniques? We must begin with a rank r solution which has a non-singular rational curve. This solution, which I will call the vacuum, will play the same role that ∂ plays in the usual construction of KP solutions. Moreover, this solution will have a vector Baker-Akhiezer function [21] which will play the role of the eigenfunction.

Thus to apply these ideas in the rank r case one considers r-vectors of finitely supported distributions rather than the scalar distributions above. All of this leads to yet another grassmannian construction for higher rank KP solutions; however this grassmannian encodes deformations of a rank r vacuum rather than the vacuum ∂. This construction is very weak compared to the others in the literature[4] (cf. [1, 17, 21]) since it is limited to the case of singular rational spectral curves. On the other hand, it has the distinct advantage that, as in the rank one case, all calculations reduce to something finite dimensional. In particular, it is always the case that:

- the τ-function can be computed as a finite dimensional determinant
- the orbit in the grassmannian is finite dimensional
- the dressing operator is just an ordinary differential operator, etc...

Thus, by using this method one is able to compute explicit higher rank solutions having arbitrary singular rational spectral curves. See, for example, [13] where this method is used to produce bispectral rings of arbitrary rank. (Bispectral rings of arbitrary rank were also produced in [5] using the other approach.)

For example, one may construct the usual 2-soliton solution of KdV by requiring that the wavefunction have the same values at $z = 1$ and $z = -1$ and also at $z = 2$ and $z = -2$. This specifies a transformation of the vacuum solution $u = 0$ which adds two nodes to the spectral curve and correspondingly two solitons to the solution. Similarly, one may begin instead with the KdV solution $u(x, t) = -2x/3t$ which is a rank two solution with non-singular rational spectral curve [10]. One may then explicitly and relatively easily construct a solution to the KP equation with the same spectral curve by imposing conditions at the points $z = \pm 1$ and $z = \pm 2$ on the vector Baker-Akhiezer function [14]. Let $\tau(x, y, t)$ be the function

$$\tau = B(-2)B(\gamma_{-1})A'(-1)A'(\gamma_{-2}) - A(\gamma_{-1})B(-2)A'(\gamma_{-2})B'(-1)$$

---

[4] Consider especially [21] which is most similar to the construction being described here.

$$+A(\gamma_{-2})B(-2)A'(\gamma_{-1})B'(-1) - A(-2)B(\gamma_{-2})A'(\gamma_{-1})B'(-1)$$
$$-A(-2)B(\gamma_{-1})A'(-1)B'(\gamma_{-2}) + A(-2)A(\gamma_{-1})B'(-1)B'(\gamma_{-2})$$
$$-A(\gamma_{-2})B(-2)A'(-1)B'(\gamma_{-1}) + A(-2)B(\gamma_{-2})A'(-1)B'(\gamma_{-1})$$
$$+e^{2y}\left(-\left(B(-2)B(\gamma_1)A'(1)A'(\gamma_{-2})\right) + A(\gamma_1)B(-2)A'(\gamma_{-2})B'(1)\right.$$
$$-A(\gamma_{-2})B(-2)A'(\gamma_1)B'(1) + A(-2)B(\gamma_{-2})A'(\gamma_1)B'(1)$$
$$+A(-2)B(\gamma_1)A'(1)B'(\gamma_{-2}) - A(-2)A(\gamma_1)B'(1)B'(\gamma_{-2})$$
$$\left.+A(\gamma_{-2})B(-2)A'(1)B'(\gamma_1) - A(-2)B(\gamma_{-2})A'(1)B'(\gamma_1)\right)$$
$$+e^{4y}\left(-\left(B(\gamma_2)A'(2)A'(\gamma_{-1})B'(-1)\right) + B(\gamma_{-1})A'(-1)A'(\gamma_2)B'(2)\right.$$
$$+A(\gamma_2)A'(\gamma_{-1})B'(-1)B'(2) - A(\gamma_{-1})A'(\gamma_2)B'(-1)B'(2)$$
$$+B(\gamma_2)A'(-1)A'(2)B'(\gamma_{-1}) - A(\gamma_2)A'(-1)B'(2)B'(\gamma_{-1})$$
$$\left.-B(\gamma_{-1})A'(-1)A'(2)B'(\gamma_2) + A(\gamma_{-1})A'(2)B'(-1)B'(\gamma_2)\right)$$
$$+e^{6y}\left(B(\gamma_2)A'(2)A'(\gamma_1)B'(1) - B(\gamma_1)A'(1)A'(\gamma_2)B'(2)\right.$$
$$-A(\gamma_2)A'(\gamma_1)B'(1)B'(2) + A(\gamma_1)A'(\gamma_2)B'(1)B'(2) - B(\gamma_2)A'(1)A'(2)B'(\gamma_1)$$
$$\left.+A(\gamma_2)A'(1)B'(2)B'(\gamma_1) + B(\gamma_1)A'(1)A'(2)B'(\gamma_2) - A(\gamma_1)A'(2)B'(1)B'(\gamma_2)\right)$$

where and A and B are the solutions to the Airy equation $f''(x) = xf(x)$ such that $A(0) = 1$, $B(0) = 0$, $A'(0) = 0$ and $B'(0) = 1$ and

$$\gamma_n = \gamma_n(x,t) := (1+\frac{3t}{2})^{-1/3}x + n(1+\frac{3t}{2})^{2/3}$$

Then the function

$$u(x,y,t) = \frac{-2x}{3t+2} + 2\frac{\partial^2}{\partial x^2}\log\tau$$

is a solution to the KP equation having this same spectral curve. Recall that this merely means that the rings of ordinary differential operators which commute with the corresponding KP operators are isomorphic. However, the ring in the soliton case has rank one whereas in the case of the "Airy" vacuum, $-2x/3t$, the ring has rank two.

6 Higher Dimensional Generalizations

A commutative ring of partial differential operators in n variables is said to be an *algebraically integrable quantum system* if the smallest set of generators of the rings has at least $n+1$ elements. In such a situation, the generators may be treated as quantum Hamiltonians and, since there are enough of them which commute, one may thoroughly analyze the system. The definitions of rank, dimension and spectral variety discussed earlier still apply, so one may still consider the special case of singular rational spectral varieties.

Some of the most interesting work in this area has been on integrable quantum systems containing a second order (Schrödinger) operator [9, 7].

Of these, the method of Veselov and Chalykh (see [7–8,23–24]) is of the greatest interest to us since it is based on the method of Krichever which we have discussed earlier. In those papers, one requires that the common eigenfunction of the operators in the ring (the wave function) should satisfy conditions of the form:

$$\partial_\alpha^j \psi \Big|_{(\alpha,z)=0} = 0.$$

That is, for some vector $\alpha \in \mathbb{C}^n$, powers of the directional derivative (in the spectral parameters z_j) applied to the wave function should be equal to zero when evaluated on the *hyperplane* perpendicular to α.

As in the $r = n = 1$ construction, this leads to a Darboux transformation of a ring of constant coefficient operators which yield the commutative ring of interest – a ring with a rational spectral variety having singularities on the hyperplanes of the conditions. However, unlike the one dimensional case, there are many added complications. In particular, one may no longer choose the "conditions" arbitrarily. In fact, the one remarkable thing about the Veselov-Chalykh construction is the way in which the conditions which yield the algebraically integrable rings are determined by root systems of lie algebras.

The recent (as yet unpublished) paper [6] makes use of lessons learned from the generalizations for $r > 1$ in order to gain new insights into the case $n > 1$. In that paper, we demonstrate that it is quite natural to generalize this technique to higher dimensions by utilizing the language and theory of $\mathcal{D}$-modules. However, since I only wish to discuss one aspect of this generalization, it will be sufficient (and more consistent with the earlier sections) to continue speaking in terms of "conditions on the wave function".

Definition 6.1: Let $q_i(z) \in \mathbb{C}[z_1,\ldots,z_n]$ be a set of distinct irreducible polynomials $(1 \le i \le N)$ and to each we associate a vector space $V_i \subset \mathbb{C}(z)[\partial]$. Then we say that the function $\psi(x,z)$, $(x,z \in \mathbb{C}^n)$ satisfies the ensemble of conditions $\mathcal{M} = (q_i, V_i)$ if

$$K\psi(x,z) = 0 \qquad \text{for all } K \in V_i \text{ and } z \in q_i^{-1}(0). \tag{6}$$

By definition, the support of $\mathcal{M}$ is the variety supp $\mathcal{M} = \cup_i q_i^{-1}(0)$.

What we would then like to do, in following the program described in this article, is to construct (in some non-trivial way, analogous to the one dimensional Darboux construction) a commutative ring of partial differential operators along with a common eigenfunction which satisfies the conditions $\mathcal{M}$. In general, it may not be possible to do this. In the case $n = 1$ this can be done by Wilson's construction. In the higher dimensional case, this is done by the Veselov-Chalykh construction for *certain* choices of $\mathcal{M}$. Since these were previously the only known examples, one may look to these examples with the hope of noticing general truths.

For example, in the case $n = 1$, of course all of the polynomials q_i factor into linear terms. Moreover, in the Veselov-Chalykh construction, since all conditions are evaluated on some collection of hyperplanes, one again has that all q_i factor into linear terms. Therefore, one might be led to conclude that this hyperplane property is a necessary part of the construction. In fact, this is not the case.

Remark: It should be emphasized that [6] drops the requirement that the ring should contain a Schrödinger operator which was present in the works of Veselov and Chalykh. That paper, and the results described below, address the more general question of extending the constructions above to partial differential operators. Therefore, these results may not be particularly useful in addressing the more specific question raised in [7].

Definition 6.2: We say that $\mathcal{M}$ is a *good* ensemble of conditions if the ring

$$A_{\mathcal{M}} = \{P \in \mathbb{C}(x)[\partial] \mid KPe^{(x,z)} = 0 \text{ for any } K(z, \underline{\partial}) \in V_i \text{ and } z \in q_i^{-1}(0)\}$$

is the left ideal generated by some element $D_{\mathcal{M}}$.

If $n = 1$ then it is clear that any $\mathcal{M}$ is good. In particular, this is merely the ring of all operators whose kernel contains a specified finite dimensional space of functions in x. It is this property of ordinary differential operators which allows for the Darboux transformation that produces the commutative ring. One may therefore view Definition 6.2 as a reformulation of the question "When can we perform a *higher dimensional Darboux transformation?*" Consequently, it is simple to see that we can associate a commutative ring of partial differential operators to any set of good conditions.

Definition 6.3: If $\mathcal{M}$ is "good" then let $\mathcal{R}^0_{\mathcal{M}}$ be the ring

$$\mathcal{R}^0_{\mathcal{M}} = \{L_0 \in \mathbb{C}[\partial] \mid D_{\mathcal{M}} L_0 \in A_{\mathcal{M}}\}.$$

Since $D_{\mathcal{M}} \mathcal{R}^0_{\mathcal{M}} \subset A_{\mathcal{M}}$, it makes sense to discuss division on the right by $D_{\mathcal{M}}$. Let

$$\mathcal{R}_{\mathcal{M}} = D_{\mathcal{M}} \mathcal{R}^0_{\mathcal{M}} D_{\mathcal{M}}^{-1} \qquad \text{and} \qquad \psi_{\mathcal{M}} = D_{\mathcal{M}} e^{(x,k)}.$$

Theorem 6.4 *For any good ensemble $\mathcal{M}$, the ring $\mathcal{R}_{\mathcal{M}}$ is an algebraically integrable quantum system with common eigenfunction $\psi_{\mathcal{M}}$ satisfying the conditions $\mathcal{M}$.*

Unfortunately, it is not simple to determine which sets of conditions are good. However, there are some very simple examples of good conditions which demonstrate the fact that there exist good conditions for arbitrarily chosen q_i. The following theorem is an immediate consequence of results proved in [6].

Theorem 6.5 *Let q_i be arbitrary distinct irreducible polynomials, m_i be positive integers and $p(\underline{\partial}) \in \mathbb{C}[\underline{\partial}]$ be some fixed constant coefficient operator. If we denote by V_i the vector space of differential operators spanned by $\underline{\partial}^{\alpha} p(\underline{\partial})$*

for all multi-indices α such that $|\alpha| < m_i$ then the ensemble $\mathcal{M} = (q_i, V_i)$ is good.

Moreover, in this case the ideal $A_{\mathcal{M}}$ is the left ideal generated by the operator

$$D_{\mathcal{M}} = p(x)q(\partial)p^{-1}(x) \qquad where \qquad q(x) = \prod_{i=1}^{N} q^{m_i}(x). \qquad (7)$$

Thus we see that using Darboux transformations for partial differential operators it is possible to construct commutative rings with rational spectral varieties having singularities along arbitrarily chosen algebraic hypersurfaces. Although this result indicates that the extension of the one dimensional construction to partial differential operators is likely to be mathematically rich, the theory is still far from being complete. In particular, the higher dimensional version of this construction is still in need of higher dimensional versions of the papers [25] and [4, 12].

Acknowledgements: The author is grateful to E. Previato (thesis advisor on [12]), M. Rothstein, J. Harnad and Yu. Berest for their advice and collaboration, and to H. Aratyn, T. Imbo, W.-Y. Keung and U. Sukhatme for organizing this interesting and enjoyable workshop.

References

[1] M.R. Adams and M.J. Bergvelt, "The Krichever Map, Vector Bundles over Algebraic Curves, and Heisenberg Algebras", Comm. Math. Phys. 154 (1993) pp. 265-305

[2] A.A. Andrianov, N.V. Borisov and M.V. Ioffe, "The Factorization Method and Quantum systems with Equivalent Energy Spectra", Physics Letters 105A, 1,2,(1984) pp. 19-22

[3] H. Aratyn, E. Nissimov and S. Pacheva, "Constrained KP Hierarchies: Additional Symmetries, Darboux-Bäcklund Solutions and Relations to Multi-matrix Models", International Journal of Modern Physics A, 12 (1997) pp. 1265–1340

[4] B. Bakalov, E. Horozov, M. Yakimov, "Bäcklund-Darboux Transformations in Sato's Grassmannian", *Serdica Math. Journal*, 22 (1996) pp. 571-588

[5] B. Bakalov, E. Horozov, M. Yakimov, "Bispectral Commutative Rings of Ordinary Differential Operators", to appear in *Comm.Math.Phys.*

[6] Yu. Berest and A. Kasman, "$\mathcal{D}$-modules and Darboux Transformations for Partial Differential Operators", in preparation.

[7] O.A. Chalykh and A.P. Veselov, "Commutative rings of partial differential operators and Lie Algebras", Comm. Math. Phys. 125 (1990) 597-611.

[8] O.A. Chalykh and A.P. Veselov, "Integrability in the theory of the Schrödinger operators and harmonic analysis", Comm. Math. Phys. 152 (1993) 29-40

[9] G.J. Heckman and E.M. Opdam, "Root systems and hypergeometric functions I", comp. Math. 64 (1987) pp. 329-352

[10] G. Latham and E. Previato, "Higher Rank Darboux Transformations", in NATO Adv. Sci, Inst. Ser. B Phys. 320, pp. 117–134, Plenum, 1994

[11] A. Kasman, "Bispectral KP Solutions and Linearization of Calogero-Moser Particle Systems", *Communications in Mathematical Physics* 172 (1995) pp. 427-448

[12] A. Kasman, "Rank r KP Solutions with Singular Rational Spectral Curves", PhD. thesis, Boston University, 1995

[13] A. Kasman and M. Rothstein, "Bispectral Darboux Transformations", *Physica D* 102, (1997) p. 159-173

[14] A. Kasman, "Darboux Transformations from n-KdV to KP", to appear in *Acta Applicandae Mathematicae*

[15] I.M. Krichever, "Methods of Algebraic Geometry in the Theory of Non-linear Equations" *Russian Math. Surveys* 32:6 (1977) pp. 185-213

[16] I.M. Krichever, "Rational Solutions of the Kadomtsev-Petviashvili Equation and Integrable Systems of N Particles on a Line" *Functional Analysis and Applications* 12 (1978) pp. 59-61

[17] I.M. Krichever and S.P. Novikov, "Holomorphic Bundles over Riemann Surfaces and the KP Equation. I." *Functional Analysis and Applications*, 12 (1978) pp. 276-286

[18] D.J. Korteweg and G. deVries, "On the Change of Form of Long Waves Advancing in a Rectangular Canal" *Philos. Mag.* 39 (1895) pp. 422-443

[19] D. Mumford, "Tata Lectures on Theta II", Birkhäuser (1984)

[20] E. Previato, "Seventy Years of Spectral Curves: 1923–1993" *Lecture Notes in Mathematics 1620*, Springer-Verlag, 1996

[21] E. Previato and G. Wilson, "Vector Bundles Over Curves and Solutions of the KP Equations" *Proc. Sympos. Pure Math.*, 49 (1989), pp. 553-569

[22] G. Segal and G. Wilson, "Loop Groups and Equations of KdV Type" *Publications Mathematiques No. 61 de l'Institut des Hautes Etudes Scientifiques* (1985) pp. 5-65

[23] A.P. Veselov, K.L. Styrkas and O.A. Chalykh, "Algebraic Integrability for the Schrödinger operators and reflections groups", Theor. Math. Phys. 94 (1993) 253-275

[24] A.P. Veselov, M.V. Feigin, O.A. Chalykh, "New Integrable Deformations of Quantum Calogero-Moser problem", Usp. Math. Nauk 51 (3) (1996) 185-186

[25] G. Wilson, "Bispectral Commutative Ordinary Differential Operators", *J. reine angew. Math.* 442 (1993) pp. 177–204

[26] G. Wilson, "Collisions of Calogero-Moser particles and an adelic Grassmannian", to appear in Invent.Math.

Lectures on the Asymptotic Expansion of a Hermitian Matrix Integral

Motohico Mulase

Department of Mathematics, University of California, Davis, CA 95616–8633
email mulase@math.ucdavis.edu

Abstract. In these lectures three different methods of computing the asymptotic expansion of a Hermitian matrix integral is presented. The first one is a combinatorial method using Feynman diagrams. This leads us to the generating function of the reciprocal of the order of the automorphism group of a tiling of a Riemann surface. The second method is based on the classical analysis of orthogonal polynomials. A rigorous asymptotic method is established, and a special case of the matrix integral is computed in terms of the Riemann ζ-function. The third method is derived from a formula for the τ-function solution to the KP equations. This method leads us to a new class of solutions of the KP equations that are *transcendental*, in the sense that they cannot be obtained by the celebrated Krichever construction and its generalizations based on algebraic geometry of vector bundles on Riemann surfaces. In each case a mathematically rigorous way of dealing with asymptotic series in an infinite number of variables is established.

Date: August 18, 1997
1991 Mathematics Subject Classification. Primary: 32G15, 57R20, 81Q30. Secondary: 14H15, 30E15, 30E20, 30F30

0 Introduction

The purpose of these lectures is to explain three different methods of calculation of the asymptotic expansion of a Hermitian matrix integral.

The first method is a combinatorial one using the technique of Feynman diagram expansion. This method leads us directly to the connection between the matrix integrals and the moduli spaces of pointed Riemann surfaces (Harer-Zagier (1986)), (Kontsevich (1992)), (Penner (1988)), (Witten (1991)). The second method is the classical asymptotic analysis of orthogonal polynomials. It allows us to compute the integral explicitly in the special case known as the Penner Model, which is related to the Euler characteristic of the moduli spaces of Riemann surfaces. We will see that the values are expressed in terms of the Riemann zeta function. Except for this special case, the integral in general reduces to a Selberg integral which is not explicitly computable. However, through the fact that the Hermitian matrix integral satisfies the KP equations, we give another expression of the asymptotic expansion as a τ-function of the KP equations.

The Hermitian matrix integral thus connects three different worlds of mathematics: the moduli theory of Riemann surfaces through combinatorics, the Riemann zeta function through classical asymptotic analysis, and the theory of integrable systems through τ-functions of the KP equations. We explain these relations in this article, however, no attempt will be made to give any conceptual or geometric explanation why the KP equations are related to the topology of moduli spaces of pointed Riemann surfaces.

Riemann's collected work is a great source of imagination to a mathematician. The Riemann theta functions were introduced in his monumental paper *Theorie der Abel'schen Functionen* that was published in Crelle's journal in 1857. Two years later he published a paper on the prime number distribution where he studied the property of the zeta function as a complex analytic function. These papers are unrelated, but we note that his proof of the functional equation of the zeta function is based on the transformation property of a Jacobi theta function with respect to the Jacobi imaginary transform $\tau \mapsto -1/\tau$. The Jacobi theta functions are the 1-dimensional version of the Riemann theta functions, and the Jacobi imaginary transform is a special case of more general modular transforms in the moduli parameters. The coincidental equivalence between the functional equation of the Riemann zeta function and the modular invariance of a theta function is mysterious. How much more did Riemann know about the relations between these two types of functions?

In the following sections we explore another relation between these two types of functions. The way we will encounter the moduli spaces of Riemann surfaces is quite different from Riemann's in the above mentioned paper of 1857. They appear very naturally in the asymptotic expansion of Hermitian matrix integrals, which can be considered as a kind of generalization of the Riemann theta functions. We know that Riemann theta functions associated with Riemann surfaces are characterized as finite-dimensional solutions to the system of KP equations (Arbarello et al (1984)), (Mulase (1984)), (Mumford (1978)), (Shiota (1986)). The matrix integrals that we will investigate in this article satisfy again the same KP equations, though this time they are truly infinite-dimensional solutions (Mulase (1994a)).

Using a combinatorial and number-theoretic method, Harer and Zagier (Harer-Zagier (1986)) obtained a formula for the Euler number of the moduli space of pointed Riemann surfaces (defined as an algebraic stack or an orbifold) in terms of the Riemann zeta function. Later an analytic method of calculating the asymptotic expansion of a special Hermitian matrix integral was proposed by Penner (Penner (1988)). He discovered that the coefficients of the asymptotic series are given in terms of special values of the Riemann zeta function. Penner's proposed computation coincides with the formula of Harer and Zagier, except for the subtle point of giving an ordering to the set of marked points or not. The calculation of the asymptotic expansion of the Penner model has been rigorously performed (Mulase (1995)). The theorem of Harer and Zagier gives an amazing relation between the Riemann zeta function and the Riemann theta functions, if we think the latter to be essentially related to the moduli spaces of Riemann surfaces.

We add to this link yet another player: the KP equations. The observation (Mulase (1994b)) that the Hermitian matrix integral is a continuum soliton solution to the KP equation is suggestive from the geometric point of view. Soliton solutions represent singular Riemann surfaces with rational double points. When we increase the number of singularities to continuum infinity, the τ-function of the soliton solution converges to a Hermitian matrix integral that has the information of the

Euler characteristic of the moduli spaces of pointed Riemann surfaces. We do not know why.

Many explicit formulas for solutions of the KP equations have been established. All these solutions are based on the one-to-one correspondence between certain class of solutions of the KP equations and a set of geometric data consisting of an arbitrary irreducible algebraic curve, which can be singular as well, and a torsion-free sheaf defined on it (Mulase (1990)). Let us call a solution to the KP equations *transcendental* if it does not correspond to any algebraic curve. How can we construct a transcendental solution, then? An answer has been obtained by an accident. It turns out that the Hermitian matrix integrals we deal with in this article are transcendental solutions of the KP equations. This is closely related to the unexpected $sl(2)$ stability condition of the points of the infinite-dimensional Grassmannian of Sato (Sato (1981)) that correspond to the matrix integrals. Again we do not have any satisfactory explanation why the KP equations, the $sl(2)$ stability condition, and the Euler characteristic of the moduli spaces of pointed Riemann surfaces are related. The last section is devoted to this topic.

The organization of the article is as follows. In Section 1 we explain the technique of the Feynman diagram expansion through a toy model. A Feynman diagram is a kind of *graph*, but the notion of the automorphism group of a Feynman diagram is different from the usual graph theoretic automorphism. This topic is carefully treated in this section. Section 2 is devoted to explaining the ribbon graph expansion of a Hermitian matrix integral. The mathematical method of dealing with asymptotic series in an infinite number of variables is also explained in this section. The Penner model is rigorously calculated in Section 3, following the idea of (Mulase (1995)). The value we obtain is the Euler characteristic of the moduli spaces of pointed Riemann surfaces calculated by Harer and Zagier, but we will not go into the moduli theory in this article. The third expression of the asymptotic expansion of the Hermitian matrix integral is computed by using the formula for the τ-function solution to the KP equations in Section 4. This solution is transcendental, which is proved in Section 5 from the $sl(2)$ stability condition of the point of the Grassmannian that corresponds to the Hermitian matrix integral. The last two sections contain our new results, including Theorem 4.1 and Theorem 5.1, which were presented in the UIC Workshop in 1997.

Acknowledgement. This article is based on the series of lectures delivered by the author as graduate courses at Kyoto University (1994, 1995), Mathematical Society of Japan Summer Institute for Youth (1995), Humboldt Universität zu Berlin (1995, 1996), and the University of California, Davis (1994, 1996). He thanks the organizers and the enthusiastic audience of these courses, in particular, Mikio and Yasuko Sato, Takahiro Shiota, and Kenji Ueno of Kyoto, Thomas Friedrich, Herbert Kurke, and Ines Quandt of Berlin, and Michael Pencava and Craig Tracy of Davis, for encouragements and valuable comments. The author's special thanks are due to Laura Loos who went through the earlier version of the lecture notes and made useful comments and suggestions that are incorporated in this article.

1   Feynman diagram expansion of a toy model

Let us start with a simple integral:

$$\int_{-\infty}^{\infty} e^{-x^2/2} dx = \sqrt{2\pi}.$$ (1.1)

According to Lord Kelvin, a mathematician is one to whom that is as obvious as that twice two makes four is to you. However, the usual proof of this formula using polar coordinates of a plane is *really trivial*, and it is hardly a good qualification for a mathematician. It is plausible that Lord Kelvin had in mind a proof using functions only in one variable and appealing to an infinite product expansion of trigonometric functions, that requires reasonably deep knowledge of function theory.

The integral we consider is a variation of (1.1):

$$Z(t) = \int_{-\infty}^{\infty} e^{-x^2/2} e^{t \cdot x^4/4!} \frac{dx}{\sqrt{2\pi}}.$$ (1.2)

We want to know the integral $Z(t)$ as a function of t. Since

$$\left| e^{t \cdot x^4/4!} \right| = e^{Re(t) \cdot x^4/4!}$$

for every $x \in \mathbb{R}$, the integral converges to make $Z(t)$ a holomorphic function in t for $Re(t) < 0$. Unfortunately there is no analytic method to give a simple closed formula like (1.1) for (1.2), so we need a different approach. Since a holomorphic function defined on a domain is completely determined by its convergent Taylor expansion at a point in the domain, we can try to find a convergent power series expansion of $Z(t)$. But here again we encounter the same problem, and the only thing we can do is restricted to the power series expansion of $Z(t)$ at $t = 0$. At a boundary point of the domain where the function is not holomorphic, there is no longer a Taylor expansion, but we still have a useful power series expansion called an *asymptotic expansion*.

Definition 1.1 *Let Ω be an open domain of the complex plane $\mathbb{C}$ having the origin 0 on its boundary, and let $h(z)$ be a holomorphic function defined on Ω. A formal power series*

$$\sum_{v=0}^{\infty} a_v z^v$$

is said to be an asymptotic expansion *of $h(z)$ on Ω at $z = 0$ if*

$$\lim_{\substack{z \to 0 \\ z \in \Omega}} \frac{h(z) - \sum_{v=0}^{m} a_v z^v}{z^{m+1}} = a_{m+1}$$ (1.3)

holds for all $m \geq 0$.

If $h(z)$ happens to be holomorphic at $z = 0$, then the Taylor series expansion of $h(z)$ at the origin is by definition an asymptotic expansion. Formula (1.3) shows that if $h(z)$ admits an asymptotic expansion, then it is unique. However, we cannot recover the original holomorphic function from its asymptotic expansion. Let us compute the asymptotic expansion of $e^{1/z}$ defined on a domain

$$\Omega_\epsilon = \{z \in \mathbb{C} | \pi/2 + \epsilon < \arg(z) < 3\pi/2 - \epsilon\}$$ (1.4)

for a small $\epsilon > 0$. Since

$$\lim_{\substack{z \to 0 \\ z \in \Omega_\epsilon}} \frac{e^{1/z} - 0}{z^{m+1}} = 0$$

for any $m \geq 0$, the asymptotic expansion of $e^{1/z}$ at the origin is the 0-series. Thus the asymptotic expansion does not recognize the difference between $e^{1/z}$ and the 0-function. We will use this fact many times in Section 3 when we compute the Penner model. This example also shows us that even when $h(z)$ is not holomorphic at $z = 0$, its asymptotic expansion can be a convergent power series.

To indicate that the asymptotic expansion of a holomorphic function is *not equal* to the original function, we use the following notation:

$$A\big(h(z)\big) = \sum_{v=0}^{\infty} a_v z^v.$$

If two holomorphic functions $h(z)$ and $f(z)$ defined on Ω have the same asymptotic expansion at $z = 0$, then we write

$$h(z) \overset{A}{\equiv} f(z).$$

Thus $0 \overset{A}{\equiv} e^{1/z}$ at $z = 0$ as holomorphic functions defined on the domain Ω_ϵ. For two holomorphic functions $f(z)$ and $g(z)$ defined on Ω admitting the asymptotic expansions at 0, we have

$$A\big(f(z) + g(z)\big) = A\big(f(z)\big) + A\big(g(z)\big)$$
$$A\big(f(z) \cdot g(z)\big) = A\big(f(z)\big) \cdot A\big(g(z)\big).$$

We note that the asymptotic expansion of a holomorphic function *does* depend on the choice of the domain Ω. For example, $e^{1/z}$ does not admit any asymptotic expansion at $z = 0$ as a holomorphic function on the right half plane. However, if

$$\Omega_1 \subset \Omega_2, \qquad 0 \in \partial\Omega_1 \cap \partial\Omega_2,$$

as in Figure 1.1, and $h(z)$ has an asymptotic expansion on Ω_2 at $z = 0$, then it also admits an asymptotic expansion on Ω_1 at $z = 0$, which is actually the same series.

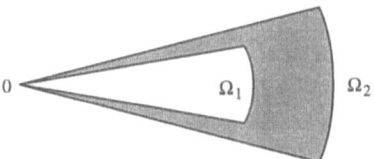

Fig. 1.1. Domains $\Omega_1 \subset \Omega_2$

We can also define the asymptotic expansion of a real analytic function: if K is an open interval of the real axis with 0 as its one of the boundary points and $h(z)$ a real analytic function on K, then the same formula (1.3), replacing Ω by K, defines the asymptotic expansion of $h(z)$ at $z = 0$.

Now let us compute the asymptotic expansion of $Z(t)$ of (1.2) as a holomorphic function defined on $\Omega_\epsilon = \{t \in \mathbb{C}| \pi/2 + \epsilon < \arg(t) < 3\pi/2 - \epsilon\}$. The Taylor expansion of the exponential gives

$$\int_{-\infty}^{\infty} e^{-x^2/2} e^{t \cdot x^4/4!} \frac{dx}{\sqrt{2\pi}} = \int_{-\infty}^{\infty} e^{-x^2/2} \sum_{v=0}^{\infty} \frac{1}{(4!)^v \cdot v!} \cdot x^{4v} \cdot t^v \cdot \frac{dx}{\sqrt{2\pi}}.$$

The infinite integral and the infinite sum we have here are not interchangeable. But let's just interchange them and see what happens:

$$\sum_{v=0}^{\infty} \frac{1}{(4!)^v \cdot v!} \left(\int_{-\infty}^{\infty} e^{-x^2/2} \cdot x^{4v} \cdot \frac{dx}{\sqrt{2\pi}} \right) t^v. \tag{1.5}$$

Note that this is a well-defined formal power series in t because the integral

$$\int_{-\infty}^{\infty} e^{-x^2/2} \cdot x^{4v} \cdot \frac{dx}{\sqrt{2\pi}}$$

converges.

Lemma 1.1 *The formal power series* (1.5) *gives the asymptotic expansion of* $Z(t)$:

$$\mathcal{A} \left(\int_{-\infty}^{\infty} e^{-x^2/2} \sum_{v=0}^{\infty} \frac{1}{(4!)^v \cdot v!} \cdot x^{4v} \cdot t^v \cdot \frac{dx}{\sqrt{2\pi}} \right)$$

$$= \sum_{v=0}^{\infty} \frac{t^v}{(4!)^v \cdot v!} \left(\int_{-\infty}^{\infty} e^{-x^2/2} \cdot x^{4v} \cdot \frac{dx}{\sqrt{2\pi}} \right).$$

Although we cannot get an *equality* by interchanging the integral and the sum because the power series expansion of the integrand of Lemma 1.1 is not uniformly convergent on the infinite interval $(-\infty, \infty)$, at least we obtain a formula which is correct in one direction.

Proof. Using the linearity of the integral, we have

$$\int_{-\infty}^{\infty} e^{-x^2/2} \sum_{v=0}^{\infty} \frac{1}{(4!)^v \cdot v!} \cdot x^{4v} \cdot t^v \cdot \frac{dx}{\sqrt{2\pi}} - \sum_{v=0}^{m} \frac{t^v}{(4!)^v \cdot v!} \int_{-\infty}^{\infty} e^{-x^2/2} \cdot x^{4v} \cdot \frac{dx}{\sqrt{2\pi}}$$

$$= \int_{-\infty}^{\infty} e^{-x^2/2} \sum_{v=m+1}^{\infty} \frac{1}{(4!)^v \cdot v!} \cdot x^{4v} \cdot t^v \cdot \frac{dx}{\sqrt{2\pi}}$$

$$= t^{m+1} \int_{-\infty}^{\infty} e^{-x^2/2} \sum_{a=0}^{\infty} \frac{1}{(4!)^{m+1+a} \cdot (m+1+a)!} \cdot x^{4(m+1+a)} \cdot t^a \cdot \frac{dx}{\sqrt{2\pi}}.$$

As long as t stays in $\Omega_\epsilon = \{t \in \mathbb{C} | \pi/2 + \epsilon < \arg(t) < 3\pi/2 - \epsilon\}$, we can divide the above expression by t^{m+1} and take the limit $t \to 0$, because the integral converges. The result is the $(m+1)$-th coefficient of the asymptotic expansion, which proves the claim. $\square$

How can we calculate the coefficient of the expansion? The standard technique is the following:

$$\int_{-\infty}^{\infty} e^{-x^2/2} \cdot x^{4v} \cdot \frac{dx}{\sqrt{2\pi}} = \int_{-\infty}^{\infty} e^{-x^2/2} \cdot \left(\frac{d}{dy} \right)^{4v} e^{xy} \Bigg|_{y=0} \cdot \frac{dx}{\sqrt{2\pi}}$$

$$= \left(\frac{d}{dy} \right)^{4v} \int_{-\infty}^{\infty} e^{-x^2/2} \cdot e^{xy} \cdot \frac{dx}{\sqrt{2\pi}} \Bigg|_{y=0}$$

$$= \left(\frac{d}{dy}\right)^{4v} \int_{-\infty}^{\infty} e^{-(x-y)^2/2} \cdot e^{y^2/2} \cdot \frac{dx}{\sqrt{2\pi}}\Bigg|_{y=0}$$

$$= \left(\frac{d}{dy}\right)^{4v} e^{y^2/2}\Bigg|_{y=0},$$

where we have used the translational invariance of the integral (1.1). Note that the integration is reduced to a differentiation. All we need now is a Taylor coefficient of the exponential function $e^{y^2/2}$, from which we obtain

$$\int_{-\infty}^{\infty} e^{-x^2/2} \cdot x^{4v} \cdot \frac{dx}{\sqrt{2\pi}} = \left(\frac{d}{dy}\right)^{4v} e^{y^2/2}\Bigg|_{y=0} = \frac{(4v)!}{(2v)! \cdot 2^{2v}} = (4v-1)!!, \qquad (1.6)$$

where the double factorial is defined by

$$(2n-1)!! = (2n-1) \cdot (2n-3) \cdot (2n-5) \cdots 5 \cdot 3 \cdot 1.$$

The quantity (1.6) has a combinatorial meaning. Let us denote the differential operator d/dy by a dot •. We have $4v$ dots attacking the fort $e^{y^2/2}$. Since y is set equal to 0 after the operation, if only one dot attacks the fort, the result would be just 0:

$$\frac{d}{dy} e^{y^2/2}\Bigg|_{y=0} = y\, e^{y^2/2}\big|_{y=0} = 0.$$

To obtain a nonzero result, the dots have to attack the fort by pairs:

$$\left(\frac{d}{dy}\right)^2 e^{y^2/2}\Bigg|_{y=0} = y^2\, e^{y^2/2}\big|_{y=0} + e^{y^2/2}\big|_{y=0} = 1.$$

Noting that the result we get by the paired attack is 1, we conclude that the value of the integral (or the differentiation) (1.6) is equal to

The number of ways of making $2v$ pairs out of $4v$ dots

$$= \binom{4v}{2}\binom{4v-2}{2}\binom{4v-4}{2}\cdots\binom{4}{2}\binom{2}{2}\bigg/(2v)!$$

$$= \frac{4v(4v-1)}{2} \cdot \frac{(4v-2)(4v-3)}{2} \cdots \frac{4\cdot 3}{2} \cdot \frac{2\cdot 1}{2}\bigg/(2v)!$$

$$= \frac{(4v)!}{(2v)! \cdot 2^{2v}}.$$

These pairs can be visualized by a diagram like Figure 1.2. Let us call such a diagram a *pairing scheme*. Thus (1.6) gives the number of pairing schemes of $4v$ dots. An example of a pairing scheme of $8 = 4 \times 2$ dots is given in Figure 1.2.

The coefficient of the asymptotic expansion of Lemma 1.1 has an extra factor of $1/(4!)^v \cdot v!$. How can we interpret it combinatorially? Here enters the idea of Feynman diagrams. The $4v$ dots are grouped into v sets of 4 dots. Let us replace each set of 4 dots by a *cross*, identifying the four dots with the four endpoints of the cross. Then the pairing scheme changes into a *Feynman diagram*, as shown in Figure 1.3, by connecting the endpoints according to the pairing rules. This is an example of a *graph*. We use this word for a CW complex like Figure 1.3 in this

Fig. 1.2. Pairing Scheme

article. A graph $\Gamma = (V, E, i)$ consists of a finite set V of *vertices*, a finite set E of *edges*, and the incidence relation i of vertices and edges. The number of half-edges coming out of a vertex of a graph is called the *degree* of the vertex. A *degree d* graph is a graph whose vertices have the same degree d. A degree 3 graph is also called a *trivalent* graph. The *order* of the graph Γ is the number of vertices $|V|$ of Γ.

When we make the Feynman diagram Γ from a pairing scheme, we consider the center of a cross as a vertex and a pairing of dots as an edge of the graph Γ. Figure 1.3 is thus considered as a degree 4 graph of order 2.

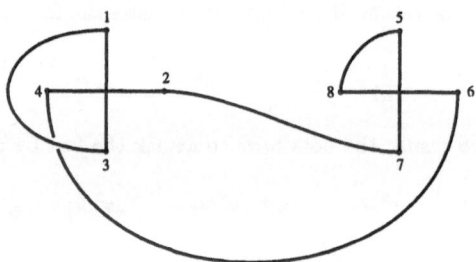

Fig. 1.3. Feynman Diagram

As a graph we can interchange the v crosses freely, and in each cross we can place the four edges in any way we want, as long as *the strings are attached*. The degrees of freedom for these moves are exactly $(4!)^v \cdot v!$. Thus we (tentatively) conclude that *the v-th coefficient of the asymptotic expansion of the integral $Z(t)$ is the number of degree 4 graphs of order v.* As an example, let us compute the simplest case $v = 1$. From the above considerations, the number of degree 4 graphs with one vertex should be

$$\frac{(4 - 1)!!}{4!} = \frac{1}{8}.$$

But this is impossible! What went wrong?

The number of different pairing schemes of 4 dots is three, as in Figure 1.4. When we factored out (1.6) by $(4!)^v \cdot v!$, we assumed that interchanging the v crosses and renumbering each edge of a cross would lead to a different pairing scheme that still corresponds to the same graph. In other words, we assumed that the group $\mathfrak{S}_v \rtimes (\mathfrak{S}_4)^v$ acts on the set of all pairing schemes freely, where $\mathfrak{S}_n$ denotes the permutation group of n letters, and the product is the semi-direct product of two factors with $(\mathfrak{S}_4)^v$ as its normal subgroup. But as we see clearly from the above example, some pairing schemes are stable under the action of non-trivial permutations. The isotropy group that stabilizes any of the three pairing schemes of Figure 1.4 is $(\mathbb{Z}/2\mathbb{Z})^3$.

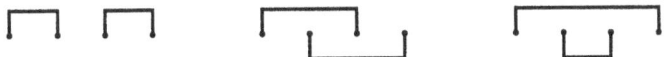

Fig. 1.4. Pairing Schemes of 4 Dots

Since our graphs are constructed from pairing schemes, we define the *automorphism group* in the following manner:

Definition 1.2 *The automorphism group* $\mathrm{Aut}(\Gamma)$ *of a graph* Γ *is the isotropy subgroup* I_P *of* $\mathfrak{S}_v \rtimes (\mathfrak{S}_4)^v$ *that preserves the original pairing scheme* P. *If pairing schemes* P *and* P' *correspond to* Γ, *then the isotropy groups* I_P *and* $I_{P'}$ *are conjugate to one another in* $\mathfrak{S}_v \rtimes (\mathfrak{S}_4)^v$. *Therefore, as an abstract group,* $\mathrm{Aut}(\Gamma)$ *is well-defined.*

Remark. Our definition of $\mathrm{Aut}(\Gamma)$ *does not* coincide with the traditional graph theoretic definition of automorphism.

The correct interpretation of $1/8$ is then $1/|\mathrm{Aut}(\Gamma)|$, where Γ in this case is the degree 4 graph with only one vertex, and we denote by $|\mathrm{Aut}(\Gamma)|$ the order of the group. More generally, we can interpret the formal power series (1.5) as a summation over the set of all pairing schemes modulo the group $\mathfrak{S}_v \rtimes (\mathfrak{S}_4)^v$, which is equivalent to the set of all degree 4 graphs. The contribution of a graph Γ is modified by the weight of $1/|\mathrm{Aut}(\Gamma)|$. Summarizing, we have established: *The asymptotic expansion of the integral* $Z(t)$ *is given by*

$$\mathcal{A}\left(\int_{-\infty}^{\infty} e^{-x^2/2} e^{t \cdot x^4/4!}\, \frac{dx}{\sqrt{2\pi}}\right) = \sum_{v=0}^{\infty} \left(\sum_{\substack{\text{degree 4 graph} \\ \Gamma \text{ of order } v}} \frac{1}{|\mathrm{Aut}(\Gamma)|} \right) \cdot t^v.$$

Since the number of degree 4 graphs with a fixed number of vertices is finite, the right hand side of the above formula is a well-defined element of the power series ring $\mathbb{Q}[[t]]$. The degree of each vertex of the graph is 4, which is due to the power 4 in the exponent of the integral. The same argument thus establishes

Theorem 1.1 *The asymptotic formula*

$$\mathcal{A}\left(\int_{-\infty}^{\infty} e^{-x^2/2} e^{t \cdot x^{2j}/(2j)!}\, \frac{dx}{\sqrt{2\pi}}\right) = \sum_{v=0}^{\infty} \left(\sum_{\substack{\text{degree } 2j \text{ graph} \\ \Gamma \text{ of order } v}} \frac{1}{|\mathrm{Aut}(\Gamma)|} \right) \cdot t^v \in \mathbb{Q}[[t]]$$

holds for an arbitrary $j \geq 2$.

We can consider more general graphs with the integral

$$Z(t_1, t_2, \cdots, t_{2m}) = \int_{-\infty}^{\infty} e^{-x^2/2} \exp\left(\sum_{j=1}^{2m} \frac{t_j}{j!} x^j\right) \frac{dx}{\sqrt{2\pi}}, \qquad (1.7)$$

where $m \geq 2$ is an integer. The integral converges if t_{2m} is in the domain Ω_ϵ of (1.4) and determines a holomorphic function on

$$(t_1, t_2, \cdots, t_{2m}) \in \mathbb{C}^{2m-1} \times \Omega_\epsilon.$$

We can expand $Z(t_1, t_2, \cdots, t_{2m})$ as a Taylor series in $(t_1, t_2, \cdots, t_{2m-1}) \in \mathbb{C}^{2m-1}$ and as an asymptotic series in $t_{2m} \in \Omega_\epsilon$ at the origin. Fix a value of $t_{2m} \in \Omega_\epsilon$. Then

$$\exp\left(t_{2m} \cdot x^{2m}/(2m)!\right)$$

acts as a uniformizing factor so that the power series expansion of the integrand in terms of x converges uniformly on $(-\infty, \infty)$ for all values of $t_1, t_2, \cdots, t_{2m-1} \in \mathbb{C}$. Therefore, we can interchange the infinite integral and the infinite sums:

$$Z(t_1, t_2, \cdots, t_{2m}) = \int_{-\infty}^{\infty} e^{-x^2/2} \exp\left(\sum_{j=1}^{2m} \frac{t_j}{j!} x^j\right) \frac{dx}{\sqrt{2\pi}}$$

$$= \int_{-\infty}^{\infty} e^{-x^2/2} \exp\left(\frac{t_1}{1!} x\right) \cdots \exp\left(\frac{t_{2m-1}}{(2m-1)!} x^{2m-1}\right) \cdot \exp\left(\frac{t_{2m}}{(2m)!} x^{2m}\right) \frac{dx}{\sqrt{2\pi}}$$

$$= \int_{-\infty}^{\infty} e^{-x^2/2} \left(\sum_{v_1=0}^{\infty} \frac{t_1^{v_1}}{v_1! \cdot (1!)^{v_1}}\right) \cdots \left(\sum_{v_{2m-1}=0}^{\infty} \frac{t_{2m-1}^{v_{2m-1}}}{v_{2m-1}! \cdot ((2m-1)!)^{v_{2m-1}}}\right)$$

$$\times \left(\sum_{v_{2m}=0}^{\infty} \frac{t_{2m}^{v_{2m}}}{v_{2m}! \cdot ((2m)!)^{v_{2m}}}\right) x^{v_1 + 2v_2 + \cdots + (2m)v_{2m}} \frac{dx}{\sqrt{2\pi}}$$

$$= \left(\sum_{v_1=0}^{\infty} \frac{t_1^{v_1}}{v_1! \cdot (1!)^{v_1}}\right) \cdots \left(\sum_{v_{2m-1}=0}^{\infty} \frac{t_{2m-1}^{v_{2m-1}}}{v_{2m-1}! \cdot ((2m-1)!)^{v_{2m-1}}}\right)$$

$$\times \int_{-\infty}^{\infty} e^{-x^2/2} \left(\sum_{v_{2m}=0}^{\infty} \frac{t_{2m}^{v_{2m}}}{v_{2m}! \cdot ((2m)!)^{v_{2m}}}\right) x^{v_1 + 2v_2 + \cdots + (2m)v_{2m}} \frac{dx}{\sqrt{2\pi}}.$$

We already know that

$$A\left(\int_{-\infty}^{\infty} e^{-x^2/2} \left(\sum_{v_{2m}=0}^{\infty} \frac{t_{2m}^{v_{2m}}}{v_{2m}! \cdot ((2m)!)^{v_{2m}}}\right) x^{v_1 + 2v_2 + \cdots + (2m)v_{2m}} \frac{dx}{\sqrt{2\pi}}\right)$$

$$= \sum_{v_{2m}=0}^{\infty} \frac{t_{2m}^{v_{2m}}}{v_{2m}! \cdot ((2m)!)^{v_{2m}}} \int_{-\infty}^{\infty} e^{-x^2/2} x^{v_1 + 2v_2 + \cdots + (2m)v_{2m}} \frac{dx}{\sqrt{2\pi}}$$

at $t_{2m} = 0$ when the top integral is considered to be a holomorphic function in $t_{2m} \in \Omega_\epsilon$. Therefore, we have

$$A\left(\int_{-\infty}^{\infty} e^{-x^2/2} \exp\left(\sum_{j=1}^{2m} \frac{t_j}{j!} x^j\right) \frac{dx}{\sqrt{2\pi}}\right)$$

$$= \sum_{v_1=0}^{\infty} \frac{t_1^{v_1}}{v_1! \cdot (1!)^{v_1}} \cdots \sum_{v_{2m}=0}^{\infty} \frac{t_{2m}^{v_{2m}}}{v_{2m}! \cdot ((2m)!)^{v_{2m}}} \int_{-\infty}^{\infty} e^{-x^2/2} x^{v_1 + 2v_2 + \cdots + (2m)v_{2m}} \frac{dx}{\sqrt{2\pi}}.$$

$$(1.8)$$

We now apply the Feynman diagram expansion to the above integral. First, we have

$$\int_{-\infty}^{\infty} e^{-x^2/2} x^{v_1+2v_2+\cdots+(2m)v_{2m}} \frac{dx}{\sqrt{2\pi}} = \left(\frac{d}{dy}\right)^{v_1+2v_2+\cdots+(2m)v_{2m}} e^{y^2/2}\Bigg|_{y=0}.$$

The pairing scheme of the dot diagram has v_1 sets of single dot, v_2 sets of double dots, $\cdots$, and v_{2m} sets of $2m$ dots. Passing to a Feynman diagram, we have a graph with v_j vertices of degree j for $j = 1, 2, \cdots, 2m$. Thus

Theorem 1.2 *We have the following asymptotic formula:*

$$A\left(\int_{-\infty}^{\infty} e^{-x^2/2} \exp\left(\sum_{j=1}^{2m} \frac{t_j}{j!} x^j\right) \frac{dx}{\sqrt{2\pi}}\right) = \sum_{\substack{Graph\ \Gamma\ with \\ vertices\ of\ degree\ \leq 2m}} \frac{1}{|\mathrm{Aut}(\Gamma)|} \cdot \prod_{j=1}^{2m} t_j^{v_j(\Gamma)},$$

where $v_j(\Gamma)$ denotes the number of vertices of degree j in Γ.

Note that the asymptotic series is a well-defined element of the formal power series ring

$$\mathbb{Q}[[t_1, t_2, \cdots, t_{2m}]],$$

because there are only finitely many graphs for given numbers $v_1(\Gamma)$, $v_2(\Gamma)$, $\cdots$, $v_{2m}(\Gamma)$.

Definition 1.3 *The automorphism group of Γ is defined as the isotropy subgroup of*

$$\prod_{j=1}^{2m} \mathfrak{S}_{v_j} \ltimes \mathfrak{S}_j^{v_j}$$

that stabilizes the pairing scheme corresponding to Γ.

As before, the definition of $\mathrm{Aut}(\Gamma)$ as an abstract group does not depend on the particular choice of the pairing scheme corresponding to the graph.

Let us now consider the relation between general graphs and connected graphs. Let a_v be the number of arbitrary degree j graphs of order v and c_v the number of *connected* degree j graphs of order v, where $j \geq 1$ is a fixed number. From Figure 1.5, it is obvious that

$$a_v = \sum_{n_1+2n_2+3n_3+\cdots=v} \frac{c_1^{n_1} \cdot c_2^{n_2} \cdot c_3^{n_3} \cdots}{n_1! \cdot n_2! \cdot n_3! \cdots}, \tag{1.9}$$

where n_i is the number of connected components with i vertices in a given graph. Formula (1.9) is equivalent to a simple functional relation in terms of generating functions:

$$\sum_{v=0}^{\infty} a_v t_j^v = \exp\left(\sum_{v=1}^{\infty} c_v t_j^v\right), \tag{1.10}$$

where we use the convention that $a_0 = 1$ and $c_0 = 0$.

In a more general case, let $\mathbf{v} = (v_1, v_2, \cdots, v_{2m})$ and

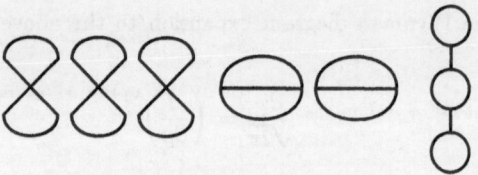

Fig. 1.5. A Disconnected Graph

$$t^{\mathbf{v}} = \prod_{j=1}^{2m} t_j^{v_j}.$$

We denote by $c_{\mathbf{v}}$ the number of connected graphs with v_j vertices of degree j, where $1 \le j \le 2m$, and by $a_{\mathbf{v}}$ the number of all graphs with v_j vertices of degree j. Then we have

$$\sum_{\mathbf{v}} a_{\mathbf{v}} t^{\mathbf{v}} = \exp\left(\sum_{\mathbf{v}} c_{\mathbf{v}} t^{\mathbf{v}}\right)$$

$$= 1 + \sum_{\mathbf{v}} c_{\mathbf{v}} t^{\mathbf{v}} + \frac{1}{2!}\left(\sum_{\mathbf{v}} c_{\mathbf{v}} t^{\mathbf{v}}\right)^2 + \frac{1}{3!}\left(\sum_{\mathbf{v}} c_{\mathbf{v}} t^{\mathbf{v}}\right)^3 + \cdots, \tag{1.11}$$

where the n-th term of the right hand side counts the number of graphs consisting of n connected components. The factor $1/(n!)$ means that we can permute the n connected components without changing the original graph.

The case we are considering is slightly more complicated because the generating function we have in Theorem 1.2 counts the number of graphs with weight $1/|\text{Aut}(\Gamma)|$. The automorphism group of a graph Γ consisting of n connected components $\Gamma_1, \cdots, \Gamma_n$ is the semi-direct product

$$\text{Aut}(\Gamma) = \mathfrak{S}_n \ltimes \prod_{j=1}^{n} \text{Aut}(\Gamma_j). \tag{1.12}$$

Therefore, we have

$$\frac{1}{|\text{Aut}(\Gamma)|} = \frac{1}{n!} \cdot \prod_{j=1}^{n} \frac{1}{|\text{Aut}(\Gamma_j)|}.$$

Note that the right hand side is the product of n factors following the key factor $1/(n!)$. It shows, therefore, that the exponential formula (1.11) connecting connected graphs and general graphs also holds in the case we are investigating. Thus we have established

Theorem 1.3 *The asymptotic series involving only connected graphs is given by*

$$\log \mathcal{A}\left(\int_{-\infty}^{\infty} e^{-x^2/2} \exp\left(\sum_{j=1}^{2m} \frac{t_j}{j!} x^j\right) \frac{dx}{\sqrt{2\pi}}\right)$$

$$= \sum_{\substack{\textit{Connected graph } \Gamma \textit{ with} \\ \textit{vertices of degree } \le 2m}} \frac{1}{|\text{Aut}(\Gamma)|} \cdot \prod_{j=1}^{2m} t_j^{v_j(\Gamma)}.$$

We note here that the function log is *not* considered as an analytic function. It is applied to the formal power series appearing in the right hand side of Theorem 1.2 as the inverse power series of $\exp(z)$ defined by

$$\log(1 - z) = -\sum_{n=1}^{\infty} \frac{1}{n} z^n.$$

2   Matrix integrals and ribbon graphs/fatgraphs

Let $\mathcal{H}_n$ denote the space of all $n \times n$ Hermitian matrices. It is an n^2-dimensional Euclidean space with a metric

$$\sqrt{\text{trace}(X - Y)^2}, \qquad X, Y \in \mathcal{H}_n.$$

The standard volume form on $\mathcal{H}_n$, which is compatible with the above metric, is given by

$$d\mu(X) = dx_{11} \wedge dx_{22} \wedge \cdots \wedge dx_{nn} \wedge \left(\bigwedge_{i<j} d(Re x_{ij}) \wedge d(Im x_{ij}) \right)$$

for $X = [x_{ij}] \in \mathcal{H}_n$. It is important to note that the metric and the volume form of $\mathcal{H}_n$ are invariant under the conjugation $X \longmapsto UXU^{-1}$ by a unitary matrix $U \in U(n)$. The main subject of this section is the *Hermitian matrix integral*

$$Z_n(t, m) = \int_{\mathcal{H}_n} \exp\left(-\frac{1}{2} \text{trace}(X^2) \right) \exp\left(\text{trace} \sum_{j=3}^{2m} \frac{t_j}{j} X^j \right) \frac{d\mu(X)}{N}, \qquad (2.1)$$

where

$$N = \int_{\mathcal{H}_n} \exp\left(-\frac{1}{2} \text{trace}(X^2) \right) d\mu(X) = 2^{n/2} \cdot \pi^{n^2/2} \qquad (2.2)$$

is a normalization constant to make $Z_n(0, m) = 1$. We note that $Z_n(t, m)$ is a holomorphic function in $(t_3, t_4, \cdots, t_{2m-1}) \in \mathbb{C}^{2m-3}$ and

$$t_{2m} \in \Omega_\epsilon = \{t \in \mathbb{C} | \pi/2 + \epsilon < \arg(t) < 3\pi/2 - \epsilon\}$$

($\epsilon > 0$), because the dominating term $\text{trace}(X^{2m})$ is positive definite on $\mathcal{H}_n$. Thus we can expand $Z_n(t, m)$ as a convergent power series in $t_3, t_4, \cdots, t_{2m-1}$ about 0, and as an asymptotic series in t_{2m} at $t_{2m} = 0$.

We also note here that we do not include the t_1 and t_2 terms in the integral because of our interests in topology, which will become clearer as we proceed. From the point of view of graphs, we do not allow degree 1 and 2 vertices in this section.

Corresponding to the fact that the integral (2.1) has richer structure than (1.8), the Feynman diagrams appearing in the asymptotic expansion of $Z_n(t, m)$ have more information than just a graph as in Theorem 1.2. As we are going to see below, the new information we have from the Hermitian matrix integral is that *the graph is drawn on a compact oriented surface*. Suppose we have such a graph drawn on an oriented surface, as in Figure 2.1.

Locally at each vertex of the graph, the orientation of the surface gives rise to a cyclic order of the edges coming out of the vertex, as shown in Figure 2.2.

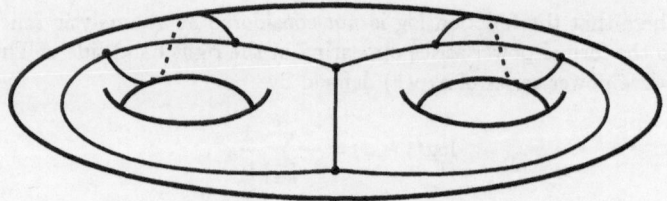

Fig. 2.1. Graph on a Surface

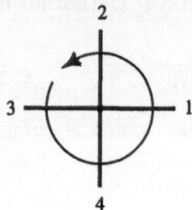

Fig. 2.2. Cyclic Order of Edges

A graph drawn on a surface thus gives a graph with a cyclic order of edges at each vertex. An example, that is corresponding to Figure 2.1, is shown in Figure 2.3.

Note that two circles are reversed in Figure 2.3, corresponding to the fact that two edges of the graph of Figure 2.1 go around the back side of the surface.

Conversely, suppose we have a connected graph Γ_{rib} with a cyclic order of edges assigned to each vertex. To indicate that we have the extra information of cyclic order at each vertex, we use Γ_{rib} and distinguish it from the underlying graph Γ. We can construct a compact oriented surface $C(\Gamma_{rib})$ canonically such that the graph Γ is drawn on it, as follows. First, the graph around each vertex can be drawn on a positively oriented plane that is compatible with the cyclic order. Next we *fatten* the local part of the graph into a crossroad of multiple intersection. The orientation of the plane defines an orientation on each sidewalk of the crossroad, as in Figure 2.4.

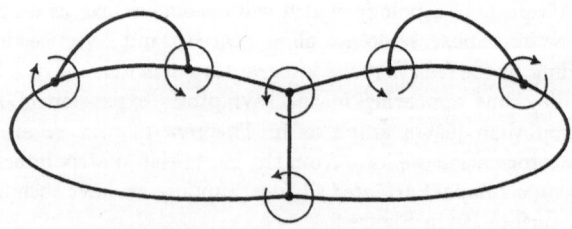

Fig. 2.3. Ribbon Graph

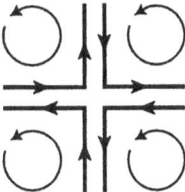

Fig. 2.4. Oriented Crossroad

The roads are connected to the other parts of the graph, with matching orientation on the sidewalks. Then we obtain an oriented surface with boundary. Figure 2.5 shows such a surface with boundary.

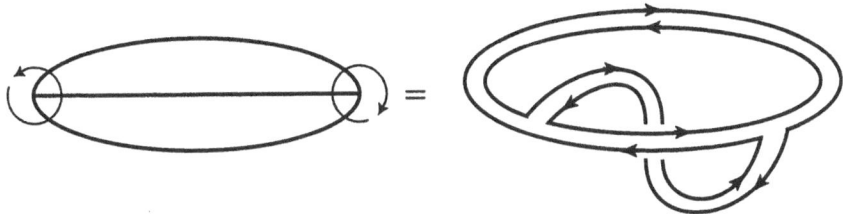

Fig. 2.5. Ribbon Graph and Surface with Boundary

Let $b(\Gamma_{rib})$ denote the number of boundary components of this oriented surface (= fattened graph) made out of Γ_{rib}. From the construction, each boundary component has a unique orientation compatible with that of the fattened graph. Thus a boundary component is indeed an oriented circle, which we also call a *boundary circuit*. So we can attach an oriented 2-dimensional disk to each boundary component of the fattened graph to construct a compact oriented surface $C(\Gamma_{rib})$.

Definition 2.1 *A ribbon graph (or a fatgraph) is a graph with a cyclic order of edges assigned to each vertex.*

The ribbon graph Γ_{rib} of Figure 2.5 has only one boundary component, and the resulting compact surface $C(\Gamma_{rib})$ is the 2-torus on which the underlying graph Γ is drawn (Figure 2.6).

We have shown that every graph drawn on an oriented surface is a ribbon graph, and conversely, that every connected ribbon graph Γ_{rib} gives rise to a canonical compact oriented surface $C(\Gamma_{rib})$ on which the underlying graph is drawn. The attached boundary disks and the underlying graph Γ give a cell-decomposition of $C(\Gamma_{rib})$.

Lemma 2.1 *Let Γ_{rib} be a connected ribbon graph with vertices of degree ≥ 3, and $v_j(\Gamma)$ denote the number of vertices of the underlying graph Γ of degree j. Then the genus $g(C(\Gamma_{rib}))$ of the canonical oriented surface $C(\Gamma_{rib})$ associated with Γ_{rib} is computed by the following formula:*

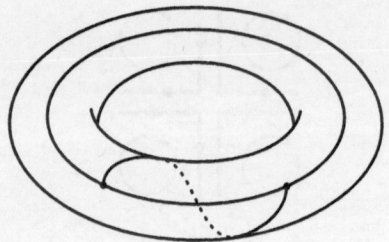

Fig. 2.6. 3-valent Graph on a Torus

$$2 - 2g(C(\Gamma_{rib})) = \sum_{j \geq 3} v_j(\Gamma) - \frac{1}{2} \sum_{j \geq 3} j \cdot v_j(\Gamma) + s(\Gamma_{rib}).$$

Proof. The total number of vertices of the cell-decomposition is given by $\sum_{j \geq 3} v_j(\Gamma)$. Since each edge is bounded by two vertices (possibly the same), the number of edges is given by $\frac{1}{2} \sum_{j \geq 3} j \cdot v_j(\Gamma)$. By construction, $b(\Gamma_{rib})$ is the number of 2-cells. Thus the Euler characteristic of a compact surface gives the above formula. $\square$

To see how ribbon graphs appear in the matrix integral, let us consider a simple example:

$$\int_{\mathcal{H}_n} \exp\left(-\frac{1}{2} \mathrm{trace}(X^2)\right) \cdot \exp\left(\frac{t}{4} \mathrm{trace} X^4\right) \frac{d\mu(X)}{N}.$$

Using the same argument as in Lemma 1.1 we can prove the asymptotic formula

$$\mathcal{A}\left(\int_{\mathcal{H}_n} \exp\left(-\frac{1}{2} \mathrm{trace}(X^2)\right) \cdot \exp\left(\frac{t}{4} \mathrm{trace} X^4\right) \frac{d\mu(X)}{N}\right)$$

$$= \sum_{v=0}^{\infty} \frac{t^v}{4! \cdot v!} \int_{\mathcal{H}_n} \exp\left(-\frac{1}{2} \mathrm{trace}(X^2)\right) \cdot (\mathrm{trace} X^4)^v \frac{d\mu(X)}{N}.$$

We need another matrix $Y = [y_{ij}] \in \mathcal{H}_n$ and a differential operator

$$\frac{\partial}{\partial Y} = \left[\frac{\partial}{\partial y_{ij}}\right]$$

to compute the asymptotic expansion of the integral.

Lemma 2.2 *For every $j > 0$ and $v > 0$, we have*

$$\left(\mathrm{trace}\left(\frac{\partial}{\partial Y}\right)^j\right)^v e^{\mathrm{trace}(X^t \cdot Y)}\bigg|_{Y=0} = \left(\mathrm{trace} X^j\right)^v. \tag{2.3}$$

Proof. Suppose that Y and X are both arbitrary complex matrices of size n. Then for each $j > 0$, we have

$$\mathrm{trace}\left(\frac{\partial}{\partial Y}\right)^j e^{\mathrm{trace}(X^t \cdot Y)}\bigg|_{Y=0}$$

$$= \sum_{i_1, i_2, i_3, \cdots, i_j = 1}^{n} \frac{\partial}{\partial y_{i_1 i_2}} \frac{\partial}{\partial y_{i_2 i_3}} \cdots \frac{\partial}{\partial y_{i_j i_1}} \exp\left(\sum_{k,\ell=1}^{n} x_{k\ell} \cdot y_{k\ell}\right)\bigg|_{Y=0}$$

$$= \sum_{i_1,i_2,i_3,\cdots,i_j=1}^{n} x_{i_1 i_2} x_{i_2 i_3} \cdots x_{i_j i_1} = \text{trace} X^j.$$

Repeating it v times, we obtain the desired formula (2.3) for general complex matrices. Certainly, the formula holds after changing coordinates:

$$\begin{cases} y_{ij} = u_{ij} + \sqrt{-1} w_{ij} & \text{for } i < j \\ y_{ji} = u_{ij} - \sqrt{-1} w_{ij} & \text{for } i < j \\ y_{ii} = u_{ii} & \end{cases} \qquad (2.4)$$

where u_{ij} and w_{ij} are complex variables. Since (2.3) is an algebraic formula, it holds for an arbitrary field of characteristic 0. In particular, (2.3) holds for real u_{ij} and w_{ij}, which proves the lemma. $\square$

Therefore, we have

$$\int_{\mathcal{H}_n} \exp\left(-\frac{1}{2}\text{trace}(X^2)\right) \cdot (\text{trace} X^4)^v \frac{d\mu(X)}{N}$$

$$= \int_{\mathcal{H}_n} \exp\left(-\frac{1}{2}\text{trace}(X^2)\right) \cdot \left(\text{trace}\left(\frac{\partial}{\partial Y}\right)^4\right)^v e^{\text{trace}(X^t \cdot Y)}\bigg|_{Y=0} \frac{d\mu(X)}{N}$$

$$= \left(\text{trace}\left(\frac{\partial}{\partial Y}\right)^4\right)^v \int_{\mathcal{H}_n} \exp\left(-\frac{1}{2}\text{trace}(X - Y^t)^2\right) \cdot e^{1/2\text{trace}(Y^t)^2}\bigg|_{Y=0} \frac{d\mu(X)}{N}$$

$$= \left(\text{trace}\left(\frac{\partial}{\partial Y}\right)^4\right)^v e^{1/2\text{trace} Y^2}\bigg|_{Y=0}$$

$$= \left(\sum_{i,j,k,\ell} \frac{\partial}{\partial y_{ij}} \frac{\partial}{\partial y_{jk}} \frac{\partial}{\partial y_{k\ell}} \frac{\partial}{\partial y_{\ell i}}\right)^v \exp\left(\frac{1}{2}\sum_{i,j} y_{ij} y_{ji}\right)\bigg|_{Y=0}.$$

The only nontrivial contribution of the differentiation comes from paired derivatives:

$$\frac{\partial}{\partial y_{ij}} \frac{\partial}{\partial y_{k\ell}} \exp\left(\frac{1}{2}\sum_{i,j} y_{ij} y_{ji}\right)\bigg|_{Y=0} = \frac{\partial}{\partial y_{ij}} y_{\ell k} = \delta_{i\ell} \cdot \delta_{jk}.$$

If we denote by $\bullet_{ij}$ the differential operator $\frac{\partial}{\partial y_{ij}}$, then we have a pairing scheme of $4v$ dots as before, and the pairing of two dots $\bullet_{ij}$ and $\bullet_{k\ell}$ contributes $\delta_{i\ell} \cdot \delta_{jk}$. Thus

$$\left(\text{trace}\left(\frac{\partial}{\partial Y}\right)^4\right)^v e^{1/2\text{trace} Y^2}\bigg|_{Y=0}$$

$$(2.5)$$

$$= \sum_{i_1,j_1,k_1,\ell_1=1}^{n} \cdots \sum_{i_v,j_v,k_v,\ell_v=1}^{n} \sum_{\substack{\text{All pairings} \\ P \text{ of } 4v \text{ dots}}} \left(\prod_{\substack{\text{All paired dots} \\ (\bullet_{ij},\bullet_{k\ell}) \text{ in } P}} \delta_{i\ell} \cdot \delta_{jk}\right).$$

A symbolic description of the contribution of parings is given in Figures 2.7.

An interpretation of Figure 2.7 in terms of Feynman Diagrams was introduced by 'tHooft ('tHooft (1974)). The set of four *indexed* dots $\bullet_{ij} \bullet_{jk} \bullet_{k\ell}\bullet_{\ell i}$ is replaced by a crossroad (Figure 2.8).

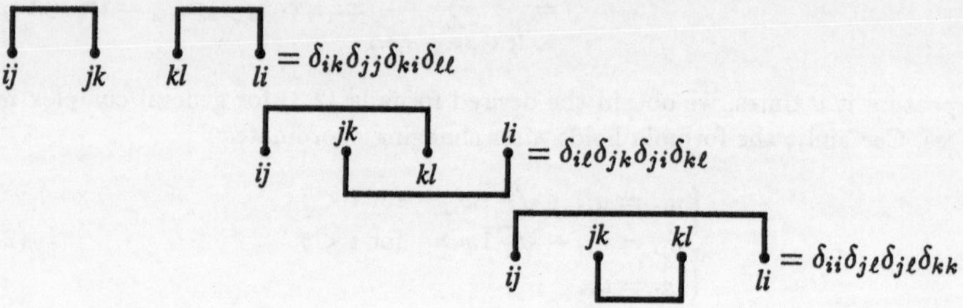

$$\overbrace{\underset{ij}{\vert}\quad\underset{jk}{\vert}}\quad\overbrace{\underset{kl}{\vert}\quad\underset{li}{\vert}} = \delta_{ik}\delta_{jj}\delta_{ki}\delta_{\ell\ell}$$

$$\overbrace{\underset{ij}{\vert}\quad\overset{jk}{\underset{\underbrace{\vert\quad\vert}_{kl}}{}}\quad\overset{li}{\vert}} = \delta_{i\ell}\delta_{jk}\delta_{ji}\delta_{k\ell}$$

$$\overset{}{\underset{ij}{\vert}\quad\overset{jk\qquad kl}{\underbrace{\vert\quad\vert}}\quad\underset{li}{\vert}} = \delta_{ii}\delta_{j\ell}\delta_{j\ell}\delta_{kk}$$

Fig. 2.7. Pairing Contribution

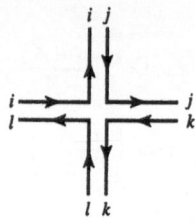

Fig. 2.8. Indexed Crossroad

Since $\bullet_{ij} = \frac{\partial}{\partial y_{ij}}$ is different from $\bullet_{ji} = \frac{\partial}{\partial y_{ji}}$, the different roles of the indices are represented by an arrow. If $\bullet_{ij}$ is connected to $\bullet_{jk}$, then it gives a contribution of $\delta_{ik} \cdot \delta_{jj}$. 'tHooft visualized this situation graphically by making a crossroad loop (Figure 2.9).

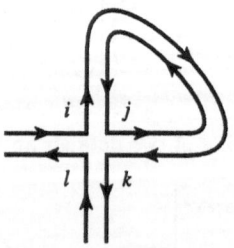

Fig. 2.9. Crossroad Loop

Note that the orientation of the sidewalks of this crossroad loop is consistent. Thus we obtain a ribbon graph, as we expected.

The passage from the pairing scheme to a ribbon graph has again some redundancy. In Section 1, the permutation group $\mathfrak{S}_j$ appeared for a vertex of degree j. This is due to the fact that a scalar monomial $x_1 x_2 \cdots x_j$ is invariant under the $\mathfrak{S}_j$-

action. In the case of matrix integrals, a monomial is of type $\text{trace}(X_1 X_2 \cdots X_j)$, which is invariant under the action of the cyclic group $\mathbb{Z}/j\mathbb{Z}$, but not under the full symmetric group $\mathfrak{S}_j$. This is the origin of the appearance of the extra cyclic order of the edges at each vertex.

Definition 2.2 *Let P be a pairing scheme of indexed dots and Γ_{rib} the corresponding ribbon graph. Then the group*

$$\prod_j \left(\mathfrak{S}_{v_j(\Gamma)} \rtimes (\mathbb{Z}/j\mathbb{Z})^{v_j(\Gamma)} \right)$$

acts on the set of all pairing schemes. As before, we define the automorphism group of a ribbon graph Γ_{rib} to be the isotropy subgroup of the above group that fixes P. As an abstract group, $\text{Aut}(\Gamma_{rib})$ does not depend on the choice of the pairing scheme of indexed dots corresponding to Γ_{rib}.

One more difference between the matrix integral and the integrals considered in Section 1 is the appearance of the size of matrix in the calculation. To illustrate this effect, let us continue our consideration of the degree 4 case with one vertex:

$$\frac{1}{4 \cdot 1!} \int_{\mathcal{H}_n} e^{-1/2 \text{trace}(X^2)} \text{trace}(X^4) \frac{d\mu(X)}{N}$$

$$= \frac{1}{4} \sum_{i,j,k,\ell=1}^{n} \left(\delta_{ik}\delta_{jj}\delta_{ki}\delta_{\ell\ell} + \delta_{i\ell}\delta_{jk}\delta_{ji}\delta_{k\ell} + \delta_{ii}\delta_{j\ell}\delta_{j\ell}\delta_{kk} \right) \qquad (2.6)$$

$$= \frac{1}{4}(n^3 + n + n^3) = \frac{1}{2}n^3 + \frac{1}{4}n.$$

As shown in Figure 2.10 there are two degree 4 ribbon graphs of order one. The one on the left has the automorphism group $\mathbb{Z}/2\mathbb{Z}$, while the second has $\mathbb{Z}/4\mathbb{Z}$.

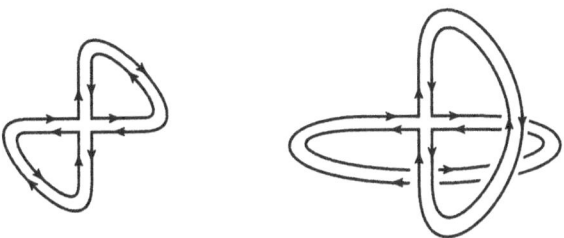

Fig. 2.10. Degree 4 Ribbon Graphs with 1 Vertex

We also note that the exponent of n in (2.6) is exactly the number of boundary components of the ribbon graph which is considered as a surface with boundary. For every $v \geq 1$, we now have

$$\frac{1}{4^v \cdot v!} \int_{\mathcal{H}_n} e^{-1/2\mathrm{trace}(X^2)} \left(\mathrm{trace}(X^4)\right)^v \frac{d\mu(X)}{N}$$

$$= \sum_{\substack{\text{degree 4 ribbon graph} \\ \Gamma_{rib} \text{ of order } v}} \frac{1}{|\mathrm{Aut}(\Gamma_{rib})|} n^{b(\Gamma_{rib})} \in \mathbb{Q}[n].$$

The same argument that we used to prove Theorem 1.2 works and we have:

Theorem 2.1 *The asymptotic expansion of the Hermitian matrix integral (2.1) is given by*

$$\mathcal{A} \left(\int_{\mathcal{H}_n} \exp\left(-\frac{1}{2}\mathrm{trace}(X^2)\right) \exp\left(\mathrm{trace} \sum_{j=3}^{2m} \frac{t_j}{j} X^j\right) \frac{d\mu(X)}{N} \right)$$

$$= \sum_{\substack{\text{Ribbon graph } \Gamma_{rib} \text{ with} \\ \text{vertices of degree } 3,4,\cdots,2m}} \frac{1}{|\mathrm{Aut}(\Gamma_{rib})|} n^{b(\Gamma_{rib})} \cdot \prod_{j=3}^{2m} t_j{}^{v_j(\Gamma)},$$

where $b(\Gamma_{rib})$ denotes the number of boundary components of the ribbon graph Γ_{rib}, and $v_j(\Gamma)$ the number of degree j vertices in the underlying graph Γ.

Here we note that for given values of $v_3(\Gamma), \cdots, v_{2m}(\Gamma)$, the number of ribbon graphs is finite. Thus the above asymptotic series belongs to $(\mathbb{Q}[n])[[t_3, t_4, \cdots, t_{2m}]]$. The relation between connected ribbon graphs and arbitrary ribbon graphs are the same as in Section 1. In particular, since (1.12) also holds for ribbon graphs, application of the logarithm gives us

Theorem 2.2

$$\log \mathcal{A} \left(\int_{\mathcal{H}_n} \exp\left(-\frac{1}{2}\mathrm{trace}(X^2)\right) \exp\left(\mathrm{trace} \sum_{j=3}^{2m} \frac{t_j}{j} X^j\right) \frac{d\mu(X)}{N} \right)$$

$$= \sum_{\substack{\text{Connected ribbon graph } \Gamma_{rib} \\ \text{with maximum degree } 2m}} \frac{1}{|\mathrm{Aut}(\Gamma_{rib})|} n^{b(\Gamma_{rib})} \cdot \prod_{j=3}^{2m} t_j{}^{v_j(\Gamma)}.$$

This formula is particularly useful, because we are interested in *connected* Riemann surfaces and only connected ribbon graphs give rise to connected surfaces. Using Lemma 2.1, we can rearrange the summation in terms of the genus of a compact oriented surface and the number of marked points on it:

$$\log \mathcal{A} \left(Z_n(t, m) \right)$$

$$= \sum_{\substack{g \geq 0, s > 0 \\ 2 - 2g - s < 0}} \left(\sum_{\substack{\text{Connected ribbon graph } \Gamma_{rib} \\ \text{with vertices of degree } 3,4,\cdots,2m, \\ \chi(\Gamma) = 2 - 2g - s, b(\Gamma_{rib}) = s}} \frac{n^s}{|\mathrm{Aut}(\Gamma_{rib})|} \cdot \prod_{j=3}^{2m} t_j{}^{v_j(\Gamma)} \right), \quad (2.7)$$

where $\chi(\Gamma)$ denotes the Euler characteristic of the underlying graph Γ. Note that Lemma 2.1 implies that

$$2 - 2g(C(\Gamma_{rib})) - b(\Gamma_{rib}) < 0$$

for every ribbon graph Γ_{rib}. Let $v(\Gamma)$ and $e(\Gamma)$ be the total number of vertices and edges of the graph Γ, respectively. Then

$$\begin{cases} \chi(\Gamma) = v(\Gamma) - e(\Gamma) \\ v(\Gamma) = v_3(\Gamma) + v_4(\Gamma) + \cdots + v_{2m}(\Gamma) \\ e(\Gamma) = \frac{1}{2}(3 \cdot v_3(\Gamma) + 4 \cdot v_4(\Gamma) + \cdots + 2m \cdot v_{2m}(\Gamma)), \end{cases} \qquad (2.8)$$

because the vertices of Γ have degree in between 3 and $2m$. Thus for every fixed g and s, the second summation of (2.7) is a finite sum, which again shows that (2.7) is an element of the formal power series ring

$$(\mathbb{Q}[n])[[t_3, t_4, \cdots, t_{2m}]].$$

The number g is of course the genus of $C(\Gamma_{rib})$. The topological type of the ribbon graph Γ_{rib} is the same as the compact surface $C(\Gamma_{rib})$ minus $b(\Gamma_{rib})$ points. The number of boundary components becomes the number of *marked* points of a Riemann surface in later sections.

Let $(\mathbb{Q}[n])[[t_3, t_4, \cdots]]$ be the formal power series ring in infinitely many variables. The *adic* topology of this ring is given by the degree

$$\deg t_j = j, \qquad j \geq 3$$

and the ideal $\mathfrak{I}_j(t)$ of $(\mathbb{Q}[n])[[t_3, t_4, \cdots]]$ generated by polynomials in $t_3, t_4, \cdots$ of degree greater than j, with coefficients in $\mathbb{Q}[n]$. We have a natural projection

$$\pi_j : (\mathbb{Q}[n])[[t_3, t_4, \cdots]] \longrightarrow (\mathbb{Q}[n])[[t_3, t_4, \cdots]]/\mathfrak{I}_j(t) = (\mathbb{Q}[n])[[t_3, \cdots, t_j]]/\mathfrak{I}_j(t).$$

For each fixed j, the projection image

$$\pi_j\big(\log \mathcal{A}(Z_n(t, m))\big) \in (\mathbb{Q}[n])[[t_3, t_4, \cdots]]/\mathfrak{I}_j(t) = (\mathbb{Q}[n])[[t_3, \cdots, t_j]]/\mathfrak{I}_j(t)$$

is stable for all $2m \geq j$. Since

$$(\mathbb{Q}[n])[[t_3, t_4, \cdots]] = \varprojlim_{j} (\mathbb{Q}[n])[[t_3, t_4, \cdots]]/\mathfrak{I}_j(t)$$

and

$$\big\{\pi_{2m}\big(\log \mathcal{A}(Z_n(t, m))\big)\big\}_{m \geq 2}$$

defines an element of the projective system, it gives a well-defined formal power series in infinitely many variables. We denote it symbolically by

$$\begin{aligned} \lim_{m \to \infty} \log \mathcal{A}(Z_n(t, m)) \\ = \big\{\pi_{2m}\big(\log \mathcal{A}(Z_n(t, m))\big)\big\}_{m \geq 2} \in (\mathbb{Q}[n])[[t_3, t_4, \cdots]]. \end{aligned} \qquad (2.9)$$

Going back to the Feynman diagram expansion (2.7), we have an equality

$$\lim_{m \to \infty} \log \mathcal{A}(Z_n(t, m))$$

$$= \sum_{\substack{g \geq 0, s > 0 \\ 2-2g-s < 0}} \left(\sum_{\substack{\text{Connected ribbon graph } \Gamma_{rib} \\ \text{with vertices of degree } \geq 3, \\ \chi(\Gamma) = 2-2g-s, \, b(\Gamma_{rib}) = s}} \frac{n^s}{|\text{Aut}(\Gamma_{rib})|} \cdot \prod_{j \geq 3} t_j^{v_j(\Gamma)} \right)$$

$$(2.10)$$

as an element of $(\mathbb{Q}[n])[[t_3, t_4, \cdots]]$. For each fixed g and s, the maximum possible valency of the ribbon graphs in the second summation is $4g + 2s - 2$. To see this, let Γ be a graph with the largest possible degree ℓ. Since the Euler characteristic of Γ is given by $2 - 2g - s = v(\Gamma) - e(\Gamma)$, the degree becomes maximum when Γ has only one vertex. Thus

$$2 - 2g - s = 1 - \frac{1}{2}\ell.$$

This shows us that the right hand side of (2.10) does not have any infinite products.

3   Asymptotic analysis of the Penner model

There are no known analytic methods to compute the matrix integral $Z_n(t, m)$ for general m. It is therefore an amazing observation of Penner that at the limit of $m \to \infty$ a certain specialization of $Z_n(t, m)$ is actually computable. In this section we study the *Penner model* and calculate its asymptotic expansion analytically.

The specialization Penner considered is the substitution

$$t_j = -(\sqrt{z})^{j-2}, \quad j = 3, 4, \cdots, 2m \tag{3.1}$$

in the matrix integral $Z_n(t, m)$, where $\sqrt{z}$ is defined for $Re(z) > 0$. The condition

$$\pi/2 + \epsilon < \arg(t_{2m}) < 3\pi/2 - \epsilon$$

for t_{2m} translates into the condition

$$|\arg(z)| < \frac{\pi}{2m - 2}. \tag{3.2}$$

Thus we have a holomorphic function

$P_n(z, m)$

$$= \int_{\mathcal{H}_n} \exp\left(-\frac{1}{2}\text{trace}(X^2)\right) \exp\left(-\sum_{j=3}^{2m} \frac{(\sqrt{z})^{j-2}}{j}\text{trace}(X^j)\right) \frac{d\mu(X)}{N}$$

$$= \int_{\mathcal{H}_n} \exp\left(-\sum_{j=2}^{2m} \frac{(\sqrt{z})^{j-2}}{j}\text{trace}(X^j)\right) \frac{d\mu(X)}{N} \tag{3.3}$$

defined on the region of the complex plane given by (3.2).

We note that the domain (3.2) still makes sense as the positive real axis when we take the limit $m \to \infty$. The quantity N is the same normalization constant as in (2.2).

The asymptotic expansion of (3.3) at $z = 0$ can be calculated by making the same substitution (3.1) in Theorem 2.1. Taking the logarithm, we obtain

$\log \mathcal{A}(P_n(z, m))$

$$= \sum_{\substack{g \geq 0, s > 0 \\ 2-2g-s < 0}} \left(\sum_{\substack{\text{Connected ribbon graph } \Gamma_{rib} \\ \text{with vertices of degree } 3,4,\cdots,2m, \\ \mathcal{X}(\Gamma)=2-2g-s, b(\Gamma_{rib})=s}} \frac{(-1)^{e(\Gamma)}}{|\text{Aut}(\Gamma_{rib})|} \right) n^s \cdot (-z)^{2g+s-2},$$

$$\tag{3.4}$$

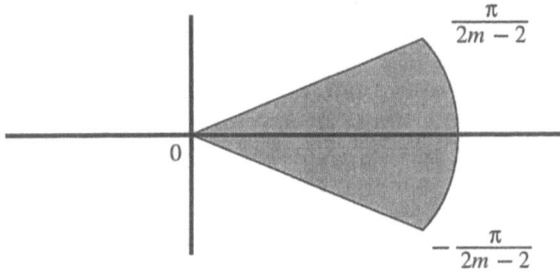

Fig. 3.1. Wedge-shape Domain

where we used (2.8) to compute

$$\prod_{j=3}^{2m} \left(-(\sqrt{z})^{j-2}\right)^{v_j(\Gamma)} = (-1)^{\Sigma_{j=3}^{2m} v_j(\Gamma)} \cdot z^{\frac{1}{2}\Sigma_{j=3}^{2m} j v_j(\Gamma) - \Sigma_{j=3}^{2m} v_j(\Gamma)}$$

$$= (-1)^{v(\Gamma)} z^{e(\Gamma)-v(\Gamma)}$$

$$= (-1)^{e(\Gamma)} (-z)^{-X(\Gamma)}.$$

Note that the right hand side of (3.4) is a well-defined element of $(\mathbb{Q}[n])[[z]]$. For every $\nu > 0$, the terms in $\log \mathcal{A}(P_n(z,m))$ of degree less than or equal to ν with respect to z are stable for all $m \geq \nu + 1$. Again by the same argument we used in Section 2, we can define an element

$$\lim_{m \to \infty} \log \mathcal{A}(P_n(z,m)) \in (\mathbb{Q}[n])[[z]].$$

Thus we have an equality

$$\lim_{m \to \infty} \log \mathcal{A}\left(\int_{\mathcal{H}_n} \exp\left(-\sum_{j=2}^{2m} \frac{(\sqrt{z})^{j-2}}{j} \operatorname{trace}(X^j)\right) \frac{d\mu(X)}{N}\right)$$

$$= \sum_{\substack{g \geq 0, s > 0 \\ 2-2g-s < 0}} \left(\sum_{\substack{\text{Connected ribbon graph } \Gamma_r \\ \text{with vertices of degree} \geq 3, \\ X(\Gamma) = 2-2g-s, b(\Gamma_{rib}) = s}} \frac{(-1)^{e(\Gamma)}}{|\operatorname{Aut}(\Gamma_{rib})|}\right) n^s \cdot (-z)^{2g+s-2}$$

$$(3.5)$$

as a well-defined element of $(\mathbb{Q}[n])[[z]]$. We recall that in (2.10) we proved that the number of ribbon graphs in the second summation for fixed g and s is finite.

Let us now compute $\lim_{m \to \infty} \log \mathcal{A}(P_n(z,m))$. The standard analytic technique to compute the Hermitian matrix integrals is the following formula. Let $f(X)$ be a function on $X \in \mathcal{H}_n$ which is invariant under the conjugation by a unitary matrix $U \in U(n)$:

$$f(X) = f(U^{-1} \cdot X \cdot U) = f(k_0, k_1, \cdots, k_{n-1}),$$

where $k_0, k_1, \cdots, k_{n-1}$ are the eigenvalues of the Hermitian matrix X. If $f(X)$ is integrable on $\mathcal{H}_n$ with respect to the measure $d\mu(X)$, then

$$\int_{\mathcal{H}_n} f(X)d\mu(X) = c(n) \cdot \int_{\mathbb{R}^n} f(k_0, k_1, \cdots, k_{n-1})\Delta(k)^2 dk_0 dk_1 \cdots dk_{n-1}, \qquad (3.6)$$

where

$$c(n) = \frac{\pi^{n(n-1)/2}}{n! \cdot (n-1)! \cdots 2! \cdot 1!}, \qquad (3.7)$$

and

$$\Delta(k) = \Delta(k_0, k_1 \cdots, k_{n-1}) = \det \begin{pmatrix} 1 & k_0 & k_0^2 & \cdots & k_0^{n-1} \\ 1 & k_1 & k_1^2 & \cdots & k_1^{n-1} \\ 1 & k_2 & k_2^2 & \cdots & k_2^{n-1} \\ \vdots & \vdots & \vdots & \ddots & \vdots \\ 1 & k_{n-1} & k_{n-1}^2 & \cdots & k_{n-1}^{n-1} \end{pmatrix} = \prod_{i>j}(k_i - k_j)$$

is the Vandermonde determinant. The proof of (3.6) goes as follows:

Let $\overset{\circ}{\mathcal{H}}_n$ denote the open dense subset of $\mathcal{H}_n$ consisting of non-singular Hermitian matrices of size n with n distinct eigenvalues. If $f(X)$ is a regular integrable function on $\mathcal{H}_n$, then

$$\int_{\mathcal{H}_n} f(X)d\mu(X) = \int_{\overset{\circ}{\mathcal{H}}_n} f(X)d\mu(X).$$

We denote by $\overset{\circ}{\mathbb{R}}^n$ the space of real diagonal matrices of all distinct, non-zero eigenvalues. Here again integration over $\mathbb{R}^n$ is equal to integration over $\overset{\circ}{\mathbb{R}}^n$. Since every Hermitian matrix is diagonalizable by a unitary matrix, we have a surjective map

$$U(n) \times \overset{\circ}{\mathbb{R}}^n \ni \left(U, \begin{bmatrix} k_0 & & \\ & \ddots & \\ & & k_{n-1} \end{bmatrix} \right) \longmapsto U \cdot \begin{bmatrix} k_0 & & \\ & \ddots & \\ & & k_{n-1} \end{bmatrix} \cdot U^{-1} \in \overset{\circ}{\mathcal{H}}_n.$$

The fiber of this map is the set of all unitary matrices that are commutative with a generic real diagonal matrix, which can be identified with the product of two subgroups

$$T^n \cdot W_n \subset U(n),$$

where $T^n \subset U(n)$ is the maximal torus of $U(n)$, and $W_n \subset U(n)$ the group of permutation matrices of size n. Note that

$$\dim U(n) = \dim \mathcal{H}_n = n^2, \quad \dim T^n = n.$$

Therefore, the induced map

$$h : U(n)/T^n \times \overset{\circ}{\mathbb{R}}^n \longrightarrow \overset{\circ}{\mathcal{H}}_n$$

is a covering map of degree $|W_n| = n!$. We need the Jacobian determinant of h. Put

$$X = [x_{ij}] = U \cdot \begin{bmatrix} k_0 & & \\ & \ddots & \\ & & k_{n-1} \end{bmatrix} \cdot U^{-1} \in \overset{\circ}{\mathcal{H}}_n,$$

and denote

$$dX = [dx_{ij}].$$

Then

$$dX = dU \cdot \begin{bmatrix} k_0 & & \\ & \ddots & \\ & & k_{n-1} \end{bmatrix} \cdot U^{-1} + U \cdot \begin{bmatrix} dk_0 & & \\ & \ddots & \\ & & dk_{n-1} \end{bmatrix} \cdot U^{-1}$$

$$- U \cdot \begin{bmatrix} k_0 & & \\ & \ddots & \\ & & k_{n-1} \end{bmatrix} \cdot U^{-1} \cdot dU \cdot U^{-1}$$

$$= U \cdot \begin{bmatrix} dk_0 & & \\ & \ddots & \\ & & dk_{n-1} \end{bmatrix} \cdot U^{-1} + [dU \cdot U^{-1}, X]$$

$$= U \cdot \left(\begin{bmatrix} dk_0 & & \\ & \ddots & \\ & & dk_{n-1} \end{bmatrix} + \left[U^{-1} \cdot dU, \begin{bmatrix} k_0 & & \\ & \ddots & \\ & & k_{n-1} \end{bmatrix} \right] \right) \cdot U^{-1}$$

$$= U \cdot \left(\begin{bmatrix} dk_0 & & \\ & \ddots & \\ & & dk_{n-1} \end{bmatrix} + [(k_j - k_i)d\omega_{ij}] \right) \cdot U^{-1},$$

where

$$U^{-1} \cdot dU = [d\omega_{ij}],$$

which is a skew Hermitian matrix. In terms of the above expression, we compute

$$d\mu(X) = dk_0 \wedge \cdots \wedge dk_{n-1} \wedge \left(\bigwedge_{i<j} (k_j - k_i)^2 Re(d\omega_{ij}) \wedge \Im(d\omega_{ij}) \right).$$

Thus the integration on $\overset{\circ}{\mathcal{H}}_n$ is separated to integration on $U(n)/T^n$ and $\overset{\circ}{\mathbb{R}}^n$. Let

$$c(n) = \frac{1}{n!} \int_{U(n)/T^n} \bigwedge_{i<j} Re(d\omega_{ij}) \wedge \Im(d\omega_{ij}).$$

Then we obtain

$$\int_{\overset{\circ}{\mathcal{H}}_n} f(X)d\mu(X) = c(n) \int_{\overset{\circ}{\mathbb{R}}^n} \Delta(k)^2 f(k_0, \cdots, k_{n-1}) dk_0 \cdots dk_{n-1}.$$

For computation of $c(n)$, we refer to, for example, Bessis-Itzykson-Zuber (Bessis et al (1980)).

Using formula (3.6), we can reduce our integral to

$$P_n(z, m) = \frac{c(n)}{N} \int_{\mathbb{R}^n} \Delta(k)^2 \prod_{i=0}^{n-1} \left(\exp\left(-\sum_{j=2}^{2m} \frac{(\sqrt{z})^{j-2}}{j} k_i^j \right) dk_i \right).$$

At this stage, one might want to compute

$$\lim_{m \to \infty} \exp\left(-\sum_{j=2}^{2m} \frac{(\sqrt{z})^{j-2}}{j} k_i^j\right) = \exp\left(\frac{1}{z}\log(1 - \sqrt{z}k_i) + \frac{k_i}{\sqrt{z}}\right)$$

$$= (1 - \sqrt{z}k_i)^{1/z} \cdot e^{k_i/\sqrt{z}}$$

$$= z^{1/z} \cdot e^{1/z} \cdot x^{1/z} \cdot e^{-x},$$

where

$$x = \frac{1 - \sqrt{z}k_i}{z}.$$

Since the above function in x is proportional to the Laguerre potential, one might expect that the integral becomes computable. However, such a substitution requires a very careful treatment. First of all, we have to justify the limit $m \to \infty$ taken *inside* the integral over the whole space. Secondly, the integral with respect to k_i is for the entire real axis, which translates to an integral in x again on the entire real line. Since the Laguerre potential is not integrable for negative x, the above formal computation cannot be justifiable inside the integral sign. What should we do, then?

The following is our key idea to compute the Penner model.

Theorem 3.1 ((Mulase (1995))) *Let* $\mathfrak{I}_\nu(z) = z^\nu \cdot \mathbb{C}[[z]]$ *denote the ideal of* $\mathbb{C}[[z]]$ *generated by* z^ν, *and*

$$\pi_\nu : \mathbb{C}[[z]] \longrightarrow \mathbb{C}[[z]]/\mathfrak{I}_\nu(z)$$

the natural projection. For an arbitrary polynomial $p(k) \in \mathbb{C}[k]$, *consider the following two asymptotic series:*

$$a(z, m) = \mathcal{A}\left(\int_{-\infty}^{\infty} p(k) \cdot \exp\left(-\sum_{j=2}^{2m} \frac{(\sqrt{z})^{j-2}}{j} k^j\right) dk\right) \in \mathbb{C}[[z]]$$

as $z \to +0$ *with* $|\arg(z)| < \frac{\pi}{2m-2}$, *and*

$$b(z) = \mathcal{A}\left(\sqrt{z}(ez)^{1/z} \int_0^{\infty} p\left(\frac{1 - zx}{\sqrt{z}}\right) \cdot x^{1/z} \cdot e^{-x} \cdot dx\right) \in \mathbb{C}[[z]]$$

as $z \to +0$ *with* $z > 0$. *Then for every* $m > 2$, *we have*

$$\pi_m\big(a(z, m)\big) = \pi_m\big(b(z)\big)$$

as an element of $\mathbb{C}[[z]]/\mathfrak{I}_m(z)$. *In other words,*

$$\lim_{m \to \infty} \mathcal{A}\left(\int_{-\infty}^{\infty} p(k) \cdot \exp\left(-\sum_{j=2}^{2m} \frac{(\sqrt{z})^{j-2}}{j} k^j\right) dk\right)$$

$$= \mathcal{A}\left(\sqrt{z}(ez)^{1/z} \int_0^{\infty} p\left(\frac{1 - zx}{\sqrt{z}}\right) \cdot x^{1/z} \cdot e^{-x} \cdot dx\right)$$

holds with respect to the $\mathfrak{I}_m(z)$-*adic topology of* $\mathbb{C}[[z]]$.

Remark. The above integrals are *never* equal as holomorphic functions in z. The limit $m \to \infty$ makes sense only for real positive z, and the equality holds only asymptotically.

Proof. Putting $y = \sqrt{z}k$, we have

$$\int_{-\infty}^{\infty} p(k) \cdot \exp\left(-\sum_{j=2}^{2m} \frac{(\sqrt{z})^{j-2}}{j} k^j\right) dk$$

$$= \frac{1}{\sqrt{z}} \int_{-\infty}^{\infty} p\left(\frac{y}{\sqrt{z}}\right) \cdot \exp\left(-\frac{1}{z}\sum_{j=2}^{2m} \frac{y^j}{j}\right) dy$$

$$= \int_{-\infty}^{\infty} d\nu(y, m),$$

where

$$d\nu(y, m) = \frac{1}{\sqrt{z}} \cdot p\left(\frac{y}{\sqrt{z}}\right) \cdot \exp\left(-\frac{1}{z}\sum_{j=2}^{2m} \frac{y^j}{j}\right) dy.$$

Let us decompose the integral into three pieces:

$$\int_{-\infty}^{\infty} d\nu(y, m) = \int_{-\infty}^{-1} d\nu(y, m) + \int_{-1}^{1} d\nu(y, m) + \int_{1}^{\infty} d\nu(y, m). \qquad (3.8)$$

Note that the polynomial $\sum_{j=2}^{2m} \frac{y^j}{j}$ of degree $2m$ takes positive values on the intervals $(-\infty, -1]$ and $[1, \infty)$. Since $p(k)$ is a polynomial, it is obvious that the asymptotic expansion of the first and the third integrals of the right hand side of (3.8) for $z \to +0$ with $z > 0$ is the 0-series. Therefore, we have

$$\int_{-\infty}^{\infty} d\nu(y, m) \overset{A}{\equiv} \int_{-1}^{1} d\nu(y, m).$$

On the interval $[-1, 1]$, if we fix a z such that $\text{Re}(z) > 0$, then the convergence

$$\lim_{m \to \infty} \exp\left(-\frac{1}{z}\sum_{j=2}^{2m} \frac{y^j}{j}\right) = (1 - y)^{1/z} \cdot e^{y/z}$$

is absolute and uniform with respect to y. Thus, for a new variable $t = 1 - y$, we have

$$\lim_{m \to \infty} \int_{-1}^{1} d\nu(y, m)$$

$$= \frac{1}{\sqrt{z}} \int_{-1}^{1} p\left(\frac{y}{\sqrt{z}}\right) (1 - y)^{1/z} e^{y/z} dy$$

$$= \frac{1}{\sqrt{z}} e^{1/z} \int_{0}^{2} p\left(\frac{1-t}{\sqrt{z}}\right) t^{1/z} e^{-t/z} dt$$

$$= \frac{1}{\sqrt{z}} e^{1/z} \int_{0}^{\infty} p\left(\frac{1-t}{\sqrt{z}}\right) t^{1/z} e^{-t/z} dt - \frac{1}{\sqrt{z}} e^{1/z} \int_{2}^{\infty} p\left(\frac{1-t}{\sqrt{z}}\right) t^{1/z} e^{-t/z} dt.$$

This last integral is

$$\frac{1}{\sqrt{z}} e^{1/z} \int_{2}^{\infty} p\left(\frac{1-t}{\sqrt{z}}\right) t^{1/z} e^{-t/z} dt = \frac{1}{\sqrt{z}} \int_{2}^{\infty} p\left(\frac{1-t}{\sqrt{z}}\right) e^{(1+\log t - t)/z} dt.$$

Since $1 + \log t - t < 0$ for $t \geq 2$, the asymptotic expansion of this integral as $z \to +0$ with $z > 0$ is the 0-series. Therefore, since the integrals do not depend on the integration variables, we have

$$
\lim_{m \to \infty} A \left(\int_{-\infty}^{\infty} p(k) \cdot \exp \left(-\sum_{j=2}^{2m} \frac{(\sqrt{z})^{j-2}}{j} k^j \right) dk \right)
$$
$$
= A \left(\frac{1}{\sqrt{z}} e^{1/z} \int_{0}^{\infty} p \left(\frac{1-t}{\sqrt{z}} \right) t^{1/z} e^{-t/z} dt \right)
$$
$$
= A \left(\sqrt{z} e^{1/z} z^{1/z} \int_{0}^{\infty} p \left(\frac{1 - zx}{\sqrt{z}} \right) x^{1/z} e^{-x} dx \right)
$$

as a formal power series in z. This completes the proof of Theorem. $\square$

By applying Theorem 3.1 for each k_i, we obtain

$$
\lim_{m \to \infty} A \left(\int_{\mathbb{R}^n} \Delta(k)^2 \cdot \prod_{i=0}^{n-1} \exp \left(-\sum_{j=2}^{2m} \frac{(\sqrt{z})^{j-2}}{j} k_i^j \right) dk_i \right)
$$
$$
= A \left(\left(\sqrt{z} e^{1/z} z^{1/z} \right)^n \int_{0}^{\infty} \cdots \int_{0}^{\infty} \Delta \left(\frac{1 - zx}{\sqrt{z}} \right)^2 \cdot \prod_{i=0}^{n-1} x_i^{1/z} e^{-x_i} dx_i \right) \qquad (3.9)
$$
$$
= A \left(\left(\sqrt{z} e^{1/z} z^{1/z} \right)^n z^{\frac{n(n-1)}{2}} \int_{0}^{\infty} \cdots \int_{0}^{\infty} \Delta(x)^2 \cdot \prod_{i=0}^{n-1} x_i^{1/z} e^{-x_i} dx_i \right),
$$

where we used the multilinear property of the Vandermonde determinant. We can use the standard technique of orthogonal polynomials to compute the above integral. Let $p_j(x)$ be a monic orthogonal polynomial in x of degree j with respect to the measure

$$
d\lambda(x) = x^{1/z} e^{-x} dx
$$

defined on $K = [0, \infty)$ for a positive $z > 0$:

$$
\int_{K} p_i(x) p_j(x) d\lambda(x) = \delta_{ij} \parallel p_j(x) \parallel^2 .
$$

Because of the multilinearity of the determinant, we have once again

$$
\Delta(x) = \det \left(x_i^j \right) = \det \left(p_j(x_i) \right).
$$

Therefore,

$$\int_{K^n} \Delta(x)^2 d\lambda(x_0) \cdots d\lambda(x_{n-1})$$

$$= \int_{K^n} \det (p_j(x_i)) \det (p_j(x_i) d\lambda(x_i))$$

$$= \int_{K^n} \sum_{\sigma \in \mathfrak{S}_n} \sum_{\tau \in \mathfrak{S}_n} \text{sign}(\sigma) \text{sign}(\tau) \prod_{i=0}^{n-1} p_{\sigma(i)}(x_i) \prod_{i=0}^{n-1} p_{\tau(i)}(x_i) d\lambda(x_i)$$

$$= \sum_{\sigma \in \mathfrak{S}_n} \sum_{\tau \in \mathfrak{S}_n} \text{sign}(\sigma) \text{sign}(\tau) \prod_{i=0}^{n-1} \int_K p_{\sigma(i)}(x) p_{\tau(i)}(x) d\lambda(x)$$

$$= \sum_{\sigma \in \mathfrak{S}_n} \text{sign}(\sigma)^2 \prod_{i=0}^{n-1} \int_K p_{\sigma(i)}(x) p_{\sigma(i)}(x) d\lambda(x)$$

$$= n! \prod_{i=0}^{n-1} \| p_i(x) \|^2 .$$

(3.10)

For a real number $z > 0$, the *Laguerre* polynomial

$$L_m^{1/z}(x) = \sum_{j=0}^{m} \binom{m+1/z}{m-j} \frac{(-1)^j}{j!} x^j = \frac{(-1)^m}{m!} x^m + \cdots$$

of degree m satisfies the orthogonality condition

$$\int_0^{\infty} L_i^{1/z}(x) L_j^{1/z}(x) e^{-x} x^{1/z} dx = \delta_{ij} \frac{(j+1/z)!}{j!}. \qquad (3.11)$$

Thus we can use

$$p_i(x) = (-1)^i \cdot i! \cdot L_i^{1/z}(x) \qquad (3.12)$$

for the computation. From (3.9)–(3.12), we have

$$\lim_{m \to \infty} \mathcal{A} \left(\int_{\mathbb{R}^n} \Delta(k)^2 \cdot \prod_{i=0}^{n-1} \exp \left(-\sum_{j=2}^{2m} \frac{(\sqrt{z})^{j-2}}{j} k_i^j \right) dk_i \right)$$

$$= \mathcal{A} \left(\left(\sqrt{z} e^{1/z} z^{1/z} \right)^n z^{\frac{n(n-1)}{2}} n! \prod_{i=0}^{n-1} i! \cdot \left(i + \frac{1}{z} \right)! \right) \qquad (3.13)$$

$$= \mathcal{A} \left((ez)^{\frac{n}{z}} \cdot z^{\frac{n^2}{2}} \cdot n! \prod_{i=0}^{n-1} i! \cdot \left(-1 + \frac{1}{z} \right)! \cdot \left(i + \frac{1}{z} \right)^{n-i} \right).$$

Applying (3.6) and (3.13) to (3.5), we conclude

$$\lim_{m\to\infty} \log \mathcal{A}\left(\frac{1}{N}\int_{\mathcal{H}_n}\exp\left(-\sum_{j=2}^{2m}\frac{(\sqrt{z})^{j-2}}{j}\mathrm{trace}(X^j)\right)d\mu(X)\right)$$

$$= \log \mathcal{A}\left(\frac{1}{N}\cdot\pi^{-\frac{n(n-1)}{2}}\cdot(ez)^{\frac{n}{z}}\cdot z^{\frac{n^2}{2}}\cdot\prod_{i=0}^{n-1}\left(-1+\frac{1}{z}\right)!\cdot\left(i+\frac{1}{z}\right)^{n-i}\right)$$

$$= \log \mathcal{A}\left(\frac{1}{N}\cdot\pi^{-\frac{n(n-1)}{2}}\cdot(ez)^{\frac{n}{z}}\cdot z^{\frac{n^2}{2}}\cdot\left(\Gamma\left(\frac{1}{z}\right)\right)^n\cdot\prod_{i=0}^{n-1}\left(i+\frac{1}{z}\right)^{n-i}\right)$$

$$= \mathrm{const} + \frac{n}{z} + \frac{n}{z}\log z + \frac{n^2}{2}\log z + n\log \mathcal{A}\left(\Gamma\left(\frac{1}{z}\right)\right) \tag{3.14}$$

$$\qquad + \sum_{i=0}^{n-1}(n-i)\log\frac{1+iz}{z}$$

$$= \mathrm{const} + \frac{n}{z} + \frac{n}{z}\log z - \frac{n}{2}\log z + n\log \mathcal{A}\left(\Gamma\left(\frac{1}{z}\right)\right)$$

$$\qquad + \sum_{r=1}^{\infty}\frac{(-1)^{r-1}}{r}\left(\sum_{i=0}^{n-1}(n-i)i^r\right)z^r.$$

Let us recall Stirling's formula:

$$\log \mathcal{A}\left(\Gamma\left(\frac{1}{z}\right)\right) = -\frac{1}{z}\log z - \frac{1}{z} + \frac{1}{2}\log z + \sum_{r=1}^{\infty}\frac{b_{2r}}{2r(2r-1)}z^{2r-1} + \mathrm{const}, \tag{3.15}$$

where b_r is the Bernoulli number defined by

$$\frac{x}{e^x-1} = \sum_{r=0}^{\infty}\frac{b_r}{r!}x^r.$$

We are not interested in the constant term (the term independent of z) of (3.15) because the asymptotic series in question, (3.5), has no constant term. We can see that substitution of (3.15) in (3.14) eliminates all the logarithmic terms as desired:

$$\lim_{m\to\infty} \log \mathcal{A}\left(\frac{1}{N}\int_{\mathcal{H}_n}\exp\left(-\sum_{j=2}^{2m}\frac{(\sqrt{z})^{j-2}}{j}\mathrm{trace}(X^j)\right)d\mu(X)\right)$$

$$= \sum_{r=1}^{\infty}\frac{b_{2r}}{2r(2r-1)}\cdot n\cdot z^{2r-1} + \sum_{r=1}^{\infty}\frac{(-1)^{r-1}}{r}\left(\sum_{i=0}^{n-1}(n-i)i^r\right)z^r.$$

Let

$$\phi_r(x) = \sum_{q=0}^{r-1}\binom{r}{q}b_q x^{r-q}$$

denote the Bernoulli polynomial. Then we have

$$\sum_{i=1}^{n-1}i^r = \frac{\phi_{r+1}(n)}{r+1}.$$

Thus for $r > 0$,

$$\sum_{i=0}^{n-1}(n-i)i^r = \frac{n\phi_{r+1}(n)}{r+1} - \frac{\phi_{r+2}(n)}{r+2}$$

$$= \sum_{q=0}^{r}\frac{1}{r+1}\binom{r+1}{q}b_q\cdot n^{r+2-q} - \sum_{q=0}^{r+1}\frac{1}{r+2}\binom{r+2}{q}b_q\cdot n^{r+2-q}$$

$$= \sum_{q=0}^{r}\frac{r!(1-q)}{q!(r+2-q)!}b_q\cdot n^{r+2-q} - b_{r+1}\cdot n.$$

Therefore, we have

$$\sum_{r=1}^{\infty}\frac{b_{2r}}{2r(2r-1)}\cdot n\cdot z^{2r-1} + \sum_{r=1}^{\infty}\frac{(-1)^{r-1}}{r}\left(\sum_{i=0}^{n-1}(n-i)i^r\right)z^r$$

$$= -\sum_{r=1}^{\infty}\frac{1}{2r}b_{2r}\cdot n\cdot z^{2r-1} + \sum_{r=1}^{\infty}\sum_{q=0}^{r}(-1)^r\frac{(r-1)!(q-1)}{q!(r+2-q)!}b_q\cdot n^{r+2-q}\cdot z^r$$

$$= -\sum_{r=1}^{\infty}\frac{1}{2r}b_{2r}\cdot n\cdot z^{2r-1} + \sum_{r=1}^{\infty}(-1)^{r-1}\frac{1}{r(r+1)(r+2)}n^{r+2}\cdot z^r \qquad (3.16)$$

$$+ \sum_{r=2}^{\infty}\sum_{q=1}^{[r/2]}(-1)^r\frac{(r-1)!(2q-1)}{(2q)!(r+2-2q)!}b_{2q}\cdot n^{r+2-2q}\cdot z^r.$$

It is time to switch the summation indices r and q to g and s as in (3.5). The first sum of the third line of (3.16) is the case when we specify a single point on a Riemann surface of arbitrary genus $g = r$. The second sum is for genus 0 case with more than two points specified. So we use $s = r+2$ for the number of points. In the third sum, $q = g \geq 0$ is the genus and $r+2-2q = s \geq 2$ is the number of points. Thus (3.16) is equal to

$$\sum_{g=1}^{\infty}\zeta(1-2g)\cdot n\cdot z^{2g-1} + \sum_{s=3}^{\infty}(-1)^{s-1}\frac{1}{s(s-1)(s-2)}n^s\cdot z^{s-2}$$

$$+ \sum_{g=1}^{\infty}\sum_{s=2}^{\infty}(-1)^{s-1}\frac{(2g+s-3)!}{(2g-2)!s!}\zeta(1-2g)\cdot n^s\cdot z^{-2+2g+s}, \qquad (3.17)$$

where we used Euler's formula

$$\zeta(1-2g) = -\frac{b_{2g}}{2g},$$

and the fact that $b_0 = 1$ and $b_{2q+1} = 0$ for $q \geq 1$. Note that the first two summations of (3.17) are actually the special cases of the third summation corresponding to $s = 1$ and $g = 0$. Thus we have established:

Theorem 3.2

$$\lim_{m\to\infty}\log\mathcal{A}\left(\frac{1}{N}\int_{\mathcal{H}_n}\exp\left(-\sum_{j=2}^{2m}\frac{(\sqrt{z})^{j-2}}{j}\text{trace}(X^j)\right)d\mu(X)\right)$$

$$= -\sum_{\substack{g\geq 0,s>0\\2-2g-s<0}}\frac{(2g+s-3)!(2g)(2g-1)}{(2g)!s!}\zeta(1-2g)\cdot n^s\cdot(-z)^{-2+2g+s}.$$

Since the asymptotic expansion is unique, from (3.5) we obtain

$$\sum_{\substack{\text{Connected ribbon graph } \Gamma_{rib} \\ \text{with vertices of degree } \geq 3, \\ \chi(\Gamma)=2-2g-s, b(\Gamma_{rib})=s}} \frac{(-1)^{e(\Gamma)}}{|\text{Aut}(\Gamma_{rib})|} = -\frac{(2g+s-3)!(2g)(2g-1)}{(2g)!s!}\zeta(1-2g) \quad (3.18)$$

for every $g \geq 0$ and $s > 0$ subject to $2 - 2g - s < 0$.

Remark. If we have taken into account the values of $c(n)$ and N in the above computation, then we will see that all the constant terms appearing in the computation automatically cancel out.

Let us examine a couple of examples.

Example 3.1 *The simplest case is $g = 0$ and $s = 3$. The underlying graph Γ of a ribbon graph Γ_{rib}, whose topological type is S^2 minus three points, should satisfy*

$$\chi(\Gamma) = v(\Gamma) - e(\Gamma) = 2 - 2g - s = -1 \quad \text{and} \quad (3.19)$$
$$3v(\Gamma) \leq 2e(\Gamma). \quad (3.20)$$

Eqn.(3.19) gives the Euler characteristic of a tri-punctured sphere, and Eqn.(3.20) states that every vertex of Γ has degree at least 3. It follows from these conditions that

$$e(\Gamma) \leq 3.$$

There are only three graphs in this case, as shown in Figure 3.2.

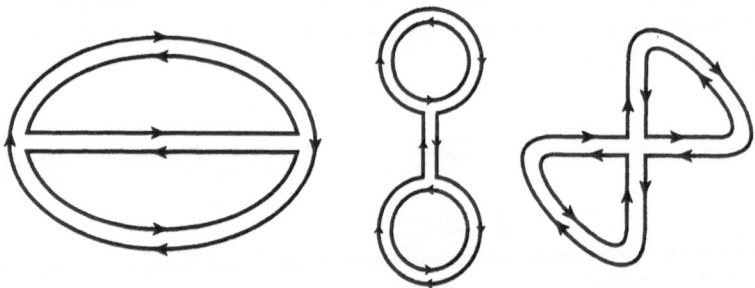

Fig. 3.2. Ribbon Graphs for $g = 0, s = 3$

The automorphism groups of these ribbon graphs are $\mathfrak{S}_2 \ltimes \mathbb{Z}/3\mathbb{Z} = \mathfrak{S}_3$, $\mathbb{Z}/2\mathbb{Z}$, and again $\mathbb{Z}/2\mathbb{Z}$, respectively. Thus the left hand side of (3.18) is

$$\frac{(-1)^3}{3!} + \frac{(-1)^3}{2} + \frac{(-1)^2}{2} = -\frac{1}{6}.$$

The right hand side is coming from the term $n^3(-z)^1$ of the second summation in (3.17). The value is, of course,

$$-\frac{1}{3(3-1)(3-2)} = -\frac{1}{6}.$$

It can be also computed from (3.18):

$$-\frac{(2g+s-3)!(2g)(2g-1)}{(2g)!s!}\zeta(1-2g) = \frac{(2g)(2g-1)b_{2g}}{(2g)!3!(2g)} = -\frac{1}{6}.$$

Example 3.2 *The next simple case is $g = s = 1$. Since the Euler characteristic condition is the same as in Example 3.1, the only possibilities are again graphs with 1 vertex and 2 edges or 2 vertices and 3 edges. There are two ribbon graphs satisfying the conditions: Figure 2.5 and the graph on the right in Figure 2.10. The first one has $\mathfrak{S}_2 \times \mathbb{Z}/3\mathbb{Z}$ as its automorphism group, which happens to be a degenerate case of the semi-direct product. The automorphism group of the second graph is $\mathbb{Z}/4\mathbb{Z}$, as noted in Section 2. Thus we have*

$$\frac{(-1)^3}{6} + \frac{(-1)^2}{4} = \frac{1}{12} = -\zeta(-1).$$

4   KP equations and matrix integrals

There are no analytic methods of evaluating the Hermitian matrix integral

$$Z_n(t,m) = \int_{\mathcal{H}_n} \exp\left(-\frac{1}{2}\text{trace}(X^2)\right) \exp\left(\text{trace}\sum_{j=3}^{2m}\frac{t_j}{j}X^j\right)\frac{d\mu(X)}{N}.$$

However, there is an interesting fact about this integral: *it satisfies the system of the KP equations.* In this section we give a proof of this fact.

To investigate the most general case, we define

$$Z_n(t,m,\phi) = \int_{\mathcal{H}_n} \exp\left(\text{trace}\sum_{j=1}^{2m}\frac{t_j}{j}X^j\right)\phi(X)\frac{d\mu(X)}{N}, \tag{4.1}$$

where $\phi(X)$ is a $U(n)$-invariant function on $\mathcal{H}_n$ which is determined by n functions $\phi_0(k), \cdots, \phi_{n-1}(k)$ in one variable in the following manner:

$$\phi(X) = \phi(k_0, k_1, \cdots k_{n-1}) = \frac{\det\begin{pmatrix} \phi_0(k_0) & \phi_1(k_0) & \cdots & \phi_{n-1}(k_0) \\ \phi_0(k_1) & \phi_1(k_1) & \cdots & \phi_{n-1}(k_1) \\ \vdots & \vdots & \ddots & \vdots \\ \phi_0(k_{n-1}) & \phi_1(k_{n-1}) & \cdots & \phi_{n-1}(k_{n-1}) \end{pmatrix}}{\Delta(k_0, k_1 \cdots, k_{n-1})}, \tag{4.2}$$

where $k_0, \cdots, k_{n-1}$ are eigenvalues of X. Unlike (2.1), we allow terms containing $t_1 X$ and $t_2 X^2$ in the integral (4.1). Using (3.6), we have

$$Z_n(t,m,\phi) = \int_{\mathcal{H}_n} \exp\left(\text{trace}\sum_{\alpha=1}^{2m}\frac{t_\alpha}{\alpha}X^\alpha\right)\phi(X)\cdot\frac{d\mu(X)}{N}$$

$$= \frac{c(n)}{N}\int_{\mathbb{R}^n}\exp\left(\sum_{i=0}^{n-1}\sum_{\alpha=1}^{2m}\frac{t_\alpha}{\alpha}k_i^\alpha\right)\Delta(k_0,\cdots,k_{n-1})\det(\phi_j(k_i))\,dk_0\cdots dk_{n-1}.$$

Here we need a simple formula. Let $\phi_0(k), \cdots, \phi_{n-1}(k)$ and $\psi_0(k), \cdots, \psi_{n-1}(k)$ be $2n$ arbitrary functions in k. Then

$$\det\left[\phi_i(k_\ell)\right]\cdot\det\left[\psi_j(k_\ell)\right]=\sum_{\sigma\in\mathfrak{S}_n}\det\left[\phi_i(k_{\sigma(j)})\cdot\psi_j(k_{\sigma(j)})\right],\tag{4.3}$$

where σ runs over all permutations of $\mathfrak{S}_n$. To prove (4.3), we calculate the left hand side by the usual product formula of the determinant. Then it becomes a summation of n^n terms. Because of the multilinearity of the determinants, only $n!$ of these terms are nonzero. Rearranging the $n!$ terms, we obtain the above formula. Using this formula for $\psi_j(k)=k^j$, we obtain

$$Z_n(t,m,\phi)$$

$$=\frac{c(n)}{N}\int_{\mathbb{R}^n}\exp\left(\sum_{i=0}^{n-1}\sum_{\alpha=1}^{2m}\frac{t_\alpha}{\alpha}k_i^\alpha\right)\sum_{\sigma\in\mathfrak{S}_n}\det\left(\phi_j(k_{\sigma(i)})k_{\sigma(i)}^i\right)dk_0\cdots dk_{n-1}$$

$$=\frac{c(n)}{N}\int_{\mathbb{R}^n}\sum_{\sigma\in\mathfrak{S}_n}\exp\left(\sum_{i=0}^{n-1}\sum_{\alpha=1}^{2m}\frac{t_\alpha}{\alpha}k_{\sigma(i)}^\alpha\right)\det\left(\phi_j(k_{\sigma(i)})k_{\sigma(i)}^i\right)dk_0\cdots dk_{n-1}$$

$$=\frac{c(n)}{N}\sum_{\sigma\in\mathfrak{S}_n}\det\left(\int_{\mathbb{R}^n}\exp\left(\sum_{\alpha=1}^{2m}\frac{t_\alpha}{\alpha}k_{\sigma(i)}^\alpha\right)\phi_j(k_{\sigma(i)})k_{\sigma(i)}^i\,dk_0\cdots dk_{n-1}\right)$$

$$=n!\cdot\frac{c(n)}{N}\det\left(\int_{-\infty}^{\infty}\exp\left(\sum_{\alpha=1}^{2m}\frac{t_\alpha}{\alpha}k_i^\alpha\right)\phi_j(k_i)k_i^i\,dk_i\right)$$

$$=n!\cdot\frac{c(n)}{N}\det\left(\int_{-\infty}^{\infty}\exp\left(\sum_{\alpha=1}^{2m}\frac{t_\alpha}{\alpha}k^\alpha\right)\phi_j(k)k^i\,dk\right).$$

The above computation makes sense as a complex analytic function in

$$(t_1,\cdots,t_{2m-1},t_{2m})\in\mathbb{C}^{2m-1}\times\{t_{2m}\in\mathbb{C}|Re(t_{2m})<0\},$$

on which the integral converges, provided that $|\phi_j(k)|$ grows slower than $\exp(k^{2m})$. To compare our t_j's with the standard time variables in the KP theory, let us set

$$T_\alpha=\frac{t_\alpha}{\alpha}.$$

Now we use the formula

$$\exp\left(\sum_{\alpha=1}^{2m}T_\alpha k^\alpha\right)=\sum_{r=0}^{\infty}p_r(T)k^r,\tag{4.4}$$

where

$$p_r(T)=\sum_{n_1+2n_2+3n_3+\cdots+(2m)n_{2m}=r}\frac{T_1^{n_1}\cdot T_2^{n_2}\cdot T_3^{n_3}\cdots T_{2m}^{n_{2m}}}{n_1!\cdot n_2!\cdot n_3!\cdots n_{2m}!}\tag{4.5}$$

is a weighted homogeneous polynomial in $\mathbb{Q}[T_1,\cdots,T_{2m}]$ of degree r. The relation (4.4) holds as an entire function in $T_1,\cdots,T_{2m}$ and k. Note that we have encountered this formula already as (1.9). From (4.4), we have

$$Z_n(t,m,\phi)=n!\cdot\frac{c(n)}{N}\det\left(\int_{-\infty}^{\infty}\sum_{r=0}^{\infty}p_r(T)k^r\phi_j(k)k^i\,dk\right)$$

$$= n! \cdot \frac{c(n)}{N} \det \left(\int_{-\infty}^{\infty} \sum_{r=0}^{\infty} p_{r-i}(T) k^r \phi_j(k) dk \right),$$

where we define $p_r(T) = 0$ for $r < 0$.

Lemma 4.1 *Let $\phi_j(k)$, $j = 0, \cdots, n-1$, be a function defined on $\mathbb{R}$ such that*

$$\int_{-\infty}^{\infty} k^r \phi_j(k) dk$$

exists for all $r \geq 0$. Then as a holomorphic function defined for $\mathrm{Re}(t_{2m}) < 0$, we have

$$\mathcal{A} \left(\int_{-\infty}^{\infty} \exp \left(\sum_{\alpha=1}^{2m} \frac{t_\alpha}{\alpha} k^\alpha \right) \phi_j(k) k^i dk \right) = \sum_{r=0}^{\infty} p_{r-i}(T) \int_{-\infty}^{\infty} k^r \phi_j(k) dk$$

as $t_{2m} \to 0$.

Proof. The argument is the same as the one we used in Section 1. We choose a fixed t_{2m} so that $\mathrm{Re}(t_{2m}) < 0$. Because of the uniform convergence of the power series expansion of the integrand, we can interchange the integral and the infinite sums for $\alpha = 1, \cdots, 2m - 1$. Using (1.8), (4.4) and (4.5), we have

$$\mathcal{A} \left(\int_{-\infty}^{\infty} \exp \left(\sum_{\alpha=1}^{2m} T_\alpha k^\alpha \right) \phi_j(k) k^i dk \right)$$

$$= \sum_{n_1=0}^{\infty} \frac{T_1^{n_1}}{n_1!} \cdots \sum_{n_{2m}=0}^{\infty} \frac{T_{2m}^{n_{2m}}}{n_{2m}!} \int_{-\infty}^{\infty} k^{i+n_1+2n_2+\cdots+(2m)n_{2m}} \phi_j(k) dk$$

$$= \sum_{r=0}^{\infty} p_r(T) \int_{-\infty}^{\infty} k^{i+r} \phi_j(k) dk.$$

□

Thus we have established

$$\mathcal{A}(Z_n(t,m,\phi)) = n! \cdot \frac{c(n)}{N} \det \left(\sum_{r=0}^{\infty} p_{r-i}(T) \int_{-\infty}^{\infty} k^r \phi_j(k) dk \right)$$

$$= \det \left(\sum_{r=0}^{\infty} p_{r-i}(T) \xi_{rj} \right), \tag{4.6}$$

where

$$\xi_{rj} = n! \cdot \frac{c(n)}{N} \int_{-\infty}^{\infty} k^r \phi_j(k) dk.$$

We recall that the determinant in (4.6) is an $n \times n$ determinant. Sato (Sato (1981)) proved that any size determinant of the form

$$\det \left(\sum_{r=0}^{\infty} p_{r-i}(T) \xi_{rj} \right) \tag{4.7}$$

satisfies the Hirota bilinear form of the KP equations. He also proved that every power series solution of the KP system should be written as (4.7), allowing certain infinite determinants. A necessary background of the KP theory can be found in (Mulase (1994a)).

We have thus proved the following theorem.

Theorem 4.1 *If $\phi_j(k)$, $j = 0, \cdots, n-1$, satisfies that*

$$\left| \int_{-\infty}^{\infty} k^r \phi_j(k) dk \right| < +\infty$$

for all $r \geq 0$, then the asymptotic expansion of the matrix integral $Z_n(t, m, \phi)$ satisfies the KP equations with respect to $T_1, T_2, \cdots, T_{2m}$. Moreover, if we choose a value of T_{2m} such that $Re(T_{2m}) < 0$ and fix it, then $Z_n(t, m, \phi)$ itself is an entire holomorphic solution to the KP equations with respect to $(T_1, T_2, \cdots, T_{2m-1}) \in \mathbb{C}^{2m-1}$. In particular,

$$u(T_1, T_2, T_3, \cdots) = \frac{\partial^2}{\partial T_1^2} \log(Z_n(t, m, \phi))$$

is a meromorphic solution to the KP equation

$$\frac{3}{4} u_{22} = \left(u_3 - \frac{1}{4} u_{111} - 3uu_1 \right)_1,$$

where u_j denotes the partial derivative of u with respect to T_j.

The formula we have just established is a continuum version of the famous Hirota soliton solution of the KP equations (Sato (1981)). The most general soliton solution of the KP equations due to Mikio and Yasuko Sato depends on $nM + M$ parameters c_{ij} and λ_i, where $0 \leq i \leq M-1$ and $0 \leq j \leq n-1$. Let

$$\eta(T, k) = \sum_{\alpha=1}^{2m} T_\alpha k^\alpha.$$

Then Sato-Sato's soliton solution is given by

$$\sum_{0 \leq i_0 < \cdots < i_{n-1} \leq M-1} \exp \left(\sum_{j=0}^{n-1} \eta(T, \lambda_{i_j}) \right)$$

$$\Delta(\lambda_{i_0}, \cdots, \lambda_{i_{n-1}}) \det \begin{pmatrix} c_{i_0 0} & \cdots & c_{i_0 n-1} \\ \vdots & & \vdots \\ c_{i_{n-1} 0} & \cdots & c_{i_{n-1} n-1} \end{pmatrix}.$$

This coincides with our $Z_n(t, m, \phi)$ if we take

$$\phi_j(k) = \sum_{i=0}^{M-1} c_{ij} \delta(k - \lambda_i).$$

Therefore, our matrix integral $Z_n(t, m, \phi)$ of (4.1) with (4.2) is indeed a *continuum soliton solution* of the KP equations.

So far we have dealt with the matrix integrals with a fixed integer m in this section. As before, we can take the limit $m \to \infty$ of these integrals, which gives formal power series solutions of the whole hierarchy of the KP equations. Note that the determinant expression of (4.6) does not have any explicit mention on the integer m. Therefore, we have obtained the third asymptotic formula for the matrix integral:

$$\lim_{m \to \infty} \mathcal{A}(Z_n(t, m, \phi)) = n! \cdot \frac{c(n)}{N} \det \left(\sum_{r=0}^{\infty} p_{r-i}(T) \int_{-\infty}^{\infty} k^r \phi_j(k) dk \right). \qquad (4.8)$$

5 Transcendental Solutions of the KP Equations and the Grassmannian

There are several different ways to construct solutions to the KP equations. The Krichever construction and its generalizations are based on the correspondence between certain points of the Grassmannian of Sato (Sato (1981)) and the algebro-geometric data consisting of an irreducible algebraic curve (possibly singular) and a torsion-free sheaf on it (Mulase (1990)). These solutions deserve to be called *algebraic*, because they carry geometric information of algebraic curves. Let us call a solution to the KP equations *transcendental* if no algebraic curve corresponds to this solution. The natural question we can ask is: how can we construct a transcendental solution?

In this section we show that the Hermitian matrix integrals we have been dealing with in the earlier sections are indeed transcendental solutions.

The technique we show that these matrix integrals are transcendental solutions is based on the observation that the points of the Grassmannian corresponding to these solutions satisfy a peculiar $sl(2)$ stability condition. Since these solutions are deeply related to the moduli theory of Riemann surfaces, the appearance of $sl(2)$ is mysteriously suggestive. At present we do not have any geometric explanation of the relation between the KP equations, the $sl(2)$ stability on the Grassmannian, and the moduli theory of pointed Riemann surfaces.

Let $V = \mathbb{C}((z))$ denote the field of formal Laurent series in one variable z. We fix its *polarization*

$$\mathbb{C}((z)) = \mathbb{C}[z^{-1}] \oplus \mathbb{C}[[z]] \cdot z. \tag{5.1}$$

For a vector subspace $W \subset V$, there is a natural map

$$\gamma_W : W \hookrightarrow V \longrightarrow V/\mathbb{C}[[z]]z \cong \mathbb{C}[z^{-1}]. \tag{5.2}$$

The infinite-dimensional Grassmannian is defined by

$$Gr = \{W \subset V \mid \gamma_W : W \longrightarrow \mathbb{C}[z^{-1}] \text{ is Fredholm of index } 0\}. \tag{5.3}$$

The *big-cell* of the Grassmannian is the subset of Gr consisting of vector subspaces $W \subset V$ such that γ_W of (5.2) is an isomorphism.

Let W be a point of the big-cell of the Grassmannian. We can choose a basis

$$\langle w_0, w_1, w_2, \cdots \rangle$$

for W such that

$$w_j = z^{-j} + \sum_{i=1}^{\infty} c_{ij} z^i, \qquad j = 0, 1, 2, \cdots. \tag{5.4}$$

The *Bosonization* is a map

$$Gr \longrightarrow \mathbb{P}(\mathbb{C}[[T_1, T_2, T_3, \cdots]]) \tag{5.5}$$

that assigns a τ-function τ_W to each point W of the Grassmannian. For a point W of the big-cell with a basis (5.4), the Bosonization has an infinite determinant expression

$$\tau_W = \det\left(p_{i-j}(T) + \sum_{\mu=1}^{\infty} p_{\mu+i}(T)c_{\mu j}\right).$$ (5.6)

The infinite determinant gives a well-defined element of $\mathbb{C}[[T_1, T_2, T_3, \cdots]]$ in the same manner as we have explained in the earlier sections. Sato's formula (4.7) gives another expression of the Bosonization map. For more detail, we refer to (Mulase (1994a)) and (Mulase (1994b)).

The *commutative stabilizer* of $W \in Gr$ is defined by

$$A_W = \{a \in \mathbb{C}((z)) \mid a \cdot W \subset W\}.$$ (5.7)

The key idea that connects the KP equations and algebraic curves is that the commutative stabilizer is the coordinate ring of an algebraic curve. If the greatest common divisor of the pole order of elements in A_W is 1, then the Bosonization τ_W of W is essentially the Riemann theta function associated with the algebraic curve C whose coordinate ring is A_W (Mulase (1984)), (Mulase (1994a)).

Definition 5.1 *A solution of the KP equations τ_W is said to be* transcendental *if*

$$A_W = \mathbb{C}.$$ (5.8)

Remark. It is known that if $A_W \neq \mathbb{C}$, then the Bosonization τ_W of W is a solution to the KP equation corresponding to a vector bundle $\mathcal{F}$ on an algebraic curve C such that

$$H^0(C, \mathcal{F}) = H^1(C, \mathcal{F}) = 0$$ (5.9)

(Mulase (1990)). Conversely, there is a solution corresponding to an arbitrary torsion-free sheaf $\mathcal{F}$ defined on an arbitrary (possibly singular) algebraic curve C satisfying (5.9). None of these solutions are transcendental.

The Hermitian matrix integral we have discussed in Section 2 gives a transcendental solution to the KP equations.

Theorem 5.1 *Choose arbitrary positive integers k and n, and let*

$$a = (a_1, a_2, \cdots, a_{2k}) \in \mathbb{C}^{2k}$$

be a complex vector such that $Re(a_{2k}) < 0$. Define a formal Laurent series

$$w_j = \sum_{r=0}^{\infty}\left(\int_{-\infty}^{\infty} \lambda^{r+j} \exp\left(\sum_{\mu=1}^{2k} a_\mu \lambda^\mu\right) d\lambda\right) z^{r+1-n} \in \mathbb{C}((z))$$ (5.10)

for $j = 0, 1, 2, \cdots, n-1$, and let

$$W(a) = \langle w_0, w_1, \cdots, w_{n-1}, z^{-n}, z^{-n-1}, \cdots\rangle \in Gr$$ (5.11)

*be a point of the Grassmannian spanned by $w_0, w_1, \cdots, w_{n-1}$ and $z^{-n}, z^{-n-1}, \cdots$.
Then the τ-function corresponding to $W(a)$ is given by the asymptotic expansion of
a Hermitian matrix integral:*

$$\tau_{W(a)} = \lim_{m\to\infty} \mathcal{A}\left(\int_{\mathcal{H}_n} \exp\left(\sum_{j=1}^{2m} T_j \text{trace}(X^j)\right) \exp\left(\sum_{\mu=1}^{2k} a_\mu \text{trace}(X^\mu)\right) dX\right),$$
(5.12)

where we take $Re(T_{2m}) < 0$ first and then let $m \to \infty$ to determine a well-defined formal power series in $\mathbb{C}[[T_1, T_2, T_3, \cdots]]$. Define a linear differential operator

$$L_i(a) = z^{1-i}\frac{d}{dz} + \frac{(3n-1) + i(n-1)}{2}z^{-i} + \sum_{\mu=1}^{2k} \mu a_\mu z^{-i-\mu} \qquad (5.13)$$

for $i = -1, 0, 1$. These differential operators satisfy the $sl(2, \mathbb{C})$ relation

$$[L_i(a), L_j(a)] = (i - j)L_{i+j}(a).$$

The point $W(a)$ of the Grassmannian satisfies the non-commutative stability condition

$$L_i(a) \cdot W(a) \subset W(a), \qquad i = -1, 0, 1. \qquad (5.14)$$

Moreover, $\tau_{W(a)}$ is a transcendental solution of the KP equations.

Proof. The function

$$\exp\left(\sum_{\mu=1}^{2k} a_\mu \mathrm{trace}(X^\mu)\right)$$

is a special case of the function $\phi(X)$ defined in (4.2). Thus the results of the previous section proves that $\tau_{W(a)}$ is a τ-function of the KP equations corresponding to the point of the Grassmannian $W(a)$.

Let us first prove that the $sl(2)$ stability condition (5.14) implies that the commutative stabilizer is trivial:

$$A_{W(a)} = \mathbb{C}.$$

Suppose $f(z) \in A_{W(a)} \subset \mathbb{C}((z))$, and let $\mathrm{ord}(f) = \nu > 0$, where we define the *pole order* by

$$\mathrm{ord}(z^{-\nu}) = \nu.$$

Since $L_{-1}(a)$ and f stabilize $W(a)$,

$$[L_{-1}(a), f] = z^2\frac{df}{dz} \in A_{W(a)}$$

also stabilizes $W(a)$. Note that

$$\mathrm{ord}([L_{-1}(a), f]) = \nu - 1.$$

Thus we can immediately conclude that

$$A_{W(a)} = \mathbb{C}[z^{-1}].$$

But then

$$L_{-1}(a) - \sum_{\mu=1}^{2k} \mu a_\mu z^{1-\mu} = z^2\frac{d}{dz} + \frac{(3n-1) - (n-1)}{2}z \qquad (5.15)$$

stabilizes $W(a)$. Since the new stabilizer (5.15) decreases the order of elements of $W(a)$ exactly by 1, $W(a)$ must have an element of arbitrary negative order. But this contradicts to the Fredholm condition of $W(a)$. This means $A_{W(a)} = \mathbb{C}$, hence $\tau_{W(a)}$ is a transcendental solution.

Now all we need is to show (5.14), which can be verified by a straightforward computation. First, we note a simple formula

$$
0 = \int_{-\infty}^{\infty} \frac{d}{d\lambda}\left(\lambda^{\alpha}\exp\left(\sum_{\mu=1}^{2k}a_{\mu}\lambda^{\mu}\right)d\lambda\right)
$$

$$
= \int_{-\infty}^{\infty}\alpha\lambda^{\alpha-1}\exp\left(\sum_{\mu=1}^{2k}a_{\mu}\lambda^{\mu}\right)d\lambda \tag{5.16}
$$

$$
+ \int_{-\infty}^{\infty}\sum_{\mu=1}^{2k}\mu a_{\mu}\lambda^{\alpha+\mu-1}\exp\left(\sum_{\mu=1}^{2k}a_{\mu}\lambda^{\mu}\right)d\lambda.
$$

Let us compute the effect of the differential operators (5.13) on the basis elements of $W(a)$. First, we have

$$
L_{-1}(a)w_j = \left(z^2\frac{d}{dz}+nz+\sum_{\mu=1}^{2k}\mu a_{\mu}z^{1-\mu}\right)\sum_{r=0}^{\infty}z^{r+1-n}\int_{-\infty}^{\infty}\lambda^{r+j}e^{\sum_{\mu=1}^{2k}a_{\mu}\lambda^{\mu}}d\lambda
$$

$$
= \sum_{r=0}^{\infty}(r+1)z^{r+2-n}\int_{-\infty}^{\infty}\lambda^{r+j}e^{\sum_{\mu=1}^{2k}a_{\mu}\lambda^{\mu}}d\lambda
$$

$$
+ \sum_{r=0}^{\infty}\sum_{\mu=1}^{2k}z^{r+2-n-\mu}\int_{-\infty}^{\infty}\lambda^{r+j}\mu a_{\mu}e^{\sum_{\mu=1}^{2k}a_{\mu}\lambda^{\mu}}d\lambda
$$

$$
= \sum_{r=0}^{\infty}rz^{r+1-n}\int_{-\infty}^{\infty}\lambda^{r+j-1}e^{\sum_{\mu=1}^{2k}a_{\mu}\lambda^{\mu}}d\lambda
$$

$$
+ \sum_{\mu=1}^{2k}\sum_{r=0}^{\mu-2}z^{r+2-n-\mu}\int_{-\infty}^{\infty}\lambda^{r+j}\mu a_{\mu}e^{\sum_{\mu=1}^{2k}a_{\mu}\lambda^{\mu}}d\lambda
$$

$$
+ \sum_{\mu=1}^{2k}\sum_{r=\mu-1}^{\infty}z^{r+2-n-\mu}\int_{-\infty}^{\infty}\lambda^{r+j}\mu a_{\mu}e^{\sum_{\mu=1}^{2k}a_{\mu}\lambda^{\mu}}d\lambda
$$

$$
= \sum_{r=0}^{\infty}rz^{r+1-n}\int_{-\infty}^{\infty}\lambda^{r+j-1}e^{\sum_{\mu=1}^{2k}a_{\mu}\lambda^{\mu}}d\lambda
$$

$$
+ \sum_{\mu=1}^{2k}\sum_{r=0}^{\mu-2}z^{r+2-n-\mu}\int_{-\infty}^{\infty}\lambda^{r+j}\mu a_{\mu}e^{\sum_{\mu=1}^{2k}a_{\mu}\lambda^{\mu}}d\lambda
$$

$$
+ \sum_{\mu=1}^{2k}\sum_{r=0}^{\infty}z^{r+1-n}\int_{-\infty}^{\infty}\lambda^{r+j+\mu-1}\mu a_{\mu}e^{\sum_{\mu=1}^{2k}a_{\mu}\lambda^{\mu}}d\lambda
$$

$$
= \sum_{r=0}^{\infty}rz^{r+1-n}\int_{-\infty}^{\infty}\lambda^{r+j-1}e^{\sum_{\mu=1}^{2k}a_{\mu}\lambda^{\mu}}d\lambda
$$

$$
+ \sum_{\mu=1}^{2k}\sum_{r=0}^{\mu-2}z^{r+2-n-\mu}\int_{-\infty}^{\infty}\lambda^{r+j}\mu a_{\mu}e^{\sum_{\mu=1}^{2k}a_{\mu}\lambda^{\mu}}d\lambda
$$

$$
- \sum_{r=0}^{\infty}z^{r+1-n}\int_{-\infty}^{\infty}(r+j)\lambda^{r+j-1}e^{\sum_{\mu=1}^{2k}a_{\mu}\lambda^{\mu}}d\lambda
$$

$$= -jw_{j-1} + \sum_{\mu=1}^{2k} \sum_{r=0}^{\mu-2} z^{r+2-n-\mu} \int_{-\infty}^{\infty} \lambda^{r+j} \mu a_\mu e^{\sum_{\mu=1}^{2k} a_\mu \lambda^\mu} d\lambda$$

$$\in W(a)$$

for all $j = 0, 1, 2, \cdots, n-1$. Note that w_{-1} does not appear in the above computation because of the combination jw_{j-1}. For the basis elements $z^{-n}, z^{-n-1}, \cdots$, we have

$$L_{-1}(a)z^{-n-i} = \left(z^2 \frac{d}{dz} + nz + \sum_{\mu=1}^{2k} \mu a_\mu z^{1-\mu} \right) z^{-n-i}$$

$$= (-i)z^{-n-i+1} + \sum_{\mu=1}^{2k} \mu a_\mu z^{1-\mu-n-i} \in W(a)$$

for all $i \geq 0$. We note that the term z^{-n+1} does not appear in this computation. Thus we conclude

$$L_{-1}(a) \cdot W(a) \subset W(a).$$

For $j = 0$, we have

$$L_0(a)w_j = \left(z\frac{d}{dz} + \frac{3n-1}{2} + \sum_{\mu=1}^{2k} \mu a_\mu z^{-\mu} \right) \sum_{r=0}^{\infty} z^{r+1-n} \int_{-\infty}^{\infty} \lambda^{r+j} e^{\sum_{\mu=1}^{2k} a_\mu \lambda^\mu} d\lambda$$

$$= \sum_{r=0}^{\infty} (r + \frac{n+1}{2}) z^{r+1-n} \int_{-\infty}^{\infty} \lambda^{r+j} e^{\sum_{\mu=1}^{2k} a_\mu \lambda^\mu} d\lambda$$

$$+ \sum_{r=0}^{\infty} \sum_{\mu=1}^{2k} z^{r+1-n-\mu} \int_{-\infty}^{\infty} \lambda^{r+j} \mu a_\mu e^{\sum_{\mu=1}^{2k} a_\mu \lambda^\mu} d\lambda$$

$$= \frac{n+1}{2} w_j + \sum_{r=0}^{\infty} r z^{r+1-n} \int_{-\infty}^{\infty} \lambda^{r+j} e^{\sum_{\mu=1}^{2k} a_\mu \lambda^\mu} d\lambda$$

$$+ \sum_{\mu=1}^{2k} \sum_{r=0}^{\mu-1} z^{r+1-n-\mu} \int_{-\infty}^{\infty} \lambda^{r+j} \mu a_\mu e^{\sum_{\mu=1}^{2k} a_\mu \lambda^\mu} d\lambda$$

$$+ \sum_{\mu=1}^{2k} \sum_{r=\mu}^{\infty} z^{r+1-n-\mu} \int_{-\infty}^{\infty} \lambda^{r+j} \mu a_\mu e^{\sum_{\mu=1}^{2k} a_\mu \lambda^\mu} d\lambda$$

$$= \frac{n+1}{2} w_j + \sum_{r=0}^{\infty} r z^{r+1-n} \int_{-\infty}^{\infty} \lambda^{r+j} e^{\sum_{\mu=1}^{2k} a_\mu \lambda^\mu} d\lambda$$

$$+ \sum_{\mu=1}^{2k} \sum_{r=0}^{\mu-1} z^{r+1-n-\mu} \int_{-\infty}^{\infty} \lambda^{r+j} \mu a_\mu e^{\sum_{\mu=1}^{2k} a_\mu \lambda^\mu} d\lambda$$

$$+ \sum_{\mu=1}^{2k} \sum_{r=0}^{\infty} z^{r+1-n} \int_{-\infty}^{\infty} \lambda^{r+j+\mu} \mu a_\mu e^{\sum_{\mu=1}^{2k} a_\mu \lambda^\mu} d\lambda$$

$$= \frac{n+1}{2} w_j + \sum_{r=0}^{\infty} r z^{r+1-n} \int_{-\infty}^{\infty} \lambda^{r+j} e^{\sum_{\mu=1}^{2k} a_\mu \lambda^\mu} d\lambda$$

$$+ \sum_{\mu=1}^{2k} \sum_{r=0}^{\mu-1} z^{r+1-n-\mu} \int_{-\infty}^{\infty} \lambda^{r+j} \mu a_\mu e^{\sum_{\mu=1}^{2k} a_\mu \lambda^\mu} d\lambda$$

$$- \sum_{r=0}^{\infty} z^{r+1-n} \int_{-\infty}^{\infty} (r+j+1)\lambda^{r+j} e^{\sum_{\mu=1}^{2k} a_\mu \lambda^\mu} d\lambda$$

$$= \left(\frac{n+1}{2} - j - 1 \right) w_j + \sum_{\mu=1}^{2k} \sum_{r=0}^{\mu-1} z^{r+1-n-\mu} \int_{-\infty}^{\infty} \lambda^{r+j} \mu a_\mu e^{\sum_{\mu=1}^{2k} a_\mu \lambda^\mu} d\lambda$$

$$\in W(a)$$

for all $j = 0, 1, 2, \cdots, n-1$. It is obvious that

$$L_0(a) \cdot z^{-n-i} \in W(a)$$

for $i \geq 0$. Finally, for $j = 1$, we have

$$L_1(a)w_j$$

$$= \left(\frac{d}{dz} + (2n-1)z^{-1} + \sum_{\mu=1}^{2k} \mu a_\mu z^{-\mu-1} \right) \sum_{r=0}^{\infty} z^{r+1-n} \int_{-\infty}^{\infty} \lambda^{r+j} e^{\sum_{\mu=1}^{2k} a_\mu \lambda^\mu} d\lambda$$

$$= \sum_{r=0}^{\infty} (r+n) z^{r-n} \int_{-\infty}^{\infty} \lambda^{r+j} e^{\sum_{\mu=1}^{2k} a_\mu \lambda^\mu} d\lambda$$

$$+ \sum_{r=0}^{\infty} \sum_{\mu=1}^{2k} z^{r-n-\mu} \int_{-\infty}^{\infty} \lambda^{r+j} \mu a_\mu e^{\sum_{\mu=1}^{2k} a_\mu \lambda^\mu} d\lambda$$

$$= \sum_{r=-1}^{\infty} (r+n+1) z^{r+1-n} \int_{-\infty}^{\infty} \lambda^{r+j+1} e^{\sum_{\mu=1}^{2k} a_\mu \lambda^\mu} d\lambda$$

$$+ \sum_{\mu=1}^{2k} \sum_{r=0}^{\mu} z^{r-n-\mu} \int_{-\infty}^{\infty} \lambda^{r+j} \mu a_\mu e^{\sum_{\mu=1}^{2k} a_\mu \lambda^\mu} d\lambda$$

$$+ \sum_{\mu=1}^{2k} \sum_{r=\mu+1}^{\infty} z^{r-n-\mu} \int_{-\infty}^{\infty} \lambda^{r+j} \mu a_\mu e^{\sum_{\mu=1}^{2k} a_\mu \lambda^\mu} d\lambda$$

$$= z^{-n} \int_{-\infty}^{\infty} \lambda^j e^{\sum_{\mu=1}^{2k} a_\mu \lambda^\mu} d\lambda + \sum_{r=0}^{\infty} (r+n+1) z^{r+1-n} \int_{-\infty}^{\infty} \lambda^{r+j+1} e^{\sum_{\mu=1}^{2k} a_\mu \lambda^\mu} d\lambda$$

$$+ \sum_{\mu=1}^{2k} \sum_{r=0}^{\mu} z^{r-n-\mu} \int_{-\infty}^{\infty} \lambda^{r+j} \mu a_\mu e^{\sum_{\mu=1}^{2k} a_\mu \lambda^\mu} d\lambda$$

$$+ \sum_{\mu=1}^{2k} \sum_{r=0}^{\infty} z^{r+1-n} \int_{-\infty}^{\infty} \lambda^{r+j+\mu+1} \mu a_\mu e^{\sum_{\mu=1}^{2k} a_\mu \lambda^\mu} d\lambda$$

$$= z^{-n} \int_{-\infty}^{\infty} \lambda^j e^{\sum_{\mu=1}^{2k} a_\mu \lambda^\mu} d\lambda + \sum_{r=0}^{\infty} (r+n+1) z^{r+1-n} \int_{-\infty}^{\infty} \lambda^{r+j+1} e^{\sum_{\mu=1}^{2k} a_\mu \lambda^\mu} d\lambda$$

$$+ \sum_{\mu=1}^{2k} \sum_{r=0}^{\mu} z^{r-n-\mu} \int_{-\infty}^{\infty} \lambda^{r+j} \mu a_\mu e^{\sum_{\mu=1}^{2k} a_\mu \lambda^\mu} d\lambda$$

$$-\sum_{r=0}^{\infty} z^{r+1-n} \int_{-\infty}^{\infty} (r+j+2)\lambda^{r+j+1} e^{\sum_{\mu=1}^{2k} a_\mu \lambda^\mu} d\lambda$$

$$= z^{-n} \int_{-\infty}^{\infty} \lambda^j e^{\sum_{\mu=1}^{2k} a_\mu \lambda^\mu} d\lambda + (n-j-1)w_{j+1}$$

$$+ \sum_{\mu=1}^{2k}\sum_{r=0}^{\mu} z^{r-n-\mu} \int_{-\infty}^{\infty} \lambda^{r+j} \mu a_\mu e^{\sum_{\mu=1}^{2k} a_\mu \lambda^\mu} d\lambda$$

$$\in W(a)$$

for all $j = 0, 1, 2, \cdots, n-1$. Note that the term w_n does not appear in the computation. It is again obvious that

$$L_1(a) \cdot z^{-n-i} \in W(a)$$

for $i \geq 0$. This completes the proof of the $sl(2)$ stability of $W(a)$, and hence we have established the theorem. □

The action of these $sl(2)$ generators on $W(a)$ is very subtle, and it does not seem to allow any generalization. For example, the above proof does not apply for the Virasoro generators $L_i(a)$ other than $i = -1, 0, 1$, although the operators $L_i(a)$ are defined for all $i \in \mathbb{Z}$ and they satisfy the Witt algebra relation

$$[L_i(a), L_j(a)] = (i-j)L_{i+j}(a)$$

for $i, j \in \mathbb{Z}$.

References

Enrico Arbarello and C. De Concini (1984): "On a Set of Equations Characterizing the Riemann Matrices", *Annals of Mathematics* **120** 119–140

D. Bessis and C. Itzykson and J. B. Zuber (1980): "Quantum Field Theory Techniques in Graphical Enumeration", *Advances in Applied Mathematics* **1** 109–157

J. Harer and D. Zagier (1986): "The Euler Characteristic of the Moduli Space of Curves", *Inventiones Mathematicae* **85** 457–485

Maxim Kontsevich (1992): "Intersection Theory on the Moduli Space of Curves and the Matrix Airy Function", *Communications in Mathematical Physics* **147** 1–23

Motohico Mulase (1984): "Cohomological Structure in Soliton Equations and Jacobian Varieties", *Journal of Differential Geometry* **19** 403–430

Motohico Mulase (1990): "Category of vector bundles on algebraic curves and infinite dimensional Grassmannians", *International Journal of Mathematics* **1** 293–342

Motohico Mulase (1994a): "Algebraic Theory of the KP Equations", *Perspectives in Mathematical Physics*, International Press Inc., ed Robert C. Penner and Shing-Tung Yau, 151–217

Motohico Mulase (1994b): "Matrix Integrals and Integrable Systems", *Topology, Geometry and Field Theory*, World Scientific Publishing Co., eds. K. Fukaya and M. Furuta and T. Kohno and D. Kotschick, 111–127

Motohico Mulase (1995): "Asymptotic Analysis of a Hermitian Matrix Integral", *International Journal of Mathematics* **6** 881–892

David Mumford (1978): "An algebro-geometric constructions of commuting operators and of solutions to the Toda lattice equations, Korteweg-de Vries equations and related nonlinear equations", *Proceedings of the International Symposium on Algebraic Geometry, Kyoto 1977*, Kinokuniya Publishers, 115–153

Robert C. Penner (1988): "Perturbation Series and the Moduli Space of Riemann Surfaces", *Journal of Differential Geometry* **27** 35–53

Mikio Sato (1981): "Soliton Equations as Dynamical Systems on an Infinite-Dimensional Grassmann Manifold", *Kokyuroku of the Research Institute for Mathematical Sciences, Kyoto University*, **439** 30–46

Takahiro Shiota (1986): "Characterization of jacobian varieties in terms of soliton equations", *Inventiones Mathematicae* **83** 333–382

G. 'tHooft (1974): "A Planer Diagram Theory for Strong Interactions", *Nuclear Physics* **B72** 461–473

Edward Witten (1991): "Two Dimensional Gravity and Intersection Theory on Moduli Space", *Surveys in Differential Geometry* **1** 243–310

Solitons and Generalized Tau-Functions for Affine Integrable Hierarchies

L.A. Ferreira[1] and J. Sánchez Guillén[2]

[1] Instituto de Física Teórica - IFT/UNESP, Rua Pamplona 145, 01405-900, São Paulo - SP, Brazil

[2] Facultad de Física, Universidad de Santiago de Compostela, 15706 Santiago de Compostela, Spain

Abstract. We consider a very large class of hierarchies of zero-curvature equations constructed from affine Kac-Moody algebras $\hat{\mathcal{G}}$. We argue that one of the basic ingredients for the appearance of soliton solutions in such theories is the existence of "vacuum solutions" corresponding to Lax operators lying in some abelian (up to central term) subalgebra of $\hat{\mathcal{G}}$. Using the dressing transformation procedure we construct the solutions in the orbit of those vacuum solutions, and conjecture that the soliton solutions correspond to some special points in those orbits. The generalized tau-function for those hierarchies are defined for integrable highest weight representations of $\hat{\mathcal{G}}$, and it applies for any level of the representation and it is independent of its realization. We illustrate our methods with the recently proposed non abelian Toda models coupled to matter fields. A very special class of such theories possess a $U(1)$ Noether charge that, under a suitable gauge fixing of the conformal symmetry, is proportional to a topological charge. That leads to a mechanism that confines the matter fields inside the solitons.

1 Introduction

The study of soliton solutions of non linear differential equations has been developed considerably in the last decades using (apparently) quite diverse methods. In spite of the great variety of types of equations considered, some basic features seem to be common to a large class of them. Attempts to unify such various aspects of basic classical Soliton Theory is clearly very important and may lead to new insights into the role of solitons in Physics and Mathematics.

Among the several methods of constructing solutions for non linear differential equations we have the Hirota method (Hirota (1972)), the dressing transformation procedure (Zakharov et al. (1979), Date et al.(1981), Semenov-Tian-Shansky (1985), Babelon et al (1992), Babelon et al. (1993)), Backlund transformations (Fordy et al. (1980), Liao et al. (1993)), the inverse scattering method (Faddeev et al. (1987)), the Leznov-Saveliev algebraic construction (Leznov et al. (1992)) and the tau-function approach (Kac et al. (1991), Date et al.(1981)). Each one of these methods have their own advantages, and the choice of one or the other depends on the particular problem and model one wants to address. However, soliton solutions are special and the theories presenting them must possess some common structures. By solitons here we mean a solution localized in space that travels without dispersion, and that keeps its form when scattered by other soliton, suffering just a shift in its position with respect to the one it would have if not for the scattering. In fact, in some theories like the Toda models which possess two dimensional Lorentz invariance, the solitons can have a particle interpretation and there are indications for the existence of a duality between the one-soliton solutions and the fundamental particles. In four dimensions, a similar duality, generalizing the electromagnetic duality, seems to exist between monopoles and gauge particles (Montonen et al. (1977)).

Pratically all theories in one and two space time dimensions, presenting soliton solutions, have a representation in terms of a zero curvature condition or Lax-Zakharov-Shabat equation. In addition, the corresponding Lax operators lie in some infinite dimensional Lie algebra. In fact, we can say that basically almost all known soliton equations are related to Kac-Moody algebras (Drinfel'd (1985), Groot et al. (1992)).

In this contribution we try to unify some aspects concerning soliton solutions. We argue that a basic ingredient for the appearence of soliton solutions is that there must exist one or several solutions, which we call "vacuum solutions", such that the Lax operators, when evaluated on them, should lie in some abelian subalgebra (up to central term) of the Kac-Moody algebra associated to the model. Such subalgebra can be written as an algebra of oscillators b_i

$$[b_i, b_j] = i \beta_i C \delta_{i+j,0} \tag{1}$$

with β_i being some complex numbers and C the central element of the Kac-Moody algebra $\hat{\mathcal{G}}$. In several cases, but not in all, (1) constitutes a Heisenberg subalgebra (Kac et al. (1985)) of $\hat{\mathcal{G}}$. In addition, we argue that the components of the Lax operators in the direction of the b_i's should be constant for the vacuum solutions.

Using the dressing transformation method one can then construct, out of a given vacuum solution, an orbit of solutions parametrized by a constant group element ρ of the Kac-Moody group. We conjecture that the soliton solutions of those theories lie on such orbits, and that they correspond to points of the orbits where ρ is the product of exponentials of eigenvectors of the constant elements defined by the Lax operators evaluated on the corresponding vacuum solution. Such observations provide not only a powerful and elegant method of constructing soliton solutions, but also allow us to connect and generalize several results known in the literature (Wilson (1984), Wilson (1993)). The so called solitonic specialization of the Leznov-Saveliev solution proposed in (Olive et al. (1993a), Olive et al. (1993b), Olive et al. (1993c)) in the context of Toda type models, can then be connected to the dressing method applied to such vacuum solutions. In addition, we believe that in several cases there is a connection with the Backlund transformation method.

We can also connect that observation with tau-functions. We define tau-functions as states in integrable highest weight representations of the Kac-

Moody algebra, lying in the orbit, under the action of the group elements performing the dressing transformations, of the highest weight state. Our definition is independent of the level of the representation and also on the way it is realised, constituting a generalization of the previous definitions of tau-functions for level one vetex operator representations (Kac et al. (1991)). The connection with the Hirota method is then made by realizing the Hirota's tau-function as projections of the tau-function on some suitable states of representation. The truncation of the Hirota's expansion is then understood in terms of the nilpotency of some operators in those integrable representations. The projections which give the components of the fields for those special solutions are related so that their conserverved densities involve actually a single function

We illustrate our methods with the recently proposed non abelian Toda models coupled to matter fields (Ferreira et al. (1996)). In fact, we discuss a very special class of such theories where the solitons play a crucial role. These models possess a $U(1)$ Noether charge that, under a suitable gauge fixing of the conformal symmetry, is proportional to a topological charge. That leads to a mechanism that confines the matter fields inside the solitons.

The paper contains results of collaboration with other authors (Ferreira et al. (1997), Ferreira et al. (1996), Ferreira et al. (1995)) and it is organized as follows. In section 2 we introduce the type of hierarchies of soliton equations we shall consider, and discuss their vacuum solutions. In section 3 we define the dressing transformations and in section 4 we construct the soliton solutions. Section 5 introduces the tau-functions and discuss how they connect to previous definitions and to the Hirota's tau functions. The non abelian Toda models are introduced in section 6, their soliton solutions constructed in section 7, and their properties discussed in sections 8 and 9. In section 10 we present in great detail, the example of the model associated to the principal gradation of $sl(2)^{(1)}$.

2   Hierarchies and Vacuum Solutions

Non-linear integrable hierarchies of equations are most conveniently discussed by associating them with a system of first-order differential equations

$$\mathcal{L}_N \Psi = 0 , \tag{2}$$

where $\mathcal{L}_N$ are Lax operators of the form

$$\mathcal{L}_N \equiv \frac{\partial}{\partial t_N} - A_N \tag{3}$$

and the variables t_N are the various "times" of the hierarchy. Then, the equivalent zero-curvature formulation is obtained through the integrability conditions of the associated linear problem (2),

$$[\mathcal{L}_N , \mathcal{L}_M] = 0 . \tag{4}$$

An equivalent way to express the relation between the solutions of the zero-curvature equations and of the associated linear problem is

$$A_N = \frac{\partial \Psi}{\partial t_N} \, \Psi^{-1} . \tag{5}$$

The class of integrable hierarchies of zero-curvature equations that will be studied here is constructed from graded Kac-Moody algebras in the following way. Consider a complex affine Kac-Moody algebra $\tilde{\mathcal{G}} = \hat{\mathcal{G}} + \mathbb{C}D$, associated to a simple finite Lie algebra $\mathcal{G}$ of rank r, and an integer gradation of its derived algebra $\hat{\mathcal{G}}$ labelled by a vector $\mathbf{s} = (s_0, s_1, \ldots, s_r)$ of $r+1$ non-negative co-prime integers such that

$$\hat{\mathcal{G}} = \bigoplus_{i \in \mathbb{Z}} \hat{\mathcal{G}}_i(\mathbf{s}) \quad \text{and} \quad [\hat{\mathcal{G}}_i(\mathbf{s}), \hat{\mathcal{G}}_j(\mathbf{s})] \subseteq \hat{\mathcal{G}}_{i+j}(\mathbf{s}) . \tag{6}$$

According to (Kac et al. (1990)), integral gradations of $\hat{\mathcal{G}}$ are labelled by a set of co-prime integers $\mathbf{s} = (s_0, s_1, \ldots s_r)$, and the grading operators are given by

$$Q_{\mathbf{s}} \equiv H_{\mathbf{s}} + N_{\mathbf{s}} D - \frac{1}{2N_{\mathbf{s}}} \mathrm{Tr}\,(H_{\mathbf{s}})^2 \, C . \tag{7}$$

Here

$$H_{\mathbf{s}} \equiv \sum_{a=1}^{r} s_a \lambda_a^v \cdot H^0 , \qquad N_{\mathbf{s}} \equiv \sum_{i=0}^{r} s_i m_i^{\psi} , \qquad \psi = \sum_{a=1}^{r} m_a^{\psi} \alpha_a , \qquad m_0^{\psi} = 1 ; \tag{8}$$

H^0 is an element of the Cartan subalgebra of $\mathcal{G}$; α_a, $a = 1, 2, \ldots r$, are its simple roots; ψ is its maximal root; m_a^{ψ} the integers in expansion $\psi = \sum_{a=1}^{r} m_a^{\psi} \alpha_a$; and λ_a^v are the fundamental co-weights satisfying the relation $\alpha_a \cdot \lambda_b^v = \delta_{ab}$.

We have in mind basically two types of integrable systems. The first one corresponds to the Generalized Drinfel'd-Sokolov Hierarchies considered in (Groot et al. (1992)), and (Hollowood et al. (1993)), which are generalizations of the KdV type hierarchies studied in (Drinfel'd (1985)). In particular, and using the parlance of the original references, we will be interested in the generalized mKdV hierarchies, whose construction can be summarised as follows (see (Groot et al. (1992)) and, especially, (Hollowood et al. (1993)) for details). Given an integer gradation $\mathbf{s}$ of $\hat{\mathcal{G}}$ and a semisimple constant element E_l of grade l with respect to $\mathbf{s}$, one defines the Lax operator

$$L \equiv \partial_x + E_l + A , \tag{9}$$

where the components of A are the fields of the hierarchy. In (Hollowood et al. (1993)), it was shown that the component of A along the central term of $\hat{\mathcal{G}}$ should not be considered as an actual degree of freedom of the hierarchy. This is the reason why these hierarchies can be equivalently formulated both in terms of affine Kac-Moody algebras or of the corresponding loop algebras. They are functions of x and of the other times of the hierarchy taking values in the subspaces of $\hat{\mathcal{G}}$ with grades ranging from 0 to $l-1$. For each element in the centre of $\mathrm{Ker}(\mathrm{ad}\, E_l)$ with positive s-grade N, one constructs a local functional of those fields, B_N, whose components take values in the subspaces $\hat{\mathcal{G}}_0(\mathbf{s}), \ldots, \hat{\mathcal{G}}_N(\mathbf{s})$. Then, B_N defines the flow equation

$$\frac{\partial L}{\partial t_N} = [B_N\,,\, L]\,, \tag{10}$$

and the resulting Lax operators $\mathcal{L}_N = \partial/\partial t_N - B_N$ commute among themselves (Groot et al. (1992)). These zero-curvature equations are continuity equations and the projections of the diagonalized connections B_N along the constant regular elements E_l are essentially the conserved densities.The abelianization procedure has a gauge ambiguity, given by negative grade conjugations, which is reflected as total derivatives (Miramontes et al. (1997)).

The second type of integrable systems corresponds to the non-abelian affine Toda theories (Leznov et al. (1992), Ferreira et al. (1995), Ferreira et al. (1996), Underwood (1993)), and a very general class of these models will be described in section 6.

An important common feature of all those hierarchies is that they possess trivial solutions which will be called "vacuum solutions". These particular solutions are singled out by the condition that the Lax operators evaluated on them lie on some abelian subalgebra of $\hat{\mathcal{G}}$, up to central terms. Then, the dressing transformation method can be used to generate an orbit of solutions out of each "vacuum". Moreover, it is generally conjectured that multi-soliton solutions lie in the resulting orbits. As a bonus, the fact that we only consider the particular subset of solutions connected with a generic vacuum allows one to perform the calculations in a very general way and, consequently, our results apply to a much broader class of hierarchies.

For a given choice of the Kac-Moody algebra $\tilde{\mathcal{G}}$ and the gradation s, let us consider Lax operators of the form (3) where the potentials can be decomposed as

$$A_N = \sum_{i=N_-}^{N_+} A_{N,i}\,, \quad \text{where} \quad A_{N,i} \in \hat{\mathcal{G}}_i(\mathbf{s}) \tag{11}$$

N_- and N_+ are non-positive and non-negative integers, respectively, and the times t_N are labelled by (positive or negative) integer numbers. The particular form of these potentials will be constrained only by the condition that the corresponding hierarchy admits vacuum solutions where they take the form

$$A_N^{(\mathrm{vac})} = \sum_{i=N_-}^{N_+} c_N^i b_i + f_N(t)\, c \equiv \varepsilon_N + f_N(t)\, C\,. \tag{12}$$

In this equation, C is the central element of $\hat{\mathcal{G}}$, and $b_i \in \hat{\mathcal{G}}_i(\mathbf{s})$ are the generators of a subalgebra $\hat{s}$ of $\hat{\mathcal{G}}$ defined by

$$\hat{s} = \{b_i \in \hat{\mathcal{G}}_i(\mathbf{s}),\ i \in E \subset \mathbb{Z} \mid [b_i, b_j] = i\,\beta_i\, C\, \delta_{i+j,0}\}, \tag{13}$$

where β_i are arbitrary (vanishing or non-vanishing) complex numbers such that $\beta_{-i} = \beta_i$, and E is some set of integers numbers. Moreover, c_N^i are also arbitrary numbers, and $f_N(t)$ are $\mathbb{C}$-functions of the times t_N that satisfy the equations

$$\frac{\partial f_N(t)}{\partial t_M} - \frac{\partial f_M(t)}{\partial t_N} = \sum_i i\, \beta_i\, c_M^i\, c_N^{-i}. \tag{14}$$

These vacuum potentials correspond to the solution of the associated linear problem given by the group element (5)

$$\Psi^{(\text{vac})} = \exp\left(\sum_N \varepsilon_N t_N + \gamma(t)\, C\right) \tag{15}$$

where the numeric function $\gamma(t)$ is a solution of the equations

$$\frac{\partial \gamma(t)}{\partial t_N} = f_N(t) + \frac{1}{2} \sum_{M,i} i\, \beta_i\, c_N^i\, c_M^{-i}\, t_M. \tag{16}$$

For generic choices of gradations, equations (14) and (16) can be solved, fixing thereby the arbitary functions in a convenient way. In fact, one can remove the ambiguity in the abelianization mentioned after (10) and express the conserved densities for our solutions (15) as total derivatives. (Miramontes et al. (1997), Freeman (1995)).

3   Dressing Transformations

In terms of the associated linear problem, one can define an important set of transformations called "dressing transformations", which take known solutions of the hierarchy to new solutions. Regarding the structure of the integrable hierarchies, these transformations have a deep meaning and, in fact, the group of dressing transformations can be viewed as the classical precursor of the quantum group symmetries (Babelon et al (1992)). Denote by $\hat{G}_-(\mathbf{s})$, $\hat{G}_+(\mathbf{s})$, and $\hat{G}_0(\mathbf{s})$ the subgroups of the Kac-Moody group $\hat{G}$ formed by exponentiating the subalgebras $\hat{\mathcal{G}}_{<0}(\mathbf{s}) \equiv \bigoplus_{i<0} \hat{\mathcal{G}}_i(\mathbf{s})$, $\hat{\mathcal{G}}_{>0}(\mathbf{s}) \equiv \bigoplus_{i>0} \hat{\mathcal{G}}_i(\mathbf{s})$, and $\hat{\mathcal{G}}_0(\mathbf{s})$, respectively. According to Wilson (Wilson (1984), Wilson (1993)), the dressing transformations can be described in the following way. Consider a solution Ψ of the linear problem (2), and let $\rho = \rho_- \rho_0 \rho_+$ be a constant element in the "big cell" of $\hat{G}$, i.e., in the subset $\hat{G}_-(\mathbf{s})\, \hat{G}_0(\mathbf{s})\, \hat{G}_+(\mathbf{s})$ of $\hat{G}$, such that

$$\Psi\, \rho\, \Psi^{-1} = (\Psi\, \rho\, \Psi^{-1})_{<0}\, (\Psi\, \rho\, \Psi^{-1})_0\, (\Psi\, \rho\, \Psi^{-1})_{>0}. \tag{17}$$

Notice that these conditions are equivalent to say that both h and $\Psi \, h \, \Psi^{-1}$ admit a generalized Gauss decomposition with respect to the gradation s . Define

$$
\begin{aligned}
\Psi^\rho &= \Theta_-^{(0)} [(\Psi \, \rho \, \Psi^{-1})_{<0}]^{-1} \, \Psi \rho \\
&= \Theta_+^{(0)} \, (\Psi \, \rho \, \Psi^{-1})_{>0} \, \Psi
\end{aligned}
\tag{18}
$$

where

$$
\Theta_-^{(0)^{-1}} \Theta_+^{(0)} = \left(\Psi \rho \Psi^{-1} \right)_0 .
\tag{19}
$$

Then, Ψ^ρ is another solution of the linear problem. In order to prove it, introduce the notation $g_\pm \equiv (\Psi \, \rho \, \Psi^{-1})_\pm$ and $\partial_N \equiv \partial / \partial t_N$, and consider

$$
\begin{aligned}
\partial_N \Psi^\rho \, \Psi^{\rho-1} &= \partial_N \Theta_-^{(0)} \, \Theta_-^{(0)^{-1}} - \Theta_-^{(0)} g_-^{-1} \, \partial_N g_- \Theta_-^{(0)^{-1}} \\
&\quad + \Theta_-^{(0)} g_-^{-1} (\partial_N \Psi \, \Psi^{-1}) g_- \Theta_-^{(0)^{-1}} \\
&= \partial_N \Theta_+^{(0)} \, \Theta_+^{(0)^{-1}} + \Theta_+^{(0)} \partial_N g_+ \, g_+^{-1} \Theta_+^{(0)^{-1}} \\
&\quad + \Theta_+^{(0)} g_+ (\partial_N \Psi \, \Psi^{-1}) g_+^{-1} \Theta_+^{(0)^{-1}}
\end{aligned}
\tag{20}
$$

Then, the first identity implies that $\partial_N \Psi^\rho \, \Psi^{\rho-1} \in \bigoplus_{i \leq N_+} \hat{\mathcal{G}}_i(\mathbf{s})$, and the second that $\partial_N \Psi^\rho \, \Psi^{\rho-1} \in \bigoplus_{i \geq N_-} \hat{\mathcal{G}}_i(\mathbf{s})$. Consequently

$$
A_N^\rho = \frac{\partial \Psi^\rho}{\partial t_N} \, (\Psi^\rho)^{-1} \in \bigoplus_{i=N_-}^{N_+} \hat{\mathcal{G}}_i(\mathbf{s}) ,
\tag{21}
$$

and, taking into account (11), it is a solution of the hierarchy of zero-curvature equations. If the fields of the hierarchy are such that $A_{N,i}$ does not span the whole subspace $\hat{\mathcal{G}}_i(\mathbf{s})$ then we have to impose further constraints on the group elements performing the dressing transformation.

For any ρ lying in the big cell of $\hat{G}$, the transformation

$$
\mathcal{D}_\rho : \Psi \mapsto \Psi^\rho , \quad \text{or} \quad A_N \mapsto A_N^\rho ,
\tag{22}
$$

is called a dressing transformation.

4   Solitons out of Vacuum Solutions

We now consider the orbit of the vacuum solution (15) under the group of dressing transformations. For any element ρ of the big cell of $\hat{G}$, let us define

$$
\Theta_+^> = \left(\Psi^{(\text{vac})} \rho \Psi^{(\text{vac})-1} \right)_{>0} , \qquad \Theta_-^\leq = \left(\Psi^{(\text{vac})} \rho \Psi^{(\text{vac})-1} \right)_{<0}^{-1} .
\tag{23}
$$

and

$$
\Theta_+ = \Theta_+^{(0)} \, \Theta_+^> , \qquad \Theta_- = \Theta_-^{(0)} \, \Theta_-^\leq ;
\tag{24}
$$

142   L.A. Ferreira and J. Sánchez Guillén

where $\Theta_+^{(0)}$ and $\Theta_-^{(0)}$ are the same as in (19), but with Ψ replaced by $\Psi^{(\text{vac})}$ defined in (15).

Now, orbit of the vacuum solution (15) can be easily constructed using Eqs. (18) and (21). Under the dressing transformation generated by ρ,

$$\Psi^{(\text{vac})} \mapsto \Psi^\rho = \Theta_- \Psi^{(\text{vac})} \rho = \Theta_+ \Psi^{(\text{vac})} , \qquad (25)$$

or, equivalently, $A_N^{(\text{vac})}$ becomes

$$A_N^h - f_N(t)\,c = \Theta_- \varepsilon_N \Theta_-^{-1} + \partial_N \Theta_- \Theta_-^{-1} \in \bigoplus_{i \leq N_+} \hat{\mathcal{G}}_i(s)$$

$$= \Theta_+ \varepsilon_N \Theta_+^{-1} + \partial_N \Theta_+ \Theta_+^{-1} \in \bigoplus_{i \geq N_-} \hat{\mathcal{G}}_i(s), \qquad (26)$$

Eqs. (24), (23) and (26) summarize the outcome of the dressing transformation method, which, starting with some vacuum solution (12), associates a solution of the zero-curvature equations (4) to each constant element ρ in the big cell of $\hat{G}$. The construction of this solution involves two steps. First, the Eqs. (26) can be understood as a local change of variables between the components of the potential A_N and some components of the group elements Θ_+ and Θ_-.

The second step consists in obtaining the value of the required components of Θ_+ and Θ_- from Eqs. (24) and (23). This is usually done by considering matrix elements of the form

$$\langle \mu \mid \Theta_-^{-1} \Theta_+ \mid \mu' \rangle = \langle \mu \mid e^{\sum_N \varepsilon_N t_N} \rho\, e^{-\sum_N \varepsilon_N t_N} \mid \mu' \rangle , \qquad (27)$$

where $\mid \mu \rangle$ and $\mid \mu' \rangle$ are vectors in a given representation of $\tilde{G}$. The appropriate set of vectors is specified by the condition that all the required components of $\Theta_-^{(\text{vac})}$ and $\Theta_+^{(\text{vac})}$ can be expressed in terms of the resulting matrix elements. It will be shown below that the required matrix elements, considered as functions of the group element ρ, constitute the generalization of the Hirota's tau-functions for these hierarchies. Moreover, Eq. (27) is the analogue of the, so called, solitonic specialization of the Leznov-Saveliev solution proposed in (Olive et al. (1993a), Olive et al. (1993b), Olive et al. (1993c), Ferreira et al. (1995), Ferreira et al. (1996)) for the affine (abelian and non-abelian) Toda theories.

Consider now the common eigenvectors of the adjoint action of the ε_N's that specify the vacuum solution (12). Then, the important class of multi-soliton solutions is conjectured to correspond to group elements ρ which are the product of exponentials of eigenvectors

$$\rho = e^{F_1} e^{F_2} \ldots e^{F_n} , \qquad [\varepsilon_N , F_k] = \omega_N^{(k)} F_k , \qquad k = 1, 2, \ldots n . \qquad (28)$$

In this case, the dependence of the solution upon the times t_N can be made quite explicit

$$\langle \mu \mid \Theta_-^{-1} \Theta_+ \mid \mu' \rangle = \langle \mu \mid \prod_{k=1}^{n} \exp(e^{\sum_N \omega_N^{(k)} t_N} F_k) \mid \mu' \rangle . \qquad (29)$$

We emphasize that not all solutions of the type (29) are soliton solutions, but we conjecture that the soliton and multi-soliton solutions are among them. The conjecture that multi-soliton solutions are associated with group elements of the form (28) naturally follows from the well known properties of the multi-soliton solutions of affine Toda equations and of hierarchies of the KdV type, and, in the sine-Gordon theory, it has been explicitly checked in Ref. (Babelon et al. (1993)). Actually, in all these cases, the multi-soliton solutions are obtained in terms of representations of the "vertex operator" type where the corresponding eigenvectors are nilpotent. Then, for each eigenvector F_k there exists a positive integer number m_k such that $(F_k)^m \neq 0$ only if $m \leq m_k$. This remarkable property simplifies the form of (29) because it implies that $e^{F_k} = 1 + F_k + \cdots + (F_k)^{m_k}/m_k!$, which provides a group-theoretical justification of Hirota's method.

An interesting feature of the dressing transformations method is the possibility of relating the solutions of different integrable equations. Consider two different integrable hierarchies whose vacuum solutions are compatible, in the sense that the corresponding vacuum Lax operators commute. Then, one can consider the original integrable equations as the restriction of a larger hierarchy of equations. Consequently, the solutions obtained through the group of dressing transformations can also be understood in terms of the solutions of the larger hierarchy, which implies certain relations among them. (see section 4 of (Ferreira et al. (1995)) for more details). The conjugation by the special elements Θ dressing the vacuum solutions can be in fact understood as the special abelianization of the Lax potentials above mentioned, which allows to express the conserved densities for these special solutions as total derivatives, depending therefore on the asymptotic values, as corresponds to their solictonic character.

We also point out that this method of constructing solitons has recently been used in the context of the constrained KP models (Aratyn et al. (1997)).

5   The Tau-Functions

According to the discussion in the previous section, the orbits generated by the group of dressing transformations acting on some vacuum provide solutions of certain integrable hierarchies of equations. Making contact with the method of Hirota, the generalized "tau-functions" that will be defined in this section constitute a new set of variables to describe those solutions. One of the characteristic properties of these variables is that they substantially simplify the task of constructing multi-soliton solutions (Ferreira et al. (1995)). The group-theoretical interpretation of this property has already been pointed out in the previous section. Tau-functions are given by certain matrix elements in a appropriate representation of the Kac-Moody Group $\hat{G}$. Moreover, the tau-functions corresponding to the multi-soliton solutions are expected to involve nilpotent elements of $\hat{G}$, which is the origin of their remarkable simple form.

The tau-function formulation of the Generalized Drinfel'd-Sokolov Hierarchies of (Groot et al. (1992)) has already been worked out in (Hollowood et al. (1993)), which, in fact, has largely inspired our approach. However, there are two important differences between our results and those of (Hollowood et al. (1993)). Firstly, our approach applies to the affine Toda equations too, and, secondly, it does not rely upon the use of (level-one) vertex operator representations.

At this point, it is worth recalling that the solutions constructed in sections 3 and 4 are completely representation-independent. In contrast, our definition of tau-functions makes use of a special class of representations of the Kac-Moody algebra $\hat{\mathcal{G}}$ called "integrable highest-weight" representations. The reason why these representations are called "integrable" is the following. For an infinite-dimensional representation, it is generally not possible to go from a representation of the algebra $\hat{\mathcal{G}}$ to a representation of the corresponding group $\hat{G}$ via the exponential map $x \mapsto e^x$. However, the construction does work if, for instance, the formal power series terminates at a certain power of x, or if the representation space admits a basis of eigenvalues of x. These conditions, applied to the Chevalley generators of $\hat{\mathcal{G}}$, single out this special type of representations.

The generalized tau-functions will be sets of matrix elements of the form indicated on the right-hand-side of (27), considered as functions of the group element ρ. They are characterized by the condition that they allow one to parameterize all the components of Θ_+ and Θ_- required to specify the solutions (26) of the zero-curvature equations (4). As we have discussed before, the tau-functions corresponding to the multi-soliton solutions are expected to have a very simple form. However, in contrast with the original method of Hirota, we cannot ensure in general that the equations of the hierarchy become simpler in terms of this new set of variables.

First, let us discuss the generalized Hirota tau-functions associated with the components of B. In equation (27), these components can be isolated by considering the vectors $\mid \mu_0 \rangle$ of an integrable highest-weight representation $L(\tilde{s})$ of $\tilde{\mathcal{G}}$ which are annihilated by all the elements in $\hat{\mathcal{G}}_{>0}(s)$, i.e., $T \mid \mu_0 \rangle = 0$ and $\langle \mu_0 \mid T' = 0$ for all $T \in \hat{\mathcal{G}}_{>0}(s)$ and $T' \in \hat{\mathcal{G}}_{<0}(s)$, respectively. Then, the corresponding tau-functions are defined as [1]

$$\tau_{\mu_0, \mu_0'}(t) = \langle \mu_0' \mid \Psi^{(\mathrm{vac})} \rho \, \Psi^{(\mathrm{vac})^{-1}} \mid \mu_0 \rangle$$
$$= \langle \mu_0' \mid e^{\sum_N \varepsilon_N t_N} \rho \, e^{-\sum_N \varepsilon_N t_N} \mid \mu_0 \rangle, \qquad (30)$$

---

[1] Since the resulting relations between tau-functions and components of the A_N's will be considered as generic changes of variables, we will not generally indicate the intrinsic dependence of the tau-functions on the group element ρ.

and, in terms of them, equation (27) becomes just

$$\langle \mu'_0 \mid B^{-1} \mid \mu_0 \rangle = \tau_{\mu_0, \mu'_0}(t) . \tag{31}$$

where we have denoted

$$B^{-1} \equiv \left(\Psi^{(\text{vac})} \rho \, \Psi^{(\text{vac})\,-1} \right)_0 \tag{32}$$

By construction, $\hat{\mathcal{G}}_0(\mathbf{s})$ always contains the central element C of the Kac-Moody algebra, but it is always possible to split the contribution of the corresponding field in (31). Let $s_q \neq 0$ and consider the subalgebra $\tilde{\mathcal{G}}^{(q)}$ of $\tilde{\mathcal{G}}$ generated by the $e_i^{\pm}$ with $i = 0, \ldots, r$ but $i \neq q$, which is a semisimple finite Lie algebra of rank r ($\tilde{\mathcal{G}}^{(q)}$ is always simple if $q = 0$). Then, $\hat{\mathcal{G}}_0(\mathbf{s}) = (\hat{\mathcal{G}}_0(\mathbf{s}) \cap \tilde{\mathcal{G}}^{(q)}) \oplus \mathbb{C} \, C$ and, correspondingly, B can be split as $B = b \, \exp(\nu \, c)$. Here, ν is the field along C, and b is a function taking values in the semisimple finite Lie group $G_0^{(q)}$ whose Lie algebra is $\hat{\mathcal{G}}_0(\mathbf{s}) \cap \tilde{\mathcal{G}}^{(q)}$. Since $\tilde{K} = \sum_{i=0}^r k_i^{\vee} \tilde{s}_i$ is the level of the representation $L(\tilde{\mathbf{s}})$, Eq. (31) is equivalent to

$$\langle \mu'_0 \mid B^{-1} \mid \mu_0 \rangle = e^{-\nu \, \tilde{K}} \langle \mu'_0 \mid b^{-1} \mid \mu_0 \rangle = \tau_{\mu_0, \mu'_0}(t) . \tag{33}$$

Moreover, it is always possible to introduce a tau-function for the field ν. Let us consider the highest-weight vector $\mid v_q \rangle$ of the fundamental representation $L(q)$, which is obviously annihilated by all the elements in $\tilde{\mathcal{G}}^{(q)}$. Therefore,

$$\langle v_q \mid B^{-1} \mid v_q \rangle = e^{-\nu \, k_q^{\vee}} = \tau_{v_q, v_q}(t) \equiv \tau_q^{(0)}(t) , \tag{34}$$

which leads to

$$\langle \mu'_0 \mid b^{-1} \mid \mu_0 \rangle = \frac{\tau_{\mu_0, \mu'_0}(t)}{\left(\tau_q^{(0)}(t) \right)^{\tilde{K}/k_q^{\vee}}} \quad \text{and} \quad \nu = -\ln \frac{\tau_q^{(0)}(t)}{k_q^{\vee}} . \tag{35}$$

Finally, recall that the vectors $\mid \mu_0 \rangle$ form a representation of the semisimple Lie group $G_0^{(q)}$. Therefore, if $L(\tilde{\mathbf{s}})$ is chosen such that this representation is faithful, Eq. (35) allows one to obtain all the components of b in terms of the generalized tau-functions τ_{μ_0, μ'_0} and $\tau_q^{(0)}$. Notice that, in this case, the definition of generalized tau-functions coincide exactly with the quantities involved in the solitonic specialization of the Leznov-Saveliev solution proposed in (Olive et al. (1993c)).

Let us now discuss the generalized tau-functions associated with the components of $\Theta^{\leq}$. Consider the gradation $\mathbf{s}$ of $\tilde{\mathcal{G}}$ involved in the definition of the integrable hierarchy. For each $s_i \neq 0$, let us consider the highest-weight vector of the fundamental representation $L(i)$ and define the (right) tau-function vector

$$\begin{aligned} \mid \tau_i^R(t) \rangle &= \Psi^{(\text{vac})} \, h \, \Psi^{(\text{vac})\,-1} \mid v_i \rangle \\ &= e^{\sum_N \varepsilon_N t_N} h \, e^{-\sum_N \varepsilon_N t_N} \mid v_i \rangle \end{aligned} \tag{36}$$

Notice that $\mid \tau_i^R(t) \rangle$ is a vector in the representation $L(i)$. Therefore, it has infinite components, and it will be shown soon that the role of the Hirota tau-functions will be played by a finite subset of them. Taking into account that $\mid v_i \rangle$ is annihilated by all the elements in $g_{>0}(s)$, equation (27) implies

$$\Theta_-^{\leq -1} B^{-1} \mid v_i \rangle = \mid \tau_i^R(t) \rangle, \quad i = 0, \ldots, r \quad \text{and} \quad s_i \neq 0. \tag{37}$$

The definition (36) is inspired by the tau-function approach of (Kac et al. (1991), Hollowood et al. (1993), Ferreira et al. (1995)). However, in (Hollowood et al. (1993)), and (Ferreira et al. (1995)), the authors consider a unique tau-function $\mid \tau_s(t) \rangle \in L(s)$. In fact, one could equally consider different tau-functions $\mid \tau_{s'}(t) \rangle$ associated with any integrable representation $L(s')$ such that $s_i' \neq 0$ if, and only if, $s_i \neq 0$. Since the highest weight state of $L(s)$ is obtained by the tensor product of the $\mid v_i \rangle$'s, as

$$\mid v_s \rangle = \bigotimes_{i=0}^{r} \left\{ \mid v_i \rangle^{\otimes s_i} \right\}. \tag{38}$$

one sees that all these choices lead to the same results, but the one presented here is the most economical.

Since, for any integrable representation, the grading operator (7) can be diagonalized acting on $L(s)$, these tau-functions vectors can be decomposed as

$$\mid \tau_i^R(t) \rangle = \sum_{-j \in \mathbb{Z}_{\leq 0}} \mid \tau_i^{R(-j)}(t) \rangle, \qquad Q_i \mid \tau_i^{R(-j)}(t) \rangle = -j \mid \tau_i^{R(-j)}(t) \rangle, \tag{39}$$

where we have used that $\Theta_-^{\leq} \in \hat{G}_{<0}(s)$ and $B \in \hat{G}_0(s)$, and Q_i indicates the derivation corresponding to the grading operator (7) with $s_j = \delta_{j,i}$. Moreover, the highest-weight vector is an eigenvector of the subalgebra $\hat{G}_0(s)$ and, consequently, of B. Therefore,

$$\mid \tau_i^{R(0)}(t) \rangle = B^{-1} \mid v_i \rangle = \tau_i^{(0)}(t) \mid v_i \rangle, \tag{40}$$

where, $\tau_i^{(0)}(t)$ is a $\mathbb{C}$-function, not a vector of $L(i)$, whose definition is [2]

$$\tau_i^{(0)}(t) = \langle v_i \mid e^{\sum_N \varepsilon_N t_N} \rho \, e^{-\sum_N \varepsilon_N t_N} \mid v_i \rangle \equiv \tau_{v_i, v_i}(t) \tag{41}$$

(compare with Eq. (33)). Therefore, Eq. (37) becomes

$$\Theta_-^{\leq -1} \mid v_i \rangle = \frac{1}{\tau_i^{(0)}(t)} \mid \tau_i^R(t) \rangle, \tag{42}$$

---

[2] To compare with (31), notice that $\mid \mu_0 \rangle = \mid v_i \rangle$ forms a one-dimensional representation of $\hat{G}_0(s)$ and, consequently, $\tau_{v_i, \mu_0'}(t)$ vanishes unless $\mid \mu_0' \rangle = \mid v_i \rangle$. Therefore, for non-abelian G_0, the required tau-functions $\tau_{\mu_0, \mu_0'}(t)$ have to involve the fundamental integrable representations $L(j)$ corresponding to $s_j = 0$, in contrast with $\mid \tau_i^R(t) \rangle$ (see Eq. (37)).

which is the generalization of the Eq. (5.1) of (Hollowood et al. (1993)) for general integrable highest-weight representations of $\tilde{\mathcal{G}}$. Eq. (42) allows one to express all the components of $\Theta_-^{\leq}$ in terms of the components of $\mid \tau_i^R(t)\rangle$ for all $i = 0, \ldots, r$ with $s_i \neq 0$ (for instance, by using the positive definite Hermitian form of $L(i)$). However, it is obvious that only a finite subset of them enter in the definition of the potentials A_N through Eq. (26).

In exactly the same way, one can introduce another set of "left" tau-function vectors through

$$\langle \tau_i^L(t) \mid = \langle v_i \mid \Psi^{(\text{vac})} h \Psi^{(\text{vac})-1} , \tag{43}$$

which leads to

$$\langle v_i \mid \Theta_+^{\geq} = \langle \tau_i^L(t) \mid \frac{1}{\tau_i^{(0)}(t)} , \tag{44}$$

and allows one to express all the components of $\Theta_+^{\geq}$ in terms of the components of $\langle \tau_i^L(t) \mid$ for all $i = 0, \ldots, r$ with $s_i \neq 0$.

Summarising, the generalized Hirota tau-functions of these hierarchies consist of the subset of functions $\tau_{\mu_0,\mu_0'}$ and of components of $\mid \tau_i^R\rangle$ and $\langle \tau_i^L \mid$ required to parameterize all the components of the potentials A_N in Eq. (26). Then, for the multi-soliton solutions corresponding to the group element ρ specified in (28), their truncated power series expansion follows from the possible nilpotency of the eigenvectors F_k in these representations. For instance, if $n = 1$ in (28) and $F_1^m \mid \mu_0\rangle = F_1^m \mid v_i\rangle = 0$ unless $m \leq m_1$, then

$$\tau_{\mu_0,\mu_0'}(t) = \tilde{\tau}_{\mu_0,\mu_0'}^0 + \tilde{\tau}_{\mu_0,\mu_0'}^1 + \ldots + \tilde{\tau}_{\mu_0,\mu_0'}^{m_1}$$

$$= \sum_{k=0}^{m_1} \frac{1}{k!} e^{k \sum_N w_N t_N} \langle \mu_0' \mid F_1^k \mid \mu_0\rangle , \quad \text{and}$$

$$\mid \tau_i^R(t)\rangle = \sum_{k=0}^{m_1} \frac{1}{k!} e^{k \sum_N w_N t_N} F_1^k \mid v_i\rangle . \tag{45}$$

Let us finally remark the special fact that for the conserved densities of the solitons appropriately defined as total derivatives, they can be easily obtained from one projection as the familiar second derivative of the logarithm, in terms, therefore of a single complex function. The mass computation in Section 8 are in a way one example of this.

6   The Example of the Non Abelian Toda Models

Consider an untwisted affine Kac-Moody algebra $\hat{\mathcal{G}}$ endowed with an integral gradation $\hat{\mathcal{G}} = \bigoplus_{n \in \mathbb{Z}} \hat{\mathcal{G}}_n$ (see (6),(7)). By an affine Kac-Moody algebra we mean a loop algebra corresponding to a finite dimensional simple Lie algebra $\mathcal{G}$ of rank r, extended by the center C and the derivation D.

Let $\mathcal{M}$ be a two dimensional manifold with local coordinates x_+ and x_-; $\hat{\mathcal{G}}$ be an affine Kac-Moody algebra corresponding to a finite dimensional complex simple Lie algebra $\mathcal{G}$ with the Lie group G; $\mathcal{A}$ be a flat connection in the trivial holomorphic principal fibre bundle $\mathcal{M} \times \hat{G} \longmapsto \mathcal{M}$. Specify the connection in such a way that its $(1,0)$-component takes values in the subspaces $\bigoplus_{n=0}^{l} \hat{\mathcal{G}}_{+n}$, and $(0,1)$-component takes values in $\bigoplus_{n=0}^{l} \hat{\mathcal{G}}_{-n}$, with l being a fixed positive integer. In other words, up to a relevant gauge tranformation, these components, satisfying the zero curvature condition

$$\partial_+ A_- - \partial_- A_+ + [A_+ , A_-] = 0, \tag{46}$$

are of the form

$$A_+ = -B\,F^+\,B^{-1}, \qquad A_- = -\partial_- B\,B^{-1} + F^-. \tag{47}$$

Here B is a mapping from $\mathcal{M}$ to the Lie group $\hat{G}_0$ with the Lie algebra $\hat{\mathcal{G}}_0$; $F^{\pm}$ are mappings to $\bigoplus_{n=1}^{l} \hat{\mathcal{G}}_{\pm n}$ of the form

$$F^+ = E_l + \sum_{m=1}^{l-1} F_m^+, \qquad F^- = E_{-l} + \sum_{m=1}^{l-1} F_m^-, \tag{48}$$

with $E_{\pm l}$ being some fixed elements of $\hat{\mathcal{G}}_{\pm l}$; and $F_m^{\pm}$, $1 \le m \le l-1$, take values in $\hat{\mathcal{G}}_{\pm m}$.

Substituting the gauge potentials (47) into (46), one gets the equations of motion

$$\partial_+ \left(\partial_- B\,B^{-1}\right) = [E_{-l}\,, B\,E_l\,B^{-1}] + \sum_{n=1}^{l-1}[F_n^-\,, B\,F_n^+\,B^{-1}], \tag{49}$$

$$\partial_- F_m^+ = [E_l\,, B^{-1}\,F_{l-m}^-\,B] + \sum_{n=1}^{l-m-1} [F_{n+m}^+\,, B^{-1}\,F_n^-\,B], \tag{50}$$

$$\partial_+ F_m^- = -[E_{-l}\,, B\,F_{l-m}^+\,B^{-1}] - \sum_{n=1}^{l-m-1} [F_{n+m}^-\,, B\,F_n^+\,B^{-1}]. \tag{51}$$

Since $Q_{\mathbf{s}}$, defined in (7), and C are in $\hat{\mathcal{G}}_0$, we parametrise B as

$$B = b\,e^{\eta\,Q_{\mathbf{s}}}\,e^{\nu\,C}, \tag{52}$$

where b is a mapping to G_0, the subgroup of $\hat{G}_0$ generated by all elements of $\hat{\mathcal{G}}_0$ other than $Q_{\mathbf{s}}$ and C. The fields η and ν correspond to the extension of the loop algebra, and, as we will show below, are responsible for making the system conformally invariant (Aratyn et al. (1991), Babelon et al. (1990)). Clearly, the order of the three factors in (52) is irrelevant, since they commute. In addition, we will use a special basis for the generators of $\hat{\mathcal{G}}_0$ such that they are all orthogonal to $Q_{\mathbf{s}}$ and C. From (7) one observes that the generators of $\hat{\mathcal{G}}_0$ are, besides C and $Q_{\mathbf{s}}$, the elements H_a^0, $a = 1,2,\ldots r$, of the Cartan subalgebra, and step operators $E^0_{\pm a}$ and $E^{\mp 1}_{\pm \beta}$, such that $\sum^r_{a=1} s_a \lambda^v_a \cdot \alpha = 0$, and $\sum^r_{a=1} s_a \lambda^v_a \cdot \beta = N_S$. There can be no step operators E^n_γ, with $\mid n \mid > 1$, as explained in appendix C of ref. (Ferreira et al. (1995)). Therefore, shifting the Cartan elements as

$$\tilde{H}^0_a = H^0_a - \frac{1}{N_S}\, \text{Tr}\left(H_S\, H^0_a\right) C = H^0_a - \frac{2}{\alpha^2_a}\frac{s_a}{N_S}C, \qquad (53)$$

one gets

$$\text{Tr}\left(C^2\right) = \text{Tr}\left(C\, \tilde{H}^0_a\right) = \text{Tr}\left(Q^2_S\right) = \text{Tr}\left(Q_S\, \tilde{H}^0_a\right) = 0, \quad \text{Tr}\left(Q_S\, C\right) = N_S,$$

$$\text{Tr}\left(\tilde{H}^0_a\, \tilde{H}^0_b\right) = \text{Tr}\left(H^0_a\, H^0_b\right) = 4\alpha_a \cdot \alpha_b / \alpha^2_a \alpha^2_b \equiv \eta_{ab}, \qquad (54)$$

for all $a, b = 1 \ldots, r$.

Here we have used $H^0_a = 2\alpha_a \cdot H^0 / \alpha^2_a$, $\text{Tr}\left(x \cdot H^0\, y \cdot H^0\right) = x \cdot y$, and $\text{Tr}\left(C\, D\right) = 1$. For more detail of such a special basis, see appendix C of ref. (Ferreira et al. (1995)).

Substituting (52) into the equations of motion (49)–(51), one has

$$\partial_+\left(\partial_- b b^{-1}\right) + \partial_+\partial_-\, \nu\, C = e^{l\eta}[E_{-l}\,,\, b\, E_l\, b^{-1}] + \sum^{l-1}_{n=1} e^{n\eta}\, [F^-_n\,,\, b\, F^+_n b^{-1}] \quad (55)$$

$$\partial_- F^+_m = e^{(l-m)\eta}\, [E_l\,,\, b^{-1}\, F^-_{l-m}\, b] + \sum^{l-m-1}_{n=1} e^{n\eta}[F^+_{m+n}\,,\, b^{-1}\, F^-_n\, b], \qquad (56)$$

$$\partial_+ F^-_m = -e^{(l-m)\eta}\, [E_{-l}\,,\, b\, F^+_{l-m}\, b^{-1}] - \sum^{l-m-1}_{n=1} e^{n\eta}[F^-_{m+n}\,,\, b\, F^+_n\, b^{-1}], \quad (57)$$

$$\partial_+\partial_-\, \eta\, Q_S = 0\,, \qquad (58)$$

where the last equation is a consenquence of the fact that D, and hence Q_S, can not be obtained as the Lie bracket of any two elements of $\hat{\mathcal{G}}$.

The structure of the vacuum of the system (55)–(58) is rather complicated. We will discuss some aspects of it below. However, there is a simple condition that guarantees the existence of static (vacuum) solutions. If the elements $E_{\pm l}$ satisfy the relation

$$[E_l\,,\, E_{-l}] = \beta C\,, \qquad \text{where} \quad \beta = \tfrac{l}{N_S}\, \text{Tr}\left(E_l\, E_{-l}\right), \qquad (59)$$

then

$$b = 1\,, \qquad F^\pm_m = 0\,, \qquad \eta = 0\,, \qquad \nu = -\beta x_+ x_-\,, \qquad (60)$$

is a (vacuum) solution of (55)–(58).

Another possibility for vacuum solutions arises when $E_{\pm l}$, $l > 1$, belong to a Heisenberg subalgebra of $\hat{\mathcal{G}}$, see (Kac et al. (1990), Kac et al. (1985)),

$$[E_M\,,\, E_N] = \text{Tr}\left(E_M E_{-M}\right) M\, \delta_{M+N,0}\, C, \qquad (61)$$

where M, N belong to some (infinite) subset $\mathbb{Z}_E$ of the integer numbers $\mathbb{Z}$. In such cases one has that

$$b = 1, \quad \eta = 0, \; F_M^\pm = c_M^\pm E_{\pm M}, \; F_m^\pm = 0, \; m \notin \mathbb{Z}_E, \; \nu = -\Omega\, x_+\, x_-, \quad (62)$$

is a solution of (55)–(58) with $c_M^\pm$ being constants, and

$$\Omega \equiv \beta + \sum_{M=1}^{l-1} \mathrm{Tr}\,(E_M E_{-M})\, M\, c_M^+\, c_M^-. \tag{63}$$

Obviously, the system (55)–(58) may have many more vacuum solutions besides (60) and (62). However, the condition (59) guarantees the existence of at least one vacuum solution. Such a fact, as we will see below, favors the existence of soliton solutions.

The models introduced above are completely characterised by the data $\{\hat{\mathcal{G}}, Q_s, l; E_{\pm l}\}$; and we have a quite large class of systems with physical properties crucially depending on a choice of those data.

Equations (55) – (58) are invariant under the conformal transformation

$$x_+ \to f(x_+), \qquad x_- \to g(x_-), \tag{64}$$

with f and g being analytic functions; and with the fields transforming as

$$b(x_+, x_-) \to \tilde{b}(\tilde{x}_+, \tilde{x}_-) = b(x_+, x_-), \tag{65}$$

$$e^{-\nu(x_+, x_-)} \to e^{-\tilde{\nu}(\tilde{x}_+, \tilde{x}_-)} = (f')^\delta\, (g')^\delta\, e^{-\nu(x_+, x_-)}, \tag{66}$$

$$e^{-\eta(x_+, x_-)} \to e^{-\tilde{\eta}(\tilde{x}_+, \tilde{x}_-)} = (f')^{1/l}\, (g')^{1/l}\, e^{-\eta(x_+, x_-)}, \tag{67}$$

$$F_m^+(x_+, x_-) \to \tilde{F}_m^+(\tilde{x}_+, \tilde{x}_-) = (f')^{-1+m/l}\, F_m^+(x_+, x_-), \tag{68}$$

$$F_m^-(x_+, x_-) \to \tilde{F}_m^-(\tilde{x}_+, \tilde{x}_-) = (g')^{-1+m/l}\, F_m^-(x_+, x_-), \tag{69}$$

where the conformal weight δ, associated to $e^{-\nu}$, is arbitrary.

Notice that the Lorentz transformation $x_\pm \to \lambda^{\mp 1} x_\pm$ is obtained from (64) by taking $f(x_+) = x_+/\lambda$ and $g(x_-) = \lambda x_-$.

Equations (55)–(58) are also invariant under the transformations

$$b(x_+, x_-) \to h_L(x_-)\, b(x_+, x_-)\, h_R(x_+), \tag{70}$$

$$F_m^+(x_+, x_-) \to h_R^{-1}(x_+)\, F_m^+(x_+, x_-)\, h_R(x_+), \tag{71}$$

$$F_m^-(x_+, x_-) \to h_L(x_-)\, F_m^-(x_+, x_-)\, h_L^{-1}(x_-), \tag{72}$$

where $h_L(x_-)$ and $h_R(x_+)$ are elements of subgroups $\mathcal{H}_0^L$ and $\mathcal{H}_0^R$ of G_0, respectively, satisfying the conditions

$$h_R(x_+)\, E_l\, h_R^{-1}(x_+) = E_l, \qquad h_L^{-1}(x_-)\, E_{-l}\, h_L(x_-) = E_{-l}. \tag{73}$$

The left and right gauge transformations commute, and so the gauge group is $\mathcal{H}_0^L \otimes \mathcal{H}_0^R$. Whenever $\mathcal{H}_0^L$ and $\mathcal{H}_0^R$ have a set of common generators, we get an important subgroup of the gauge group, namely $\mathcal{H}_D \equiv \mathcal{H}_0^L \cap \mathcal{H}_0^R$. These are global gauge transformations, where the fields are transformed under conjugation ($h_L = h_R^{-1} \equiv h_D = $ const.),

$$b \to h_D\, b h_D^{-1}\,, \qquad F_m^{\pm} \to h_D\, F_m^{\pm} h_D^{-1}\,, \tag{74}$$

and $E_{\pm l} = h_D\, E_{\pm l} h_D^{-1}$. We discuss the relevance of these transformations below.

7 Soliton Solutions

We now perform the dressing transformation, dicussed in sections 3 and 4, by taking as an initial configuration a vacuum solution of (55)–(58). As we have said, the model under consideration may have several type of vacuum solutions. However, here we will deal with the solutions of type (60) or (62).

For the vacuum solutions (62), the gauge potentials (47) become

$$A_+^{(0)} = -\mathcal{E}_+\,, \qquad A_-^{(0)} = \mathcal{E}_- + \Omega x_+ C, \tag{75}$$

with $\mathcal{E}_{\pm}$ given by

$$\mathcal{E}_{\pm} \equiv E_{\pm l} + \sum_{N=1}^{l-1} c_N^{\pm} E_{\pm N}\,, \quad \text{and so} \quad [\mathcal{E}_+, \mathcal{E}_-] = \Omega\, C; \tag{76}$$

where $c_N^{\pm}$ and Ω were introduced in (62) and (63), respectively.

They can be written as

$$A_{\pm}^{(0)} = -\partial_{\pm}\Psi^{(\mathrm{vac})}\,\Psi^{(\mathrm{vac})-1}\,, \quad \text{with} \quad \Psi^{(\mathrm{vac})} = e^{x_+\,\mathcal{E}_+}\, e^{-x_-\,\mathcal{E}_-}. \tag{77}$$

The gauge potentials for the vacuum solution (60) are obtained from (75) by taking $c_n^{\pm} = 0$. In fact, they are connected by the gauge transformation

$$A_{\pm}^{(0)} = \tilde{\Psi}^{(\mathrm{vac})} A_{\pm}^{(0)}\big|_{c_n^{\pm}=0}\,\left(\tilde{\Psi}^{(\mathrm{vac})}\right)^{-1} - \partial_{\pm}\tilde{\Psi}^{(\mathrm{vac})}\left(\tilde{\Psi}^{(\mathrm{vac})}\right)^{-1}, \tag{78}$$

with

$$\tilde{\Psi}^{(\mathrm{vac})} = \exp[x_+\,(\mathcal{E}_+ - E_l)]\exp[-x_-\,(\mathcal{E}_- - E_{-l})]. \tag{79}$$

However, in general, the vacuum solutions (60) and (62) may not be connected by any dressing transformation, and, in such a case, the existence of two elements of form (24), is not always possible. Consequently, one can have soliton solutions lying on different orbits under the dressing transformations.

In order to perform the dressing procedure, we take (75) as initial gauge potentials. Then, we obtain, under the dressing procedure, the solutions on the orbit of vacuum (62), and for $c_n^{\pm} = 0$ those on the orbit of the vacuum (60). From the structure of the dressing transformations and from the fact that the grading operator (7) is never the result of any commutation, since it contains D, it follows that the dressing transformations do not excite the field η. Therefore, from (47), (52), (75) and (26) we get

$$b\,E_l\,b^{-1} + \sum_{m=1}^{l-1} b\,F_m^+\,b^{-1} = \Theta_\pm \left(E_l + \sum_{n=1}^{l-1} c_n^+\,E_n + \Theta_\pm^{-1}\,\partial_+\Theta_\pm \right) \Theta_\pm^{-1}, \quad (80)$$

$$-\partial_-\,bb^{-1} - (\partial_-\,\nu + \Omega x_+)\,C + E_{-l} + \sum_{m=1}^{l-1} F_m^- = \qquad\qquad (81)$$

$$= \Theta_\pm \left(E_{-l} + \sum_{m=1}^{l-1} c_m^-\,E_{-m} - \Theta_\pm^{-1}\,\partial_-\Theta_\pm \right) \Theta_\pm^{-1}.$$

Note that in the above relations, the fields b, ν and $F_m^\pm$ stand for the solutions on the orbit of the vacuum solution (62). The procedure to construct the solution requires to split the above equations into the eigensubspaces of the grading operator (7). It is convenient to write

$$\Theta_+^> = \exp\left(\sum_{s>0} t^{(s)} \right), \qquad \Theta_-^\le = \exp\left(\sum_{s>0} t^{(-s)} \right), \quad \text{where } t^{(\pm s)} \in \hat{\mathcal{G}}_{\pm s} \tag{82}$$

The mappings $t^{(\pm s)}$, for each choice of ρ, are determined from (23) with Ψ being $\Psi^{(\text{vac})}$ given in (77). Then, the components of (80) and (82) in each eigensubspace, give an equation connecting the fields with $t^{(\pm s)}$. Thus the solutions for the fields b, ν and $F_m^\pm$ are determined from $t^{(\pm s)}$. Such a procedure is rather cumbersome, but fortunately, one needs to know very few $t^{(\pm s)}$'s to get the solution. For instance, taking relations (80) and (82) for Θ_+ (Θ_-) with grade components 0 and $-l$ (l and 0), one gets

$$\Theta_+^{(0)} = h_L^{-1}(x_-), \qquad \Theta_-^{(0)} = b\,e^{(\nu + \Omega x_+ + x_-)\,C}\,h_R(x_+), \tag{83}$$

with $h_L(x_-)$ and $h_R(x_+)$ defined in (73).

From (17), (24), (77) and (83) it follows that

$$\Theta_-^{\le\,-1} \left(h_L(x_-)\,b\,e^{(\nu + \Omega x_+ + x_-)\,C}\,h_R(x_+) \right)^{-1} \Theta_+^> =$$
$$e^{x_+\,\mathcal{E}_+}\,e^{-x_-\,\mathcal{E}_-}\,\rho\,e^{x_-\,\mathcal{E}_-}\,e^{-x_+\,\mathcal{E}_+}. \tag{84}$$

The space–time dependence of the r.h.s. of the above relation is given explicitly. One can extract the solutions out of (84) by taking the expectation value of its both sides between suitable states of a given representation of $\hat{\mathcal{G}}$, in a similar way to that one explained in section 4.

The solitons solutions are obtained from (84) by choosing the fixed group element ρ, characterising the dressing transformation, as the exponential of an eigenvector of $\mathcal{E}_\pm$, i.e.

$$\rho = e^V. \tag{85}$$

That is the *solitonic specialization* discussed in section 4. Indeed, if V satisfies the relations

$$[\mathcal{E}_\pm , V] = \omega_\pm V, \tag{86}$$

then (84) reads as

$$\exp\left(e^{x_+ \omega_+ - x_- \omega_-} V\right) \equiv \exp\left(e^{\gamma(x - vt)} V\right), \tag{87}$$

with $\gamma = \omega_+ + \omega_-$, and $v = (\omega_- - \omega_+)/(\omega_+ + \omega_-)$, since $x_\pm = t \pm x$.

Therefore, for each eigenvector V, expression (87) corresponds to a solution that travels with a constant velocity v without dispersion. Depending upon the properties of V, as we will see below in the examples, such solutions correspond to one–soliton solutions.

The multi–soliton solutions are obtained by taking ρ to be the product of several one–soliton ρ's, i.e.,

$$\rho = e^{V_1} e^{V_2} e^{V_3} \ldots e^{V_N}, \tag{88}$$

with each V_i satisfying $[\mathcal{E}_\pm , V_i] = \omega_\pm^i V_i$.

Notice that, under the global gauge transformations (74), the gauge potentials (47) are transformed as $A_\pm \to h_D A_\pm h_D^{-1}$. Therefore, since the potencials are pure gauge, $A_\mu = -\partial_\mu T T^{-1}$, one has $T \to h_D T$, and consequently (23) implies $\Theta_+^> \to h_D \Theta_+^> h_D^{-1}$ and $\Theta_-^\leq \to h_D \Theta_-^\leq h_D^{-1}$. Hence, with solution (84) corresponding to a fixed element ρ, a solution, obtained from that by a global gauge transformation (74), is given by (84) with the replacement

$$\rho \to h_D \rho h_D^{-1}, \tag{89}$$

if the condition $h_D \mathcal{E}_\pm h_D^{-1} = \mathcal{E}_\pm$ is satisfied. For the solutions on the orbit of the vacuum (60), that is indeed true, since $\mathcal{E}_\pm = E_{\pm l}$; see (76). For the solitonic case, one then obtains for each eigenvector V of $\mathcal{E}_\pm$, an orbit of equivalent one–soliton (or multi–soliton) solutions generated by $h_D V h_D^{-1}$.

8   Masses of Fundamental Particles and Solitons

As we have seen above, the system under consideration is conformally invariant. Therefore, since we do not have a continuum mass spectrum, its fundamental particles have to be massless. However, such a symmetry can be spontaneoulsy broken by choosing a particular constant solution for the field η, say $\eta = \eta_0$. The resulting theory is then massive. Representing the mapping B as $B \equiv \exp T$, and considering only the linear field approximation, i.e., the free part of the equations of motion (49)–(51), one gets

$$\partial_+ \partial_- T = -v_\eta [E_{-l} , [E_l , T]], \tag{90}$$

$$\partial_+ \partial_- F_m^+ = -v_\eta [E_{-l} , [E_l , F_m^+]], \tag{91}$$

$$\partial_+ \partial_- F_m^- = -v_\eta [E_{-l} , [E_l , F_m^-]], \tag{92}$$

where $v_\eta = e^{l\,\eta_0}$.

Therefore, the masses of fundamental particles in such a theory are given by the eigenvalues of the operator $[E_{-l}, [E_l, *]]$ in the subspaces $\hat{\mathcal{G}}_n$, $n = 0, \pm 1, \pm 2, \ldots \pm (l-1)$, i.e.,

$$[E_{-l}, [E_l, X]] = \lambda X. \tag{93}$$

Since $\partial_+ \partial_- = \frac{1}{4}(\partial_t^2 - \partial_x^2)$, we obtain the masses from the Klein–Gordon type equations (90) – (92) as

$$m_\lambda^2 = 4\,\lambda\,v_\eta. \tag{94}$$

That result constitute a generalization of the arguments used in the abelian and non abelian affine Toda models (Braden et al. (1989), Ferreira et al. (1995)). Of course, we are interested in those cases where the eigenvalues of the operator $[E_{-l}, [E_l, *]]$ are real and positive on the subspaces under consideration. That will be, in fact, one of the conditions we use to select the data $\{\hat{\mathcal{G}}, Q_s, l, E_{\pm l}\}$ for defining physical models through (47).

Notice that the field $e^{l\,\eta}$ plays the role of a Higgs field, since it not only spontaneously breaks the conformal symmetry, but also because its vacuum expectation value sets the mass scale of the theory. We have here the same mechanism as in non abelian affine Toda theories (Aratyn et al. (1993), Ferreira et al. (1995)).

Let us explain now, following the reasonings of (Aratyn et al. (1993)) and (Ferreira et al. (1995)), that the masses of solitons are also generated by the spontaneous breakdown of the conformal symmetry.

The energy momentum tensor of such theories is of the form (see (Ferreira et al. (1996)) for more details)

$$L_{\mu\nu} = \Theta_{\mu\nu} + S_{\mu\nu}. \tag{95}$$

where $S_{\mu\nu}$ is the improvement term

$$
\begin{aligned}
S_{\mu\nu} &\equiv -\frac{k}{l}\,\mathrm{Tr}\left(Q_s\left(\partial_\mu\left(B^{-1}\partial_\nu B\right) - g_{\mu\nu}\partial_\rho\left(B^{-1}\partial^\rho B\right)\right)\right) \\
&= -\frac{k\,N_s}{l}\left(\partial_\mu\partial_\nu - g_{\mu\nu}\partial^2\right)\nu,
\end{aligned}
\tag{96}
$$

Due to the fact we are dealing with a conformally invariant theory, $L_{\mu\nu}$ satisfies

$$\partial_- L_{++} = 0, \qquad \partial_+ L_{--} = 0, \qquad L_{+-} = L_{-+} = 0. \tag{97}$$

Even though it is not traceless, $\Theta_{\mu\nu}$ is symmetric and conserved,

$$\partial^\mu \Theta_{\mu\nu} = 0, \tag{98}$$

The energy of classical solutions are given by the space integral of the $(0,0)$ component of energy–momentum tensor $L_{\mu\nu}$. In the Lorentz frame where the classical soliton solution is static, the energy should be interpreted as the mass of the soliton. However, since the theory is conformally invariant, it has no mass scale, and the soliton mass should vanish. When the conformal symmetry is spontaneously broken by choosing a particular constant solution for the field η, we obtain a massive theory. Construct the energy–momentum tensor of such a theory as follows. Clearly, the tensor $\Theta_{\mu\nu}$, introduced in (95) and evaluated at any classical solution, satisfies (98). Therefore, the tensor defined by

$$\Theta_{\mu\nu}^{\text{broken}} \equiv \Theta_{\mu\nu}\,|_{\eta=\text{constant}}\,, \qquad (99)$$

is symmetric and conserved,

$$\partial^{\mu}\Theta_{\mu\nu}^{\text{broken}} = 0\,, \qquad (100)$$

since $\eta = $ constant is a solution of the equations of motion. Then, let the energy in the massive theory be proportional to the space integral of $\Theta_{00}^{\text{broken}}$. Using (96) and (95), we obtain the soliton mass in the form

$$\frac{M}{\sqrt{1-v^2}} \equiv -\int_{-\infty}^{\infty} dx\,\Theta_{00}^{\text{broken}} + E_{\text{vac.}} = -\frac{k\,N_{\text{s}}}{l}\partial_x\left(\nu + \Omega x_+ x_-\right)|_{-\infty}^{\infty} \quad (101)$$

because the integral of $L_{00}^{\text{red.}}$ vanishes by the above arguments. Here v is the soliton velocity in the units of the speed of the light. Notice that we have subtracted the energy $E_{\text{vac.}}$ of the vacuum solution which is, in fact, divergent. Of course, the vacuum solution is not unique, and it is not clear which one provides the absolute minimum of the energy. We will use the following prescription for the soliton mass formula. For the soliton solutions lying, under the dressing transformations, on the orbit of the vacuum solution (62), we take Ω in (101) to be that one given in (63). However, for those soliton solutions lying on the orbit of the vacuum (60), we take Ω in (101) to be equal to the parameter β introduced in (59). Such a prescription guarantees the finiteness of the soliton masses.

The soliton masses are determined solely by the behaviour at $x = \pm\infty$ of the space derivative of the field ν. That is quite a remarkable fact. In addition, as we now explain, it is very easy to obtain such a behaviour in the general case from the solitonic solutions (87). Actually this is a manifestation at the level of the energy-momentum tensor of a more general feature of all conserved densities of the solitonic solutions, expressed also by the tau-function, as above mentioned.

Consider an integral gradation of $\hat{\mathcal{G}}$, with $s_i' = \frac{\psi^2}{\alpha_i^2}s_i$, $\alpha_0 \equiv -\psi$, and s_i labeling the gradation that defines the model (55)–(58). Consider the integrable highest weight representation with highest weight state

$$|\,\lambda_{\text{s}}'\rangle = \bigotimes_{i=0}^{r} |\,\hat{\lambda}_i\rangle^{\oplus s_i'}, \qquad (102)$$

where $|\,\hat{\lambda}_i\rangle$ are the highest weight states of the fundamental representations of $\hat{\mathcal{G}}$, and $\hat{\lambda}_i$ are the corresponding fundamental weights of $\hat{\mathcal{G}}$.

Then it is possible to show (Ferreira et al. (1996)) that taking the expectation value of both sides of (84) in such state, one gets (with the gauge choice $h_L(x_-) = h_R(x_+) = 1$)

$$e^{-(\nu+\Omega x_+ + x_-)N_{\mathbf{S}}\frac{\psi^2}{2}} = \langle \lambda_{\mathbf{s}'} \mid e^{x_+ \mathcal{E}_+} e^{-x_- \mathcal{E}_-} \rho\, e^{x_- \mathcal{E}_-} e^{-x_+ \mathcal{E}_+} \mid \lambda_{\mathbf{s}'} \rangle. \quad (103)$$

Now, choosing ρ to be the exponential of an eigenvector of $\mathcal{E}_\pm$,

$$[\mathcal{E}_\pm , V] = \omega_\pm V, \quad (104)$$

we obtain a soliton solution

$$e^{-(\nu+\Omega x_+ + x_-)N_{\mathbf{S}}\frac{\psi^2}{2}} = \langle \lambda_{\mathbf{s}'} \mid e^{e^{\Gamma} V} \mid \lambda_{\mathbf{s}'} \rangle \quad (105)$$

with $\Gamma = \omega_+ x_+ - \omega_- x_- \equiv \gamma\,(x - vt)$.

Suppose V is an operator in such a representation for which there is a positive integer N'_V, such that

$$\langle \lambda_{\mathbf{s}'} \mid V^n \mid \lambda_{\mathbf{s}'} \rangle = 0 \qquad \text{for } n > N'_V. \quad (106)$$

Then the soliton mass is easily obtained from (101), where for $\gamma > 0$ ($\gamma < 0$) only the upper (lower) limit $x = \infty$ ($x = -\infty$) contributes in the integral[3] ,

$$M = \frac{2}{\psi^2}\frac{k\,N'_V}{l} \mid \gamma \mid \sqrt{1-v^2} = \frac{2}{\psi^2}\frac{2k\,N'_V}{l} \sqrt{\omega_+ \omega_-}. \quad (107)$$

Notice that we must have $\omega_+ \omega_- > 0$ in order to have the soliton velocity $v = (\omega_- - \omega_+)/(\omega_- + \omega_+)$, not exceeding the light velocity ($c = 1$). The soliton mass formula (107) has some remarkable properties. One of them concerns the relation particle–soliton in the theory, indicating some sort of duality similar to the electromagnetic duality of some four dimensional gauge theories possessing the Bogomolny (monopole) limit (Montonen et al. (1977)). As we have seen, the soliton solutions are created by the eigenvectors V of $\mathcal{E}_\pm$. From (104) one has $[\mathcal{E}_+ , [\mathcal{E}_- , V]] = \omega_+ \omega_- V$. Expanding V over the eigenvectors of the grading operator $Q_{\mathbf{s}}$ as $V = \sum_n V^{(n)}$, one observes that $[\mathcal{E}_+ , [\mathcal{E}_- , V^{(n)}]] = \omega_+ \omega_- V^{(n)}$. Therefore, if some $V^{(n)} \in \hat{\mathcal{G}}_n$, $n = 0, \pm1, \pm2, \ldots \pm (l-1)$, does not vanish, it implies that $V^{(n)}$ must be one of the eigenvectors X in (93). Then we associate a soliton with a fundamental particle. In addition, we have $\lambda \equiv \omega_+ \omega_-$, and, consequently, from (94) and (107), the masses of the corresponding soliton and fundamental particle are determined by the same eigenvalue. In fact, we have from (94) and (107), with $v_\eta = 1$, that

---

[3] We point out that the soliton mass formula (107) could be equally obtained by defining the mass through the momentum formula, instead through the energy like in (101), as $\frac{M\,v}{\sqrt{1-v^2}} \equiv \int dx \Theta_{01}^{\text{broken}}$. In this case, we do not have to subtract the vacuum momentum, since it vanishes.

$$M_{\text{sol.}} = \frac{2}{\psi^2} \frac{k}{l} \frac{N_V'}{l} \, m_\lambda^{\text{part.}}. \tag{108}$$

Of course, in the expansion of V, we may have more than one non vanishing $V^{(n)}$, with $n = 0, \pm 1, \pm 2, \ldots \pm (l-1)$. Then we would associate a one–soliton solution to more than one fundamental particle. The counting of one–soliton solutions has to be better analysed in each particular case. We discuss this issue in section 10.

9   The Matter Fields

It is clear from (65)-(69), that the massive fields associated with non vanishing grade (namely $F_m^\pm$), are chiral fields with non vanishing spins, in contrast with the Toda type fields. In fact, we show that the free equations for such fields take the form of the massive Dirac equation, as could be expected from general covariance arguments.

Consider the subspace $\hat{\mathcal{G}}_m$ for $0 < m < l$. Let $\hat{\mathcal{G}}_m^{(F)}$ be the subspace of $\hat{\mathcal{G}}_m$, generated by the eigenvectors of $[E_{-l}, [E_l, \cdot]]$ with non zero eigenvalues, i.e.,

$$\hat{\mathcal{G}}_m^{(F)} \equiv \{ T^{(m)} \in \hat{\mathcal{G}}_m \mid \lambda^{(m)} \neq 0 \}, \tag{109}$$

where $\lambda^{(m)}$ is defined as

$$[E_{-l}, [E_l, T^{(m)}]] = [E_l, [E_{-l}, T^{(m)}]] = \lambda^{(m)} \, T^{(m)}. \tag{110}$$

Decompose the subspace $\hat{\mathcal{G}}_m$, as a vector space, into the sum

$$\hat{\mathcal{G}}_m = \hat{\mathcal{G}}_m^{(F)} + \hat{\mathcal{G}}_m^{(K)}, \tag{111}$$

where $\hat{\mathcal{G}}_m^{(K)}$ is the complement of $\hat{\mathcal{G}}_m^{(F)}$ in $\hat{\mathcal{G}}_m$.

It is possible to show (see (Ferreira et al. (1996)) for details) that the subspaces $\hat{\mathcal{G}}_{-l+m}^{(F)}$ and $\hat{\mathcal{G}}_m^{(F)}$ are isomorphic. The mapping is given by the action of E_{-l} on $\hat{\mathcal{G}}_m^{(F)}$, or equivalently by the action of E_l on $\hat{\mathcal{G}}_{-l+m}^{(F)}$. Therefore, we can put in one-to-one correspondence the fields in $\hat{\mathcal{G}}_{-l+m}^{(F)}$ and $\hat{\mathcal{G}}_m^{(F)}$. Then, one can show that each pair of such fields constitute a Dirac spinor under the two dimensional Lorentz group, and their equations of motion can indeed be written in the form of (obviously not free) Dirac equations. Consequently, we interpret the massive fields associated to generators of non zero grading as matter fields

10   An Example of a Special Class of Models

There is a class of models possessing a $U(1)$ Noether current, which, under some circumstances, is proportional to a topological current. That occurs for those models where the grade l of the operator E_l, introduced in (48), is equal to the integer N_s defined in (7). In addition, it is necessary that the operators $E_{\pm N_s}$ satisfy the condition

$$z E_{-N_s} = \mu z^{-1} E_{N_s} \in \text{center of } \hat{\mathcal{G}}_0 \tag{112}$$

where, μ is some constant independent of z, and z is a complex variable used to realize the generators of the affine Kac-Moody algebra $\hat{\mathcal{G}}$, in terms of those of the finite simple Lie algebra $\mathcal{G}$ as

$$H_a^n \equiv z^n H_a, \qquad E_\alpha^n \equiv z^n E_\alpha, \qquad D \equiv z \frac{d}{dz} \tag{113}$$

That means that the "projections" of $E_{\pm N_s}$ onto $\hat{\mathcal{G}}_0$, are parallel and lie in the center of $\hat{\mathcal{G}}_0$. When those condition are satisfied it is possible to gauge fix the conformal symmetry, such that a special $U(1)$ Noether charge is proportional to a topological charge.

In this section we discuss a example where that happens. It correspond to the principal gradation of $sl(2)^{(1)}$ with $l = 2$. Let us denote by H^n, $E_\pm^n$, D and C the Chevalley basis generators of the $sl(2)^{(1)}$. The commutation relations are

$$[H^m, H^n] = 2 m C \delta_{m+n,0}, \qquad [E_+^m, E_-^n] = H^{m+n} + m C \delta_{m+n,0}, \tag{114}$$
$$[H^m, E_\pm^n] = \pm 2 E_\pm^{m+n}, \qquad [D, T^m] = m T^m, \qquad T^m \equiv H^m, E_\pm^m; \tag{115}$$

all other commutation relations are trivial. The grading operator for the principal gradation ($\mathbf{s} = (1,1)$) is $Q \equiv \frac{1}{2} H^0 + 2D$. Then the eigensubspaces are $\hat{\mathcal{G}}_0 = \{H^0, C, Q\}$, $\hat{\mathcal{G}}_{2n+1} = \{E_+^n, E_-^{n+1}\}$, with $n \in \mathbb{Z}$, and $\hat{\mathcal{G}}_{2n} = \{H^n\}$, with $n \in \{\mathbb{Z} - 0\}$.

The mapping B is parametrised as

$$B = e^{\varphi H^0} e^{\tilde{\nu} C} e^{\eta Q} = e^{\varphi \tilde{H}^0} e^{\nu C} e^{\eta Q}, \tag{116}$$

where $\tilde{H}^0 = H^0 - \frac{1}{2} C$ is the Cartan generator in the special basis introduced in (53), and so $\tilde{\nu} = \nu - \frac{1}{2}\varphi$.

In the case $l = 2$, we choose

$$E_2 \equiv m H^1, \qquad E_{-2} \equiv m H^{-1}, \tag{117}$$

where m is a constant. We then have

$$[E_2, [E_{-2}, E_\pm^n]] = 4m^2 E_\pm^n. \tag{118}$$

Therefore, each of the subspaces $\hat{\mathcal{G}}_{\pm 1}$ has two generators with the same eigenvalue $4m^2$. Following section 9 we write

$$F_1^+ = 2\sqrt{im} \left(\psi_R E_+^0 + \tilde{\psi}_R E_-^1 \right), \qquad F_1^- = 2\sqrt{im} \left(\psi_L E_+^{-1} - \tilde{\psi}_L E_-^0 \right), \tag{119}$$

and introduce the Dirac fields

$$\psi = \begin{pmatrix} \psi_R \\ \psi_L \end{pmatrix} ; \qquad \tilde{\psi} = \begin{pmatrix} \tilde{\psi}_R \\ \tilde{\psi}_L \end{pmatrix} \tag{120}$$

From (94) we obtain the masses of the particles,

$$m_\varphi = m_{\tilde{\nu}} = m_\eta = 0; \qquad m_\psi = 4m; \tag{121}$$

The equations of motion derived from (55)–(58), are

$$\partial^2 \varphi = -4m_\psi \, \overline{\tilde{\psi}} \gamma_5 e^{\eta + 2\varphi \gamma_5} \psi, \tag{122}$$

$$\partial^2 \tilde{\nu} = -2m_\psi \, \overline{\tilde{\psi}} (1 - \gamma_5) e^{\eta + 2\varphi \gamma_5} \psi - \frac{1}{2} m_\psi^2 e^{2\eta}, \tag{123}$$

$$\partial^2 \eta = 0, \tag{124}$$

$$i\gamma^\mu \partial_\mu \psi = m_\psi \, e^{\eta + 2\varphi \gamma_5} \psi, \tag{125}$$

$$i\gamma^\mu \partial_\mu \tilde{\psi} = m_\psi \, e^{\eta - 2\varphi \gamma_5} \tilde{\psi}, \tag{126}$$

where the gamma matrices are defined as

$$\gamma_0 = -i \begin{pmatrix} 0 & -1 \\ 1 & 0 \end{pmatrix}, \qquad \gamma_1 = -i \begin{pmatrix} 0 & 1 \\ 1 & 0 \end{pmatrix}, \tag{127}$$

and $\gamma_5 = \gamma_0 \gamma_1$, and $\bar{\psi} \equiv \tilde{\psi}^T \gamma_0$. Recall that $\partial^2 = \partial_t^2 - \partial_x^2$, $x_\pm = t \pm x$. The corresponding Lagrangian has the form

$$\frac{1}{k} \mathcal{L} = \frac{1}{4} \partial_\mu \varphi \, \partial^\mu \varphi + \frac{1}{4} \partial_\mu \varphi \, \partial^\mu \eta + \frac{1}{2} \partial_\mu \tilde{\nu} \, \partial^\mu \eta - \frac{1}{8} m_\psi^2 \, e^{2\eta}$$
$$+ i\bar{\psi} \gamma^\mu \partial_\mu \psi - m_\psi \, \bar{\psi} \, e^{\eta + 2\varphi \gamma_5} \psi. \tag{128}$$

It is real (for $\eta = $ real constant) if $\tilde{\psi}$ is the complex conjugate of ψ, and if φ is pure imaginary. This will be true for the soliton solution as we shall see below.

Notice that such model is invariant under the transformations

$$x_+ \leftrightarrow x_- ; \quad \psi_R \leftrightarrow \epsilon \tilde{\psi}_L; \quad \tilde{\psi}_R \leftrightarrow -\epsilon \psi_L; \quad \varphi \leftrightarrow \varphi; \quad \eta \leftrightarrow \eta; \quad \nu \leftrightarrow \nu \tag{129}$$

where $\epsilon = \pm 1$. It should be interpreted as the product CP of charge conjugation times parity. Parity alone is clearly violated.

The generator $H^0 \in \hat{\mathcal{G}}_0$ commutes with $E_{\pm 2}$, and, therefore, the gauge symmetry (70)–(72) of the model is $U(1)_L \otimes U(1)_R$,

$$h_L(x_-) = e^{\xi_- (x_-) H^0}, \qquad h_R(x_+) = e^{\xi_+ (x_+) H^0}. \tag{130}$$

Since the genereators of $U(1)_L$ and $U(1)_R$ are the same, we have the global gauge transfomations (74) generated by $h_D \equiv h_L = h_R^{-1} \equiv e^{i\theta \, H^0/2}$ ($\theta = $ const.). The fields are transformed as

$$\psi \rightarrow e^{i\theta} \psi \quad \tilde{\psi} \rightarrow e^{-i\theta} \tilde{\psi} \quad \varphi \rightarrow \varphi, \quad \tilde{\nu} \rightarrow \tilde{\nu}, \quad \eta \rightarrow \eta; \tag{131}$$

and the corresponding Noether current is

$$J^\mu = \bar{\psi}\gamma^\mu\psi, \qquad \partial_\mu J^\mu = 0. \tag{132}$$

The fields ψ and $\tilde{\psi}$ have charges 1 and -1, respectively; and φ, $\tilde{\nu}$ and η have charge zero.

Let us next see how the general arguments given above concerning Noether and topological charges apply here. The topological current and charges are

$$j^\mu = \frac{1}{2\pi i}\epsilon^{\mu\nu}\partial_\nu\varphi, \qquad Q_{\text{topol.}} \equiv \int dx\, j^0, \tag{133}$$

Indeed, the Lagrangian (128) has infinitely many degenerate vacua due to the invariance under $\varphi \to \varphi + i\pi$. Making use of the field equations, one easily verifies that

$$\partial_\mu\left[i\bar{\psi}\gamma_5\gamma^\mu\psi + \frac{1}{2}\partial^\mu\varphi\right] = 0 \tag{134}$$

Combining this equation with the conservation of the vector current $\bar{\psi}\gamma^\mu\psi$, one deduces that there exist two charges defined by

$$\mathcal{J} = -i\tilde{\psi}_R\psi_R + \frac{1}{2}\partial_+\varphi, \quad \bar{\mathcal{J}} = i\tilde{\psi}_L\psi_L + \frac{1}{2}\partial_-\varphi$$

which satisfy $\partial_-\mathcal{J} = 0$, $\partial_+\bar{\mathcal{J}} = 0$. We now make a "gauging fixing" of the conformal symmetry by choosing $\mathcal{J} = \bar{\mathcal{J}} = 0$. We call it a "gauging fixing", because any values of $\mathcal{J}$ and $\bar{\mathcal{J}}$ can be transformed to zero by a conformal transformation. This gives, altogether,

$$\frac{1}{2\pi i}\epsilon^{\mu\nu}\partial_\nu\varphi = \frac{1}{\pi}\bar{\psi}\gamma^\mu\psi, \tag{135}$$

so that the topological and Noether currents are proportional. As discussed at the beginning of this section, that is a consequence of the fact that $E_{\pm 2}$ satisfies (112).

Let us turn to the Noether charge which here is simply the fermion number. It should be defined such that it satisfies the Poisson bracket relation

$$i\{\psi, Q_{\text{Noether}}\}_{\text{P.B.}} = \psi \tag{136}$$

Since the coupling constant k was taken as an overall factor, this is satisfied by

$$Q_{\text{Noether}} = k\int dx\, \bar{\psi}\gamma^0\psi \tag{137}$$

so that

$$Q_{\text{topol.}} = \frac{1}{k\pi}Q_{\text{Noether}} \tag{138}$$

As argued in general, this means that k should only take discrete values as expected, since our actions are related with the one of WZNW.

Let us now construct the soliton solutions. The operators $E_{\pm 2}$ given in (117), lie in the homogeneous Heisenberg subalgebra generated by H^n, with the commutation relations (114). Such a subalgebra has no generators of grade ± 1 for the principal gradation. Therefore, the model under consideration has no vacuum solutions of type (62). Then, from (76), we get

$$\mathcal{E}_{\pm} = E_{\pm 2} = m\, H^{\pm 1}. \tag{139}$$

We perform the dressing transformation starting from the vacuum solution (60), namely

$$\varphi = \eta = \psi = \tilde{\psi} = 0\,, \qquad ; \quad \tilde{\nu} = -\frac{1}{8} m_\psi^2 x_+ x_- \equiv \nu_0. \tag{140}$$

Now, let $\mid \hat{\lambda}_0 \rangle$ and $\mid \hat{\lambda}_1 \rangle$ be the highest weight states of two fundamental representations of the affine Kac–Moody algebra $sl(2)^{(1)}$, respectively the scalar and spinor ones. Then, from (84) with $\eta = 0$, we obtain the solutions on the orbit of the vacuum (140),

$$e^{-\varphi} = \frac{\langle \hat{\lambda}_1 \mid G \mid \hat{\lambda}_1 \rangle}{\langle \hat{\lambda}_0 \mid G \mid \hat{\lambda}_0 \rangle}, \qquad e^{-(\tilde{\nu}-\nu_0)} = \langle \hat{\lambda}_0 \mid G \mid \hat{\lambda}_0 \rangle,$$

$$\psi_R = \sqrt{\frac{m}{i}}\, \frac{\langle \hat{\lambda}_0 \mid E_-^1\, G \mid \hat{\lambda}_0 \rangle}{\langle \hat{\lambda}_0 \mid G \mid \hat{\lambda}_0 \rangle}, \qquad \tilde{\psi}_R = -\sqrt{\frac{m}{i}}\, \frac{\langle \hat{\lambda}_1 \mid E_+^0\, G \mid \hat{\lambda}_1 \rangle}{\langle \hat{\lambda}_1 \mid G \mid \hat{\lambda}_1 \rangle}$$

$$\psi_L = -\sqrt{\frac{m}{i}}\, \frac{\langle \hat{\lambda}_1 \mid G\, E_-^0 \mid \hat{\lambda}_1 \rangle}{\langle \hat{\lambda}_1 \mid G \mid \hat{\lambda}_1 \rangle}, \qquad \tilde{\psi}_L = -\sqrt{\frac{m}{i}}\, \frac{\langle \hat{\lambda}_0 \mid G\, E_+^{-1} \mid \hat{\lambda}_0 \rangle}{\langle \hat{\lambda}_0 \mid G \mid \hat{\lambda}_0 \rangle} \tag{141}$$

where

$$G \equiv e^{x_+ \mathcal{E}_+}\, e^{-x_- \mathcal{E}_-}\, \rho\, e^{x_- \mathcal{E}_-}\, e^{-x_+ \mathcal{E}_+}. \tag{142}$$

In order to get the soliton solutions, we choose the fixed mapping ρ to be an exponentiation of an eigenvector of $\mathcal{E}_\pm$ (solitonic specialization); namely, $\rho = e^V$, with $[\mathcal{E}_\pm\,, V] = \omega_\pm V$. Therefore,

$$G = \exp\left(e^\Gamma V\right) \quad \text{with} \quad \Gamma = \omega_+ x_+ - \omega_- x_- \equiv \gamma\,(x - v t). \tag{143}$$

In this case the eigenvectors of $\mathcal{E}_\pm$ are

$$V_\pm(z) = \sum_{n \in \mathbb{Z}} z^{-n}\, E_\pm^n. \tag{144}$$

Indeed,

$$[\mathcal{E}_+\,, V_\pm(z)] = \pm 2mz\, V_\pm(z) \equiv \omega_+^\pm V_\pm(z), \tag{145}$$

$$[\mathcal{E}_-\,, V_\pm(z)] = \pm \frac{2m}{z}\, V_\pm(z) \equiv \omega_-^\pm V_\pm(z). \tag{146}$$

The solution, associated with $V_+(z)$, is

$$\nu = \nu_0, \quad \varphi = \tilde{\psi} = 0, \quad \psi = \sqrt{\frac{m}{i}} e^{\Gamma} \begin{pmatrix} z \\ -1 \end{pmatrix}; \tag{147}$$

while those, associated with $V_-(z)$, is given by

$$\nu = \nu_0, \quad \varphi = \psi = 0, \quad \tilde{\psi} = -\sqrt{\frac{m}{i}} e^{-\Gamma} \begin{pmatrix} 1 \\ 1/z \end{pmatrix}, \tag{148}$$

where

$$\Gamma = 2m(zx_+ - \frac{1}{z} x_-) \equiv \gamma (x - vt). \tag{149}$$

The masses of these solutions are obtained from (107). Here the relevant state $| \lambda_{s'} \rangle$ in (106) is

$$| \lambda_{s'} \rangle = | \hat{\lambda}_0 \rangle \otimes | \hat{\lambda}_1 \rangle. \tag{150}$$

Using level one vertex operators, one can verify that

$$\langle \hat{\lambda}_i | (V_{\pm}(z))^n | \hat{\lambda}_i \rangle = 0, \quad \text{for } n \geq 1 \text{ and } i = 0, 1. \tag{151}$$

Therefore, $N_V' = 0$ in (106), and from (107) one gets that the masses of the solutions (147) and (148) vanish. Such solutions correspond to the objects which travel with velocities $v = \pm (1 - z^2) / (1 + z^2)$; and keeping $z^2 > 0$, one has $| v | < 1$. Therefore, these solutions cannot be interpreted as solitons (particles), since they would correspond to massless particles traveling with velocity smaller that light velocity. We should interpret them as vacuum configurations, since they have the same energy as vacuum (140).

The true soliton solutions of the system are constructed as follows. Notice that $V_+(z)$ and $V_-(-z)$ have the same eigenvalues. Therefore, any linear combination of them, leads to solutions traveling with a constant velocity without dispersion. So, we let

$$V(a_{\pm}, z) \equiv \sqrt{i} (a_+ V_+(z) + a_- V_-(-z)); \tag{152}$$

$$[\mathcal{E}_+ , V(a_{\pm}, z)] = 2mz \, V(a_{\pm}, z), \quad [\mathcal{E}_- , V(a_{\pm}, z)] = \frac{2m}{z} V(a_{\pm}, z), \tag{153}$$

and so $\omega_+ = 2mz$ and $\omega_- = \frac{2m}{z}$. The particular factor $\sqrt{i}$ is chosen such that the reality condition will be obeyed with $a_- = a_+^*$. Again, using level one vertex operators, one can verify that[4]

$$\langle \lambda_{s'} | V(a_{\pm}, z)^n | \lambda_{s'} \rangle = 0 \quad \text{for } n > 4. \tag{154}$$

Therefore, $N_V' = 4$ in (106), and from (107) with $\psi^2 = 2$, and ψ being the highest root of $sl(2)$, one gets that the mass of such solutions is

$$M = 8k \, m = 2 \, k \, m_{\psi}, \tag{155}$$

---

[4] Notice that the truncation occurs for powers greater than 4, and not 2, because $| \lambda_{s'} \rangle$ lies in the tensor product representation, see (150)

where k is the coupling constant appearing in the Lagrangian (128). The solutions generated by (152), have two parameters, namely $a_\pm$. One parameter is always present, because one can scale an eigenvector of $\mathcal{E}_\pm$ without changing the width γ and velocity v of the soliton, obtained from the eigenvectors $\omega_\pm$; see (87). However, in this case, the second parameter comes from a symmetry. As we have pointed out in (89), associated to the fixed element $\rho = e^{V(a_\pm, z)}$, we have an orbit of equivalent solutions due to the global transformations (131),

$$V(a_\pm, z) \to \sqrt{i}\left(a_+ e^{i\theta} V_+(z) + a_- e^{-i\theta} V_-(-z)\right). \tag{156}$$

The explicit form of the solutions generated by (152), is obtained using (141),

$$\varphi = \log\left(\frac{1 + i\sigma e^{2\Gamma}}{1 - i\sigma e^{2\Gamma}}\right), \tag{157}$$

$$\tilde{\nu} = -\log\left(1 + i\sigma e^{2\Gamma}\right) - \frac{1}{8} m_\psi^2 x_+ x_-, \tag{158}$$

$$\eta = 0; \tag{159}$$

and

$$\psi = a_+ \sqrt{m}\, e^{\Gamma}\left(\frac{\frac{z}{1+i\sigma e^{2\Gamma}}}{\frac{-1}{1-i\sigma e^{2\Gamma}}}\right), \qquad \tilde{\psi} = a_- \sqrt{m}\, e^{\Gamma}\left(\frac{\frac{z}{1-i\sigma e^{2\Gamma}}}{\frac{-1}{1+i\sigma e^{2\Gamma}}}\right); \tag{160}$$

where Γ is given in (149), and $\sigma = a_+ a_- z/4$. Keeping m and z real, we have the mass M of the soliton, from (155), real and positive, and also the parameters γ and v (149) are real. The reality condition is obeyed if $a_- = a_+^*$, as anticipated. At this point, it is useful to re-express the expressions just given in terms of the physical parameters of the soliton. Using equations (127) and (121) and one deduces that

$$\gamma = m_\psi\Big/\sqrt{1 - v^2}, \quad z = \sqrt{(1 - v)/(1 + v)}. \tag{161}$$

Moreover, since $a_\pm$ are complex conjugate, we may write

$$a_\pm = e^{\pm i\theta} 2\sqrt{\frac{\sigma}{z}}. \tag{162}$$

The dependence upon space-time appears only through $\sqrt{\sigma} \exp(\Gamma)$. We will write [5]

$$\sqrt{\sigma}\, e^{\Gamma} = \exp((\gamma(x - x_0 - vt)) \tag{163}$$

where x_0 is the position of the soliton at time zero. Then we have

$$\varphi = 2i \arctan\left(\exp\left(2m_\psi (x - x_0 - vt)\Big/\sqrt{1 - v^2}\right)\right), \tag{164}$$

which is the sine–Gordon soliton. The Dirac fields are given by

---

[5] by convention, we choose σ to be positive

$$\psi = e^{i\theta}\sqrt{m_\psi}\,e^{m_\psi(x-x_0-vt)/\sqrt{1-v^2}}\left(\begin{array}{c}\left(\frac{1-v}{1+v}\right)^{1/4}\frac{1}{1+ie^{2m_\psi(x-x_0-vt)/\sqrt{1-v^2}}}\\ -\left(\frac{1+v}{1-v}\right)^{1/4}\frac{1}{1-ie^{2m_\psi(x-x_0-vt)/\sqrt{1-v^2}}}\end{array}\right),$$

$$(165)$$

and $\widetilde{\psi}$ is the complex conjugate of ψ. Thus the only parameters are the soliton mass and velocity, together with the angle θ which reflects the global invariance (131). Notice that the sign of the tolopogical charge can be reversed by reversing the sign of z. Therefore, the solutions (164)–(165) contain the sine–Gordon soliton and anti–soliton.

Finally, we come to the very important feature of the present model already mentioned above in general, namely it is clear from the explicit expressions Eqs.165 that ψ vanishes exponentially when $x - x_0 \to \pm\infty$, so that the Dirac field is confined inside the soliton. That this must be true is of course a general consequence of Eq.135 which may be verified directly on the explicit solution. This phenomenon has been much studied for electron phonon systems. Models of a similar type describe the electron self-localization in quasi-one-dimensional dielectrics (for recent reviews see (Brazovskii et al. (1994)), (Heeger et al. (1988))). At low temperature these systems go over to dielectric states characterized by charge density waves which can be constructed on the basis of the Peierls model. The continuous limits are described by Lagrangians similar to Eq.128. Discussing this important issue is beyond the scope of the present article, so we will not dwell upon it here. Let us simply recall that the typical example of the polyacteline molecule was much discussed in connection with fermion number fractionization (Jackiw et al. (1976)). Clearly, on the other hand one may regard our soliton solution a sort of one dimensional bag model for QCD. In this connection let us note that, if we introduce the two-by-two matrix $U = \exp(\eta + 2\varphi\gamma_5)$, we may rewrite the Lagrangian Eq.128 as

$$\mathcal{L} = \frac{1}{16}\left\{\mathrm{tr}\left[U^{-1}\partial_\mu U\frac{1+\gamma_5}{2}U^{-1}\partial^\mu U\right] - \frac{1}{2}\mathrm{tr}\left[U^{-1}\partial_\mu U\right]\mathrm{tr}\left[U^{-1}\partial^\mu U\right]\right\}$$

$$+i\bar{\psi}\gamma_\mu\partial_\mu\psi - \bar{\psi}U\psi - \frac{m_\psi^2}{8}\det(U), \qquad (166)$$

which is similar to a two-dimensional version of the low energy effective action for QCD (see e.g. (Chodos et al. (1984))).

11 Conclusions

From the common description based on zero curvature equations, a unified treatment has been given of the procedures to obtain solutions of nonlinear equations for a large class of integrable hierarchies including the generalized KdV and non-abelian Toda coupled to matter. Interesting examples of them have been discussed in detail, taking advantage of the simplifications achieved, and the same procedure can be equally well applied to more complex equations. The role of the tau functions has been clarified, as specific matrix elements in special integrable highest weight representations (for any level). The results are a further step towards stablishing the conjecture that all multisoliton solutions lie in the orbit of the vacuum, generated by dressing transformations, which are now clearly related to the tau functions and solitonic specialization. These results should be very useful for the classification of integrable theories in two dimensions and for the generalization to the quantum case and to higher dimensions.

References

H. Aratyn, L.A. Ferreira, J.F. Gomes and A.H. Zimerman, *Phys. Lett.* **254B** (1991) 372.

H. Aratyn, C.P. Constantinidis, L.A. Ferreira, J.F. Gomes and A.H. Zimerman, *Nucl. Phys.* **B406** (1993) 727.

H. Aratyn, L.A. Ferreira, J.F. Gomes and A.H. Zimerman, *Vertex Operators and Solitons of Constrained KP Hierarchies*, contribution to this volume

O. Babelon and L. Bonora, *Phys. Lett.* **244B** (1990) , 220.

O. Babelon and D. Bernard, Phys. Lett. **B260** 81; Commun. Math. Phys. **149** (1992) 279-306, hep-th/9111036.

O. Babelon and D. Bernard, Int. J. Mod. Phys. **A8** (1993) 507-543, hep-th/9206002.

H.W. Braden, E. Corrigan, P.E. Dorey and R. Sasaki, *Phys. Lett.* **227B** (1989) 411, *Nucl. Phys.* **B338** (1990) 689; M.D. Freeman *Phys. Lett.* **261B** (1991) 57; A. Fring, H.C. Liao and D.I. Olive, *Phys. Lett.* **266B** (1991) 82-86.

S.A. Brazovskii, N. N. Kirova, *Sov. Phys. Rev. A Phys.* **5** (1994) 99.

A. Chodos, E. Hadjimichael, C. Tze, *Solitons in Nuclear and Elementary Particle Physics*, Proceedings of the Lewes Workshop, 1984, World Scientific.

E.Date, M. Jimbo, M. Kashiwara, T. Miwa, *Proc. Japan. Acad.* **57A** (1981) 3806; *Physica* **D4** (1982) 343; Publ. RIMS Kyoto University **18**, (1982) 1077.

V.G. Drinfel'd and V.V. Sokolov, J. Sov. Math. **30** (1985) 1975.

L.D. Faddeev and L.A. Takhtajan, *Hamiltonian Methods in the Theory of Solitons*, Springer, 1987.

L.A. Ferreira, J.L. Miramontes and J. Sánchez Guillén, *Nucl. Phys.* **B449** (1995) 631-679, hep-th/9412127.

L.A. Ferreira, J-L. Gervais, J. Sánchez Guillén and M.V. Saveliev, *Affine Toda Systems Coupled to Matter Fields*, Nucl. Phys. **B470** (1996) 236, hep-th/9512105.

L.A. Ferreira, J.L. Miramontes and J. Sánchez Guillén, *J. Math. Phys.* **38** (1997) 2, hep-th/9606066.

A.P. Fordy and J. Gibbons, *Commun. Math. Phys.* **77** (1980) 21. P.J. Caudrey, J.D. Gibbon, J.C. Eilbeck and R.K. Bullough, *Phys. Rev. Lett.* **30**, (1973), 237. G.L. Lamb Jr., *Elements of Soliton Theory*, Wiley, New York, 1980. G.L. Lamb, *Backlund transformations at the turn of the century*. In: Lecture Notes in Mathematics, Vol. 515, pp. 69-79, Springer, 1976.

M. D.Freeman *Nucl. Phys.* **B433** (1995) 657.

M.F. de Groot, T.J. Hollowood, and J.L. Miramontes, Commun. Math. Phys.**145** (1992) 57; N.J. Burroughs, M.F. de Groot, T.J. Hollowood, and J.L. Miramontes, Commun. Math. Phys.**153** (1993) 187; Phys. Lett. **B 277** (1992) 89.

A.J. Heeger, S. Kivelson, J.R. Schrieffer, W.-P. Wu, Rev. Mod. Phys. **60** (1988) 782.

R. Hirota, *Direct methods in soliton theory*, in "Soliton" (R.K. Bullough and P.S. Caudrey, eds.), (1980) 157.; J. Phys. Soc. Japan **33** (1972) 1459.

T.J. Hollowood and J.L. Miramontes, Commun. Math. Phys.**157** (1993) 99.

R. Jackiw, C. Rebbi, *Phys. Rev. D* **13** (1976) 3398; J. Goldstone, F. Wilczek, *Phys. Rev. Lett.* **47** (1981) 986.

V.G. Kac and D.H. Peterson, *112 constructions of the basic representation of the loop group of* E_8. In: Symp. on Anomalies, Geometry and Topology, W.A. Bardeen and A.R. White (eds.), World Scientific, Singapore, 1985.

V.G. Kac, *Infinite dimensional Lie algebras*, Third Edition, Cambridge University Press, Cambridge, 1990.

V.G. Kac and M. Wakimoto, *Exceptional hierarchies of soliton equations*, in "Proceedings of Symposia in Pure Mathematics", Vol. **49** (1989) 191.

A.N. Leznov and M.V. Saveliev, *Group Theoretical Methods for Integration of Non-Linear Dynamical Systems*, Progress in Physics Series, v. 15, Birkhaüser-Verlag, Basel, 1992.

H.C. Liao, D. Olive and N. Turok, *Phys. Lett.* **298B** (1993) 95-102.

J.L.Miramontes, private communication and in preparation with C.R.Fernandez-Pousa.

C. Montonen and D.I. Olive, *Phys. Lett.* **72B** (1977) 117
P. Goddard, J. Nuyts and D.I. Olive, *Nucl. Phys.* **B125** (1977) 1 D.I. Olive, *Magnetic monopoles and electromagnetic duality conjectures* in *Monopoles in quantum field theory* eds. N. S. Craigie, P. Goddard and W. Nahm (World Scientific, Singapore, 1982) p. 157; D. Olive, *Exact electromagnetic duality*, hep-th/9508089, to appear in the Proc. of ICTP Conf. on Recent Developments in Statistical Mechanics and Quantum Field Theory, Trieste, 10-12 Apr. 1995. N. Seiberg and E. Witten *Nucl. Phys.* **B426** (1994) 19-52, hep-th/9407087; C. Vafa and E. Witten *Nucl. Phys.* **B431** (1994) 3-77, hep-th/9408074.

D. Olive, N. Turok and J.W.R. Underwood, Nucl. Phys. **B401** (1993) 663-697.

D. Olive, N. Turok and J.W.R. Underwood, Nucl. Phys. **B409** (1993) 509-546, hep-th/9305160.

D. Olive, M.V. Saveliev and J.W.R. Underwood, Phys. Lett. **B311** (1993) 117-122, hep-th/9212123.

M. Semenov-Tian-Shansky, *Functional Analysis and Its Application* **17** (1983) 259; Publ. RIMS Kyoto Univ. **21** (1985) 1237.

J.W.R. Underwood, *Aspects of Non Abelian Toda Theories*, hep-th/9304156.

G. Wilson, Phil. Trans. R. Soc. Lond. **A 315** (1985) 383; *Habillage et fonctions* τ C. R. Acad. Sc. Paris **299 (I)** (1984) 587.

G. Wilson, *The τ-Functions of the gAKNS Equations*, in "Verdier memorial conference on integrable systems" (O. Babelon, P. Cartier, and Y. Kosmann-Schwarzbach, eds.), Birkhauser (1993) 131-145.

V.E. Zakharov, A.B. Shabat, *Functional Analysis and Its Application* **13** (1979) 166.

Constrained and Rational Reductions of the KP Hierarchy

Gerard Helminck and Johan van de Leur *

Faculty of Applied Mathematics,
University of Twente,
P.O.Box 217, 7500 AE Enschede, The Netherlands

Abstract. We consider Segal and Wilson's description of the KP hierarchy on the Hilbert–Schmidt Grassmannian. In this setting we show that the (vector) constrained KP hierarchy and Krichever and Dickey's rational reductions of the KP hierarchy are the same. To prove this, we use some results on elementary Bäcklund–Darboux transformations.

1 Introduction to the KP Hierarchy

The KP hierarchy consists of a tower of nonlinear differential equations in infinitely many variables $\{t_n | n \geq 1\}$. It is named after the simplest nontrivial equation in this tower, the Kadomtsev–Petviashvili equation:

$$\frac{3}{4}\frac{\partial^2 v}{\partial t_2^2} = \frac{\partial}{\partial t_1}\left(\frac{\partial v}{\partial t_3} - 3v\frac{\partial v}{\partial t_1} - \frac{1}{4}\frac{\partial^3 v}{\partial t_1^3}\right). \tag{1}$$

The solutions of these equations belong to a ring of functions R that is stable under the operators $\partial_n = \frac{\partial}{\partial t_n}$. The form in which one presents these equations, is the so-called Lax form. This is an equality between operators in $\partial = \partial_1$ of a specific nature. It is instructive to digress for a moment on this form. We start with a simple example. To v in R we associate the differential operators

$$\mathcal{L} = \partial^2 + 2v \quad \text{and} \quad \mathcal{P} = \partial^3 + 3v\partial + \frac{3}{2}\frac{\partial v}{\partial t_1}. \tag{2}$$

Both $\mathcal{L}$ and $\mathcal{P}$ are regarded as operators acting on R and as such we can speak of the commutant of these operators. A direct computation shows

$$[\mathcal{P}, \mathcal{L}] = \frac{1}{2}\frac{\partial^3 v}{\partial t_1^3} + 6\frac{\partial v}{\partial t_1}v.$$

In particular, $[\mathcal{P}, \mathcal{L}]$ is a zero-th order operator in ∂. Thus we see that the condition that v is a solution of the Korteweg–de Vries (KdV) equation,

---

* JvdL is financially supported by the Netherlands Organization for Scientific Research (NWO).

$$\frac{\partial v}{\partial t_3} = \frac{1}{4}\frac{\partial^3 v}{\partial t_1^3} + 3v\frac{\partial v}{\partial t_1}, \tag{3}$$

is equivalent to the following condition for the Schrödinger operator $\mathcal{L}$:

$$\frac{\partial \mathcal{L}}{\partial t_3} = 0 \cdot \partial + 2\frac{\partial v}{\partial t_3} = [\mathcal{P}, \mathcal{L}]. \tag{4}$$

The relation (4) is called the Lax form of the KdV equation. Equation (4) gives rise e.g. to the following questions:

(Q1) Is there a systematic way to come up with $\mathcal{P}$ once one has $\mathcal{L}$?
(Q2) Are there more differential operators in ∂ such that the commutator with $\mathcal{L}$ has order 0?
(Q3) What can be done for other operators besides the Schrödinger operator?

The first two questions lead in a natural way to an extension of $R[\partial] = \{\sum_{i=0}^{n} a_i \partial^i |\ a_i \in R\}$. By adding, namely, suitable integral operators to the ring, it becomes possible to take roots of differential operators like the Schrödinger operator and this framework provides the answer to these questions. Consider the ring $R[\partial, \partial^{-1})$ of pseudodifferential operators with coefficients in R. It consists of all expressions

$$\sum_{i=-\infty}^{N} a_i \partial^i \quad , \quad a_i \in R \quad \text{for all } i,$$

that are added in an obvious way and multiplied according to

$$\partial^j \circ a\partial^i = \sum_{k=0}^{\infty} \binom{j}{k} \partial^k (a)\partial^{i+j-k}.$$

Each operator $P = \sum p_j \partial^j$ decomposes as $P = P_+ + P_-$ with $P_+ = \sum_{j \geq 0} p_j \partial^j$ its differential operator part and $P_- = \sum_{j < 0} p_j \partial^j$ its integral operator part. We denote by $Res_{\partial} P = p_{-1}$ the *residue* of P. For each $\mathcal{L} \in R[\partial]$ of the form

$$\mathcal{L} = \partial^k + \sum_{i=0}^{k-2} a_i \partial^i, \tag{5}$$

one shows recursively that there is a unique operator $L \in R[\partial, \partial^{-1})$ of the form

$$L = \partial + \sum_{j<0} \ell_j \partial^j \quad , \ell_j \in R \quad \text{for all } j < 0. \tag{6}$$

such that $\mathcal{L} = L^k$ and consequently one denotes L also as $\mathcal{L}^{1/k}$. Any element in $R[\partial, \partial^{-1})$ of the form (6) is called a *Lax operator*. In particular, the operator $\mathcal{P}$ equals $(\mathcal{L}^{3/2})_+$ and this answers question Q1. Since $\mathcal{L}$ and $\mathcal{L}^{1/2}$ commute, we have for all $i \geq 1$

$$[(\mathcal{L}^{i/2})_+, \mathcal{L}] = -[(\mathcal{L}^{i/2})_-, \mathcal{L}]. \tag{7}$$

As the left-hand side of (7) belongs to $R[\partial]$ and the right-hand side is of order ≤ 0, we see that all $\{(\mathcal{L}^{i/2})_+ | i \geq 1\}$ are examples of elements in $R[\partial]$ whose commutator with $\mathcal{L}$ is of order zero. This answers question Q2. Hence it makes sense to consider the system of equations

$$\partial_n(\mathcal{L}) = [(\mathcal{L}^{n/2})_+, \mathcal{L}], \quad n \geq 1,$$

which is called the KdV hierarchy. Of course, for all even n this gives $\partial_n(v) = 0$. Likewise one can consider for each differential operator $\mathcal{L}$ of the form (5), the equations

$$\partial_n(\mathcal{L}) = [(\mathcal{L}^{n/k})_+, \mathcal{L}], \quad n \geq 1. \tag{8}$$

This system of equations is called the k-th Gelfand–Dickey hierarchy and shows that the Schrödinger operator is not an isolated case (this answers question Q3). It is not difficult to show that the equations (8) for $\mathcal{L}$ are equivalent to the following ones for $L = \mathcal{L}^{1/k}$

$$\partial_n(L) = \sum_{j<0} \partial_n(\ell_j)\partial^j = [(L^n)_+, L], \quad n \geq 1. \tag{9}$$

Thus we have come to a system of equations, called the *Lax equations* for a Lax operator L that includes all Gelfand–Dickey hierarchies. Because of the reason mentioned before, it is called the KP hierarchy.

Equation (9) has at least the trivial solution $L = \partial$ and can be seen as the compatibility equation of the linear system

$$L\psi = z\psi \quad \text{and} \quad \partial_n(\psi) = (L^n)_+(\psi) \tag{10}$$

One needs a context in which the actions of (10) make sense and that allows you to derive (9) from (10). For the trivial solution (10) becomes

$$\partial\psi = z\psi \quad \text{and} \quad \partial_n\psi = z^n\psi \quad \text{for all } n \geq 1.$$

Hence, the function $\gamma(t) = \exp(\sum_{i \geq 1} t_i z^i)$ is a solution. The space M of so-called *oscillating functions* is a space for which we can make sense of (10) and that can be seen as a collection of perturbations of this solution. It is defined as

$$M = \{(\sum_{j \leq N} a_j z^j) e^{\sum t_i z^i} \mid a_i \in R, \text{ for all } i\}.$$

Expressing $e^{\sum t_i z^i}$ in terms of the elementary Schur functions, i.e.,

$$e^{\sum t_i z^i} = \sum_{k=0}^{\infty} p_k(t) z^k$$

one notices that the product

$$\left(\sum_{j \le N} a_j z^j\right)e^{\sum t_i z^i} = \sum_{\ell \in \mathbb{Z}}\left(\sum_{k=0}^{\infty} a_{\ell-k} p_k\right)z^\ell$$

is still formal. To make sense of this expression as an infinite series in z and z^{-1}, the coefficients $\sum_{k=0}^{\infty} a_{\ell-k} p_k$ for all $\ell \in \mathbb{Z}$ have to be well defined functions of $t = (t_n)$. In that light it is natural to have a context such that for each element t

$$\sum_{j \le N} |a_j(t)|^2 < \infty \quad \text{and} \quad \sum_{k=0}^{\infty} |p_k(t)|^2 < \infty.$$

Such a context has been given in Segal and Wilson (1985) and will be considered in section 2.

The space M becomes a $R[\partial, \partial^{-1})$-module by the natural extension of the actions

$$b\{(\textstyle\sum_j a_j z^j)e^{\sum t_i z^i}\} = (\textstyle\sum_j ba_j z^j)e^{\sum t_i z^i}$$
$$\partial\{(\textstyle\sum_j a_j z^j)e^{\sum t_i z^i}\} = (\textstyle\sum_j \partial(a_j)z^j + \textstyle\sum_j a_j z^{j+1})e^{\sum t_i z^i}.$$

It is even a free $R[\partial, \partial^{-1})$-module, since we have

$$(\textstyle\sum p_j \partial^j)e^{\sum t_i z^i} = (\textstyle\sum p_j z^j)e^{\sum t_i z^i}.$$

An element ψ in M is called an *oscillating function of type* z^ℓ, if it has the form

$$\psi(z) = \psi(t, z) = \{z^\ell + \sum_{j<\ell} \alpha_j z^j\}e^{\sum t_i z^i}.$$

The fact that M is a free $R[\partial, \partial^{-1})$-module, permits you to show that each oscillating function of type z^ℓ that satisfies (10) gives you a solution of (9). This function is then called a *wavefunction* of the KP-hierarchy. Notice that we can write $\psi(z) = P(\partial)e^{\sum t_i z^i}$. It is not difficult to see that the equations (9) for $L = P\partial P^{-1}$ are equivalent to the Sato–Wilson equations for P:

$$\partial_n(P)P^{-1} = -(P\partial^n P^{-1})_-. \tag{11}$$

On $R[\partial, \partial^{-1})$ we have an anti-algebra morphism called *taking the adjoint*. The adjoint of $P = \sum p_i \partial^i$ is given by

$$P^* = \sum_i (-\partial)^i p_i.$$

Using this, we associate to an oscillating function (resp. wavefunction) $\psi(z)$ its *adjoint oscillating function* (resp. *wavefunction*) $\psi^*(z) = P(\partial)^{*-1}e^{-\sum t_i z^i}$. The adjoint wavefunction satisfies

$$L^*\psi^* = z\psi^* \quad \text{and} \quad \partial_n(\psi^*) = -(L^n)^*_+(\psi^*). \tag{12}$$

Moreover, an oscillating function is a wavefunction if and only if $\psi(t, z)$ and its adjoint oscillating function $\psi^*(t, z)$ satisfy the famous bilinear identity (see Date et al. (1983)):

$$Res_{z=0}\psi(t, z)\psi^*(s, z) = 0. \tag{13}$$

2   The Segal–Wilson Grassmannian

An analytic approach to construct wavefunctions of the KP-hierarchy is considered in Segal and Wilson (1985). Segal and Wilson consider the Hilbert space

$$H = \{\sum_{n \in \mathbb{Z}} a_n z^n \mid a_n \in \mathbb{C}, \sum_{n \in \mathbb{Z}} \mid a_n \mid^2 < \infty\},$$

with decomposition $H = H_+ \oplus H_-$, where

$$H_+ = \{\sum_{n \geq 0} a_n z^n \in H\} \quad \text{and} \quad H_- = \{\sum_{n < 0} a_n z^n \in H\}$$

and inner product $< \cdot \mid \cdot >$ given by

$$< \sum_{n \in \mathbb{Z}} a_n z^n \mid \sum_{m \in \mathbb{Z}} b_m z^m >= \sum_{n \in \mathbb{Z}} a_n \overline{b_n}.$$

To this decomposition is associated the Grassmannian $Gr(H)$ consisting of all closed subspaces W of H such that the orthogonal projection $p_+ : W \to H_+$ is Fredholm and the orthogonal projection $p_- : W \to H_-$ is Hilbert-Schmidt. The connected components of $Gr(H)$ are given by

$$Gr^{(\ell)}(H) = \{W \in Gr(H) \mid p_+ : z^{-\ell}W \to H_+ \text{ has index zero}\}.$$

On each of these components we have a natural action by multiplication of the group of commuting flows

$$\Gamma_+ = \{\exp(\sum_{i \geq 1} t_i z^i) \mid t_i \in \mathbb{C}, \sum \mid t_i \mid (1 + \epsilon)^i < \infty \quad \text{for some } \epsilon > 0\}.$$

Now we take for R the ring of meromorphic functions on Γ_+ and for ∂_n the partial derivative w.r.t. the parameter t_n of Γ_+. For each $W \in Gr^{(\ell)}(H)$, let Γ_+^W be given by

$$\Gamma_+^W = \{\gamma(t) = \exp(\sum_{i=1}^{\infty} t_i z^i) \in \Gamma_+ \mid p_+ : \gamma^{-1}z^{-\ell}W \to H_+ \text{ is a bijection}\}.$$

In a similar way as in Segal and Wilson (1985), one shows that Γ_+^W is nonempty. For $\gamma \in \Gamma_+^W$, let $P_W(t, z)$ be z^ℓ times the inverse image of 1 under the projection $p_+ : \gamma^{-1}z^{-\ell}W \to H_+$. Then we associate to W an oscillating function $\psi_W(t, z) = P_W(t, z)e^{\sum_{i=1}^{\infty} t_i z^i}$ of type z^ℓ, which has the properties that it is defined on a dense open subset of Γ_+ and that it takes values in W. Moreover, it is known that the range of ψ_W spans a dense subspace of W. A crucial property of the elements $\gamma(t)$ in this dense open subset of Γ_+ is that

$$W \cap (z^\ell H_+)^\perp \gamma(t) = \{0\}.$$

By exploiting this property, one shows that ψ_W is a wavefunction for the KP hierarchy. Hence, if we write $\psi_W(t,z) = P_W(t, \partial) \cdot e^{\sum t_i z^i}$ with $P_W \in R[\partial, \partial^{-1}]$, then $L_W = P_W \partial P_W^{-1}$ is a solution of the KP-hierarchy. Each component of $Gr(H)$ generates in this way the same set of solutions of the KP-hierarchy, so it would suffice, as is done in Segal and Wilson (1985), to consider only $Gr^{(0)}(H)$. However, we need to consider here all components.

Let $W \in Gr^{(\ell)}(H)$, then $W^\perp$ is a closed subspace of H, moreover p_- : $W^\perp \to H_-$ is a Fredholm operator of index $-\ell$ and $p_+ : W^\perp \to H_+$ is a Hilbert–Schmidt operator. Interchanging the role of H_+ and H_-, we see that $W^\perp$ is a plane in the adjoint Grassmannian $Gr^*(H)$ consisting of planes U for which p_- is a Fredholm operator and p_+ is a Hilbert–Schmidt operator. We will now show that the closure of the linear span of all $\overline{z\psi_W^*(t,z)}$ is in fact $W^\perp$. Since the wavefunction ψ_W and its adjoint wavefunction ψ_W^* satisfy the bilinear identity (13), one has that

$$< \psi_W(t,z) | \overline{z\psi_W^*(s,z)} >= 0.$$

Which means that $\overline{z\psi_W^*(s,z)} \in W^\perp$ for all relevant s. Let U be the closure of the linear span of all $\overline{z\psi_W^*(s,z)}$, then $U \in Gr^*(H)$ and moreover $p_- : U \to H_-$ has index $-\ell$. Hence $U = W^\perp$ (see Helminck and Post (1988) for more details). The following proposition is now obvious.

Proposition 2.1 *let V, W be planes in $Gr(H)$. Then $V \subset W$ (and hence $W^\perp \subset V^\perp$) if and only if*

$$< \psi_V(t,z) | \overline{z\psi_W^*(s,z)} >= 0.$$

3   Elementary Bäcklund–Darboux Transformations

Let ψ be a wavefunction of type z^ℓ, not necessarily in the Segal–Wilson setting. Like in Oevel and Schief (1993), we define elementary Bäcklund–Darboux transformations that produce new wavefunctions of type $z^{\ell \pm 1}$. For $q, r \in R$, one considers the first order differential operators $q\partial q^{-1}$ and $r^{-1}\partial r$. An obvious question now is: When are $q\partial q^{-1}\psi$ and $r^{-1}\partial^{-1}r\psi$ again wavefunctions? The answer is given by the following theorem:

Theorem 3.1 *Let ψ be a wavefunction of type z^ℓ and let ψ^* be its adjoint wave function. Then the following holds:*
(a) Let q and $r \in R$ satisfy

$$\partial_n(q) = (L^n)_+(q), \tag{14}$$
$$\partial_n(r) = -(L^n)^*_+(r) \quad \text{for all } n = 1, 2, \ldots \tag{15}$$

Then $q\partial q^{-1}\psi$ (resp. $r^{-1}\partial^{-1}r\psi$) is a wavefunction of type $z^{\ell+1}$ (resp. $z^{\ell-1}$) and $-q^{-1}\partial^{-1}q\psi^*$ (resp. $-r\partial r^{-1}\psi^*$) is its adjoint wavefunction.

(b) If $q\partial q^{-1}\psi$ (resp. $r^{-1}\partial^{-1}r\psi$) is again a wavefunction and hence $-q^{-1}\partial^{-1}q\psi^*$ (resp. $-r\partial r^{-1}\psi^*$) its adjoint wavefunction, then one can find an up to a scalar multiple unique $\hat{q}$ (resp. $\hat{r}$) $\in R$, such that

$$q\partial q^{-1} = \hat{q}\partial\hat{q}^{-1} \quad (\text{resp. } r^{-1}\partial r = \hat{r}^{-1}\partial\hat{r}),$$

which satisfies the equations (14) (resp. (15)).

The proof of this theorem will be given in a forthcoming publication, it uses e.g. the following identity (which can be proved using (11) and some elementary identities of residues of pseudodifferential operators, see Helminck and van de Leur (1997))

$$\begin{aligned}
\partial_n(q\partial q^{-1}P)(q\partial q^{-1}P)^{-1} &= -(q\partial q^{-1}P\partial^n(q\partial q^{-1}P)^{-1})_- \\
&\quad +q^{-1}\partial_n(q) - q\partial q^{-2}\partial_n(q)q\partial^{-1}q^{-1} \\
&\quad -q^{-1}(L^n)_+(q) + q\partial q^{-1}(L^n)_+(q)\partial^{-1}q^{-1}.
\end{aligned} \tag{16}$$

For (b) one needs the Zakharov–Shabat equations for the Lax operator L, which follow from the equations (9).

In the Segal–Wilson setting one has the following consequence of Theorem 3.1.

Corollary 3.1 Let W be a plane in $Gr(H)$, L_W the corresponding Lax operator and let ψ_W and ψ^*_W be the corresponding wavefunction and its adjoint. Then there exists a function q (resp. r) $\in R$, defined on an open dense subset of Γ_+ which satisfies (14) (resp. (15)) for $L = L_W$, if and only if there exists a plane W_q (resp. W_r) in $Gr(H)$ satisfying the codimension 1 inclusion:

$$W_q \subset W \quad (\text{resp. } W \subset W_r) \tag{17}$$

such that the corresponding (adjoint) wavefunctions are connected by $\psi_{W_q} = q\partial q^{-1}\psi_W$ and $\psi^*_{W_q} = -q^{-1}\partial^{-1}q\psi^*_W$ (resp. $\psi_{W_r} = r^{-1}\partial^{-1}r\psi_W$ and $\psi^*_{W_r} = -r\partial r^{-1}\psi^*_W$).

Clearly the Bäcklund–Darboux operators of Theorem 3.1 are invertible. Moreover, considering elementary properties of the residue of pseudodifferential operators, one cane prove the following useful lemma.

Lemma 3.1 Let q, $r \in R$, then q (resp. r) satisfies (14) (resp. (15)) if and only if q^{-1} (resp. r^{-1}) satisfies

$$\begin{aligned}
\partial_n(q^{-1}) &= -(q\partial q^{-1}L^n q\partial^{-1}q^{-1})^*_+(q^{-1}), \\
\partial_n(r^{-1}) &= (r^{-1}\partial^{-1}rL^n r^{-1}\partial r)_+(r^{-1}), \quad \text{respectively.}
\end{aligned}$$

We will use the above results in the proof of the main theorem of this paper.

4 Reductions of the KP Hierarchy

There exist many reductions of the KP hierarchy. The most famous ones are
the ones described in section 1, viz., the ones related to the Gelfand-Dickey
hierarchies. Recall, that this is a subsystem of the KP-hierarchy consisting
of all solutions L that are the k-th root of a differential operator. This gives
solutions of the KP-hierarchy that do not depend on the $\{t_{kn}$, with $n \geq
1\}$. Those operators satisfy the condition $L^k = (L^k)_+$. These k-th Gelfand–
Dickey reductions contain the Korteweg–de Vries (KdV) equation (3) for
$k = 2$ and the Boussinesq equation for $k = 3$:

$$3\partial_2^2(u) + \partial_1^4(u) + 6u(\partial_1(u))^2 = 0.$$

Another reduction, the so-called (vector) constrained KP hierarchies, were
introduced as natural generalizations of the Gelfand–Dickey hierarchies. Instead of assuming that L^k has no integral operator part, one allows an $(L^k)_-$
of a very simple form, viz.,

$$L^k = (L^k)_+ + \sum_{j=1}^{m} q_j \partial^{-1} r_j, \tag{18}$$

such that the following conditions on the functions q_j and r_j hold:

$$\partial_n(q_j) = (L^n)_+(q_j), \tag{19}$$
$$\partial_n(r_j) = -(L^n)_+^*(r_j) \quad \text{for all } n \geq 1. \tag{20}$$

Many interesting integrable hierarchies e.g. AKNS, Yajima–Oikawa and Melnikov hierarchies appear amongst these constrained families. For $k = 1$ one
finds the following equations:

$$\partial_2(q_j) = \partial_1^2(q_j) + 2q_j \sum_{i=1}^{m} q_i r_i, \quad \partial_2(r_j) = -\partial_1^2(r_j) - 2r_j \sum_{i=1}^{m} q_i r_i. \tag{21}$$

If one moreover assumes that $q_j = r_j^*$ then (21) becomes the vector non-linear
Schrödinger equation. For $k = 2$, $m = 1$ one finds;

$$\partial_2(u) = \partial_1(qr), \quad \partial_2(q) = \partial_1^2(q) + 2uq, \quad \partial_2(r) = -\partial_1^2(r) - 2uq, \tag{22}$$

which is the Yajima–Oikawa system. For more information on the (vector
constrained hierarchy we refer the reader to the references at the end of this
paper.

Finally we discuss the rational reductions of the KP hierarchy introduced
in Krichever (1995) and Dickey (1995b). In this case one assumes that the
Lax operator L satisfies

$$L^k = L_2^{-1} L_1, \tag{23}$$

where L_1 and L_2 are coprime differential operators of order $k + m$ and m,
respectively. This generalizes the Gelfand–Dickey reduction in a different way.

It can be shown (see Krichever (1995)) that in this case the Lax equations
(9) are equivalent to

$$\partial_i L_1 = ((L_1 L_2^{-1})^{i/k})_+ L_1 - L_1 ((L_2^{-1} L_1)^{i/k})_+, \qquad (24)$$

$$\partial_i L_2 = ((L_1 L_2^{-1})^{i/k})_+ L_2 - L_2 ((L_2^{-1} L_1)^{i/k})_+. \qquad (25)$$

For $k = m = 1$ this reduction gives the AKNS hierarchy. As the first equations
one finds (21) with $m = 1$.

It is well known that, among the solutions coming from the Segal-Wilson
Grassmannian, the ones that satisfy the k-th Gelfand-Dickey hierarchy are
exactly characterized by

$$z^k W \subset W. \qquad (26)$$

In the next section we consider a generalization of this condition and we show
that that condition describes both generalizations of the Gelfand–Dickey reduction.

5   An Extension of the Condition $z^k W \subset W$

In this section we consider, for each k and m in $\mathbb{N} = \{0, 1, 2, \ldots\}$, $k \neq 0$
subspaces W in $Gr(H)$ that satisfy the m-Vector k-Constrained ($mVkC$)-
condition:

There is a subspace W' of W of codimension m such that $z^k(W') \subset W$.
(27)

This is a natural generalization of the condition that describes inside $Gr(H)$
the solutions of the k-th Gelfand-Dickey hierarchy. We will associate to each
W, satisfying the $mVkC$-condition, $2m$ functions $\{q_j \mid 1 \le j \le m\}$ and
$\{r_j \mid 1 \le j \le m\}$ which satisfy the equations (18), (19), (20) for $L = L_W$.
For details we refer to Helminck and van de Leur (1997).

Take any W in $Gr^{(\ell)}$ that satisfies the $mVkC$-condition. It is no restriction to assume that the m occurring in (27) is optimal, i.e. there is a basis
$\{u_1, \ldots, u_m\}$ of the orthocomplement of W' in W such that

$$(\text{Span}\{z^k u_1, \ldots, z^k u_m\}) \cap W = \{0\}.$$

It is no restriction to assume that the vectors $\{z^k(u_1), \ldots, z^k(u_m)\}$ are an
orthonormal basis of the orthocomplement of W in $z^k W + W$. To the space
W we associate the subspaces

$$W_j = W \oplus \mathbb{C} z^k u_j, 1 \le j \le m.$$

Clearly the W_j all belong to $Gr^{(\ell-1)}(H)$ and hence, they have wavefunctions
ψ_{W_j} of type $z^{\ell-1}$. Recall that $\psi_{W_j}(t, z)$ is well-defined for all t belonging to
the open dense subset

$$\Gamma_+^{W_j} = \{\gamma(t) = \exp(\textstyle\sum t_i z^i) \in \Gamma_+ \mid p_+ : \gamma^{-1} z^{-\ell+1} W_j \to H_+ \text{ is a bijection}\}.$$

On $\Gamma_+^{W_j}$ we consider the function

$$s_j(t) = <\psi_{W_j}(t,z) \mid z^k u_j>. \tag{28}$$

Since the vectors $\{\psi_{W_j}(t,z) \mid t \in \Gamma_+^{W_j}\}$ are lying dense in W_j and m was assumed to be optimal, the functions $\{s_j\}$ do not vanish. Denote by $r_j = s_j^{-1}$, then on a dense open subset of Γ_+, there is defined the function

$$\varphi_j = \frac{1}{s_j}\psi_{W_j} = r_j\psi_{W_j}. \tag{29}$$

It takes values in W_j and has moreover the following useful property

$$\varphi_j(t) - z^k u_j \in W, \tag{30}$$

for all t in a dense open subset of Γ_+. This property is a consequence of the facts that $\varphi_j(t) - z^k u_j$ is by construction orthogonal to $z^k u_j$ and that W is the orthocomplement of $\mathbb{C}z^k u_j$ inside W_j. In Zhang (1996), similar functions φ_j are introduced, only not using the geometry, but as solutions of a certain system of differential equations. In particular, we can dispose of the condition (a) in the Proposition of Zhang (1996). Thus we have obtained m functions r_j.

We define the q_j as follows

$$q_j(t) = <\psi_W(t,z) \mid u_j>. $$

Because m is optimal, the functions q_j are non-zero on an open dense subset of Γ_+. Since u_j does not depend on t, (10) implies the equations (19) for all q_j.

First we derive an expression for $(L_W^k)_-(\psi_W)$. Thereto we consider

$$\Phi(t) = z^k\psi_W - (L_W^k)_+(\psi_W) - \sum_{j=1}^m q_j\varphi_j. \tag{31}$$

Since φ_j takes values in W_j, the function $(L_W^k)_+(\psi_W)$ does so in the space W and $z^k\psi_W$ in $z^k W$. Hence we have that $\Phi(t)$ belongs to $W + z^k W$ for all relevant t. By construction we have that for all $j, 1 \le j \le m, \Phi(t)$ is orthogonal to $z^k u_j$, hence $\Phi(t)$ even belongs to W. From the form of the φ_j, we see that on an open dense set of Γ_+ one has

$$\Phi(t) = \{\sum_{s \ge 0} c_s z^{\ell-1-s}\}e^{\sum t_i z^i}. $$

Since $W \cap (z^\ell H_+)^\perp \gamma(t) = \{0\}$, we arrive at

$$z^k\psi_W - (L_W^k)_+(\psi_W) = \sum_{j=1}^m q_j\varphi_j. \tag{32}$$

This equation is part of the system of differential equations for the φ_j as used in [Z]. Recall that φ_j has the form

$$\varphi_j = \{r_j z^{\ell-1} + \text{ lower order terms in } z\}e^{\sum t_i z^i}.$$

Hence,

$$\frac{\partial \varphi_j}{\partial x} = \frac{\partial \varphi_j}{\partial t_1} = \{r_j z^{\ell} + \text{ lower order terms }\}e^{\sum t_i z^i}.$$

On the other hand we know that $\varphi_j(t) - z^k u_j$ belongs to W for all t. Thus also $\frac{\partial \varphi_i}{\partial x}(t)$ belongs to W. In W we have that

$$\frac{\partial \varphi_j}{\partial x} - r_j \psi_W = \{\sum_{s \geq 0} \alpha_s z^{\ell-1-s}\}e^{\sum t_i z^i} \in (z^{\ell} H_+)^{\perp} \gamma$$

and this has to be zero. By definition we have $\varphi_j = r_j \psi_{W_j}$ and differentiation w.r.t. x gives

$$\psi_W = \frac{1}{r_j}\partial(r_j \psi_{W_j}) = (r_j^{-1}\partial r_j)(\psi_{W_j}). \tag{33}$$

Consequently, we have that $\phi_j = r_j \psi_{W_j} = \partial^{-1} r_j \psi_W$. We substitute this in equation (32) and obtain

$$(L_W^k)_{-}(\psi_W) = \{\sum_{j=1}^{m} q_j \partial^{-1} r_j\}\psi_W. \tag{34}$$

Since the pseudodifferential operators act freely on wavefunctions, we see that L_W and the functions q_j and r_j are exactly connected by equation (18) and we can conclude that L_W, the q_j and the r_j form a solution of the m-vector k-constrained KP-hierarchy.

What remains to be shown are the differential equations (20) for the r_j. Recall that $r_j = s_j^{-1}$, hence from (28) and (33) one deduces that s_j satisfies

$$\partial_n(s_j) = (L_{W_j}^n)_{+}s_j = (r_j^{-1}\partial^{-1} r_j L_W^n r_j^{-1}\partial r_j)_{+}s_j.$$

Now apply Lemma 3.1 to this equation and one obtains the equations (20).

Remark 5.1 *Notice that, using the terminology of Oevel and Schief (1993), Aratyn (1997) and Aratyn et al. (1997d), ϕ_j is the squared eigenfunction potential of ψ_W and r_j.*

Next we will show that the $mVkC$-condition is also connected to the rational reductions introduced in Krichever (1995) and Dickey (1995b). For that purpose we assume that W is a plane in $Gr(H)$ that satisfies the $mVkC$-condition, where we choose m to be as minimal as is possible for that plane. Let $L_W = P_W \partial P_W^{-1}$, with P_W of the form

$$P_W = \partial^{\ell} + \sum_{j<\ell} p_j \partial^j = \{1 + \sum_{s<0} p_{\ell+s}\partial^s\}\partial^{\ell}, \tag{35}$$

be the corresponding solution of the KP hierarchy and let $W^1 \subset W$ be the subspace of codimension m such that $W_1 = z^k W^1 \subset W$. Notice first that W_1 is a subspace of W and $z^k W$ of codimension $k+m$ and m, respectively. Hence using Corollary 3.1 iteratively, there exist differential operators L_1 and L_2 of order $k + m$ and m, respectively, such that

$$L_1 \psi_W = \psi_{W_1}, \quad L_2 z^k \psi_W = \psi_{W_1} \tag{36}$$

and that ψ_{W_1} is again a wavefunction. From (36) one immediately deduces that

$$L_W^k = L_2^{-1} L_1. \tag{37}$$

Now use the following lemma:

Lemma 5.1 *Let $L = P\partial^k P^{-1}$ be a pseudodifferential operator of order k and let L_1 and L_2 be differential operators of order $k + m$ and m, respectively, such that $L = L_2^{-1} L_1$. Then one has the following identities:*

$$L_1 (L_2^{-1} L_1)^{i/k} = (L_1 L_2^{-1})^{i/k} L_1, \quad L_2 (L_2^{-1} L_1)^{i/k} = (L_1 L_2^{-1})^{i/k} L_2.$$

Its proof is straightforward (see e.g. Helminck and van de Leur (1997)). Since both ψ_W and ψ_{W_1} are wavefunctions that are connected by equations (36), we find, using (37) and Lemma 5.1, that

$$L_W = (L_2^{-1} L_1)^{1/k} \quad \text{and} \quad L_{W_1} = L_1 (L_2^{-1} L_1)^{1/k} L_1^{-1} = (L_1 L_2^{-1})^{1/k}. \tag{38}$$

Hence

$$\partial_i \psi_{W_1} = ((L_1 L_2^{-1})^{i/k})_+ \psi_{W_1} = ((L_1 L_2^{-1})^{i/k})_+ L_1 \psi_W$$

and on the other hand it is also equal to

$$\partial_i (L_1 \psi_W) = \partial_i (L_1) \psi_W + L_1 ((L_2^{-1} L_1)^{i/k})_+ \psi_W.$$

From these relations one deduces exactly the equations (24). One obtains (25) in a similar fashion (see Helminck and van de Leur (1997). Hence we have obtained the rational reductions of the KP hierarchy. It is not difficult to see that our operators must be coprime, since we have chosen our m to be minimal.

So we have shown in this section that the $mVkC$-condition leads both to the (vector) constrained KP hierarchy and to the rational reductions of KP.

6   The Main Theorem

We will now prove the converse of the result from the foregoing section and thus come to the main theorem:

Theorem 6.1 *Let W be a plane in $Gr(H)$ and let L_W be the corresponding solution of the KP-hierarchy. Then for $m, k \in \mathbb{N}$, $k \neq 0$ and m optimal, the following 3 conditions are equivalent*

(a) The space W satisfies the $mVkC$-condition:

There is a subspace W' of W of codimension m such that $z^k(W') \subset W$.

(b) There exist functions $\{q_j \mid 1 \le j \le m\}$ and $\{r_j \mid 1 \le j \le m\}$ defined on an open dense subset of Γ_+ such that the following conditions are fulfilled:
(i) $\partial_n(q_j) = (L_W^n)_+(q_j)$ for all $n \ge 1$,
(ii) $\partial_n(r_j) = -(L_W^n)_+^*(r_j)$ for all $n \ge 1$,
(iii) $L_W^k = (L_W^k)_+ + \sum_{j=1}^{m} q_j \partial^{-1} r_j$.

(c) There exist coprime differential operators L_1 and L_2 of order $k+m$ and m, respectively, such that the following conditions are fulfilled:
(i) $L_W^k = L_2^{-1} L_1$
(ii) $\partial_i L_1 = ((L_1 L_2^{-1})^{i/k})_+ L_1 - L_1((L_2^{-1} L_1)^{i/k})_+,$
(iii) $\partial_i L_2 = ((L_1 L_2^{-1})^{i/k})_+ L_2 - L_2((L_2^{-1} L_1)^{i/k})_+.$

Proof. We have already shown that (a) implies (b) and (c). So assume first that condition (b) holds. Moreover assume that W in $Gr^{(\ell)}(H)$. Recall that there is a unique pseudodifferential operator P_W of the form (35), such that $\psi_W = P_W(e^{\sum t_i z^i})$. This P_W satisfies the Sato-Wilson equations (11).

Next we consider for each $j, 1 \le j \le m$, the operators Q_j and R_j defined by

$$Q_j := q_j \partial q_j^{-1} \quad \text{and} \quad R_j = r_j^{-1} \partial^{-1} r_j. \tag{39}$$

Since all q_j and r_j satisfy (19), (20), respectively, Q_j and R_j are elementary Bäcklund–Darboux transformations. Applying Theorem 3.1, one obtains that the oscillating functions

$$\psi_{Q_j} = q_j \partial q_j^{-1} \cdot \psi_W \quad \text{and} \quad \psi_{R_j} = r_j^{-1} \partial^{-1} r_j \cdot \psi_W. \tag{40}$$

of type $z^{\ell+1}$, respectively $z^{\ell-1}$, are wavefunctions of planes W_{Q_j} and W_{R_j}. Moreover we have the following codimension 1 inclusions:

$$W_{Q_j} \subset W \quad \text{and} \quad W \subset W_{R_j}.$$

The relation (b) (iii) leads to

$$L_W^k(\psi_W) = z^k \psi_W$$
$$= (L_W^k)_+(\psi_W) + \sum_{j=1}^{m} q_j \partial^{-1} r_j \psi_W$$
$$= (L_W^k)_+(\psi_W) + \sum_{\substack{j \\ r_j \ne 0}} q_j r_j \psi_{R_j}.$$

Thus we see with the usual density argument that

$$z^k W \subset W + \sum_j W_{R_j} = \sum_j W_{R_j} = \tilde{W}.$$

Since each W has codimension one in W_{R_j}, we see that the codimension of W in $\tilde{W}$ is $\leq m$. As one assumes in the conditions of the Theorem that m is optimal, this codimension is exactly m. Let W_1 be the orthocomplement of W in $\tilde{W}$ and $p_1 : H \to W_1$ the orthogonal projection on W_1. Inside W we consider

$$W^1 = \{w \in W \mid p_1(z^k w) = 0\}.$$

Since $\dim(W_1) = m$, we see that W^1 is a subspace of W of codimension m and by construction $z^k W^1 \subset W$. This completes the proof (b)$\Rightarrow$(a).

Next assume that (c) holds. Let ψ_1 be the oscillating function $L_1 \psi_W$, then by using Lemma 5.1:

$$(L_1 L_2^{-1})^{1/k} \psi_1 = (L_1 L_2^{-1})^{1/k} L_1 \psi_W = L_1 (L_2^{-1} L_1)^{1/k} \psi_W = z L_1 \psi_W = z \psi_1.$$

Now consider

$$\begin{aligned}
\partial_i \psi_1 &= \partial_i(L_1) \psi_W + L_1 \partial_i \psi_W \\
&= (((L_1 L_2^{-1})^{i/k})_+ L_1 - L_1 ((L_2^{-1} L_1)^{i/k})_+ + L_1 ((L_2^{-1} L_1)^{i/k})_+) \psi_W \\
&= ((L_1 L_2^{-1})^{i/k})_+ L_1 \psi_W \\
&= ((L_1 L_2^{-1})^{i/k})_+ \psi_1.
\end{aligned}$$

Hence ψ_1 is again a wavefunction of the KP hierarchy. If we let W_1 be the closure of the span of the $\psi_1(t, z)$ then $\psi_{W_1} = \psi_1$. Since $z^k \psi_W$ is also a wavefunction,

$$L_2 z^k \psi_W = \psi_{W_1}.$$

Thus we see with the usual density argument that

$$\begin{aligned}
W_1 &\subset z^k W \quad \text{of codimension } m, \\
W_1 &\subset W \quad \text{of codimension } k + m.
\end{aligned} \tag{41}$$

Hence $W^1 = z^{-k} W_1$ is a subset of W of codimension m such that $z^k W^1 \subset W$. Since our differential operators are coprime, one cannot find lower order operators M_1 and M_2 such that $L_W = M_2^{-1} M_1$. Therefore there is no smaller subspace W_1 and no smaller m such that (41) is satisfied. $\qquad\square$

As a consequence of this, we obtain that in the Segal–Wilson setting, the vector constrained KP hierarchy and rational reduction of the KP hierarchy define the same reduction of the KP hierarchy.

Acknowledgements. JvdL would like to thank the participants of the workshop, especially Henrik Aratyn, Maarten Bergvelt, John Harnad, Manuel Mañas and Alex Kasman for useful discussions. We both want to thank Henrik Aratyn for sending us Aratyn (1997). In this article, published in these proceedings, he presents a different proof that the vector constrained KP and the rational reduction's of KP describe the same hierarchies. His proof is based on kernels of differential operators and properties of Wronskians and is quite different from the proof given in this paper.

References

Aratyn H. (1997): The constrained KP hierarchy as ratio of differential operators. in the proceedings of this workshop

Aratyn H. (1995): Integrable Lax hierarchies, their symmetry reductions and multi-matrix models. hep-th 9503211

Aratyn H., Ferreira L., Gomes J.F., Zimerman A.H. (1997a): Constrained KP models as integrable matrix hierarchies. Journ. Math. Phys. **38**,1559– (hep-th 9509096)

Aratyn H., Gomes J.F., Zimerman A.H. (1995a): Affine Lie algebraic origin of constrained KP hierarchies. Journ. Math. Phys **36**, 3419– (hep-th 9408104)

Aratyn H., Nissimov E., Pacheva S. (1997b): Virasoro symmetry of constrained KP hierarchies. Phys. Letters **228A**, 164– (hep-th 9602068)

Aratyn H., Nissimov E., Pacheva S. (1997c): Constrained KP hierarchies: Additional symmetries, Darboux-Bäcklund solutions and relations to multi-matrix models. Int. J. Mod. Phys. **A12**, 1265–1340. (hep-th 9607234)

Aratyn H., Nissimov E., Pacheva S. (1997d): Methods of squared eigenfunction potentials in integrable hierarchies of KP type. solv-int 9701017

Aratyn H., Nissimov E., Pacheva S., Zimerman A.H. (1995b): Two-matrix string model as constrained (2+1)-dimensional integrable system. Int.J. Mod Phys. **A10**, 2537– (hep-th 9407017)

Cheng Y. (1992): Constraints of the Kadomtsev-Petviashvili hierarchy. Journ. Math.Phys. **33**, 3747–3782

Cheng Y. (1995): Modifying the KP, the nth constrained KP hierarchies and their Hamiltonian structures. Commun. Math. Phys. **171**, 661–682

Cheng Y., Strampp W., Zhang Y-J. (1995a): Bilinear Bäcklund transformations for the KP and k-constrained KP hierarchy. Acta Appl. Math. **41**, 341–348

Cheng Y., Strampp W., Zhang B. (1995ab): Constraints of theKP hierarchy and multilinear forms. Commun. Math. Phys. **168**, 117–135

Cheng Y., Zhang Y-J. (1994): Bilinear equations for the constrained KP hierarchy. Inverse Problems **10**. L11–L17

Date E., Jimbo M., Kashiwara M., Miwa T. (1983): Transformation groups for soliton equations. in: *Nonlinear integral systems – classical theory and quantum theory* eds. Jimbo M. and Miwa T., World Scientific, 39–120

Dickey L.A. (1995a): On the constrained KP. Letters Math. Phys **34**, 379–384

Dickey L.A. (1995b): On the constrained KP hierarchy II. Letters Math. Phys **35**, 229–236

Dickey L., Strampp W. (1996): On new identities for KP Baker functions and their application to constrained hierarchies, preprint.

Helminck G.F., Leur J.W. van de (1997): An analytic description of the vector constrained KP hierarchy. preprint solv-int 9706004

Helminck G.F., Post G.F. (1988): Geometrical interpretation of the bilinear equations for the KP hierarchy. Letters Math. Phys. **16**, 359–364

Krichever I. (1995): General rational reductions of the KP hierarchy and their symmetries. Funct. Anal. Appl. **29**, 75–80

Leur J. van de (1996a): A geometrical interpretation of the constrained KP hierarchy. preprint

Leur J. van de (1996b): The vector constrained KP hierarchy and Sato's Grassmannian. preprint to appear in Journ. Geom. Phys. (q-alg 9609001)

Liu Q.P. (1996) Bi-hamiltonian structures of coupled AKNS hierarchy and coupled Yajima–Oikawa hierarchy. Journ. Math. Phys., 2307–2314

Loris I., Willox R. (1996): Bilinear form and solutions of the k-constrained Kadomtsev–Petviashvili hierarchy. preprint.

Loris I., Willox R. (1997): On solutions of constrained KP equations. Journ. Math. Phys. **38**, 283–291

Mas J., Ramos E. (1995): The constrained KP hierarchy and the generalised Miura transformation. q-alg 9501009

Oevel W., Schief W. (1993): Darboux theorems and the KP hierarchy. in: *applications of analytic and geometric Methods in Differential Equations* (Proceedings of the NATO Advanced Research Workshop, Exeter, 14–17 July 1992, UK) P.A. Clarkson (ed.), Kluwer Publ., 193–206

Oevel W. Strampp W. (1993): Constrained KP hierarchies and bi–hamiltonian structures. Commun. Math. Phys. **157**, 51–81

Oevel W. Strampp W. (1995): Wronskian solutions of the constrained Kadomtsev–Petviashvili hierarchy. Journ. Math. Phys. **37**, 6213–6219

Orlov A.Yu. (19): Symmetries for unifying different soliton systems into a single integrable hierarchy. preprint IINS/Oce04/03

Orlov A.Yu. (1991): Volterra operator algebra Zero curvature representation. Universality of KP. in: *Nonlinear Processes in physics*, proceeding of the III Potsdam-V Kiev Workshop at Clarkson Univ., Potsdam , N.Y., USA, eds. A.S. Fokas, D.J. Kaup, A.C. Newell and V.E. Zakharov, Springer series in Nonlinear Dynamics, Springer Verlag, Berlin, 126–131

Sato M. (1981): Soliton equations as dynamical systems on infinite dimensional Grassmann manifolds. Res. Inst. Math. Sci. Kokyuroku **439** 30–46

Sidorenko J., Strampp W. (1993): Multicomponent integrable reductions in the Kadomtsev–Petviashvilli hierarchy. Journ. Math Phys.**34**, 1429–1446

Segal G., Wilson G. (1985): Loop groups and equations of KdV type. Publ. Math IHES **63**, 1–64

Zhang Y-J. (1996): On Segal-Wilson's construction for the τ-functions of the constrained KP hierarchies. Letters Math. Physics **36**, 1–15

Zhang Y.-J, Cheng Y. (1994): Solutions for the vector k-constrained KP hierarchy. Journ. Math. Phys. **35**, 5869–5884

Matrix Membranes and Integrability

Cosmas Zachos[1], David Fairlie[2], and Thomas Curtright[3]

[1] High Energy Physics Division, Argonne National Laboratory, Argonne, IL 60439-4815, USA   zachos@hep.anl.gov
[2] Department of Mathematical Sciences, University of Durham, Durham, DH1 3LE, UK   David.Fairlie@durham.ac.uk
[3] Department of Physics, University of Miami, Box 248046, Coral Gables, FL 33124, USA   curtright@phyvax.ir.Miami.edu

Abstract. This is a pedagogical digest of results reported in [Curtright, Fairlie, & Zachos 1997], and an explicit implementation of Euler's construction for the solution of the Poisson Bracket dual Nahm equation. But it does not cover 9 and 10-dimensional systems, and subsequent progress on them [Fairlie 1997]. Cubic interactions are considered in 3 and 7 space dimensions, respectively, for bosonic membranes in Poisson Bracket form. Their symmetries and vacuum configurations are explored. Their associated first order equations are transformed to Nahm's equations, and are hence seen to be integrable, for the 3-dimensional case, by virtue of the explicit Lax pair provided. Most constructions introduced also apply to matrix commutator or Moyal Bracket analogs.

1   Introduction

A proposal for non-perturbative formulation of M-theory [Banks et al. 1997] has encouraged a reappraisal of matrix membrane theory [Collins & Tucker 1976, Hoppe 1982]. Symmetry features of membranes and their connection to matrix models [Hoppe 1982, Floratos et al. 1989, Fairlie, Fletcher, & Zachos 1989, Floratos 1989, Fairlie, Fletcher, & Zachos 1990] have been appreciated for quite some time. Effectively, infinite-N quantum mechanics matrix models (presented as a restriction of SU(∞) Yang-Mills theories) amount to membranes, by virtue of the connection between SU(N) and area-preserving diffeomorphisms (*Sdiff*) generated by Poisson Brackets: in these, "color" algebra indices Fourier-transform to "membrane" sheet coordinates. The two are underlain and linked by *Moyal Brackets*, the universal associative generalization of Poisson Brackets.

Below, we care to introduce novel Poisson Bracket interactions for a bosonic membrane embedded in 3-space

$$\mathcal{L}_{IPB} = \frac{1}{3}\epsilon^{\mu\nu\kappa}X^{\mu}\{X^{\nu},\ X^{\kappa}\}\ , \tag{1}$$

which are restrictions of the Moyal Bracket generalization

$$\mathcal{L}_{IMB} = \frac{1}{3}\epsilon^{\mu\nu\kappa}X^{\mu}\{\{X^{\nu},\ X^{\kappa}\}\}\ , \tag{2}$$

which, in turn, also encompasses the plain matrix commutator term

$$\mathcal{L}_{IC} = \frac{1}{3}\epsilon^{\mu\nu\kappa}X^{\mu}[X^{\nu},\ X^{\kappa}]\ . \tag{3}$$

The structure of (1) may be recognized as that of the interaction term $\epsilon_{ijl}\phi^{i}\epsilon^{\mu\nu}\partial_{\mu}\phi^{j}\partial_{\nu}\phi^{j}$ of the 2-dimensional SO(3) pseudodual chiral σ-model of Zakharov & Mikhailov (1978)—this is a limit of the WZWN interaction term, where the integer WZWN coefficient goes to infinity while the coupling goes to zero, such that the product of the integer with the cube of the coupling is kept constant [Curtright & Zachos 1995]. (N.B. Contrast to the interaction of a different model [Plebanski et al. 1996], with derivative structure 0-2-2, which could be regarded as a large-N limit of the pseudodual interaction exemplified above by SO(3).)

Alternatively, the structure of (3) is linked to what remains of the gauge theory instanton density,

$$K^{0} = \epsilon^{\mu\nu\kappa}\mathrm{Tr}A_{\mu}\left(\partial_{\nu}A_{\kappa} - \frac{1}{4}[A_{\nu}, A_{\kappa}]\right), \tag{4}$$

in the standard space-invariant limit (where the first term vanishes).

There is some formal resemblance to membrane interaction terms introduced in Zaikov (1991) (in that case a quartic in the X^{μ}s), which, in turn, reflect the symplectic twist of topological terms of Biran et al. (1987) for self-dual membranes. But, unlike those interactions, the cubic terms considered here do not posit full Lorentz invariance beyond 3-rotational invariance: they are merely being considered as quantum mechanical systems with internal symmetry. One may therefore expect this fact to complicate supersymmetrization.

We also succeed in introducing analogous trilinear interactions for membranes embedded in 7-space, which evince similarly interesting properties.

In what follows, after a brief review of some matrix membrane technology, we explore the symmetry features of the new terms, and the remarkable symmetry of the corresponding vacuum configurations; we describe classical configurations of the Nahm type, which we find to be integrable in 3d, as in the conventional membrane models. When we interchange the rôle of dependent and independent variables of the 3d PB Nahm equations, we detail how these "dual" equations are solved by Euler's construction, based on harmonic scalar functions. Our discussion will concentrate on Poisson Brackets, but the majority of our results carry over to the Moyal Bracket and matrix commutator cases, by dint of the underlying formal analogy.

However, even though mostly integrable first-order equations are studied here, it should be borne in mind that the behaviour of the generic solutions to the second-order equations of motion for such systems is often chaotic. For example, in the case of QM matrix models (Yang-Mills on a finite gauge group, with fields dependent only upon time), characteristic features of chaotic behaviour were observed on the solutions of the *second-order* equations of motion [Matinyan et al. 1981]. Still, it is not known whether this ergodicity persists in the large N limit, i.e. the corresponding PB system.

Subtler topological considerations of special features for various membranes are not addressed here. Nontrivial boundary terms, e.g. of the type linked to D-branes, are also not considered.

2   Review of Brackets, Matrix Commutators, and Matrix Membrane Actions

Poisson Brackets, Moyal Brackets, and matrix commutators are inter-related antisymmetric derivative operations, sharing similar properties, such as integration by parts, associativity (hence comportance to the Jacobi identity), suitable Leibniz chain rules, etc. They are all representable as commutators of associative operators. Much of their technology is reviewed in [Moyal 1949; Fairlie, Fletcher, & Zachos 1989; Fairlie, Fletcher, & Zachos 1990; Hoppe 1990].

Poisson Brackets act on the "classical phase-space" of Fourier-transformed color variables, with membrane coordinates $\xi = \alpha,\ \beta$,

$$\{X^\mu,\ X^\nu\} = \frac{\partial X^\mu}{\partial \alpha}\frac{\partial X^\nu}{\partial \beta} - \frac{\partial X^\mu}{\partial \beta}\frac{\partial X^\nu}{\partial \alpha}\ . \tag{5}$$

This may be effectively regarded as the infinitesimal canonical transformation on the coordinates ξ of X^ν, generated by $\nabla X^\mu \times \nabla$, s.t. $(\alpha, \beta) \mapsto (\alpha - \partial X_\mu/\partial \beta,\ \beta + \partial X_\mu/\partial \alpha)$, which preserves the membrane area element $d\alpha d\beta$. This element is referred to as a *symplectic form* and the class of transformations that leaves it invariant specifies a symplectic geometry; the area preserving diffeomorphisms are known as *Sdiff*.

PBs correspond to $N \to \infty$ matrix commutators [Hoppe 1982]. However, there is a generalization which covers both finite and infinite N. The virtually unique associative generalization of PBs is the Moyal Bracket [Moyal 1949],

$$\{\{X^\mu,\ X^\nu\}\} = \frac{1}{\lambda}\sin\left(\lambda\frac{\partial}{\partial\alpha}\frac{\partial}{\partial\beta'} - \lambda\frac{\partial}{\partial\beta}\frac{\partial}{\partial\alpha'}\right)\ X^\mu(\xi)X^\nu(\xi')\bigg|_{\xi'=\xi}\ . \tag{6}$$

For $\lambda = 2\pi/N$, Fairlie, Fletcher, & Zachos 1989 demonstrate that the Moyal Bracket is essentially equivalent to the commutator of SU(N) matrices—or subalgebras of SU(N), depending on the topology of the corresponding membrane surface involved in the Fourier-transform of the color indices [Fairlie, Fletcher, & Zachos 1990; Kim & Rey 1997].

In the limit $\lambda \to 0$, the Moyal Bracket goes to the PB, i.e. λ may be thought of as $\hbar$. Thus, PBs are seen to represent the infinite N limit. This type of identification was first noted without benefit of the MB construction by Hoppe 1982 on a spherical membrane surface; the foregoing MB limit argument was first formulated on the torus [Fairlie, Fletcher, & Zachos 1989], but readily extends to other topologies [Fairlie, Fletcher, & Zachos 1989; Fairlie, Fletcher, & Zachos 1990; Kim & Rey 1997].

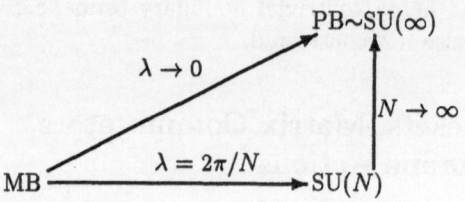

It would suffice to simply treat these most general Moyal Brackets; but their technical manipulations are sometimes more involved and less familiar, so that we also cover the matrix commutator and PB special cases as well, redundancy outweighed by pedagogy.

Floratos et al. 1989 utilize the abovementioned identification of $SU(\infty)$ with *Sdiff* on a 2-sphere, to take the large N limit of $SU(N)$ gauge theory and produce membranes. This procedure was found to be more transparent on the torus [Fairlie, Fletcher, & Zachos 1990]: the Lie algebra indices Fourier-conjugate to surface coordinates, and the fields are rescaled Fourier transforms of the original $SU(N)$ fields. The group composition rule for them is given by the PBs and the group trace by surface integration,

$$[A_\mu, A_\nu] \mapsto \{a_\mu, a_\nu\} ; \tag{7}$$

$$F_{\mu\nu} = \partial_\mu A_\nu - \partial_\nu A_\mu + [A_\mu, A_\nu] \mapsto f_{\mu\nu}(\alpha, \beta) = \partial_\mu a_\nu - \partial_\nu a_\mu + \{a_\mu, a_\nu\} ; \tag{8}$$

$$\mathrm{Tr} F^{\mu\nu} F_{\mu\nu} \mapsto -\frac{N^3}{64\pi^4} \int d\alpha d\beta \, f^{\mu\nu}(\alpha, \beta) \, f_{\mu\nu}(\alpha, \beta) . \tag{9}$$

But the large N limit need *not* really be taken to produce sheet actions. The Lagrangian with the Moyal Bracket supplanting the Poisson Bracket is itself a gauge-invariant theory, provided that the gauge transformation also involves the Moyal instead of the Poisson Bracket:

$$\delta a_\mu = \partial_\mu \Lambda - \{\{\Lambda, a_\mu\}\} , \tag{10}$$

and hence, by virtue of the Jacobi identity,

$$\delta f_{\mu\nu} = -\{\{\Lambda, f_{\mu\nu}\}\} . \tag{11}$$

Color invariance then follows,

$$\delta \int d\alpha d\beta \, f^{\mu\nu} f_{\mu\nu} = -2 \int d\alpha d\beta \, f^{\mu\nu} \{\{\Lambda, f_{\mu\nu}\}\} = 0 . \tag{12}$$

The relevant manipulations are specified in Fairlie, Fletcher, & Zachos 1990: the last equality is evident by integrations by parts, where the surface term is discarded—or nonexistent if the color membrane surface is closed[1]. For $\lambda = 2\pi/N$, this is equivalent to a conventional $SU(N)$ commutator gauge theory.

Consider, with Hoppe 1982, the $SU(\infty)$ Yang-Mills lagrangian; and trivialize all space dependence (through dimensional reduction), leaving only time dependence, while preserving all the color-Fourier-space (membrane coordinates $\xi = \alpha, \beta$) dependence of the gauge fields, which are now denoted $X^\mu(t, \alpha, \beta)$. Fix the gauge to $X^0 = 0$, and consider μ, ν to henceforth only range over spacelike values.

The Yang-Mills lagrangian density now reduces to the bosonic membrane lagrangian density

$$\mathcal{L}_{PB} = \frac{1}{2}(\partial_t X^\mu)^2 - \frac{1}{4}\{X^\mu, X^\nu\}^2. \tag{13}$$

The PB is also the determinant of the tangents to the membrane, so that the conventional "potential term" was identified in Fairlie, Fletcher, & Zachos 1990 as the Schild-Eguchi string lagrangian density [Schild 1977] (sheet area squared instead of area), $\{X_\mu, X_\nu\}\{X_\mu, X_\nu\}$. It can be seen that the equations of motion of such a string action contain those of Nambu's action.

Note that, fixing the gauge $X_0 = 0$ preserves the global color invariance, i.e. with a time-independent parameter $\Lambda(\alpha, \beta)$. The action is then invariant under

$$\delta X^\mu = \{\Lambda, X^\mu\}. \tag{14}$$

By Noether's theorem, this implies the time invariance of the color charge,

$$\mathcal{Q}_\Lambda = \int d\alpha d\beta \, \Lambda(\alpha, \beta) \, \{\partial_t X^\mu, X^\mu\}. \tag{15}$$

The same also works for the Moyal case [Fairlie, Fletcher, & Zachos 1990]. The corresponding Moyal Schild-Eguchi term was utilized to yield a "star-product-membrane" [Hoppe 1990],

$$\mathcal{L}_{MB} = \frac{1}{2}(\partial_t X^\mu)^2 - \frac{1}{4}\{\{X^\mu, X^\nu\}\}^2, \tag{16}$$

invariant under

$$\delta X^\mu = \{\{\Lambda, X^\mu\}\}. \tag{17}$$

As argued, this includes the commutator case (QM matrix model),

$$\mathcal{L}_C = \frac{1}{2}(\partial_t X^\mu)^2 - \frac{1}{4}[X^\mu, X^\nu]^2, \tag{18}$$

invariant under

$$\delta X^\mu = [\Lambda, X^\mu]. \tag{19}$$

---

[1] But note this topological term may be nontrivial for D-membranes.

3   Cubic Terms for 3 Dimensions

By suitable integrations by parts, it is straightforward to check that the cubic terms (1,2,3) in the respective actions, $\int dt d\alpha d\beta \, \mathcal{L}$, are 3-rotational invariant, as well as time-translation invariant and translation symmetric. They are also global color invariant, as specified above.

Now, further consider a plain mass term in the action[2],

$$\mathcal{L}_{3dPB} = \frac{1}{2}(\partial_t X^\mu)^2 - \frac{1}{4}\{X^\mu,\ X^\nu\}^2 - \frac{m}{2}\epsilon^{\mu\nu\kappa}X^\mu\{X^\nu,\ X^\kappa\} - \frac{m^2}{2}(X^\mu)^2. \quad (20)$$

The second order equation of motion,

$$\partial_t^2 X^\mu = -m^2 X^\mu - \frac{3m}{2}\epsilon^{\mu\nu\kappa}\{X^\nu,\ X^\kappa\} - \{X^\nu, \{X^\mu,\ X^\nu\}\}, \quad (21)$$

follows not only from extremizing the action, but also results from a first-order equation of the Nahm (self-dual) type [Nahm 1983], albeit complex,

$$\partial_t X^\mu = imX^\mu + \frac{i}{2}\epsilon^{\mu\nu\kappa}\{X^\nu,\ X^\kappa\}. \quad (22)$$

These equations hold for PBs, as well as for MBs and matrix commutators.

For solutions of this first-order equation, the conserved energy vanishes. In general, however, such solutions are not real, and do not provide absolute minima for the action—the reader may consider the simple harmonic oscillator to illustrate the point. Nonetheless, the lagrangian density can be expressed as a sum of evocative squares with positive relative signs, since the potential in (20) is such a sum,

$$\mathcal{L}_{3dPB} = \frac{1}{2}(\partial_t X^\mu)^2 - \frac{1}{2}\left(mX^\mu + \frac{1}{2}\epsilon^{\mu\nu\kappa}\{X^\nu,\ X^\kappa\}\right)^2. \quad (23)$$

By integration by parts in the action $\int dt d\alpha d\beta \, \mathcal{L}_{3dPB}$, the lagrangian density itself can then be altered to

$$\mathcal{L}_{3dPB} \cong -\frac{1}{2}\left(i\partial_t X^\mu + mX^\mu + \frac{1}{2}\epsilon^{\mu\nu\kappa}\{X^\nu,\ X^\kappa\}\right)^2, \quad (24)$$

just like the conventional bosonic membrane lagrangian density[3]. Naturally, the complex-conjugate versions of the above are equally valid.

---

[2] N.B. Of the type that may arise as a remnant of space gradients in compactified dimensions.

[3] The congruence symbol, $\cong$, denotes equivalence up to surface terms, which, e.g., vanish for a closed surface; again, consideration of D-membranes would proceed separately.

4 Vacuum Configurations and their Symmetry

The minimum of the conventional matrix membrane trough potential favors alignment of the dynamical variables X^μs. The mass parameter introduced above parameterizes a partial trough symmetry breaking[4], but does not lift "dilation" invariance, seen as follows.

The static (t-independent) minima for the action (vacuum configurations) are solutions of

$$mX^\mu + \frac{1}{2}\epsilon^{\mu\nu\kappa}\{X^\nu,\ X^\kappa\} = 0 \ . \tag{25}$$

The previously considered case, $m = 0$, is easily solved by "color-parallel" configurations. But for $m \neq 0$, the static solutions must lie on a 2-sphere, since from the previous equation

$$X^\mu \frac{\partial X^\mu}{\partial \alpha} = 0 = X^\mu \frac{\partial X^\mu}{\partial \beta} \ , \tag{26}$$

so

$$X^\mu X^\mu = R^2 \ , \tag{27}$$

an unspecified constant[5]. However, from (25), note that both m and also R, the scale of the X^μs, can be absorbed in the membrane coordinates ξ and will not be specified by the solution of (25).

Indeed, solving for one coordinate component on this sphere, say

$$X(Y, Z) = \pm\sqrt{R^2 - Y^2 - Z^2} \ , \tag{28}$$

reduces the three equations (25) to one. Namely,

$$\{Z, Y\} = m\sqrt{R^2 - Y^2 - Z^2} \ , \tag{29}$$

on the positive X branch ($m \mapsto -m$ on the negative X branch). This last equation is solved by

$$Z = \alpha \ , \qquad Y = \sqrt{R^2 - \alpha^2}\sin(m\beta) \ . \tag{30}$$

One can then interpret $m\beta$ as the usual azimuthal angle around the Z-axis. Hence, $-\pi/2 \le m\beta \le \pi/2$ and $-R \le \alpha \le R$ covers the $X \ge 0$ hemisphere completely. The other hemisphere is covered completely by the negative X branch. Since R is not fixed, it amounts to an unlifted residual trough dilation degeneracy.

All static solutions are connected to this explicit one by rescaling R and exploiting the equation's area-preserving diffeomorphism invariance for $\xi = (\alpha, \beta)$.

---

[4] Curtright & McCarty (1989) have considered a system equivalent to the limit of constant $X^3 =$ in this model.

[5] Hence $\epsilon^{\mu\nu\kappa}X^\mu\{X^\nu,\ X^\kappa\} = -2mR^2$.

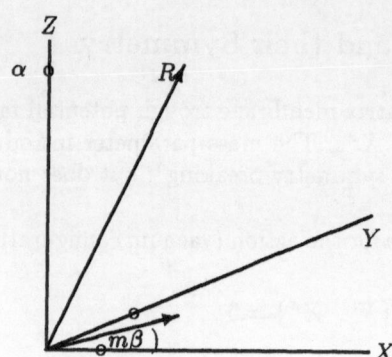

5   Nahm's Equation and its Lax Pairs

The first order equation, (22), simplifies upon changing variables to

$$\tau = \frac{e^{imt}}{m}\,,\qquad\qquad X^\mu = e^{imt}Y^\mu\,, \tag{31}$$

and reduces to the conventional PB version [Ward 1990, Floratos & Leontaris 1989] of Nahm's equation [Nahm 1983],

$$\partial_\tau Y^\mu = \frac{1}{2}\epsilon^{\mu\nu\kappa}\{Y^\nu,\,Y^\kappa\}\,. \tag{32}$$

This one does have real solutions, and can be linearized by Ward's transformation [Ward 1990]. However, the action (23) does not reduce to the conventional one upon these transformations[6].

Moreover, note

$$\partial_\tau Y^\mu \partial_\tau Y^\mu = \frac{\partial(Y^1,Y^2,Y^3)}{\partial(\tau,\alpha,\beta)}\,, \tag{33}$$

$$\partial_\tau Y^\mu \partial_\xi Y^\mu = 0\,. \tag{34}$$

One may further utilize the cube root of unity, $\omega = \exp(2\pi i/3)$, to recast[7] (32),

$$L \equiv \omega Y^1 + \omega^2 Y^2 + Y^3\,,\quad \overline{L} = \omega^2 Y^1 + \omega Y^2 + Y^3\,,\quad M \equiv Y^1 + Y^2 + Y^3\,, \tag{35}$$

$$\omega(\omega - 1)\,\partial_\tau L = \{M\,,\,L\}\,,\qquad \omega(\omega - 1)\,\partial_\tau \overline{L} = -\{M\,,\,\overline{L}\}\,, \tag{36}$$

$$\omega(\omega - 1)\,\partial_\tau M = \{L\,,\overline{L}\}\,, \tag{37}$$

which thus yields an infinite number of complex time-invariants,

$$Q_n = \int d\alpha d\beta\; L^n\,, \tag{38}$$

---

[6] Likewise, the second order equations of motion only reduce to $\partial_\tau^2 Y^\mu = \{Y^\nu,\{Y^\mu,Y^\nu\}\} + \frac{3}{\tau}(\partial_\tau Y^\mu - \frac{\epsilon^{\mu\nu\kappa}}{2}\{Y^\nu,Y^\kappa\})$.

[7] N.B. $\omega(\omega - 1)$ is pure imaginary.

for arbitrary integer power n, as the time derivative of the integrand is a surface term. (This is in complete analogy with the standard case of commutators.) The link to classical integrability is discussed next.

Equations (36,37) amount to one complex and one real equation, but these are known to be further capable of compacting into just one by virtue of an arbitrary real spectral parameter ζ, e.g. as introduced in Floratos & Leontaris 1989:

$$H \equiv \frac{i}{\sqrt{2}\omega(\omega - 1)} \left(\zeta L - \frac{\overline{L}}{\zeta} \right) , \qquad K \equiv i\sqrt{2}M + \zeta L + \frac{\overline{L}}{\zeta} , \qquad (39)$$

$$\partial_\tau K = \{H, K\} . \qquad (40)$$

Likewise, this Lax pair[8], analogous to Hitchin 1983, leads to a one-parameter family of time-invariants[9],

$$Q_n(\zeta) = \int d\alpha d\beta \ K^n , \qquad (41)$$

conserved for all n and ζ.

It also yields the usual Lax isospectral flow. PBs (and, *mutatis mutandis*, MBs) can be recast into commutators of suitable associative operators

$$\mathcal{K} \equiv \nabla K \times \nabla, \qquad \mathcal{H} \equiv \nabla H \times \nabla, \qquad \nabla \equiv \left(\frac{\partial}{\partial \alpha}, \frac{\partial}{\partial \beta} \right) , \qquad (42)$$

such that:

$$\partial_\tau \mathcal{K} = \mathcal{H}\mathcal{K} - \mathcal{K}\mathcal{H} . \qquad (43)$$

As a consequence, the spectrum of $\mathcal{K}$ is preserved upon time evolution by the (pure imaginary) $\mathcal{H}$:

$$\partial_\tau \psi = \mathcal{H}\psi , \qquad (44)$$

since time-differentiating

$$\mathcal{K}\psi = \lambda\psi \qquad (45)$$

and applying (43) yields

$$(\partial_\tau \lambda) \ \psi = (\partial_\tau \mathcal{K})\psi + \mathcal{K}\partial_\tau\psi - \lambda \ \partial_\tau\psi = 0 . \qquad (46)$$

This isospectral flow then provides integrability for (22), as in the case of the matrix commutator Nahm equation.

The discussion so far also carries over to plain matrix commutators or Moyal Brackets as well, with suitable adaptations for the associative operators involved, e.g. a $\star$-product structure [Fairlie, Fletcher, & Zachos 1989; Hoppe 1990],

---

[8] Note the wave solutions $H = \alpha$, $K = f(\beta + \tau)$.

[9] L. Dickey calls our attention to the parameter introduced in the Lax pair for the generalized Euler equations by S. Manakov, Funct Anal Appl **10** (1976) 328, which parallels that in Hitchin's Lax pair for the Nahm equations.

$$\mathbb{K} \equiv \frac{1}{2i\lambda} \, e^{i\lambda \nabla' \times \nabla} \, K(\xi') \Big|_{\xi'=\xi} . \tag{47}$$

For the rest of this section, we shall restrict our attention to the PB case only. Ward 1990 has solved the PB Nahm equation (32) implicitly through twistor linearization. Another solution procedure for (32) may be found by interchanging the rôles of dependent and independent variables: the equations then take the dual form

$$\frac{\partial \tau}{\partial Y^1} = \frac{\partial \alpha}{\partial Y^2} \frac{\partial \beta}{\partial Y^3} - \frac{\partial \alpha}{\partial Y^3} \frac{\partial \beta}{\partial Y^2} , \tag{48}$$

together with cyclic permutations, i.e.

$$\partial_\mu \tau = \epsilon^{\mu\nu\rho} \partial_\nu \alpha \, \partial_\rho \beta . \tag{49}$$

Cross-differentiation produces integrability conditions

$$\partial_\mu (\partial_\kappa \alpha \, \partial_\mu \beta - \partial_\mu \alpha \, \partial_\kappa \beta) = 0 . \tag{50}$$

Another evident consistency condition is

$$\partial^2 \tau = 0 . \tag{51}$$

In fact, any harmonic function $\tau(Y^1, Y^2, Y^3)$, readily yields α, β. Euler appreciated (1770) that any continuous differentiable divergenceless vector field may be represented locally as a cross product of two gradients [Ericksen 1960]. The problem of solving the inverse Nahm equation (49) then reduces to Euler's construction, given an arbitrary harmonic function τ.

To determine scalar fields α and β, note that

$$\nabla \alpha \cdot \nabla \tau = 0 = \nabla \beta \cdot \nabla \tau , \tag{52}$$

so that $\nabla \tau$ lies on surfaces of constant α and β. One may then choose the surface of constant α through $\nabla \tau$ arbitrarily, and further choose an arbitrary continuous vector field $\mathbf{V}$ s.t. $\mathbf{V} \cdot \nabla \alpha \neq 0$. One may thus integrate on the constant-α surface to obtain β:

$$\beta = \int d\mathbf{Y} \, \cdot \frac{\nabla \tau \times \mathbf{V}}{\mathbf{V} \cdot \nabla \alpha} . \tag{53}$$

Evidently, solutions of these dual equations

$$\partial_\mu f = \epsilon^{\mu\nu\rho} \partial_\nu g \, \partial_\rho h , \tag{54}$$

for $f(Y^1, Y^2, Y^3)$ produce constants of the motion $\int d^2\xi \, f$, beyond those already found by the Lax procedure, for the *original* equation (32), in illustration of a phenomenon noted in Fairlie & Strachan (1996), as it is straightforward to verify that

$$\frac{df}{d\tau} = \{g, h\} . \tag{55}$$

From the foregoing discussion, f need only solve Laplace's equation (51): any harmonic function $f(Y^1, Y^2, Y^3)$ yields a conserved density for (32) by also satisfying (54). To belabor the point, by virtue of Helmholtz's theorem, a divergenceless 3-vector ∇f is representable as a curl of another vector $\mathbf{A}$. On the other hand, an arbitrary 3-vector can also be represented in terms of three scalars ("Monge potentials") by means of the non-unique "Clebsch decomposition" of that vector as $\mathbf{A} = g\nabla h + \nabla u$, in a trivial extension of Euler's construction just outlined [Ericksen 1960].

6 Membrane Embedding in 7 Dimensions

Remarkably, the same type of cubic interaction term may also be introduced for a membrane embedded in 7 space dimensions. An antisymmetric, self-dual 4-tensor in 8 dimensions, $f_{\mu\nu\rho\sigma}$ was invoked by Corrigan et al. 1982 as an 8-dimensional analog of the 4-dimensional fully antisymmetric tensor $\epsilon_{\mu\nu\rho\sigma}$. Some useful technology for the manipulation of this tensor (which has 35 nonzero components, is linked to Cayley's octonionic structure constants, and is invariant under a particular $SO(7)$ subgroup of $SO(8)$) can be found in Dündarer et al. (1984); in particular, the identity

$$f^{0\mu\nu\kappa} f^{0\mu\lambda\rho} = f^{\nu\kappa\lambda\rho} + \delta^{\nu\lambda}\delta^{\kappa\rho} - \delta^{\kappa\lambda}\delta^{\nu\rho} \ . \tag{56}$$

By analogy with (32), we postulate a first-order equation[10],

$$\partial_\tau Y^\mu - \frac{f^{0\mu\nu\kappa}}{2}\{Y^\nu, \ Y^\kappa\} = 0 \ . \tag{57}$$

The indices run from $\mu = 1$ to $\mu = 7$, since we are working in a gauge where $Y^0 = 0$. The second order equation arising from iteration of (57), by virtue of the above identity, as well as the Jacobi identity, is

$$\partial_{\tau\tau} Y^\mu = -\{\{Y^\mu, \ Y^\nu\}, \ Y^\nu\} \ . \tag{58}$$

This arises from the lagrangian density

$$\mathcal{L}_{7dPB} = \frac{1}{2}\left(\partial_\tau Y^\mu\right)^2 + \frac{1}{4}\{Y^\mu, \ Y^\nu\}^2 \ . \tag{59}$$

As in the 3-dimensional case, this action reduces to a sum of squares with positive relative signs, up to a mere surface term,

$$\frac{1}{2}\left(\partial_\tau Y^\mu - \frac{f^{0\mu\nu\kappa}}{2}\{Y^\nu, \ Y^\kappa\}\right)^2 = \mathcal{L}_{7dPB} - f^{0\mu\nu\kappa}\partial_\tau Y^\mu \partial_\alpha Y^\nu \partial_\beta Y^\kappa \cong \mathcal{L}_{7dPB}. \tag{60}$$

In this lagrangian density, apparent extra terms $f^{\mu\nu\rho\kappa}\{Y^\mu, \ Y^\nu\}\{Y^\rho, \ Y^\kappa\}$ induced by the identity (56) have, in fact, vanished, by virtue of the identity,

---

[10] At present, we are not in a position to comment on its integrability or lack thereof.

$$\{f,\ g\}\{h,\ k\} + \{f,\ h\}\{k,\ g\} + \{f,\ k\}\{g,\ h\} \equiv 0 \ , \tag{61}$$

which holds for Poisson Brackets on a 2-dimensional phase-space—but not for matrix commutators nor Moyal Brackets[11]. This cancellation works at the level of the lagrangian density for the PB case.

However, note that even for ordinary matrices the corresponding term would vanish in the traced lagrangian, by the cyclicity of the trace pitted against full antisymmetry,

$$f^{\mu\nu\kappa\rho}\mathrm{Tr}X^\mu X^\nu X^\kappa X^\rho = 0 \ . \tag{62}$$

Likewise, the corresponding interaction for Moyal Brackets,

$$f^{\mu\nu\kappa\rho} \int d^2\xi \ \{\{X^\mu, X^\nu\}\}\{\{X^\kappa, X^\rho\}\} \ , \tag{63}$$

is forced by associativity to reduce to a surface term, vanishing unless there are contributions from surface boundaries or D-membrane topological numbers involved. (Shortcuts for the manipulation of such expressions underlain by $\star$-products can be found, e.g., in Hoppe 1990.) The cross terms involving time derivatives are expressible as divergences, as in the 3-dimensional case, and hence may give rise to topological contributions.

As a result, (57) is the Bogomol'nyi minimum of the action of (59) with the bottomless potential.

As in the case of 3-space, the conventional membrane signs can now be considered (for energy bounded below), and a symmetry breaking term m introduced, to yield

$$\mathcal{L}_{7dPB} \cong -\frac{1}{2}\left(-i\partial_t X^\mu + mX^\mu + \frac{f^{0\mu\nu\kappa}}{2}\{X^\nu,\ X^\kappa\}\right)^2 . \tag{64}$$

This model likewise has 7-space rotational invariance, and its vacuum configurations are, correspondingly, 2-surfaces lying on the spatial 6-sphere embedded in 7-space: $X^\mu X^\mu = R^2$. But, in addition, because of (56), these surfaces on the sphere also satisfy the trilinear constraint

$$f^{\lambda\mu\nu\kappa}X^\mu\{X^\nu,\ X^\kappa\} = 0 \ , \tag{65}$$

for $\lambda \neq 0$. (For $\lambda = 0$ this trilinear is $-2mR^2$.)

---

[11] D. Fairlie and A. Sudbery (1988, unpublished). It follows from $\epsilon^{[jk}\delta^{l]m} = 0$, whence $\epsilon^{[jk}\epsilon^{l]m} = 0$ for these membrane symplectic coordinates.

References

Banks, T., Fischler, W., Shenker, S., and Susskind, L. (1997): Phys Rev **D55** 5112

Biran, B., Floratos, E., and Savvidi, G. (1987): Phys Lett **B198** 329

Collins, P., and Tucker, R. (1976): Nucl Phys **B112** 150

Corrigan, E., Devchand, C., Fairlie, D., and Nuyts, J. (1982): Nucl Phys **B214** 452

Curtright, T., Fairlie, D., and Zachos, C. (1997): Phys Lett **B405** 37

Curtright, T. and McCarty, T. (1989): ICTP'89 talk (unpublished), (http://phyvax.ir.miami.edu:8001/curtright/ictp89.html); and the latter's University of Florida Thesis (unpublished)

Curtright, T., and Zachos, C. (1995): in *PASCOS'94*, K. C. Wali (ed), (World Scientific), pp 381-390, (hep-th/9407044)

Dündarer, R., Gürsey, F., and Tze, C-H. (1984): J Math Phys **25** 1496

Ericksen, J. (1960): Appendices 32, 34, & 35 in Vol. III/1 of the *Encyclopedia of Physics* (*Handbuch der Physik*), S. Flügge (ed), (Springer, Berlin) pp 822-829

Fairlie, D. (1997): hep-th/9707190

Fairlie, D., Fletcher, P., and Zachos, C. (1989): Phys Lett **B218** 203; D. Fairlie and C. Zachos, Phys Lett **B224** (1989) 101

Fairlie, D., Fletcher, P., and Zachos, C. (1990): J Math Phys **31** 1088

Fairlie, D., and Strachan, I. (1996): Physica **D90** 1

Floratos, E., (1989): Phys Lett **B228** 335

Floratos, E., Iliopoulos, J., and Tiktopoulos, G. (1989): Phys Lett **B217** 285; also see B. de Wit, J. Hoppe, and H. Nicolai, Nucl Phys **B305 [FS23]** (1988) 545; J. Hoppe, Int J Mod Phys **A4** (1989) 5235; E. Bergshoeff, E. Sezgin, Y. Tanii, and P. Townsend, Ann Phys **199** (1990) 340

Floratos, E., and Leontaris, G. (1989): Phys Lett **B223** 153

Hitchin, M. (1983): Comm Math Phys **89** 145

Hoppe, J. (1982): M.I.T. Ph.D. Thesis; also in Elem Part Res J (Kyoto) **83** no.3 (1989/90)

Hoppe, J. (1990): Phys Lett **B250** 44

Kim, N., and Rey, S-J. (1997): hep-th/9701139

Matinyan, S., Savvidy, G., and Ter-Arutunian Savvidy, N. (1981): Sov Phys JETP **53** 421; JETP Lett **34** (1981) 590; S. Matinyan, E. Prokhorenko, and G. Savvidy, Nucl Phys **B298** (1988) 414

Moyal, J. (1949): Proc Camb Phil Soc **45** 99; further see J. Vey, Comment Math Helvetici **50** (1975) 412; T. Jordan and E. Sudarshan, Rev Mod Phys **33** (1961) 515; D. Fairlie, Proc Camb Phil Soc **60** (1964) 581; D. Fairlie and C. Manogue, J Phys **A24** (1991) 3807

Nahm, W. (1983): in *Group Theoretical Methods in Physics: XIth International Colloquium, Istanbul, 1982*, M. Serdaroğlu and E. İnönü (eds), (Springer Lecture Notes in Physics **180**, Berlin) pp 456-466

Plebański, J., Przanowski, M., and García-Compeán, H. (1996): Mod Phys Lett **A11** 663

Schild, A. (1977): Phys Rev **D16** 1722; aso see T. Eguchi, Phys Rev Lett **44** (1980) 126

Ward, R. (1990): Phys Lett **B234** 81

Zaikov, R. (1991): Phys Lett **B266** 303

Zakharov, V., and Mikhailov, A. (1978): Sov Phys JETP **47** 1017;
    T. Curtright and C. Zachos, Phys Rev **D49** (1994) 5408

Vertex Operators and Solitons of Constrained KP Hierarchies

H. Aratyn[1], L.A. Ferreira[2], J.F. Gomes[2] and A.H. Zimerman[2]

[1] Department of Physics, University of Illinois at Chicago, 845 W. Taylor St., Chicago, IL 60607-7059
[2] Instituto de Física Teórica - IFT/UNESP, Rua Pamplona 145, 01405-900, São Paulo - SP, Brazil

Abstract. We construct the vertex operator representation for the Affine Kac-Moody $\hat{sl}(M + K + 1)$ algebra, which is relevant for the construction of the soliton solutions of the constrained KP hierarchies. The oscillators involved in the vertex operator construction are provided by the Heisenberg subalgebras of $\hat{sl}(M + K + 1)$ realized in the unconventional gradations. The well-known limiting cases are the homogeneous Heisenberg subalgebra of $\hat{sl}(M + 1)$ and the principal Heisenberg subalgebra of $\hat{sl}(K + 1)$. The explicit example of $M = K = 1$ is discussed in detail and the corresponding soliton solutions and tau-functions are given.

1 Introduction

We consider a generalized Drinfeld-Sokolov hierarchy (Drinfel'd and Sokolov (1985), de Groot, et al (1992) , Burroughs et al (1993)), based on a Kac-Moody algebra $\hat{\mathcal{G}} = \hat{sl}(M + K + 1)$ and defined by the following matrix eigenvalue problem:

$$L\Psi = 0 \quad ; \quad L \equiv (D - A - E) \quad ; \quad D \equiv I\frac{\partial}{\partial x} \tag{1.1}$$

The constant E is a non-regular and semisimple element of $\hat{\mathcal{G}}$ given by

$$E = \sum_{a=1}^{K} E^{(0)}_{\alpha_{M+a}} + E^{(1)}_{-(\alpha_{M+1}+\cdots+\alpha_{M+K})} \tag{1.2}$$

The potential A in (1.1) contains the dynamical variables of the model, namely q_i, r_i, U_a and ν, and is given by (Aratyn et al (1997a), McIntosh (1993)):

$$A = \sum_{i=1}^{M} (q_i P_i + r_i P_{-i}) + \sum_{a=1}^{K} U_{M+a}(\alpha_{M+a} \cdot H^{(0)}) + \nu\,\hat{c} \tag{1.3}$$

where $P_{\pm i} = E^{(0)}_{\pm(\alpha_i + \alpha_{i+1} + \ldots + \alpha_M)}$, $i = 1, 2, \ldots, M$, and $\hat{c}$ is a central element of $\hat{\mathcal{G}}$. We are using the Cartan-Weyl basis for $\hat{\mathcal{G}}$ with generators $H_i^{(n)}$ and

$E_\alpha^{(n)}$, with $i = 1, 2, \ldots \operatorname{rank} \hat{\mathcal{G}}$, $n \in \mathbb{Z}$ being the eigenvalues of the standard derivation d of $\hat{\mathcal{G}}$, and α being roots of the finite algebra $sl(M + K + 1)$, associated to the affine Kac-Moody algebra $\hat{\mathcal{G}} = \hat{sl}(M + K + 1)$.

The integral gradations of $\hat{\mathcal{G}}$ (Kac and Peterson (1985)) play an important role in the construction of integrable hierarchies (de Groot, et al (1992)). They are labeled by sets of $\operatorname{rank} \hat{\mathcal{G}} + 1$ co-prime non negative integers $s = (s_0, s_1, \ldots, s_r)$. The corresponding grading operators, in the case of $\hat{sl}(M + K + 1)$, are given by

$$Q_s \equiv \sum_{a=1}^{M+K} s_a \lambda_a \cdot H^{(0)} + d \sum_{i=0}^{M+K} s_i \qquad (1.4)$$

where λ_a are the fundamental weights of $sl(M+K+1)$. The relevant gradation to the hierarchy associated to (1.1) is

$$s_{cKP} = (1, \underbrace{0, \ldots, 0}_{M}, \underbrace{1, \ldots, 1}_{K}) ; \quad \leftrightarrow \quad Q_{cKP} \equiv \sum_{a=1}^{K} \lambda_{M+a} \cdot H^{(0)} + (K + 1)d$$

$$(1.5)$$

Notice that E and A have grades 1 and 0 respectively, w.r.t. (1.5).

The gradation (1.5) interpolates between the principal $s_{principal} = (1, \ldots, 1)$ and the homogeneous one $s_{homogeneous} = (1, 0 \ldots, 0)$. As it is well-known these two limits define the KdV (de Groot, et al (1992)) and AKNS hierarchies (Fordy and Kulish (1983), Aratyn et al (1995)), respectively with

$$E_{KdV} = E_{-(\alpha_1 + \cdots + \alpha_K)}^{(1)} + \sum_{a=1}^{K} E_{\alpha_a}^{(0)} \qquad (1.6)$$

and

$$E_{AKNS} = \lambda_M H^{(1)} \qquad (1.7)$$

The fact that E is semisimple means it decomposes $\hat{\mathcal{G}}$ into the kernel and image of its adjoint action, i.e. $\hat{\mathcal{G}} = \operatorname{Ker} (\operatorname{ad} E) + \operatorname{Im} (\operatorname{ad} E)$. Then one can gauge transform L into Ker (ad E), $L \to L_0 \equiv U L U^{-1}$, with U being an exponentiation of generators of negative s_{cKP}-grade of $\hat{\mathcal{G}}$. One introduces a flow equation for each element $b^{(N)}$, in the center of Ker (ad E), with positive s_{cKP}-grade N, as follows

$$\frac{dL}{dt_N} = \frac{dA}{dt_N} \equiv [L, B_N] \qquad (1.8)$$

where $B_N \equiv \left(U^{-1} b^{(N)} U \right)_{s_{cKP} \geq 0}$. We shall choose $b^{(1)} \equiv E$, $B_1 \equiv E + A$ and $t_1 \equiv x$.

We are interested in discussing the soliton solutions for the model defined by equation (1.1), following the method of (Ferreira et al (1997)). According to (Ferreira et al (1997)) the basic ingredient for the appearance of soliton solutions is that there must exist one or several "vacuum solutions", such that the Lax operators, when evaluated on them, should lie in some abelian subalgebra of $\hat{\mathcal{G}}$, and in addition the corresponding components of it should be constant. Accordingly, one requires

$$B_N^{(\text{vac.})} \equiv \varepsilon_N = \sum_{i=0}^{N} c_N^i \, a_i ; \qquad [a_i, a_j] = i \, \beta_i \, \delta_{i+j,0} . \tag{1.9}$$

where, β_i is some constant, and the oscillators a_i have s$_{\text{cKP}}$-grade i. Then one can write

$$B_N^{(\text{vac.})} = \frac{\partial \Psi^{(\text{vac.})}}{\partial t_N} \, \Psi^{-1\,(\text{vac.})} ; \qquad \text{with} \qquad \Psi^{(\text{vac.})} = \exp\left(\sum_N \varepsilon_N \, t_N \right) \tag{1.10}$$

The soliton solutions are constructed using the dressing transformation method as follows. Choose a constant element h which is an exponentiation of the generators of $\hat{\mathcal{G}}$, and perform the generalized Gauss decomposition

$$\Psi^{(\text{vac.})} \, h \, \Psi^{-1\,(\text{vac.})} = \left(\Psi^{(\text{vac.})} \, h \, \Psi^{-1\,(\text{vac.})} \right)_{\text{s}_{\text{cKP}}<0} \left(\Psi^{(\text{vac.})} \, h \, \Psi^{-1\,(\text{vac.})} \right)_{\text{s}_{\text{cKP}}=0} \times$$
$$\times \left(\Psi^{(\text{vac.})} \, h \, \Psi^{-1\,(\text{vac.})} \right)_{\text{s}_{\text{cKP}}>0} . \tag{1.11}$$

Define the group element

$$\Psi^h \equiv \left(\left(\Psi^{(\text{vac.})} \, h \, \Psi^{-1\,(\text{vac.})} \right)_{\text{s}_{\text{cKP}}<0} \right)^{-1} \Psi^{(\text{vac.})} \, h \tag{1.12}$$
$$= \left(\Psi^{(\text{vac.})} \, h \, \Psi^{-1\,(\text{vac.})} \right)_{\text{s}_{\text{cKP}}=0} \left(\Psi^{(\text{vac.})} \, h \, \Psi^{-1\,(\text{vac.})} \right)_{\text{s}_{\text{cKP}}>0} \Psi^{(\text{vac.})}$$

Then, one can easily verify that

$$B_N^h = \frac{\partial \Psi^h}{\partial t_N} \, \Psi^{h-1} \tag{1.13}$$

has the same grading structure as B_N. Since B_N^h is a flat connection it automatically satisfies the Lax (or flow) equation, and therefore by equating B_N to B_N^h one gets a solution for the hierarchy for each choice of h.

The soliton solutions correspond to those h's which are products of exponentials of eigenvectors of the oscillators a_i's, i.e.

$$h = e^{F_1} \, e^{F_2} \cdots e^{F_n} , \qquad [\varepsilon_N , F_k] = \omega_N^{(k)} \, F_k , \qquad k = 1, 2, \dots, n . \tag{1.14}$$

Notice, that the F_k's do not have to be eigenvectors of all ε_N's. The soliton is a solution of a given flow equation, associated to a time t_N, only if the F_k's are eigenvectors of the corresponding ε_N and also ε_1.

We also define, following (Ferreira et al (1997)), the generalized Hirota tau-function as

$$\tau_{\mu_0,\mu_0'}(t) = \langle \mu_0' \mid \Psi^{(\text{vac})} \, h \, \Psi^{(\text{vac})^{-1}} \mid \mu_0 \rangle$$

$$= \langle \mu_0' \mid e^{\sum_N \varepsilon_N t_N} \, h \, e^{-\sum_N \varepsilon_N t_N} \mid \mu_0 \rangle, \qquad (1.15)$$

where $|\mu_0\rangle$ and $|\mu_0'\rangle$ are suitably chosen states in an integrable highest weight representation of $\hat{\mathcal{G}}$, which are annihilated by all generators with positive s_{cKP}-grade.

The truncation of the Hirota's formal expansion is now understood from the nilpotency of the operators F_k's. Such property is most easily verified using the vertex operator construction of the integrable highest weight representation of $\hat{\mathcal{G}}$. In this paper, we give the explicit construction of the vertex operator representation relevant for the soliton solutions of the hierarchy (1.1).

For pedagogical reasons, in section 2 we review the construction for the pure homogeneous and pure principal gradations corresponding to untwisted and twisted vertex operators related to AKNS and m-KDV hierarchies respectively.

In section 3 we present, following the same line of thought, the vertex functions construction for the intermediate gradation hierarchy. In section 4 we present the vertex operator algebra in terms of OPE (Operator Product Expansion) and show that it, in fact, reproduces the original $SL(M+K+1)$ Kac-Moody algebra. In section 5 we specialize for $K = M = 1$ and discuss the resulting tau functions (Aratyn et al (1997b)).

2   Vertex Operator Construction for the Homogeneous and Principal Gradations

Let us first consider the general Kac-Moody algebra in the Cartan-Weyl basis

$$[H_i^m, H_j^n] = m\, \delta_{m+n,0}\, \delta_{i,j} \qquad i,j = 1,\dots,\text{rank } \mathcal{G} \qquad (2.16)$$

$$[H_i^m, E_\alpha^n] = (\alpha)^i E_\alpha^{m+n} \qquad\qquad (2.17)$$

$$[E_\alpha^m, E_\beta^n] = \begin{cases} \epsilon(\alpha,\beta) E_{\alpha+\beta}^{m+n} & \text{if } \alpha+\beta \text{ is a root} \\ H_\alpha^{m+n} & \text{if } \alpha+\beta = 0 \\ 0 & \text{otherwise} \end{cases} \qquad (2.18)$$

where $(H_i^m)^\dagger = H_i^{-m}$, $(E_\alpha^m)^\dagger = E_\alpha^{-m}$. From eqns ((2.16), (2.17) and (2.18)) we will define an affine Heisenberg subalgebra satisfying

$$[\mathcal{A}_i^m, (\mathcal{A}_j^n)^\dagger] = m\, \delta_{m,n}\, \delta_{i,j} \qquad i,j = 1,\dots, rank\ \mathcal{G} \qquad (2.19)$$

2.1   Homogeneous Gradation

We associate $\mathcal{A}_i^m = H_i^m$ and $(\mathcal{A}_i^m)^\dagger = (H_i^m)^\dagger$. Define now the rank $\mathcal{G}$ dimensional Fubini-Veneziano field (see (Goddard et al (1986)))

$$Q^i(z) = i \sum_{n>0} \frac{A_i^n z^{-n}}{n}$$

$$(Q^i)^\dagger(z) = -i \sum_{n>0} \frac{(A_i^n)^\dagger z^n}{n}$$

$$Q_0^i(z) = q^i - ip^i \ln z \tag{2.20}$$

where $[q^i, p^j] = i\delta_{i,j}$ and $p^i = H_i^0$.

The construction of step operators satisfying (2.17) and (2.18) is provided by the vertex operator

$$V^\alpha(z) = e^{i\alpha Q^\dagger(z)} e^{i\alpha Q_0} e^{i\alpha Q(z)} = z^{\frac{\alpha^2}{2}} e^{i\alpha Q^\dagger(z)} e^{i\alpha q} e^{\alpha p \ln z} e^{i\alpha Q(z)} \tag{2.21}$$

where the vector α denotes a root of the algebra $\mathcal{G}$. It follows that

$$H^i(z_1) V^\alpha(z_2) =: H^i(z_1) V^\alpha(z_2) : + \frac{(\alpha)^i V^\alpha(z_1)}{z_1 - z_2} \tag{2.22}$$

Eqn. ((2.18)) is equivalent to the product

$$V^\alpha(z_1) V^\beta(z_2) =: V^\alpha(z_1) V^\beta(z_2) : z_1^{\frac{\alpha^2}{2}} z_2^{\frac{\beta^2}{2}} (z_1 - z_2)^{\alpha\beta} \tag{2.23}$$

where $: \ldots :$ denotes normal ordering in the sense that A_n are moved to the right of $A^\dagger_n$ and p to the right of q for α, β roots of $\mathcal{G}$.

2.2 Principal Gradation

In this case the affine Heisenberg subalgebra is defined by

$$A^a_{a+n(K+1)} = \sum_{i=1}^{K+1-a} E^{(n)}_{\alpha_i + \alpha_{i+1} + \cdots + \alpha_{i+a-1}} + \sum_{i=1}^{a} E^{(n+1)}_{-(\alpha_i + \alpha_{i+1} + \cdots + \alpha_{i+K-a})} \tag{2.24}$$

with $a = 1, 2, \cdots, K$. It is straightforward to verify that

$$[A^a_{a+m(K+1)}, A^{b\dagger}_{b+n(K+1)}] = (a + m(K+1))\delta_{m,n} \tag{2.25}$$

For instance, take $g = SL(2)$ with

$$A^1_{2n+1} = E^{(n)}_\alpha + E^{(n+1)}_{-\alpha}. \tag{2.26}$$

Equation (2.25) is then easily verified.

We now define the rank $\mathcal{G}$ dimensional Fubini-Veneziano field

$$Q^a = i \sum_{n=0}^{\infty} \frac{A^a_{a+n(K+1)} z^{-a-n(K+1)}}{a + n(k+1)}$$

$$Q^{a\dagger} = -i \sum_{n=0}^{\infty} \frac{A^{a\dagger}_{a+n(K+1)} z^{a+n(K+1)}}{a + n(K+1)} \tag{2.27}$$

Notice that in this case the zero modes are absent.

We now seek within the Kac-Moody algebra $\hat{sl}(K+1)$ the eigenvalues and eigenfunctions of the affine Heisenberg subalgebra defined in (2.25). We therefore find that

$$\left[\mathcal{A}^a_{a+n(K+1)} , F_{b,l}(z) \right] = \omega^{-al} \left(\omega^{ab} - 1 \right) z^{a+(K+1)n} F_{b,l}(z) \tag{2.28}$$

with $a, b = 1, \ldots, K$, $l = 1, \ldots, K, K+1$, and where $\omega = exp\frac{2\pi i}{K+1}$, $F_{b,l}$ is a linear combination of Kac-Moody generators of $\hat{sl}(K+1)$ and will be given explicitly in the next section for a more general case. The eigenvalues in equation (1.14) define the roots of $sl(K+1)$. Our task is now to find a set of simple roots lying in a K dimensional complex space such that any root defined in (1.14) may be written as a integer linear combination with all coefficients being positive (or negative). The following set of simple roots satisfy our requirements:

$$\alpha_{(K)_i} = \left(\omega^{i-1}(\omega - 1), \omega^{2(i-1)}(\omega^2 - 1), \ldots, \omega^{K(i-1)}(\omega^K - 1) \right) \tag{2.29}$$

for $i = 1, \ldots, K$. Define now the vertex operator satisfying the eigenvalue equation (1.14) with eigenvalue α to be

$$V^\alpha(z) = e^{i\alpha^* Q(z)^\dagger} e^{i\alpha Q(z)} \tag{2.30}$$

The OPE of vertices (2.30) is a straightforward calculations yielding

$$V^\alpha(z_1) V^\beta(z_2) =: V^\alpha(z_1) V^\beta(z_2) : \prod_{p=1}^{K+1} (1 - \frac{z_2 \omega^{-p}}{z_1})^{\sum_{c=1}^{k} \frac{\alpha^c \beta^{c*} \omega^{pc}}{K+1}} \tag{2.31}$$

Since the simple roots in (2.29) satisfy the same addition table as the simple roots of $sl(K+1)$ the product of vertices (3.58) closes into the OPE algebra of $\hat{sl}(K+1)$.

3   Vertex Construction for the Intermediate Gradation

In this section we shall discuss the vertex functions for the model defined by eqn. (1.1) with the constant element E lying in the Kac-Moody algebra $sl(M+K+1)$ given by (1.2) and gradation given by (1.5). Consider the Heisenberg algebra associated to the center of Ker (ad E) and consisting of :
1) <u>Homogeneous part of $\hat{sl}(M)$</u>

$$B_i^{(n)} = \alpha_i \cdot H^{(n)} \quad , \quad i = 1, 2, \ldots, M - 1 \tag{3.32}$$

2) <u>Principal part of $\hat{sl}(K+1)$</u>

$$\mathcal{A}^a_{a+n(K+1)} = \sum_{i=1}^{K+1-a} E^{(n)}_{\alpha_i+M+\alpha_{i+M+1}+\cdots+\alpha_{i+M+a-1}} \tag{3.33}$$

$$+ \sum_{i=1}^{a} E^{(n+1)}_{-(\alpha_{i+M}+\alpha_{i+M+1}+\cdots+\alpha_{i+M+K-a})} \quad , \quad a = 1, 2, \ldots, K$$

3) "A border part"

$$\mathcal{A}^0_{n(K+1)} = \sqrt{\frac{M+K+1}{M}} \lambda_M . H^{(n)} - \frac{K}{2} \sqrt{\frac{M}{M+K+1}} \, \delta_{n,0} \tag{3.34}$$

The above elements of the Heisenberg subalgebra enter the oscillator algebra relations (we put $c = 1$):

$$\left[\mathcal{A}^a_{a+n(K+1)} , \mathcal{A}^{b\dagger}_{b+m(K+1)} \right] = (a + (K+1)n) \, \delta_{n,m} \delta^{a,b} \quad ; \quad a, b = 1, \cdots, K \tag{3.35}$$

with $\mathcal{A}^{a\dagger}_{a+m(K+1)} = \mathcal{A}^{K-a+1}_{-a-m(K+1)}$,

$$\left[\mathcal{A}^0_{n(K+1)} , \mathcal{A}^{0\dagger}_{m(K+1)} \right] = (K+1) \, n \, \delta_{n,m} \quad ; \quad \left[\mathcal{A}^0_n , \mathcal{A}^{a\dagger}_m \right] = 0 \tag{3.36}$$

and

$$\left[\mathcal{B}^{(m)}_i , \mathcal{B}^{(n)}_j \right] = K_{i,j} \, m \, \delta_{m+n,0} \quad ; \quad i, j = 1, \ldots, M-1 \tag{3.37}$$

where $K_{i,j}$ is the Cartan matrix of $sl(M)$.

Instead of $\mathcal{B}^{(m)}_i$, it is more convenient to work with

$$\mathcal{K}^{(n)}_i = \frac{\sum_{p=1}^{i} p \, \mathcal{B}^{(n)}_p}{N_i} \quad ; \quad N_i \equiv \sqrt{i(i+1)} \tag{3.38}$$

such that $\mathrm{Tr} \left(\mathcal{K}^{(n)}_i \mathcal{K}^{(m)}_j \right) = \delta_{ij} \, \delta_{m+n,0}$ and

$$\left[\mathcal{K}^{(m)}_i , \mathcal{K}^n_j \right] = m \, \delta_{m+n,0} \, \delta_{ij} \quad ; \quad i, j = 1, \ldots, M-1 \tag{3.39}$$

To summarize, we had parametrized the Heisenberg subalgebra in terms of elements:

$$\mathcal{A}^a_{a+n(K+1)} \ , \ \mathcal{A}^0_{n(K+1)} \ , \ \mathcal{K}^{(n)}_i \quad ; \quad a = 1, 2, \ldots, K \quad ; \quad i = 1, 2, \ldots, M-1 \tag{3.40}$$

which for $n = 0$ constitute a Cartan subalgebra of $sl(M + K + 1)$.

We work with the eigenstates of this subalgebra having the following form

$$F_{a,l} = \frac{\hat{c}}{(w^a - 1)} + \sum_{n \in Z} z^{-n(K+1)} \sum_{i=1}^{K} H_{M+i}^{(n)} \sum_{p=1}^{i} w^{a(i-p)}$$

$$+ \sum_{b=1}^{K} \sum_{n \in Z} w^{bl} z^{-(b+n(K+1))} \left(\sum_{i=1}^{K+1-b} w^{a(i-1)} E_{\alpha_{i+M}+\alpha_{i+M+1}+\cdots+\alpha_{i+M+b-1}}^{(n)} \right.$$

$$\left. + \sum_{i=1}^{b} w^{a(i+K-b)} E_{-(\alpha_{i+M}+\alpha_{i+M+1}+\cdots+\alpha_{i+M+K-b})}^{(n+1)} \right)$$

where $a = 1, \ldots, K$, $l = 1, \ldots, K, K+1$ and ω is a non-trivial $K+1$-th root of unity ($\omega^{K+1} = 1$).

$$\bar{F}_{r,l} = \sum_{n \in \mathbb{Z}} z^{-n(K+1)} \sum_{p=0}^{K} w^{pl} z^{-p} E_{\alpha_r + \cdots + \alpha_{M+p}}^{(n)}, \qquad (3.41)$$

with $r = 1, \ldots, M$, $l = 1, \ldots, K, K+1$, and

$$F_{\alpha_j} = \sum_{n \in \mathbb{Z}} z^{-n} E_{\alpha_j}^{(n)} \quad ; \quad j = 1, \ldots, M-1 \qquad (3.42)$$

One finds the following form of algebraic relations for the eigenstates of the Heisenberg subalgebra:

$$\left[\mathcal{A}_{a+n(K+1)}^a, F_{b,l}(z) \right] = \omega^{-al} \left(\omega^{ab} - 1 \right) z^{a+(K+1)n} F_{b,l}(z)$$

$$\left[\mathcal{A}_{n(K+1)}^0, F_{b,l}(z) \right] = 0$$

$$\left[\mathcal{K}_i^{(n)}, F_{b,l}(z) \right] = 0 \qquad (3.43)$$

where $a, b = 1, \ldots, K$ and $l = 1, \ldots, K, K+1$,

$$\left[\mathcal{A}_{a+n(K+1)}^a, \bar{F}_{r,l}(z) \right] = \omega^{-al} z^{a+n(K+1)} \bar{F}_{r,l}(z)$$

$$\left[\mathcal{A}_{n(K+1)}^0, \bar{F}_{r,l}(z) \right] = \sqrt{\frac{K+1}{\lambda_M^2}} z^{n(K+1)} \bar{F}_{r,l}(z)$$

$$\left[\mathcal{K}_i^{(n)}, \bar{F}_{r,l}(z) \right] = \frac{1}{N_i} \left(\sum_{p=1}^{i} \delta_{r,p} - \delta_{r,i+1} \right) z^{n(K+1)} \bar{F}_{r,l}(z) \quad (3.44)$$

where $N_i = \sqrt{i(i+1)}$, $r = 1, \ldots, M$ and $l = 1, \ldots, K, K+1$.

Furthermore, we also have

$$\left[\mathcal{A}_{a+n(K+1)}^a, F_{\alpha_j}(z) \right] = 0$$

$$\left[\mathcal{A}_{n(K+1)}^0, F_{\alpha_j}(z) \right] = 0 \qquad (3.45)$$

$$\left[\mathcal{K}_i^{(n)}, F_{\alpha_j}(z) \right] = \frac{1}{N_i} \left(\sum_{p=1}^{i} 2p \delta_{p,j} - p \delta_{p,j-1} - p \delta_{p,j+1} \right) z^n F_{\alpha_j}(z)$$

where $i, j = 1, \ldots, M - 1$.

Since, $F_{j,l}, \bar{F}_{r,l}$ and F_{α_j} are step operators associated with the Cartan subalgebra defined by the Heisenberg subalgebra they correspond to the roots of $sl(M + K + 1)$. The simple roots defining these steps operators are:

$$
\begin{aligned}
\alpha_i &= \left(\alpha_{(M)_i}, 0, 0_{(K)}\right), \quad i = 1, \ldots, M - 1 \\
&= \left(0_{(M-2)}, \frac{1 - M}{N_{M-1}}, \sqrt{\frac{K + 1}{\lambda_M^2}}, 1_{(K)}\right) \quad for \ \ i = M \\
&= \left(0_{(M-1)}, 0, \alpha_{(K)_j}\right), \quad for \ \ i = M + j, \ \ j = 1, \ldots, K \quad (3.46)
\end{aligned}
$$

where $0_{(N)}$ denotes the N-dimensional null vector, $1_{(N)}$ is the N-dimensional vector with N unit components. The root vectors $\alpha_{(M)_i}$ are

$$
\alpha_{(M)_1} = \left(\frac{2}{N_1}, 0_{(M-2)}\right)
$$

$$
\alpha_{(M)_2} = \left(\frac{-1}{N_1}, \frac{3}{N_2}, 0_{(M-3)}\right)
$$

$$
\vdots \quad \vdots \quad \vdots
$$

$$
\alpha_{(M)_i} = \left(0_{(i-2)}, -\frac{(i - 1)}{N_{i-1}}, \frac{(i + 1)}{N_i}, 0_{(M-1-i)}\right), i = 3, \ldots, M - 1 \ (3.47)
$$

and $\alpha_{(K)_i}$ are given in eqn. (2.29).

The step operators associated to the simple roots of (3.47), (2.29) are nothing but the eigenstates of the Heisenberg algebra (3.40) and the association is

$$
\begin{aligned}
E_{\alpha_i} &\leftrightarrow F_{\alpha_i} \ (= E_{\alpha_i} \ of \ sl(M)) \ \ ; \ \ i = 1, 2, \ldots, M - 1 \quad (3.48) \\
E_{\alpha_M} &\leftrightarrow \bar{F}_{M, K+1} \quad (3.49) \\
E_{\alpha_{M+a}} &\leftrightarrow F_{1, K - a + 2} \ \ ; \ \ a = 1, 2, \ldots, K \quad (3.50)
\end{aligned}
$$

The remaining eigenstates are associated to the general roots of the form $\alpha = \alpha_i + \ldots + \alpha_j$. For instance $F_{2,0} \leftrightarrow E_{\alpha_{M+1} + \alpha_{M+2}}$ with $\alpha_{M+1} + \alpha_{M+2} = (0, \ldots, 0, 0, \omega^2 - 1, \ldots, \omega^{2K} - 1)$.

Define now the Fubini-Veneziano operators

$$
Q_0^i(z) = q^i - i p^i \ln z \ ; \ Q^i(z) = i \sum_{n=1}^{\infty} \frac{K_i^{(n)} z^{-n}}{n} \ ; \ i = 1, \ldots, M - 1 (3.51)
$$

$$
Q_0^M(z) = q^M - i p^M \ln z; \quad Q^M(z) = i \sum_{n=1}^{\infty} \frac{A_{n(K+1)}^0 z^{-n(K+1)}}{n(K + 1)} \quad (3.52)
$$

$$
Q^{M+a}(z) = i \sum_{n=0}^{\infty} \frac{A_{a+n(K+1)}^a z^{a+n(K+1)}}{a + n(K + 1)} \ \ ; \ \ a = 1, 2, \ldots, K \quad (3.53)
$$

where p_M is equal to $\mathcal{A}^0_{n=0}$ from expression (3.34) and the zero modes satisfy $\left[p^i, q^j\right] = -i\delta^{ij}$.

The corresponding conjugated Fubini-Veneziano operators $Q^\dagger(z)$ are obtained from (3.51)-(3.53) by taking into consideration rules $\mathcal{K}_i^{(n)\dagger} = \mathcal{K}_i^{(-n)}$, $\left(\mathcal{A}^0_{n(K+1)}\right)^\dagger = \mathcal{A}^0_{-n(K+1)}$, $\left(\mathcal{A}^a_{a+n(K+1)}\right)^\dagger = \mathcal{A}^a_{a-n(K+1)}$ as well as $z^\dagger = z^{-1}$.

The total number of Q's equals $M + K$ which is the rank of $sl(M + K + 1)$.

Putting together the simple root structure from eqs.(3.47), (2.29) with the Fubini-Veneziano operators enables us to write down a compact expression for the general vertex operator in the normal ordered form:

$$V^\alpha(z) \equiv z^{\frac{1}{2}(\alpha_{(M)})^2} \times \tag{3.54}$$
$$\times \exp\left(i\alpha^* \cdot \mathbf{Q}^\dagger(z)\right) \exp\left(i\alpha \cdot \mathbf{q}\right) \exp\left(\alpha \cdot \mathbf{p} \ln z\right) \exp\left(i\alpha \cdot \mathbf{Q}(z)\right)$$

where $(\alpha_{(M)})^2 = \sum_{j=1}^M (\alpha^j_{(M)})^2$ i.e. the sum of squares of components in the $M - 1$ subspace, and

$$\mathbf{Q} = (Q_1, Q_2, \ldots, Q_M, Q_{M+1}, \ldots, Q_{M+K}) \tag{3.55}$$
$$\mathbf{q} = (q_1, q_2, \ldots, q_M, 0_{(K)}) \tag{3.56}$$
$$\mathbf{p} = (p_1, p_2, \ldots, p_M, 0_{(K)}) \tag{3.57}$$

The product of two vertex operators is found to be:

$$V^\alpha(z_1)V^\beta(z_2) =: V^\alpha(z_1)V^\beta(z_2) : z_1^{\frac{1}{2}\sum_{j=1}^M\left((\alpha^j)^2+2\alpha^j\beta^j\right)} z_2^{\frac{1}{2}\sum_{j=1}^M \beta^j\beta^j}$$
$$\times \left(1 - \frac{z_2}{z_1}\right)^{\sum_{j=1}^{M-1}\left(\alpha^j\beta^j+(\alpha^M\beta^M/(K+1))+\sum_{a=1}^K(\alpha^{M+a}\beta^{M+a*}/(K+1))\right)}$$
$$\times \prod_{p=1}^K \left(1 - \frac{z_2}{z_1}\omega^{-p}\right)^{(\alpha^M\beta^M/(K+1))+\sum_{a=1}^K \omega^{al}(\alpha^{M+a}\beta^{M+a*}/(K+1))} \tag{3.58}$$

where the $*$ stands for complex conjugation. Similarly for the more complicated products of vertices.

It is also straightforward to see that the square of a vertex vanishes. In order to consider the product of several vertex operators we first define the generalized scalar product for $M + K$ component vectors

$$(a \odot b)_p = \delta_{p,0} \sum_{i=1}^{M-1} a_i b_i^* + \frac{a_i b_i^*}{K+1} + \sum_{i=1}^K \frac{a_{M+i} b_{M+i}^*}{K+1}\omega^{ip} \tag{3.59}$$

With this new notation, eqn ((3.58)) becomes

$$V^\alpha(z_1)V^\beta(z_2) =: V^\alpha(z_1)V^\beta(z_2) : z_1^{\frac{1}{2}(\alpha+\beta)^2}\left(\frac{z_2}{z_1}\right)^{\frac{1}{2}\beta^2}\prod_{p=0}^K \left(1 - \frac{z_2}{z_1}\omega^{-p}\right)^{(\alpha\odot\beta)_p}$$
$$\tag{3.60}$$

The general formula for the product of several vertices is therefore given as

$$V^{\gamma_1}(z_1)\cdots V^{\gamma_N}(z_N) = \, : V^{\gamma_1}(z_1)\cdots V^{\gamma_N}(z_N): \, \prod_{m=1}^{N-1} z_m^{\gamma_m(\gamma_{m+1}+\gamma_{m+2}+\cdots+\gamma_N)}$$

$$\times \prod_{r=2}^{N}\prod_{i<r}\prod_{p=0}^{K}\left(1-\frac{z_r\omega^{-p}}{z_{r-i}}\right)^{\gamma_r\odot\gamma_{r-i}} \tag{3.61}$$

4 The OPE Algebra of Vertex Operators

We have already proposed a set of $M+K$ simple roots satisfying the same addition properties of those of $sl(M+K+1)$. We now use the product of vertices given in (3.58) to obtain the OPE to show that, in fact, it reproduces the algebra $\hat{sl}(M+K+1)$. Let us associate the Kac-Moody currents to the vertex operators via

$$e_\alpha^{(M)} = V^\alpha(z) \tag{4.62}$$

for roots in the $sl(M)$ sector, i.e. $\alpha = (\alpha_{(M)}, 0, 0_{(K)})$ where $\alpha_{(M)}$ denote a root in the pure $sl(M)$ subalgebra,

$$e_\alpha^{(K)} = \frac{V^\alpha(z)}{|1-\omega^\#|} \tag{4.63}$$

for roots in the $sl(K+1)$ sector, i.e. $\alpha = (0_{(M)}, 0, \alpha_{(K)})$ where $\alpha_{(K)}$ is a root in the $sl(K+1)$ sector and $\#$ denote the number of simple roots in $\alpha_{(K)}$,

$$e_\alpha^{(MK)} = \frac{V^\alpha(z)}{(K+1)^{\frac{1}{2M}}} \tag{4.64}$$

for roots of the form $\alpha = (\alpha_{(M)}, \alpha_{(MK)}, \alpha_{(K)})$, where $\alpha_{(MK)}$ denote the M-th component of the root α.

It follows from the complex structure of the simple roots given in the previous section, that the OPE algebra displaying the most singular part of the product of vertices is given by

$$V^\alpha(z_1)V^\beta(z_2) = \begin{cases} \frac{z_1 V^{\alpha+\beta}(z_1)}{z_1-z_2} & \text{if } \alpha+\beta \text{ is a root} \\ \frac{z_1\alpha\hat{H}(z_1)}{z_1-z_2} - z_1\frac{d}{dz_1}\left(\frac{z_1}{z_1-z_2}\right) & \text{if } \alpha+\beta = 0 \\ 0 & \text{otherwise} \end{cases} \tag{4.65}$$

5 Special Case M=K=1; Tau Functions

We consider now the special case of $M = K = 1$ with $\hat{\mathcal{G}} = \hat{sl}(3)$ and the Lax matrix operator from (1.1) with A from (1.3) and E from (1.2) being given by (see (Aratyn et al (1997b)))

$$L = D - \begin{pmatrix} 0 & q & 0 \\ r & U_2 & 1 \\ 0 & \lambda & -U_2 \end{pmatrix} - \nu \hat{c} \qquad (5.66)$$

where λ is the usual loop parameter. We now describe two different two-soliton solutions obtained from the above vertex construction.

The first one has $r = 0$ and $q \neq 0$ and equal to

$$q = -\sqrt{2}\, z_2\, e^{(t_3 z_2{}^3 + t z_2{}^2 + x z_2)} \left(1 + \frac{1}{2} e^{(-2x z_1 - 2t_3 z_1{}^3)} \left(\frac{z_1 + z_2}{z_1 - z_2} \right) \right) / \tau_0^{(0)} \qquad (5.67)$$

with the tau-functions $\tau_0^{(0)}$ and $\tau_2^{(0)}$

$$\tau_0^{(0)} = 1 - \frac{1}{2} e^{(-2x z_1 - 2t_3 z_1{}^3)} \quad ; \quad \tau_2^{(0)} = 1 + \frac{1}{2} e^{(-2x z_1 - 2t_3 z_1{}^3)} \qquad (5.68)$$

from which we obtain U_2 and ν by using:

$$U_2 = -\partial_x \ln \left(\frac{\tau_0^{(0)}}{\tau_2^{(0)}} \right) \quad ; \quad \nu = -\partial_x \ln \left(\tau_0^{(0)} \right) \qquad (5.69)$$

Another two-soliton solution for which this time both $q \neq 0$ and $r \neq 0$ is:

$$\tau_\sigma^{(0)} = 1 + (-1)^{(\sigma/2)}\, 2\, \frac{z_1{}^{1+\sigma/2}\, z_2{}^{2-\sigma/2}\, e^{(x z_1 + t z_1{}^2 + t_3 z_1{}^3 + x z_2 - t z_2{}^2 + t_3 z_2{}^3)}}{(z_1 - z_2)(z_1 + z_2)^2} \qquad (5.70)$$

with $\sigma = 0, 2$, and

$$r = \frac{\sqrt{2}\, z_2\, e^{(-t z_2{}^2 + x z_2 + t_3 z_2{}^3)}}{\tau_2^{(0)}} \quad ; \quad q = \frac{\sqrt{2}\, z_1\, e^{(t z_1{}^2 + x z_1 + t_3 z_1{}^3)}}{\tau_0^{(0)}} \qquad (5.71)$$

and again U_2 and ν can be obtained from (5.69).

In the above examples we only kept the times t_n with $n \leq 3$ for which we verified validity of the relevant evolution equations.

The novel feature of the above soliton solutions is that they mix exponentials $\exp\left(\sum_{n=1}^{\infty} t_n z_j^n\right)$ which represent a typical time dependence for the KP solutions with pure KdV time dependence of the type $\exp\left(\sum_{n=0}^{\infty} t_{2n+1} z_j^{2n+1}\right)$ involving only odd times (Aratyn et al (1997b)).

6   APPENDIX

We now provide some useful relations extensively used in obtaining the formulas given in the text:

$$\sum_{n=0}^{\infty} \frac{x^{a+(K+1)n}}{a + (K+1)n} = \frac{-1}{K+1} \sum_{p=1}^{K+1} \omega^{ap} ln(1 - x\omega^{-p}) \quad ; \quad a = 1, \ldots, K \qquad (6.72)$$

for
$$\omega^{K+1} = 1 \ . \tag{6.73}$$

It also follows that
$$1 + \omega + \cdots + \omega^K = 0 \tag{6.74}$$

and
$$(1 - x^{K+1}) = \prod_{p=1}^{K+1} (1 - \omega^p x) \tag{6.75}$$

from where we obtain, after using L'Hopital's rule

$$K + 1 = \prod_{p=1}^{K} (1 - \omega^p) \tag{6.76}$$

and

$$\sum_{p=1}^{K} \frac{1}{1 - \omega^p} = \frac{K}{2} \tag{6.77}$$

Acknowledgements A.H.Z. would like to thanks UIC and FAPESP for the hospitality and financial support. L.A.F., J.F.G. and A.H.Z. acknowledge partial financial support from CNPq. H.A.'s work was supported in part by the U.S. Department of Energy Grant No. DE-FG02-84ER40173.

References

Aratyn H., Ferreira L.A., Gomes J.F. and Zimerman A.H. (1997): *Jour. Math. Phys.* **38** 1559 (hep-th/9509096)

Aratyn H., Ferreira L.A., Gomes J.F. and Zimerman A.H. (1997): "Solitons from dressing in an Algebraic aproach to the constrained KP hierarchy", solv-int/9709004

Aratyn, H., Gomes, J.F. and Zimerman A.H. (1995): *Journ. Math. Phys.* **36** 3419 (hep-th/9408104)

Burroughs N.J., de Groot M.F., Hollowood T.J. and Miramontes J.L. (1993): *Commun. Math. Phys.* **153** 187 (hep-th/9109014); *Phys. Lett.* **B277** (1992) 89 (hep-th/9110024)

Drinfel'd V.G. and Sokolov V.V. (1985): *Journal of Soviet Math.* **30** 1975; Soviet Math. Dokl. **23** (1981) 457.

Ferreira L.A., Miramontes J.L. and Sánchez Guillén J. (1997): *Jour. Math. Phys.* **38** 882 (hep-th/9606066), see also Ferreira L.A. and Sánchez Guillén J., *Solitons and Generalized Tau-Functions for Affine Integrable Hierarchies*, in this volume

Fordy A.P and Kulish P.P. (1983): *Commun. Math. Phys.* **89** 427; A.P. Fordy, in *Soliton Theory: a Survey of Results*, (ed. A.P. Fordy) University Press, Manchester (1990), pg. 315

Goddard P. and Olive D. (1986): *Int. J. Mod. Phys.* **A1** 303 de Groot M.F., Hollowood T.J. and Miramontes J.L. (1992): *Commun. Math. Phys.* **145** 57

Kac V.G. and Peterson D.H. (1985): in *Symposium on Anomalies, Geometry and Topology*, W.A. Bardeen and A.R. White (eds.), Singapore, World Scientific pgs. 276-298; Kac V.G. *Infinite Dimensional Lie Algebras (3rd ed.)*, Cambridge University Press, Cambridge (1990).

McIntosh I. (1993): *Journ. Math. Phys.* **34** 5159

Part II

Supersymmetric Integrable Models

Part II

Supersymmetric
Integrable Models

Zero Curvature Formalism in Superspace

Henrik Aratyn[1], Ashok Das[2], Constantin Rasinariu[1] and A.H. Zimerman[3]

[1] University of Illinois at Chicago, Department of Physics, Chicago, IL 60607
[2] University of Rochester, Department of Physics, Rochester, NY 14627
[3] Instituto de Física Teórica - IFT/UNESP, Rua Pamplona 145, 01405-900, São Paulo - SP, Brazil

Abstract. We discuss the generalization of the Drinfeld-Sokolov formalism to superspace in the case of supersymmetric integrable systems. We bring out the connection between this and the zero curvature formulations, for these systems, in components. Various examples are worked out in detail. We show how the symmetric space techniques can be profitably used to solve the zero curvature conditions and to obtain the hierarchy of equations in the case of sAKNS hierarchy. The dynamical equations are formulated also as the Cartan-Maurer equations in superspace and we show how the zero curvature condition can be solved, alternately, by expanding potentials according to the grading charge. We also discuss the sTB-B hierarchy to bring out some open questions.

1 Introduction

The bosonic integrable systems have been quite well studied from various points of view [1,2]. For example, it is well known that a bosonic integrable system can be represented in the form of a scalar Lax equation [3,4]

$$\frac{\partial L}{\partial t} = [B, L] \tag{1}$$

where the Lax pair, L and B, are, in general, pseudo-differential operators with scalar coefficient functions. Thus, for example, we know that the n-th generalized KdV hierarchy is described in terms of the Lax pair

$$L = \partial^n + u_{n-2}(x)\partial^{n-2} + u_{n-3}(x)\partial^{n-3} + \ldots + u_0(x)$$

$$B = (L^{\frac{k}{n}})_+ \qquad k \neq mn \tag{2}$$

where $\partial = \frac{\partial}{\partial x}$ while $u_0(x), \ldots, u_{n-2}(x)$ represent the dynamical variables of the system (time dependence is suppressed) and $(\ )_+$ denotes the part of the pseudo-differential operator with nonnegative powers. The dynamical equations are given by

$$\frac{\partial L}{\partial t} = [(L^{\frac{k}{n}})_+, L] \tag{3}$$

The scalar Lax equation, Eq. (1), can be easily seen as the compatibility condition for the system of linear equations

$$L(t)\psi = \lambda\psi$$
$$\frac{\partial\psi}{\partial t} = B\psi \tag{4}$$

Namely, the constancy of the spectral parameter, λ, leads to the Lax equation (1) as a consistency condition for the two equations. As a concrete example, let us consider the KdV equation (The KdV hierarchy is obtained from (2) by putting $n = 2$.)

$$L = \partial^2 + u(x)$$
$$B = (L^{3/2})_+ = \partial^3 + \frac{3}{4}(\partial u + u\partial) \tag{5}$$

so that the Lax equation (1) yields the KdV equation

$$\frac{\partial u}{\partial t} = \frac{3}{2}u\frac{\partial u}{\partial x} + \frac{1}{4}\frac{\partial^3 u}{\partial x^3} \tag{6}$$

Through a redefinition of variables, this can be brought to the standard form

$$\frac{\partial u}{\partial t} = u\frac{\partial u}{\partial x} + \frac{\partial^3 u}{\partial x^3}$$

However, this is not relevant for our discussions. The corresponding linear equations, which are the basis of the inverse scattering theory, are given by

$$(\partial^2 + u)\psi = \lambda\psi$$
$$\frac{\partial\psi}{\partial t} = (\partial^3 + \frac{3}{4}(\partial u + u\partial))\psi \tag{7}$$

There exists an alternate but equivalent description of bosonic integrable models, which is commonly known as the zero curvature formulation [5]. It is known that a bosonic integrable equation can be expressed as a vanishing curvature for potentials belonging to some Lie algebra, namely, for A_0, $A_1 \in G$, the dynamical equations can be written as

$$\partial_t A_1 - \partial_x A_0 - [A_0, A_1] = 0 \tag{8}$$

The Lie algebra, G, is known to be related to the second Hamiltonian structure of the theory. For KdV, which is the earliest system studied in this framework, for example, A_0, $A_1 \in$ SL(2,**R**) and can be chosen to have the form

$$A_1 = \begin{pmatrix} \sqrt{\lambda} & -u \\ 1 & -\sqrt{\lambda} \end{pmatrix}$$

$$\tag{9}$$

$$A_0 = \begin{pmatrix} \frac{1}{2}C_x + \sqrt{\lambda}C & -\frac{1}{2}C_{xx} - uC - \sqrt{\lambda}C_x \\ C & -\frac{1}{2}C_x - \sqrt{\lambda}C \end{pmatrix}$$

(The subscript 'x' denotes derivative with respect to 'x'.) and it can be easily checked that the zero curvature condition,

$$\partial_t \mathcal{A}_1 - \partial_x \mathcal{A}_0 - [\mathcal{A}_0, \mathcal{A}_1] = 0$$

leads to the KdV hierarchy. (For $\lambda = 0$ and $C = \frac{1}{2}u$, this reproduces Eq. (6).) We note here that SL(2,$\mathbf{R}$) is simply the finite dimensional subalgebra of the Virasoro algebra which corresponds to the second Hamiltonian structure of the KdV equation (hierarchy). Let us also note that the zero curvature condition of (8) can, in general, be thought of as the consistency condition for the linear equations

$$\partial_x \Psi = \mathcal{A}_1 \Psi$$
$$\partial_t \Psi = \mathcal{A}_0 \Psi \tag{10}$$

The difference from (4), however, lies in the fact that (10) represents a set of matrix equations and more importantly, the potentials, $\mathcal{A}_0$ and $\mathcal{A}_1$, are matrix functions and not operators.

The work of Drinfeld and Sokolov [6,7], among other things, clarified the connection between these two descriptions of bosonic integrable models. In fact, the work of Drinfeld-Sokolov gives a systematic procedure for constructing the zero curvature formulation given the scalar Lax equation and *vice versa*. To see how the method works, let us consider the example of the n-th generalized KdV hierarchy described in Eqs. (2)-(3). The linear equation, in this case, has the form

$$L\psi = \lambda\psi$$

or, $$(\partial^n + u_{n-2}\partial^{n-2} + u_{n-3}\partial^{n-3} + \ldots + u_0)\psi = \lambda\psi \tag{11}$$

This is a n-th order partial differential equation and can be easily expressed as n-coupled first order equations. Thus, for example, let us define

$$\psi_1 = (\partial\psi_2)$$
$$\psi_2 = (\partial\psi_3)$$
$$\vdots$$
$$\psi_{n-1} = (\partial\psi_n) = (\partial\psi) \tag{12}$$

It is clear that in terms of these new variables, Eq. (11) can be written as a first order equation

$$(\partial\psi_1) + u_{n-2}\psi_2 + u_{n-3}\psi_3 + \ldots + (u_0 - \lambda)\psi_n = 0 \tag{13}$$

In fact, defining a n-component matrix wave function

$$\Psi = \begin{pmatrix} \psi_1 \\ \psi_2 \\ \vdots \\ \psi_n \end{pmatrix} \tag{14}$$

we see that the system of equations (12)-(13) which is equivalent to Eq. (11) can be written as a linear, first order matrix equation

$$\partial_x \Psi = \mathcal{A}_1 \Psi \tag{15}$$

where we can easily identify

$$\mathcal{A}_1 = \begin{pmatrix} 0 & -u_{n-2} & -u_{n-3} & \cdots & -(u_0 - \lambda) \\ 1 & 0 & 0 & \cdots & 0 \\ 0 & 1 & 0 & \cdots & 0 \\ \vdots & & 0 & & \vdots \\ \vdots & \vdots & & & \vdots \\ 0 & 0 & 0 & \cdots 1 & 0 \end{pmatrix} \tag{16}$$

Similarly, the time evolution equation (4) can also be written in the form

$$\partial_t \Psi = \mathcal{A}_0 \Psi \tag{17}$$

However, the form of $\mathcal{A}_0$ is, in general, much more complicated and, for simplicity, we simply note here that it can be systematically obtained. Thus, we see that starting from the scalar Lax equation and the associated linear problem, we have constructed a matrix linear equation of the form (10). The consistency of these equations, as we have noted earlier, leads to the zero curvature formulation and we note from the upper triangular structure of the potential in (16) (as well as $\mathcal{A}_0$ in (17) which we have not explicitly written) that $\mathcal{A}_0$, $\mathcal{A}_1 \in \mathrm{SL}(n, \mathbf{R})$. This shows how one can obtain the zero curvature formulation starting from the scalar Lax equation. It is also straightforward to show that given a zero curvature formulation, we can systematically obtain the scalar Lax equation as well.

As a specific example, let us consider the KdV hierarchy (which corresponds to the case $n = 2$ in (11)) for which we obtain from Eq. (16)

$$\mathcal{A}_1 = \begin{pmatrix} 0 & -(u - \lambda) \\ 1 & 0 \end{pmatrix} \tag{18}$$

where we have identified $u_0 = u$. However, we note that the potential in Eq. (18) is different from the one discussed earlier in Eq. (9). On the other hand, it is easy to show that the two potentials are related by a similarity transformation, namely,

$$\begin{pmatrix} 0 & -(u - \lambda) \\ 1 & 0 \end{pmatrix} = S \begin{pmatrix} \sqrt{\lambda} & -u \\ 1 & -\sqrt{\lambda} \end{pmatrix} S^{-1} \tag{19}$$

where

$$S = \begin{pmatrix} 1 & -\sqrt{\lambda} \\ 0 & 1 \end{pmatrix} \tag{20}$$

This brings out an important feature of the zero curvature formulation. Namely, the potentials are unique only up to similarity transformations. We note that the difference in the two potentials in Eqs. (9) and (18) is mainly in how the spectral parameter, λ, is extracted. In Eq. (18), the spectral parameter has the form

$$\lambda \begin{pmatrix} 0 & 1 \\ 0 & 0 \end{pmatrix} = \lambda \sigma_+ \qquad (21)$$

while in Eq. (9), it takes the form

$$\sqrt{\lambda} \begin{pmatrix} 1 & 0 \\ 0 & -1 \end{pmatrix} = \sqrt{\lambda}\, \sigma_3 \qquad (22)$$

We recognize that σ_+ is a singular element of $SL(2,\mathbf{R})$ and its adjoint action is nilpotent while σ_3 has a semi-simple adjoint action. Extracting the spectral parameter from $\mathcal{A}_1$, we note that a linear matrix equation can be written as

$$(\partial_x - \mathcal{A}_1 - \lambda E)\Psi = 0 \qquad (23)$$

When E corresponds to a maximally nonregular element of the Lie algebra (in other words, it has the largest possible degeneracy of eigenvalues), the linear equations are known to generate the AKNS hierarchy of equations [8]. These encompass almost all of the known bosonic integrable models. In discussing the supersymmetric integrable systems later, we will restrict ourselves to super AKNS (sAKNS) hierarchies mainly because powerful techniques based on the properties of symmetric spaces become available in such a case.

Although a lot is known about bosonic integrable models, there remains a lot to be studied in the case of supersymmetric integrable systems. For example, the supersymmetric integrable systems are naturally described by Lax operators and Lax equations in superspace [9-11]. On the other hand, apart from a few isolated works [12-13], the zero curvature formulation for most supersymmetric integrable systems exists only in components [14-17]. Thus, a zero curvature formulation of the supersymmetric integrable systems as well as the generalization of the Drinfeld-Sokolov formalism in superspace is lacking so far. Nor is the connection between the "possible" zero curvature formulation in superspace and that in components clear. Thus, for example, it is known that the supersymmetric KdV equation is described by the Lax equation

$$\frac{\partial L}{\partial t} = [B, L] \qquad (24)$$

where

$$L = D^4 + D\phi$$

$$B = (L^{3/2})_+ = D^6 + \frac{3}{4}(D^3\phi + D\phi D^2)$$

$$\phi = \psi + \theta u = \text{ fermionic superfield}$$

$$D = \frac{\partial}{\partial \theta} + \theta \frac{\partial}{\partial x} \Rightarrow D^2 = \partial \qquad (25)$$

whereas the zero curvature formulation is known in components with the potentials $\mathcal{A}_0$, $\mathcal{A}_1 \in OSp(2|1)$ [14-15] (We give only $\mathcal{A}_1$ for simplicity.)

$$\mathcal{A}_1 = \begin{pmatrix} 0 & 1 & 0 \\ -u & 0 & \psi \\ \psi & 0 & 0 \end{pmatrix} \tag{26}$$

The graded algebra $OSp(2|1)$ is expected since this is the finite dimensional subalgebra of the superconformal algebra which happens to be the second Hamiltonian structure of the supersymmetric KdV equation.

Similarly, it is also known that the supersymmetric TB (Two Boson) hierarchy is described by the Lax pair [18]

$$L = D^2 - (D\phi_0) + D^{-1}\phi_1$$
$$B = (L^2)_{\geq 1} \tag{27}$$

where

$$\phi_0 = \psi_0 + \theta J_0$$
$$\phi_1 = \psi_1 + \theta J_1 \tag{28}$$

and the nonstandard Lax equation

$$\frac{\partial L}{\partial t} = [B, L] = [(L^2)_{\geq 1}, L] \tag{29}$$

The supersymmetric TB equation is known to yield the susy KdV, susy mKdV, susy NLS, etc. equations under appropriate field redefinitions/reductions. Therefore, it is a more meaningful system to study. Once again, its zero curvature formulation is known only in components 17] where $\mathcal{A}_0$, $\mathcal{A}_1 \in OSp(2|2) \sim SL(2|1)$ with (The details of the algebra of $SL(2|1)$ are given in the Appendix.)

$$\mathcal{A}_1 = \begin{pmatrix} J_0 + \lambda & -1 & 0 \\ J_1 & 0 & -\psi_1 \\ \psi_0' - \psi_1 & 0 & J_0 + \lambda \end{pmatrix} \tag{30}$$

Here prime denotes a derivative with respect to 'x'. The graded algebra $OSp(2|2)$ is again expected since it is the finite subalgebra of the $N = 2$ superconformal algebra which is the second Hamiltonian structure of the susy TB equation.

In this talk, we would generalize the Drinfeld-Sokolov formalism to superspace and, thereby, obtain the zero curvature formulation in superspace. We will clarify the connection between the zero curvature formulations in superspace and in components. For the susy AKNS hierarchies [19,20], we will show how the symmetric space techniques can be used to obtain the equations of the hierarchy directly. We will also obtain the Cartan-Maurer equations in superspace which will lead to a manifestly covariant zero curvature formulation in the superspace. We describe how the zero curvature condition can be solved by expanding the potentials according to the grading charge. Finally, we would end with some puzzling features associated with the zero curvature formulation of the susy TB-B equation [21]. The details of the algebra SL(2|1) and other technical results such as the vertex operators and soliton dressing are given in the Appendix.

2 Zero Curvature Formulation in Superspace

In this section, we will generalize the Drinfeld-Sokolov formalism to super-space. Namely, we would obtain the zero curvature formulation in superspace starting from the scalar Lax equation.

2.1 Susy KdV

To begin with, let us consider the susy KdV equation for which the Lax pair is given in Eq. (25). Thus, the linear equation can be written as $(\lambda = 0)$

$$L\chi = 0$$
$$\text{or,} \qquad (D^4 + D\phi)\chi = 0$$
$$\text{or,} \qquad (D^3 + \phi)\chi = 0 \tag{31}$$

where we assume that χ is a bosonic superfield wave function. As in Eq. (12), let us now define

$$\chi_1 = (D\chi_2)$$
$$\chi_2 = (D\chi_3) = (D\chi) \tag{32}$$

so that Eq. (31) can be written as

$$(D\chi_1) + \phi\chi_3 = 0 \tag{33}$$

Now introducing a matrix wave function

$$\Psi = \begin{pmatrix} \chi_1 \\ \chi_3 \\ \chi_2 \end{pmatrix} \tag{34}$$

we note that it is naturally graded with the first two elements bosonic and the last fermionic. Furthermore, the system of Eqs. (32)-(33) can be written in the matrix form

$$D\Psi = D \begin{pmatrix} \chi_1 \\ \chi_3 \\ \chi_2 \end{pmatrix} = \begin{pmatrix} 0 & -\phi & 0 \\ 0 & 0 & 1 \\ 1 & 0 & 0 \end{pmatrix} \begin{pmatrix} \chi_1 \\ \chi_3 \\ \chi_2 \end{pmatrix} \tag{35}$$

which immediately yields

$$\partial_x \Psi = D^2 \Psi = \mathcal{A}_1 \Psi \tag{36}$$

with

$$\mathcal{A}_1 = \begin{pmatrix} 0 & -(D\phi) & \phi \\ 1 & 0 & 0 \\ 0 & -\phi & 0 \end{pmatrix} \tag{37}$$

Similarly, we can obtain the potential $\mathcal{A}_0$ from the second linear equation (time evolution equation). However, we will not go into determining $\mathcal{A}_0$ except for noting that $\mathcal{A}_0$, $\mathcal{A}_1 \in \mathrm{OSp}(2|1)$ and lead to the zero curvature condition in superspace starting from the scalar Lax equation. It is straightforward to convince oneself that the converse is also true, namely, given the potentials $\mathcal{A}_0$, $\mathcal{A}_1$ and the zero curvature condition, we can construct the linear equations and the associated scalar Lax equation.

We note that the potential $\mathcal{A}_1$ (and also $\mathcal{A}_0$) are graded matrix functions of the superfield ϕ. Thus, we can expand and write

$$\mathcal{A}_1 = a_1 + \theta \tilde{a}_1 \tag{38}$$

where, clearly, the θ-independent part can be identified from Eq. (37) with

$$a_1 = \begin{pmatrix} 0 & -u & \psi \\ 1 & 0 & 0 \\ 0 & -\psi & 0 \end{pmatrix} \tag{39}$$

It is now straightforward to compare and see that a_1 coincides with the component potential in Eq. (26) (up to a supertransposition. It is worth noting here that supertransposition of a graded matrix introduces additional negative signs depending on the degree of the matrix [22].) Thus, it becomes clear that the θ-independent terms of the potentials in superspace would correspond to the component potentials.

2.2   Susy TB Hierarchy

Let us next look at the zero curvature formulation for the susy TB hierarchy [18] which can directly lead to the zero curvature formulation of various other equations upon appropriate field redefinitions/reductions. The Lax pair for the system is given in Eq. (27). The linear equation has the form

$$L\chi = (D^2 - (D\phi_0) + D^{-1}\phi_1)\chi = \lambda\chi \tag{40}$$

where we assume that χ represents a bosonic superfield wave function. Let us now define, in analogy with Eq. (12),

$$\chi_1 = \chi$$
$$\chi_2 = (D^{-1}\phi_1\chi_1)$$
$$\chi_3 = (D\chi_1) \tag{41}$$

so that Eq. (40) takes the form

$$(D^2 - (D\phi_0) - \lambda)\chi_1 + \chi_2 = 0 = D\chi_3 - ((D\phi_0) + \lambda)\chi_1 + \chi_2 \qquad (42)$$

If we now define a graded matrix function (χ_1, χ_2 bosonic; χ_3 fermionic)

$$\Psi = \begin{pmatrix} \chi_1 \\ \chi_2 \\ \chi_3 \end{pmatrix} \qquad (43)$$

then, from Eqs. (41) and (42), we see that we can write

$$D\Psi = \begin{pmatrix} 0 & 0 & 1 \\ \phi_1 & 0 & 0 \\ (D\phi_0) + \lambda & -1 & 0 \end{pmatrix} \Psi \qquad (44)$$

It now follows from this that

$$\partial_x \Psi = D^2 \Psi = \mathcal{A}_1 \Psi$$

where

$$(45)$$

$$\mathcal{A}_1 = \begin{pmatrix} (D\phi_0) + \lambda & -1 & 0 \\ (D\phi_1) & 0 & -\phi_1 \\ (D^2\phi_0) - \phi_1 & 0 & (D\phi_0) + \lambda \end{pmatrix}$$

Clearly $\mathcal{A}_1 \in \mathrm{OSp}(2|2) \sim \mathrm{SL}(2|1)$. Defining

$$\partial_t \Psi = \mathcal{A}_0 \Psi = \begin{pmatrix} A & B & C \\ E & F & G \\ H & J & (A+F) \end{pmatrix} \qquad (46)$$

it is straightforward to check that the zero curvature condition

$$\partial_t \mathcal{A}_1 - \partial_x \mathcal{A}_0 - [\mathcal{A}_0, \mathcal{A}_1] = 0$$

leads to the susy TB hierarchy if

$$A = (DC) - B((D\phi_0) + \lambda) + B_x$$
$$E = (DC_x) - (D(B\phi_1)) + \phi_1 C$$
$$F = (DC)$$
$$G = B\phi_1 - C_x$$
$$H = (DB_x) + C_x - (D(B((D\phi_0) + \lambda))) - ((D\phi_0) + \lambda)C + B\phi_1$$
$$J = ((DB) + C) \qquad (47)$$

This defines the zero curvature formulation of the susy TB hierarchy in superspace starting from the Lax equation. We note that $\mathcal{A}_0$, $\mathcal{A}_1 \in \mathrm{OSp}(2|2) \sim \mathrm{SL}(2|1)$.

If we now write, as in Eq. (38),

$$\mathcal{A}_1 = a_1 + \theta \tilde{a}_1$$

it follows that the θ-independent component is given by

$$a_1 = \begin{pmatrix} J_0 + \lambda & -1 & 0 \\ J_1 & 0 & -\psi_1 \\ \psi_0' - \psi_1 & 0 & J_0 + \lambda \end{pmatrix} \tag{48}$$

This is seen to coincide with Eq. (30). We note here that the susy KdV equation corresponds to the second equation in the susy TB hierarchy when we set ϕ_0 equal to zero. In this case, we see from Eq. (48) that

$$a_1(\phi_0 = 0, \lambda = 0) = \begin{pmatrix} 0 & -1 & 0 \\ J_1 & 0 & -\psi_1 \\ -\psi_1 & 0 & 0 \end{pmatrix} \tag{49}$$

This coincides with Eq. (26) (up to an overall negative sign) if we identify $J_1 = u$ and $\psi_1 = \psi$.

2.3   Susy AKNS Hierarchy

As I have mentioned earlier, like the AKNS hierarchy, the susy AKNS hierarchy [19-20] yields most of the supersymmetric integrable models upon appropriate field redefinitions/reductions. Here I will work out the zero curvature formulation for this system in some detail for the $N = 1$ supersymmetry case without any (bosonic) global symmetry, for simplicity, although the general case can be worked out as easily. The Lax pair, in this case, is known to be

$$L = D^2 + \phi D^{-1} \psi$$
$$B = (L^n)_+ \tag{50}$$

so that the equations of the hierarchy are given by

$$\frac{\partial L}{\partial t} = [B, L] = [(L^n)_+, L] \tag{51}$$

Here ϕ and ψ are bosonic and fermionic superfields respectively.

The linear equation, in this case, takes the form

$$L\chi = (D^2 + \phi D^{-1} \psi)\chi = \lambda \chi \tag{52}$$

where we assume, as before, that χ is a bosonic superfield wave function. Let us now define

$$\chi_1 = \chi$$
$$\chi_2 = (D^{-1} \psi \chi_1)$$
$$\chi_3 = (D\chi_1) \tag{53}$$

so that the linear equation, in terms of these variables, becomes

$$(D^2 - \lambda)\chi_1 + \phi\chi_2 = 0 = D\chi_3 - \lambda\chi_1 + \phi\chi_2 \tag{54}$$

But, we also note from Eqs. (53) and (54) that if we define a matrix wave function

$$\Psi = \begin{pmatrix} \chi_1 \\ \chi_2 \\ \chi_3 \end{pmatrix} \tag{55}$$

Then, we can write Eqs. (53)-(54) in the matrix form

$$D\Psi = \begin{pmatrix} 0 & 0 & 1 \\ \psi & 0 & 0 \\ \lambda & -\phi & 0 \end{pmatrix} \Psi \tag{56}$$

It now follows from this that

$$\partial_x \Psi = D^2 \Psi = \mathcal{A}_1 \Psi$$

where

$$\tag{57}$$

$$\mathcal{A}_1 = \begin{pmatrix} \lambda & -\phi & 0 \\ (D\psi) & 0 & -\psi \\ -\phi\psi & -(D\phi) & \lambda \end{pmatrix}$$

We note that Ψ is a graded matrix wave function with χ_1, χ_2 bosonic superfields while χ_3 is a fermionic superfield. Correspondingly, we note that $\mathcal{A}_1 \in$ SL(2|1). On the other hand, if we had considered a susy AKNS problem with N extended supersymmetries and m (bosonic) global symmetries, then, it is straightforward to show that the corresponding graded wave function would have the dimension

$$\dim[\Psi] = (2^N + m(2^N - 1)) \tag{58}$$

The corresponding potential, $\mathcal{A}_1 \in$ SL$((m+1)2^{N-1}|(m+1)2^{N-1} - m)$ (This would be the largest possible graded algebra that the potentials can belong to.). For $m = 1$, we see that we can identify $\mathcal{A}_1 \in$ SL$(n+1|n)$ where $n = 2^N - 1$. This shows how the graded linear algebras naturally arise in the zero curvature formulation of susy AKNS hierarchies.

Let us next define the linear equation for time evolution to be

$$\partial_t \Psi = \mathcal{A}_0 \Psi \tag{59}$$

with

$$\mathcal{A}_0 = \begin{pmatrix} A & B & C \\ E + C\psi & F & G \\ H + \lambda C & J - \phi C & A + F \end{pmatrix} \tag{60}$$

Then, it is simple to check that the consistency of Eqs. (56) and (59) would give rise to the zero curvature condition which would describe the susy AKNS hierarchy if

$$E = -(DG)$$
$$F = (DC)$$
$$H = (DA) + B\psi$$
$$J = (DB)$$
$$C_x = -G\phi + B\psi$$
$$A_x = \phi(DG) - B(D\psi) + 2C\phi\psi \tag{61}$$

Here we have assumed that B and G are the two independent superfields (bosonic and fermionic respectively) which would describe the hierarchy of equations. Incidentally, restricting to the θ-independent terms in the above potentials would again lead to the zero curvature formulation of the susy AKNS hierarchy in components, which was not known earlier.

These examples show how the Drinfeld-Sokolov formalism can be generalized to superspace. Although we have described how we can obtain the zero curvature formulation in superspace starting from the scalar Lax equation, the converse is also equally straightforward.

3   Symmetric Space Techniques

The algebras $SL(n)$ and $SL(n+1|n)$ allow for symmetric spaces and, in such a case, one can directly obtain the hierarchy of equations without having to explicitly determine $\mathcal{A}_0$ which is usually the hardest part of the calculation. In this section, we will show how these powerful methods can be used to obtain the hierarchy of susy AKNS equations [23].

To begin with, let us recapitulate some of the notions associated with symmetric spaces. Let us consider a general (graded) Lie algebra G and a specific element E in it. The adjoint map is defined to be

$$ad_E(X) = [E, X] \tag{62}$$

where X is any element of the algebra. We can now denote the kernel and the image of this map to be

$$K = \ker(ad_E)$$
$$M = \mathrm{Im}(ad_E) \tag{63}$$

It is clear, then, that

$$[K, K] \subset K \ ; \ [K, M] \subset M \ ; \ [M, M] \subset K \tag{64}$$

If for some element, E, we can write

$$G = K + M \tag{65}$$

then, we say that the algebra, G, allows for a symmetric space description.

Let us now consider Eqs. (56) and (57) and note that, if we extract the spectral parameter explicitly, we can write the linear equation as

$$(\partial_x - \lambda E - \mathcal{A}_1)\Psi = 0 \qquad (66)$$

where

$$E = \begin{pmatrix} 1 & 0 & 0 \\ 0 & 0 & 0 \\ 0 & 0 & 1 \end{pmatrix} \qquad \mathcal{A}_1 = \begin{pmatrix} 0 & -\phi & 0 \\ (D\psi) & 0 & -\psi \\ -\phi\psi & -(D\phi) & 0 \end{pmatrix} \qquad (67)$$

It is straightforward to check that

$$K = \ker(ad_E) = \begin{pmatrix} b_1 & 0 & f_1 \\ 0 & b_2 & 0 \\ f_2 & 0 & b_1 + b_2 \end{pmatrix}$$

$$M = \mathrm{Im}(ad_E) = \begin{pmatrix} 0 & b_1 & 0 \\ b_2 & 0 & f_1 \\ 0 & f_2 & 0 \end{pmatrix} \qquad (68)$$

where b and f represent respectively the bosonic and the fermionic elements of the graded matrix. Furthermore, it is clear from Eq. (68) that any element of SL(2|1) can be written as the sum $K + M$. Thus, SL(2|1) allows for a symmetric space description.

The linear equations can be written, with the spectral parameter explicitly taken out, as

$$(\partial_x - \lambda E - \mathcal{A}_1)\Psi = 0$$
$$(\partial_t - \mathcal{A}_0)\Psi = 0 \qquad (69)$$

We note that, in general, $\mathcal{A}_1$ will have components both in the K as well as M directions, namely, we can write

$$\mathcal{A}_1 = \mathcal{A}_1^K + \mathcal{A}_1^M \qquad (70)$$

In fact, we see from Eq. (67) that the term $(-\phi\psi)$ in the left hand corner belongs to the K direction – without this, $\mathcal{A}_1$ would be completely along the M direction. Let us next make a gauge transformation which would remove the K-component of $\mathcal{A}_1$. Namely, let

$$\bar{\Psi} = G^{-1}\Psi$$

$$\bar{\mathcal{A}}_\mu = G^{-1}\mathcal{A}_\mu G + G^{-1}\partial_\mu G$$

$$G = \exp(-\int^x \mathcal{A}_1^K \, dx') \qquad (71)$$

In terms of these variables, the linear equations (69) would take the form

$$(\partial_x - \lambda E - \overline{\mathcal{A}}_1^M)\overline{\Psi} = 0$$

$$(\partial_t - \overline{\mathcal{A}}_0^K - \overline{\mathcal{A}}_0^M)\overline{\Psi} = 0 \tag{72}$$

We note that while the transformation (71) is designed to remove the K-component of $\mathcal{A}_1$, in general, $\mathcal{A}_0$ would have both the components as is denoted in Eq. (72). These components would be, in general, complicated (may be even nonlocal), however, we would see that we do not need to know the explicit form of $\overline{\mathcal{A}}_0$ to obtain the hierarchy of equations.

If we ignore the "overlines" in the variables in Eq. (72) for simplicity, we note that the consistency of the two equations would give rise to the zero curvature condition which would now have components along the independent K and M directions each of which has to vanish. Thus, we will have ($\mathcal{A}_1$ is along M direction which we do not show explicitly.)

$$K: \qquad \partial_x \mathcal{A}_0^K - [\mathcal{A}_1, \mathcal{A}_0^M] = 0$$

$$M: \qquad \partial_t \mathcal{A}_1 - \partial_x \mathcal{A}_0^M + \lambda[E, \mathcal{A}_0^M] + [\mathcal{A}_1, \mathcal{A}_0^K] = 0 \tag{73}$$

The first equation is a constraint equation which can be readily solved to give

$$\mathcal{A}_0^K = (\partial^{-1}[\mathcal{A}_1, \mathcal{A}_0^M]) + \lambda^n E$$

$$= \partial^{-1}(ad_{\mathcal{A}_1}(\mathcal{A}_0^M)) + \lambda^n E \tag{74}$$

We note that the second term in Eq. (74) is a constant of integration necessary if we are interested in the n-th equation of the hierarchy.

Substituting the solution of Eq. (74) in the second of the equations in (73), we obtain

$$\partial_t \mathcal{A}_1 - \partial_x \mathcal{A}_0^M + \lambda[E, (\mathcal{A}_0^M - \lambda^{n-1}\mathcal{A}_1)] + [\mathcal{A}_1, \partial^{-1}([\mathcal{A}_1, \mathcal{A}_0^M])] = 0 \tag{75}$$

Since $\mathcal{A}_1$ contains only the dynamical variables (the spectral parameter has been explicitly taken out), Eq. (75) would make sense only if $\mathcal{A}_0^M$ has a dependence on λ. Writing an expansion in the standard form

$$\mathcal{A}_0^M = \sum_{i=0}^{n-1} \lambda^i \mathcal{A}_{0(i)}^M \tag{76}$$

and substituting this back into Eq. (75), we note that matching of different powers of λ leads to

$$\lambda^n: \qquad \mathcal{A}_{0(n-1)}^M = \mathcal{A}_1 \tag{77}$$

$$\lambda^i: \qquad -\partial_x \mathcal{A}_{0(i)}^M + [E, \mathcal{A}_{0(i-1)}^M] + [\mathcal{A}_1, \partial^{-1}([\mathcal{A}_1, \mathcal{A}_{0(i)}^M])] = 0$$

$$\text{or,} \qquad ad_E(\mathcal{A}_{0(i-1)}^M) = \partial_x \mathcal{A}_{0(i)}^M - ad_{\mathcal{A}_1}(\partial^{-1} ad_{\mathcal{A}_1}(\mathcal{A}_{0(i)}^M))$$

$$\text{or,} \qquad \mathcal{A}_{0(i-1)}^M = ad_E(\partial_x \mathcal{A}_{0(i)}^M - ad_{\mathcal{A}_1}\partial^{-1} ad_{\mathcal{A}_1}(\mathcal{A}_{0(i)}^M)) = R\, \mathcal{A}_{0(i)}^M \tag{78}$$

where we have used the fact that ad_E is idempotent on the space of M for E in Eq. (67). Eq. (78) defines a recursion relation between the successive coefficients of expansion in (76) with the recursion operator given by

$$R = ad_E(\partial - ad_{A_1}\partial^{-1}ad_{A_1}) \tag{79}$$

Once E and A_1 are given, the recursion operator is uniquely determined by Eq. (79).

The dynamical equations are obtained by looking at the coefficients of λ^0 terms in (75) which give

$$\partial_t A_1 - \partial_x A_{0(0)}^M + [A_1, \partial^{-1}([A_1, A_{0(0)}^M])] = 0$$

$$\text{or,} \quad ad_E(\partial_t A_1) = RA_{0(0)}^M = R^2 A_{0(1)}^M = \ldots = R^n A_{0(n-1)}^M = R^n A_1 \tag{80}$$

where we have used the identification in Eq. (77). This shows that once E and A_1 are known, we can determine the recursion operator R as well as any equation of the hierarchy directly without having to solve for A_0 explicitly.

4   Cartan-Maurer Equation in Superspace

In section 3, we showed how the scalar Lax equation naturally leads to the zero curvature condition in superspace. The description, however, is not quite complete. This is because, unlike the bosonic manifold, the superspace is labelled by (t, x, θ). Consequently, the Cartan-Maurer form (connection) should have three independent components and not two as we have assumed. Thus, for a manifestly covariant zero curvature equation in superspace (for the case that we are discussing), we must have three linear equations of the form

$$\partial_x \Psi = A_1 \Psi = A_x \Psi$$
$$\partial_t \Psi = A_0 \Psi = A_t \Psi$$
$$\partial_\theta \Psi = A_\theta \Psi \tag{81}$$

The consistency of these equations would lead to a manifestly covariant zero curvature condition which would be equivalent to three conditions. The main question is whether supersymmetric integrable systems can be expressed in the form of Eq. (81) and, therefore, in terms of a manifestly covariant zero curvature condition in superspace.

The answer to the above question is in the positive. For example, for the susy AKNS hierarchy, we can choose A_1 and A_0 as before in Eqs. (57) and (60). The θ-equation is new and it is straightforward to check that if we choose

$$A_\theta = \begin{pmatrix} -\theta\lambda & \theta\phi & 1 \\ -\theta(D\psi) + \psi & 0 & \theta\psi \\ \theta\phi\psi + \lambda & \theta(D\phi) - \phi & -\theta\lambda \end{pmatrix} \tag{82}$$

Then, the zero curvature conditions following from Eq. (81) generate the susy AKNS hierarchies. There are several comments to be made here. First, the form of $\mathcal{A}_\theta$ is noncovariant simply because $\partial_\theta = \frac{\partial}{\partial\theta}$ is not a covariant derivative in superspace. Second, the form of Eq. (82) appears to have come out in a mysterious manner. However, for any system, the θ-equation can be systematically obtained from the x-equation simply as a square root because (see Eqs. (35), (44) and (56))

$$\partial_x = D^2 \tag{83}$$

We must also note here that the potentials $\mathcal{A}_1$, $\mathcal{A}_0$ and $\mathcal{A}_\theta$ are graded matrices involving anti-commuting elements. Consequently, caution must be used in defining the curvature (e.g. ∂_θ neither commutes nor anti-commutes with the potentials). When the curvatures are carefully defined, then, the vanishing of these leads to three conditions. As we have seen before

$$F_{xt} = 0 \tag{84}$$

leads to the hierarchy of equations. The other equations, on the other hand, generate the supersymmetry transformations for the various fields and serve as consistency conditions as well. In that sense, this formulation is more like the formulation of the BRS transformations as vanishing curvatures along the θ-directions only [24-26]. This, therefore, completes the derivation of the zero curvature condition both in the conventional form as well as in the manifestly covariant form in superspace.

5   Alternate Derivation of the Hierarchy

It is clear from the discussion of the earlier sections that the standard derivation of the hierarchy from the zero curvature equation follows from an expansion in the spectral parameter (see, for example, Eq. (76)). Here we give an alternate derivation of the hierarchy from the zero curvature condition by expanding in the grading of the potentials.

Let us consider the susy AKNS as a specific example. From Eq. (81), let us look at the θ and t equations.

$$(D - F - \mathcal{A})\Psi = 0$$
$$(\partial_t - \mathcal{A}_0)\Psi = 0 \tag{85}$$

where, as before,

$$D = \frac{\partial}{\partial\theta} + \theta\frac{\partial}{\partial x}$$

and from Eq. (82) (see also Eq. (57)) we note that we have separated the potential as

$$F = \begin{pmatrix} 0 & 0 & 1 \\ 0 & 0 & 0 \\ \lambda & 0 & 0 \end{pmatrix}$$

$$A = \begin{pmatrix} 0 & 0 & 0 \\ \psi & 0 & 0 \\ 0 & -\phi & 0 \end{pmatrix} \tag{86}$$

We note from the definitions in the Appendix that we can write these in terms of the generators of $SL(2|1)$ or $\widehat{SL}(2|1)$ as

$$F = F_{\alpha'} + \lambda F_{-\alpha'} = F_{\alpha'}(0) + F_{-\alpha'}(1)$$
$$A = \psi E_{-\alpha'''} - \phi F_{\alpha''} = \psi E_{-\alpha'''}(0) - \phi F_{\alpha''}(0) \tag{87}$$

We can define a grading operator for the elements of $\widehat{SL}(2|1)$ as

$$Q = \alpha'' \cdot H(0) + 2d = \alpha'' \cdot H(0) + 2\lambda \frac{d}{d\lambda} \tag{88}$$

which immediately gives

$$[Q, \alpha' \cdot H(n)] = 2n\alpha' \cdot H(n)$$
$$[Q, \alpha'' \cdot H(n)] = 2n\alpha'' \cdot H(n)$$
$$[Q, E_{\pm\alpha'''}(n)] = (2n \pm 1)E_{\pm\alpha'''}(n)$$
$$[Q, F_{\pm\alpha'}(n)] = (2n \pm 1)F_{\pm\alpha'}(n)$$
$$[Q, F_{\pm\alpha''}(n)] = 2nF_{\pm\alpha''}(n) \tag{89}$$

In other words, each generator of the algebra has a unique grading corresponding to this grading operator (charge operator). Furthermore, this grading is additive in the sense that if $A^{(q_1)}$ and $B^{(q_2)}$ have grading q_1 and q_2 respectively, then, $[A^{(q_1)}, B^{(q_2)}\}$ would have a grading $(q_1 + q_2)$ as can be easily checked with the use of (super) Jacobi identities. We note here that the grading of F, with this definition, is 1 whereas A does not have a unique grading, rather

$$A = A^{(-1)} + A^{(0)} \tag{90}$$

where

$$A^{(-1)} = \psi E_{-\alpha'''}(0)$$
$$A^{(0)} = -\phi F_{-\alpha''}(0) \tag{91}$$

The compatibility condition for the two equations in Eq. (85) can be easily seen to be

$$\partial_t A - (DA_0) - \tilde{A}_0(F + A) + (F + A)A_0 = 0 \tag{92}$$

where as we have emphasized earlier (see comment after Eq. (83)) A_0 is an even graded matrix containing even and odd superfields and, consequently, some of the elements change sign when D is commuted past A_0. $\tilde{A}_0$ is, therefore, the same matrix as A_0 with additional negaive signs for the fermionic superfield terms. Eq. (92) is not in a standard zero curvature form which involves (anti) commutators. However, we note that if we were to decompose the potentials along the directions of the bosonic and fermionic generators, we can write Eq. (92) equivalently as two conditions, namely,

$$\partial_t \mathcal{A}_B - (D\mathcal{A}_{0B}) - [\mathcal{A}_{0B}, \mathcal{A}_B] + \{\mathcal{A}_{0F}, F\} + \{\mathcal{A}_{0F}, \mathcal{A}_F\} = 0$$
$$\partial_t \mathcal{A}_F - (D\mathcal{A}_{0F}) - [\mathcal{A}_{0B}, F] - [\mathcal{A}_{0B}, \mathcal{A}_F] + [\mathcal{A}_{0F}, \mathcal{A}_B] = 0 \qquad (93)$$

where $\mathcal{A}_B$ and $\mathcal{A}_{0B}$ correspond to the potentials along the directions of the bosonic generators while $\mathcal{A}_F$ and $\mathcal{A}_{0F}$ are along the direction of the fermionic generators.

As we have emphasized earlier, the (graded) bracket between two elements have a unique grading charge. Thus, the two equations in (93) can be decomposed according to the grading charge. As we have seen before the grading charge of F is 1 while $\mathcal{A}$ which contains the dynamical variables has a mixed grading (see Eq. (90)-(91)). Correspondingly, the dynamical equations are given by

$$\partial_t \mathcal{A}_F^{(0)} - (D\mathcal{A}_{0F}^{(0)}) - [\mathcal{A}_{0B}^{(-1)}, F] - [\mathcal{A}_{0B}^{(0)}, \mathcal{A}_F^{(0)}] + [\mathcal{A}_{0F}^{(1)}, \mathcal{A}_B^{(-1)}] = 0$$
$$\partial_t \mathcal{A}_B^{(0)} - (D\mathcal{A}_{0B}^{(0)}) - [\mathcal{A}_{0B}^{(0)}, \mathcal{A}_B^{(-1)}] + \{\mathcal{A}_{0F}^{(-2)}, F\} + \{\mathcal{A}_{0F}^{(-1)}, \mathcal{A}_F^{(0)}\} = 0 \quad (94)$$

In addition, of course, we have the constraint equations

$$-(D\mathcal{A}_{0B}^{(0)}) - [\mathcal{A}_{0B}^{(1)}, \mathcal{A}_B^{(-1)}] + \{\mathcal{A}_{0F}^{(-1)}, F\} + \{\mathcal{A}_{0F}^{(0)}, \mathcal{A}_F^{(0)}\} = 0$$
$$-(D\mathcal{A}_{0F}^{(-1)}) - [\mathcal{A}_{0B}^{(-2)}, F] - [\mathcal{A}_{0B}^{(-1)}, \mathcal{A}_F^{(0)}] + [\mathcal{A}_{0F}^{(0)}, \mathcal{A}_B^{(-1)}] = 0 \qquad (95)$$

as well as

$$-(D\mathcal{A}_{0B}^{(i)}) - [\mathcal{A}_{0B}^{(i+1)}, \mathcal{A}_B^{(-1)}] + \{\mathcal{A}_{0F}^{(i-1)}, F\} + \{\mathcal{A}_{0F}^{(i)}, \mathcal{A}_F^{(0)}\} = 0$$
$$-(D\mathcal{A}_{0F}^{(i)}) - [\mathcal{A}_{0B}^{(i-1)}, F] - [\mathcal{A}_{0B}^{(i)}, \mathcal{A}_F^{(0)}] + [\mathcal{A}_{0F}^{(i+1)}, \mathcal{A}_B^{(-1)}] = 0 \qquad (96)$$

for $i \neq 0, -1$.

This is an infinite sequence of equations. However, for a given equation of the hierarchy, one can truncate the series in the following way. For example, for the n-th equation of the hierarchy, we can choose

$$\mathcal{A}_0^{(2n)} = \alpha' \cdot H(n) \qquad \mathcal{A}_0^{(i)} = 0 \qquad \text{for} \qquad i \geq 2n + 1 \qquad (97)$$

This is consistent because by choice $[\mathcal{A}_0^{(2n)}] = 0$. Every potential with a lower grading will have an increased canonical dimension as is clear from Eq. (96) so that

$$[\mathcal{A}_0^{(2n-m)}] = \frac{m}{2} \qquad (98)$$

Here we are using $[D] = \frac{1}{2}$. Furthermore, for this system, we can identify $[\phi] = \frac{1}{2}$ and $[\psi] = 1$. It follows now, from Eq. (98), that the n-th order dynamical equations in (94) would have the correct canonical dimensions.

Each potential of a lower grading can be obtained by solving the constraint equations (96) recursively and the form of the potentials can be checked from the requirement of canonical dimensionality in (98) as well as from the canonical dimensions of ϕ and ψ. Through this one can obtain the true dynamical equations of the hierarchy with the potentials also truncating for gradings below a fixed negative grading.

Let us illustrate this with the example of the sAKNS equations which is the 2nd equation of the hierarchy. Correspondingly, according to our discussion in Eq. (97), we can choose

$$A_0^{(4)} = \alpha' \cdot H(2) \tag{99}$$

The potentials can now be solved to give

$$
\begin{aligned}
A_0^{(i)} &= 0 \qquad \text{for} \qquad i \geq 5 \\
A_0^{(4)} &= \alpha' \cdot H(2) \\
A_0^{(3)} &= -\phi E_{\alpha'''}(1) \\
A_0^{(2)} &= -(D\phi)F_{\alpha''}(1) - \psi F_{-\alpha''}(1) \\
A_0^{(1)} &= -(D^2\phi)E_{\alpha'''}(0) + (D\psi)E_{-\alpha'''}(1) - \phi\psi F_{\alpha'}(0) \\
A_0^{(0)} &= (D\phi\psi)\alpha'' \cdot H(0) + \phi(D\psi)\alpha' \cdot H(0) \\
&\quad -((D^3\phi) + \phi^2\psi)F_{\alpha''}(0) + (D^2\psi)F_{-\alpha''}(0) \\
A_0^{(-1)} &= -(D^3\psi)E_{\alpha'''}(0) + (-(D^2\phi)\psi + (D\phi)(D\psi) + \phi(D^2\psi))F_{-\alpha'}(0) \\
A_0^{(-i)} &= 0 \qquad \text{for} \qquad i \geq 2 \tag{100}
\end{aligned}
$$

This choice of potentials satisfies equations (95)-(96) and gives a dynamical equations

$$
\begin{aligned}
\phi_t &= (D^4\phi) + 2\phi(D\phi\psi) \\
\psi_t &= -(D^4\psi) - 2\psi(D\phi\psi) \tag{101}
\end{aligned}
$$

which are the sAKNS equations.

In order not to give the impression that this method of decomposing according to grading charges works only when one of the equations is fermionic, let us note that if we look at the two bosonic equations in (81), we can write (see Eq. (66)-(67))

$$
\begin{aligned}
(\partial_x - E(1) - A_1)\Psi &= 0 \\
(\partial_t - A_0)\Psi &= 0 \tag{102}
\end{aligned}
$$

where we recognize that

$$
E(1) = \lambda \begin{pmatrix} 1 & 0 & 0 \\ 0 & 0 & 0 \\ 0 & 0 & 1 \end{pmatrix} = F^2 = \alpha' \cdot H^{(1)} \tag{103}
$$

In this case, the appropriate grading operator is

$$Q = d = \lambda \frac{d}{d\lambda} \tag{104}$$

which leads to a homogeneous grading with

$$[Q, E(n)] = n E(n) \tag{105}$$

Furthermore, we note from Eq. (67) that we can write

$$\begin{aligned}
\mathcal{A}_1 = & - \phi E_{\alpha'''}(0) + (D\psi) E_{-\alpha'''}(0) \\
& - \phi\psi F_{\alpha'}(0) - (D\phi) F_{\alpha''}(0) - \psi F_{-\alpha''}(0)
\end{aligned} \tag{106}$$

The compatibility condition between the two equations in (102) leads to

$$\partial_t \mathcal{A}_1 - \partial_x \mathcal{A}_0 - [\mathcal{A}_0, E(1) + \mathcal{A}_1] = 0 \tag{107}$$

Once again, it is straightforward to see that we can decompose this equation according to grading charges (with respect to homogeneous grading of Eq. (104)) and that with the choice

$$\mathcal{A}_0^{(2)} = \alpha' \cdot H(2) \qquad \mathcal{A}_0^{(i)} = 0 \qquad \text{for} \qquad i \geq 3 \tag{108}$$

One can solve for the constraint equations which can consistently truncate the potentials and lead to the sAKNS equations as the dynamical equations.

6   The sTB-B Hierarchy

In this section, we will point out some puzzling features associated with the zero curvature formulation of what is known as the sTB-B hierarchy [21]. But, before I define the sTB-B hierarchy, let me summarize some of the features that we have obtained so far. We have seen that the (graded) Lie algebra associated with the zero curvature formulation is related to the finite dimensional subalgebra of the second Hamiltonian structure for a given dynamical system. We have also seen that once we have the zero curvature formulation in superspace, restricting the potentials to the θ-independent components yields the potentials for the zero curvature formulation in components.

Let us next look at the hierarchy of equations obtained from the Lax pair

$$\begin{aligned}
L &= D^2 - (D\phi_0) + D^{-2}(D\phi_1) \\
B &= (L^n)_{\geq 1}
\end{aligned} \tag{109}$$

and the Lax equation

$$\frac{\partial L}{\partial t} = [(L^n)_{\geq 1}, L] \tag{110}$$

We note that the Lax operator L in (109) is only slightly different from that in Eq. (27). The nonstandard Lax equation in (110), therefore, leads to a different supersymmetriztion of the TB hierarchy and is known as the sTB-B hierarchy. We note that if we were to restrict fields appropriately, it leads to a new supersymmetrization of the KdV equation known as the sKdV-B equation and it is this system of equations that naturally arise in the study of supersymmetric matrix models.

The peculiarity of the sTB-B (and, therefore, the sKdV-B) hierarchy is that its Hamiltonian structures are odd (fermionic). It is, therefore, interesting to ask if this system would have a zero curvature description and if so what is the associated algebra. To this end, we write the linear equation

$$L\chi = (D^2 - (D\phi_0) + D^{-2}(D\phi_1))\chi = \lambda\chi \qquad (111)$$

We can define, as before,

$$\begin{aligned} \chi_1 &= \chi \\ \chi_2 &= (D^{-2}(D\phi_1)\chi_1) \end{aligned} \qquad (112)$$

so that the linear equation (87) takes the form

$$(D^2 - (D\phi_0) - \lambda)\chi_1 + \chi_2 = 0 \qquad (113)$$

We note from Eqs. (112) and (113) that if we define the matrix wave function

$$\Psi = \begin{pmatrix} \chi_1 \\ \chi_2 \end{pmatrix} \qquad (114)$$

Then, we can write

$$\partial_x \Psi = D^2 \Psi = \mathcal{A}_1 \Psi \qquad (115)$$

where

$$\mathcal{A}_1 = \begin{pmatrix} \lambda + (D\phi_0) & -1 \\ (D\phi_1) & 0 \end{pmatrix} \qquad (116)$$

Similarly, we can write the time evolution equation and determine the form of $\mathcal{A}_0$ which would give rise to the sTB-B hierarchy of equations. However, we will not go into the details of that for simplicity. Rather, let us note the following features of this system. First, since both χ_1 and χ_2 in Eq. (112) are bosonic superfields, the matrix wave function Ψ is not graded. Second, from the structure of the potential in Eq. (116), we see that $\mathcal{A}_1 \in \mathrm{SL}(2) \oplus U(1)$. This is exactly the same algebra which arises in the zero curvature formulation of the TB hierarchy and does not, in any way, seem to be related to the second Hamiltonian structure (which is fermionic) of the sTB-B hierarchy. Furthermore, if we write

$$\mathcal{A}_1 = a_1 + \theta \tilde{a}_1 \qquad (117)$$

Then, from Eq. (116) it is clear that

$$a_1 = \begin{pmatrix} \lambda + J_0 & -1 \\ J_1 & 0 \end{pmatrix} \qquad (118)$$

and, consequently, the θ-independent component does not contain any fermion and does not lead to potentials which would give a zero curvature formulation of the supersymmetric system in components. (In fact, the θ-independent components give the zero curvature formulation for the TB hierarchy.) This is, in fact, very different from what we have seen in other supersymmetric systems so far.

On the other hand, it is interesting to point out that there does exist a zero curvature formulation for the sTB-B hierarchy in components. In this case, the potentials, say $\mathcal{A}_1$ has the form

$$\mathcal{A}_1 = \begin{pmatrix} \lambda + J_0 & -1 & 0 & 0 \\ J_1 & 0 & 0 & 0 \\ \psi_0' & 0 & \lambda + J_0 & -1 \\ \psi_1' & 0 & J_1 & 0 \end{pmatrix} \tag{119}$$

This is a graded matrix and appears to belong to OSp(2|2). However, it is not clear how to obtain it from the zero curvature formulation in superspace. The proper identification of the symmetry algebra as well as its connection to the second Hamiltonian structure remains an open question.

7   Conclusion

We have shown how the Drinfeld-Sokolov method can be generalized to superspace in the case of supersymmetric integrable models. We have shown how, in the case of the susy AKNS hierarchy, the symmetric space techniques can be used to obtain the equations in the hierarchy directly. We have formulated the manifestly covariant Cartan-Maurer equations for supersymmetric integrable systems and given an alternate derivation of the equations of the hierarchy from the zero curvature condition by decomposing the potential according to the grading charges. Finally, we have pointed out some puzzling features and open questions associated with systems such as the sTB-B hierarchy.

This work was supported in part by U.S. Department of Energy Grants DE-FG02-84ER40173 and DE-FG02-91ER40685 as well as by NSF-INT-9602559.

Appendix

Algebra of SL(2|1) and $\widehat{SL}(2|1)$:

The superalgebra (graded algebra) SL(2|1) is isomorphic to the graded algebra OSp(2|2) and corresponds to the $N = 2$ supersymmetric extension of the Lie algebra SL(2). There are eight generators of the algebra SL(2|1), four of which are bosonic and the other four fermionic. The four bosonic generators, $E_{\alpha'''}$, $E_{-\alpha'''}$, H_1 and H_2 define the Lie algebra SL(2) $\oplus$ U(1) and the fermionic generators are represented as $F_{\alpha'}$, $F_{-\alpha'}$, $F_{\alpha''}$, $F_{-\alpha''}$. Here α' and α'' denote two simple roots such that, because of the indefinite metric structure,

$$(\alpha', \alpha') = 0 = (\alpha'', \alpha'')$$
$$(\alpha', \alpha'') = 1$$

We also have

$$\alpha''' = \alpha' + \alpha'' \; ; \; (\alpha''', \alpha''') = 2$$

The generators have a 3×3 graded matrix representation (fundamental representation) in the Cartan-Weyl basis given by

$$H_1 = \begin{pmatrix} \frac{1}{2} & 0 & 0 \\ 0 & -\frac{1}{2} & 0 \\ 0 & 0 & 0 \end{pmatrix} \qquad H_2 = \begin{pmatrix} \frac{1}{2} & 0 & 0 \\ 0 & \frac{1}{2} & 0 \\ 0 & 0 & 1 \end{pmatrix}$$

$$E_{\alpha'''} = \begin{pmatrix} 0 & 1 & 0 \\ 0 & 0 & 0 \\ 0 & 0 & 0 \end{pmatrix} \qquad E_{-\alpha'''} = \begin{pmatrix} 0 & 0 & 0 \\ 1 & 0 & 0 \\ 0 & 0 & 0 \end{pmatrix}$$

$$F_{\alpha'} = \begin{pmatrix} 0 & 0 & 1 \\ 0 & 0 & 0 \\ 0 & 0 & 0 \end{pmatrix} \qquad F_{-\alpha'} = \begin{pmatrix} 0 & 0 & 0 \\ 0 & 0 & 0 \\ 1 & 0 & 0 \end{pmatrix}$$

$$F_{\alpha''} = \begin{pmatrix} 0 & 0 & 0 \\ 0 & 0 & 0 \\ 0 & 1 & 0 \end{pmatrix} \qquad F_{-\alpha''} = \begin{pmatrix} 0 & 0 & 0 \\ 0 & 0 & 1 \\ 0 & 0 & 0 \end{pmatrix}$$

Note that all the generators are super traceless as they should be.
    We can define the Cartan elements (only two are independent)

$$\alpha' \cdot H = H_1 + H_2 \; ; \; \alpha'' \cdot H = H_1 - H_2 \; ; \; \alpha''' \cdot H = 2H_1$$

It is easy to check from the form of the generators that they satisfy the following nontrivial (anti) commutation relations.

$$[\alpha' \cdot H, E_{\pm\alpha'''}] = \pm E_{\pm\alpha'''} \quad ; \quad [\alpha' \cdot H, F_{\pm\alpha''}] = \pm F_{\pm\alpha''}$$

$$[\alpha'' \cdot H, E_{\pm\alpha'''}] = \pm E_{\pm\alpha'''} \quad ; \quad [\alpha'' \cdot H, F_{\pm\alpha'}] = \pm F_{\pm\alpha'}$$

$$[\alpha' \cdot H, F_{\pm\alpha'}] = 0 \quad ; \quad [\alpha'' \cdot H, F_{\pm\alpha''}] = 0$$

$$[E_{\pm\alpha'''}, F_{\pm\alpha''}] = [E_{\pm\alpha'''}, F_{\pm\alpha'}] = 0$$

$$\{F_{\pm\alpha''}, F_{\pm\alpha''}\} = \{F_{\pm\alpha'}, F_{\pm\alpha'}\} = 0$$

$$[E_{\alpha'''}, E_{-\alpha'''}] = \alpha''' \cdot H \quad ; \quad \{F_{\alpha'}, F_{-\alpha'}\} = \alpha' \cdot H$$

$$[E_{\pm\alpha'''}, F_{\mp\alpha'}] = \mp F_{\pm\alpha''} \quad ; \quad [E_{\pm\alpha'''}, F_{\mp\alpha''}] = \pm F_{\pm\alpha'}$$

$$\{F_{\alpha''}, F_{-\alpha''}\} = -\alpha'' \cdot H \quad ; \quad \{F_{\pm\alpha''}, F_{\mp\alpha'}\} = 0$$

$$\{F_{\pm\alpha'}, F_{\pm\alpha''}\} = E_{\pm\alpha'''}$$

Given the graded algebra SL(2|1), we can define the affine algebra (loop algebra) $\widehat{SL}(2|1)$ as follows.

$$\widehat{SL}(2|1) = SL(2|1) \otimes C[\lambda, \lambda^{-1}] \oplus \hat{c}$$

$$[a(n), b(m)\} = [a, b\}(n + m) + n\delta_{n+m} \, \text{str}(ab)\hat{c}$$

where λ denotes the loop parameter and $\hat{c}$ a central extension of the algebra. Furthermore, the supertrace of a graded matrix (3×3, in this case) is defined to be

$$\text{str}\,(M) = m_{11} + m_{22} - m_{33}$$

With these, the affine algebra $\widehat{SL}(2|1)$ can now be obtained from the algebra of SL(2|1) in a straightforward manner. A few examples of such brackets are

$$[\alpha' \cdot H(n), \alpha'' \cdot H(m)] = n\delta_{n+m}\hat{c}$$

$$[E_{\alpha'''}(n), E_{-\alpha'''}(m)] = \alpha''' \cdot H(n + m) + n\delta_{n+m}\hat{c}$$

$$\{F_{\alpha''}(n), F_{-\alpha''}(m)\} = -\alpha'' \cdot H(n + m) - n\delta_{n+m}\hat{c}$$

Vertex Operator Representation for the Superalgebra $sl(2|1)$:

Let us introduce two even roots e_i $i = 1, 2$ and one odd f and use them to represent α', α'' as

$$\alpha' = e_1 - f \quad ; \quad \alpha'' = -(e_2 - f) \tag{120}$$

so that $\alpha''' = e_1 - e_2$.

Following [27] we define the vertex operators associated to the fermionic roots $\pm(e_i - f)$, $i = 1, 2$ by

$$E\left(\pm(e_i - f), z\right) = U\left(\pm e_i, z\right) c_{\pm e_i} U\left(\mp f, z\right) \tag{121}$$

and to the bosonic root by

$$E\left(\pm(e_1 - e_2), z\right) = U\left(\pm e_1, z\right) c_{\pm e_1} U\left(\mp e_2, z\right) c_{\pm e_2} \tag{122}$$

where $c_{\pm e_i}$ are cocycle operators and

$$U\left(\pm e_i, z\right) = z^{\frac{1}{2}} : e^{\pm i e_i Q^i(z)} : \tag{123}$$

$$Q^i(z) = q^i - i p^i \ln z + i \sum_{n \neq 0} \frac{\alpha_n^i z^{-n}}{n} \tag{124}$$

$$\left[\alpha_m^i, \alpha_n^j\right] = m \, \delta_{m+n,0} \, \delta_{i,j} \quad [q^i, p^j] = i \delta_{i,j}$$

For the fermionic vertex opertors we have

$$U\left(f, z\right) = z^{\frac{1}{2}} : e^{\varphi(z)} :: e^{-\chi(z)} : \tag{125}$$

$$U\left(-f, z\right) = z^{\frac{1}{2}} : e^{-\varphi(z)} :: \partial_z e^{\chi(z)} : \tag{126}$$

$$\varphi(z) = j_q - j_0 \ln z + i \sum_{n \neq 0} \frac{j_n z^{-n}}{n} \tag{127}$$

$$\chi(z) = h_q - h_0 \ln z + i \sum_{n \neq 0} \frac{h_n z^{-n}}{n} \tag{128}$$

with

$$[j_m, j_n] = -m \, \delta_{m+n,0} \quad ; \quad [h_m, h_n] = m \, \delta_{m+n,0} \tag{129}$$

Neglecting for the moment the cocycles (Klein factors) we have a following dictionary between simple roots operators and vertex operators:

$$E_{\alpha_{III}}(z) = \sum_n E_{\alpha_{III}}(n) z^{-n} = z^{\frac{1}{2}} z^{\frac{1}{2}} : e^{i\left(e_1 Q^1(z) - e_2 Q^2(z)\right)} :$$

$$E_{-\alpha_{III}}(z) = \sum_n E_{-\alpha_{III}}(n) z^{-n} = z : e^{-i\left(e_1 Q^1(z) - e_2 Q^2(z)\right)} : \tag{130}$$

$$F_{\alpha_{II}}(z) = \sum_n F_{\alpha_{II}}(n) z^{-n} = z : e^{-i e_2 Q^2(z)} :: e^{\varphi(z)} :: e^{-\chi(z)} :$$

$$F_{-\alpha_{II}}(z) = \sum_n F_{-\alpha_{II}}(n) z^{-n} = z : e^{i e_2 Q^2(z)} :: e^{-\varphi(z)} :: \partial_z e^{-\chi(z)} : \tag{131}$$

etc.

Dressing and the Soliton Solutions:

The dressing technique (see [28]) reproduces the nontrivial part $E(1) + \mathcal{A}_1^M$ of the Lax matrix operator from eq. (72) $\left(\partial_x \cdot I - E(1) - \mathcal{A}_1^M\right) \chi = 0$ by the gauge transformations involving generators of positive and negative gradings applied to the semisimple element $E(1)$:

$$E(1) + \mathcal{A}_1^M = \Theta \, E(1) \, \Theta^{-1} + \left(\partial_x \Theta\right) \Theta^{-1} \tag{132}$$

$$E(1) + \mathcal{A}_1^M = \left(B^{-1} \Gamma\right) E(1) \left(\Gamma^{-1} B\right) + \left(\partial_x B^{-1} \Gamma\right) \left(\Gamma^{-1} B\right) \tag{133}$$

where $B^{-1}\Gamma$ contains positive terms and Θ is an expansion in the terms of negative grading such that $\Theta = \exp\left(\sum_{l<0} \theta^{(l)}\right) = 1 + \theta^{(-1)} + \ldots$. From expressions (132) and (133) we obtain two alternative formulas for the same term

$$\mathcal{A}_1^M = \begin{pmatrix} 0 & \Phi & 0 \\ \bar{\Psi} & 0 & \Psi \\ 0 & \bar{\Phi} & 0 \end{pmatrix} \tag{134}$$

(see equation (72)) of grade 0:

$$\mathcal{A}_1^M = -\left[E(1), \theta^{(-1)} \right] \quad \text{or} \quad \mathcal{A}_1^M = -B^{-1} \left(\partial_x B \right) \tag{135}$$

The term $\theta^{(-1)}$ of grade -1 can be expanded as

$$\theta^{(-1)} = -\Phi\, E_{\alpha'''}(-1) + \bar{\Psi}\, E_{-\alpha'''}(-1) + \Psi\, F_{-\alpha''}(-1) - \bar{\Phi}\, F_{\alpha''}(-1) \tag{136}$$

Realization of the dynamical variables $\Phi, \Psi, \bar{\Psi}, \bar{\Phi}$ in terms of the expectation values of product of the vertex operators is possible due to existence of alternative expression for the tau-function vectors:

$$|\tau_0\rangle = (1 - \theta^{-1}) |\lambda_0\rangle \tau_0^{(0)} \tag{137}$$

or

$$|\tau_0\rangle = \Psi^{(\text{vac})}\, h\, \Psi^{(\text{vac})^{-1}} |\lambda_0\rangle \tag{138}$$

with $\tau_0^{(0)} = \langle\lambda_0|\Psi^{(\text{vac})}\, h\, \Psi^{(\text{vac})^{-1}} |\lambda_0\rangle$ and the constant group elements h which are the product of exponentials of the vertex operators (see below). Furthermore: $\Psi^{(\text{vac})} = \exp\left(\sum_{N=1}^\infty t_N b^{(N)}\right)$ in terms of the elements of the Heisenberg subalgebra.

In this way we find from (136) and (137)-(138) :

$$\Phi = \langle\lambda_0| E_{-\alpha'''}(1)|\tau_0\rangle / \tau_0^{(0)} \quad ; \quad \bar{\Psi} = -\langle\lambda_0| E_{\alpha'''}(1)|\tau_0\rangle / \tau_0^{(0)} \tag{139}$$

$$\Psi = -\langle\lambda_0| F_{\alpha''}(1)|\tau_0\rangle / \tau_0^{(0)} \quad ; \quad \bar{\Phi} = \langle\lambda_0| F_{-\alpha''}(1)|\tau_0\rangle / \tau_0^{(0)} \tag{140}$$

Take

$$h = (1 + E_{-\alpha'''}(z_1)) \, (1 + E_{\alpha'''}(z_2)) \, (1 + F_{\alpha''}(z_3)) \tag{141}$$

then

$$\tau_0^{(0)}(z1, z2, x, t) = \langle\lambda_0|\Psi^{(\text{vac})}\, h\, \Psi^{(\text{vac})^{-1}} |\lambda_0\rangle$$
$$= 1 + \frac{z1\, z2\, e^{(-x\, z1 - t\, z1^2 + x\, z2 + t\, z2^2)}}{(z1 - z2)^2} \tag{142}$$

and

$$\Phi(z1, z2, z3, x, t) = \frac{z2\, e^{(x\, z2 + t\, z2^2)}}{\tau_0^{(0)}(z1, z2, x, t)} \tag{143}$$

$$\bar{\Psi}(z1, z2, z3, x, t) = -\frac{z1\, e^{(-x\, z1 - t\, z1^2)}}{\tau_0^{(0)}(z1, z2, x, t)} \quad ; \quad \Psi(z1, z2, z3, x, t) = 0 \tag{144}$$

$$\bar{\Phi}(z1, z2, z3, x, t) = \frac{z3\, e^{(x\, z3 + t\, z3^2)}}{\tau_0^{(0)}(z1, z2, x, t)} \times \tag{145}$$

$$\times \left(1 + \frac{z1\, z2\, (z2 - z3)\, e^{(-x\, z1 - t\, z1^2 + x\, z2 + t\, z2^2)}}{(z1 - z2)^2\, (z1 - z3)} \right)$$

For

$$h = (1 + E_{-\alpha'''}(z_1))\, (1 + E_{\alpha'''}(z_2))\, (1 + F_{-\alpha''}(z_4)) \tag{146}$$

we find again (142) and

$$\Phi(z1, z2, x, t) = \frac{z2\, e^{(x\, z2 + t\, z2^2)}}{\tau_0^{(0)}(z1, z2, x, t)} \tag{147}$$

$$\bar{\Psi}(z1, z2, x, t) = -\frac{z1\, e^{(-x\, z1 - t\, z1^2)}}{\tau_0^{(0)}(z1, z2, x, t)} \quad ; \quad \bar{\Phi}(z1, z2, z4, x, t) = 0 \tag{148}$$

$$\Psi(z1, z2, z4, x, t) = \frac{z4\, e^{(x\, z4 + t\, z4^2)}}{\tau_0^{(0)}(z1, z2, z3, x, t)} \times \tag{149}$$

$$\times \left(1 + \frac{z1\, z2\, (z1 - z4)\, e^{(-x\, z1 - t\, z1^2 + x\, z2 + t\, z2^2)}}{(z1 - z2)^2\, (z2 - z4)} \right)$$

These equations are the soliton solutions of

$$\partial_2 \Phi = \Phi'' - 2\Phi^2 \bar{\Psi} - 2\Phi\Psi\bar{\Phi} \tag{150}$$

$$\partial_2 \Psi = -\Psi'' + 2\bar{\Psi}\,\Phi\,\Psi \tag{151}$$

$$\partial_2 \bar{\Psi} = -\bar{\Psi}'' + 2\bar{\Psi}^2 \Phi + 2\Psi\bar{\Psi}\bar{\Phi} \tag{152}$$

$$\partial_2 \bar{\Phi} = \bar{\Phi}'' - 2\Phi\bar{\Phi}\bar{\Psi} \tag{153}$$

following from equations (72),(73) for the matrix $\mathcal{A}_1^M$ from equation (134). Note, that the terms "fermion" $\times$ "fermion" in eqs. (150)-(153) are absent for these solutions.

References

1. Faddeev, L.D., Takhtajan, L.A., *Hamiltonian Methods in the Theory of Solitons*, Springer-Verlag, Berlin (1987).
2. Das, A., *Integrable Models*, World Scientific, Singapore (1989).
3. Lax, P.D., Comm. Pure Appl. Math **21**, 467 (1968); *ibid* **28** (1975) 141.
4. Gelfand, I.M., Dickey, L.A., Russ. Math Surveys **30**, 77 (1975).
5. Chern, S.S., Peng, C-K., Manuscripta Mathematica **28**, 207 (1979).
6. Drinfeld, V.G., Sokolov, V.V., J. Sov. Math. **30**, 1975 (1985); Sov. Math. Dokl. **23**, 457 (1981).
7. Wilson, G., Ergod. Th. and Dynam. Sys. **1**, 361 (1981).
8. A.P. Fordy and P.P. Kulish, Comm. Math. Phys. **89**, 427 (1983); H. Aratyn, J.F. Gomes and A.H. Zimerman, J. Math. Phys.**36**, 3419 (1995)
9. Manin, Y., Radul, A.O., Comm. Math. Phys. **98**, 65 (1985).
10. Mathieu, P., J. Math. Phys. **29**, 2499 (1988).
11. Kupershmidt, B., *Elements of Superintegrable Systems: basic techniques and results*, Kluwer Acad. Publ. (1987).
12. Inami, T., Kanno, H., Comm. Math. Phys. **136**, 519 (1991).
13. Toppan, F., Int. J. Mod. Phys. **A11**, 3257 (1996).
14. Gürses, M., Oguz, Ö, Phys. Lett. **108A**, 437 (1985).
15. Das, A., Roy, S., J. Math. Phys. **31**, 2145 (1990).
16. Das, A., Huang, W-J., Roy, S., Int. J. Mod. Phys. **A7**, 3447 (1992); *ibid* **A7**, 4293 (1992).
17. Das, A., Roy, S., Mod. Phys. Lett. **A11**, 1317 (1996).
18. Brunelli, J.C., Das, A., Phys. Lett. **337B**, 303 (1994); *ibid* **354B**, 307 (1995); Int. J. Mod. Phys. **A10**, 4563 (1995).
19. Aratyn, H., Rasinariu, C., Phys. Lett. **391B**, 99 (1997).
20. Popowicz, Z., J. Physics **A29**, 1281 (1996).
21. Brunelli, J.C., Das, A., hep-th/9704126 to be published in Phys. Lett. B.
22. Cornwell, J.F., *Group Theory in Physics, vol. 3*, Acad. Press (1989).
23. Aratyn, H., Das, A., Rasinariu, C., hep-th/9704119 to be published in Mod. Phys. Lett. A.
24. Fujikawa, K., Prog. Theor. Phys. **59**, 2045 (1977).
25. Bonora, L., Tonin, M., Phys. Lett. **B98**, 48 (1981).
26. Braga, N.R.F., Das, A., Nuc. Phys. **B442**, 655 (1995).
27. Frapat, L. ,Sciarrino A. and Sorba, P., Journ. of Math. Physics **30**, 2984 (1989)
28. Ferreira L.A. and Sánchez Guillén J., "Solitons and Generalized Tau-Functions for Affine Integrable Hierarchies", in this volume; Aratyn, H., Ferreira, L.A., Gomes, J.F. and Zimerman, A.H. "Vertex Operators and Solitons of Constrained KP Hierarchies", in this volume

N=2 KdV Hierarchies and Classical r-Matrix

Laurent Gallot

Laboratoire de Physique Théorique ENSLAPP
URA 14-36 du CNRS, associée à l'ENS de Lyon et au Lapp
Groupe de Lyon: ENS de Lyon, 46 allée d'Italie, 69364 Lyon, France

Abstract. We review the construction on an associative algebra $\mathcal{G}$ of the linear and quadratic Poisson brackets associated with a classical r-matrix. We then give, in the light of this construction, the formulation in extended superspace of an N=2 supersymmetric KP hierarchy with two quadratic hamiltonian structures, each one leading to a different series of N=2 n^{th} KdV hierarchies. We also give the formulation of an N=2 Dym hierarchy.

1 Introduction

There has been in the past years an important activity in the study of N=2 supersymmetric integrable hierarchies, especially generalizations of KP and KdV. The most usual method used to obtain a Lax formulation for these systems is the AKS procedure applied to the algebra of N=1 or N=2 pseudo-differential operators (ΨDOs). It has been shown in [Popowicz (1993b)][Delduc and Gallot (1996)] that one should not use the algebra of ordinary N=2 ΨDOs to construct Lax pairs for the N=2 KP and KdV hierarchies, but rather the algebra $\mathcal{C}$ of N=2 chirality preserving ΨDOs.

In section 2, we give a short review of the construction of the linear and quadratic Poisson structures associated with a classical r-matrix given by the AKS procedure. The examples of bosonic KP, non-standard KP and Dym hierarchies are given.

In section 3, we expose the construction of N=2 supersymmetric hierarchies in the light of the results of section 2. First, following [Delduc and Gallot (1996)], an N=2 KP is obtained by means of the AKS procedure on $\mathcal{C}$. The r-matrix in this case is not skew-symmetric and the hierarchy is hamiltonian with respect to a linear and two quadratic Poisson structures. Each quadratic bracket leads to different reductions of N=2 KP. In particular, two distinct N=2 n^{th} KdV hierarchies are found. Second, we present the construction of a new N=2 Dym hierarchy.

2 The AKS Scheme

This section is devoted to a reminder of well known results about the linear and quadratic Poisson structures for Lax equations of the type

$$\partial_t L = [R(L^k), L] \qquad (1)$$

where R is a classical r-matrix obtained through the AKS procedure. These results are illustrated with the examples of KP type hierarchies and their reductions.

2.1   Classical r-matrix and linear structure

Let $\mathcal{G}$ be an associative algebra equipped with a trace, denoted by tr, defining a non degenerate invariant bilinear form $\langle,\rangle$ by $\forall x, y \in \mathcal{G}$ $\langle x, y \rangle = \text{tr}(xy)$. The adjoint A^* of an endomorphism A with respect to the trace is defined by $\forall x, y \in \mathcal{G}$ $\text{tr}\,(A^*(x)y) = \text{tr}\,(xA(y))$. We shall consider Lax equations on an element $L \in \mathcal{G}$, containing the phase space, and Poisson brackets on linear functionals of L, *i.e.* functionals of the type $l_X(L) = \text{tr}(LX)$ where $X \in \mathcal{G}$ does not depend on the phase space.

The AKS scheme relies on the existence of a classical r-matrix associated with a splitting of the algebra into the direct sum

$$\mathcal{G} = \mathcal{A} \oplus \mathcal{B} \qquad (2)$$

where $\mathcal{A}$ and $\mathcal{B}$ are Lie subalgebras: $[\mathcal{A}, \mathcal{A}] \subset \mathcal{A}$, $[\mathcal{B}, \mathcal{B}] \subset \mathcal{B}$. Let $P_{\mathcal{A}}$, resp. $P_{\mathcal{B}}$, be the projector on $\mathcal{A}$, resp. $\mathcal{B}$, along $\mathcal{B}$, resp. $\mathcal{A}$ such that $P_{\mathcal{A}} + P_{\mathcal{B}} = \mathbf{1}_{\mathcal{G}}$. Hence the endomorphism of $\mathcal{G}$ defined by

$$R = \frac{1}{2}\left(P_{\mathcal{A}} - P_{\mathcal{B}}\right) \qquad (3)$$

is a classical r-matrix [Semenov (1983)], *i.e.* it satisfies the classical modified Yang-Baxter equation (mYBe) [1] :

$$R([R(X), Y] + [X, R(Y)]) = [R(X), R(Y)] + \frac{1}{4}[X, Y]. \qquad (4)$$

This is a sufficient condition for the R-commutator

$$[X, Y]_R = [R(X), Y] + [X, R(Y)] \qquad (5)$$

to satisfy the Jacobi identity. The algebra $\mathcal{G}$ equipped with the commutator $[,]_R$ is called a Baxter algebra. As a consequence of the mYBe, one can define on linear functionals of $L \in \mathcal{G}$ the following linear Poisson structure or R-bracket:

$$\{l_X, l_Y\}_R(L) = \text{tr}\,(L[X, Y]_R). \qquad (6)$$

This structure is extended to general functionals of L by using the Leibniz rule. The Lax equations (1) are hamiltonian with respect to this Poisson bracket and may be written as

$$\partial_{t_k} l_X(L) = \{l_X, \mathcal{H}_{k+1}\}_R(L) \qquad (7)$$

---

[1] This type of r-matrices is not the most general one [Semenov (1983)]

with the hamiltonians $\mathcal{H}_j = \frac{1}{j}\mathrm{tr}\left(L^j\right)$. Since R satisfies the mYBe, the flows (1) commute. Accordingly, the hamiltonians are in involution for the R-bracket

$$\{\mathcal{H}_p, \mathcal{H}_q\}_R = 0. \tag{8}$$

2.2   The Adler-Gelfand-Dickey bracket

Let us now assume that in the splitting (2) of $\mathcal{G}$, $\mathcal{A}$ and $\mathcal{B}$ are associative subalgebras. Let us also assume that the associated r-matrix R is skew-symmetric, *i.e.* $R = -R^*$. In other words $\mathcal{A}$ and $\mathcal{B}$ are isotropic. Then one can define the quadratic Poisson structure

$$\{l_X, l_Y\}_{GD}(L) = \mathrm{tr}\left(LX\,R(LY) - XL\,R(YL)\right). \tag{9}$$

This structure was first written by Adler [1979] but the Jacobi identity was obtained by Gelfand and Dickey [Dickey (1991)]. The corresponding bracket on the group G with Lie algebra $\mathcal{G}$ is called the Sklyanin bracket. The Lax equations (1) are hamiltonian with respect to this bracket:

$$\partial_{t_k} l_X(L) = \{l_X, \mathcal{H}_k\}_{GD}(L) \tag{10}$$

and, owing to the commutativity of flows, the hamiltonians are in involution

$$\{\mathcal{H}_p, \mathcal{H}_q\}_{GD} = 0. \tag{11}$$

The linear bracket is a linearization of the quadratic one, that is to say

$$\{l_X, l_Y\}_{GD}(L + z\mathbf{1}) = \{\mathbf{l_X}, \mathbf{l_Y}\}_{GD}(\mathbf{L}) + \mathbf{z}\{\mathbf{l_X}, \mathbf{l_Y}\}_{\mathbf{R}}(\mathbf{L}) \tag{12}$$

and, as a consequence, both brackets are compatible.

2.3   *abcd* structures

The preceding result is no longer valid when the r-matrix is not skew symmetric. In that case, the right hand side of (9) is not skew symmetric and definitely cannot stand for a Poisson structure. However, Freidel and Maillet [1991] have given existence conditions for not one but two quadratic Poisson structures in the case when R is not skew-symmetric [2] . Let us consider four endomorphisms of $\mathcal{G}$, a, b, c and d, satisfying the conditions
i) $a^* = -a$, $b^* = c$, $d^* = -d$
ii)

---

[2] The work of Freidel and Maillet amounts to a generalization of the Sklyanin bracket in the non skew-symmetric case. For the case of an associative algebra, see also [Li and Parmentier (1989)][Suris (1993)]

$$a([a(X), Y] + [X, a(Y)]) = [a(X), a(Y)] + \frac{1}{4}[X, Y] \qquad (13)$$

$$d([d(X), Y] + [X, d(Y)]) = [d(X), d(Y)] + \frac{1}{4}[X, Y] \qquad (14)$$

$$b([a(X), Y] + [X, a(Y)]) + [b(X), b(Y)] = 0 \qquad (15)$$

$$c([d(X), Y] + [X, d(Y)]) + [c(X), c(Y)] = 0. \qquad (16)$$

iii) $R = a - c = d - b$.

Then there exists two quadratic Poisson structures associated with the quadruplet (a, b, c, d) defined by

$$\{l_X, l_Y\}^a_{(2)}(L) = tr\left(LXa(LY) + XLb(LY) - LXc(YL) - XLd(YL)\right) \quad (17)$$

$$\{l_X, l_Y\}^b_{(2)}(L) = tr\left(LXd(LY) + XLc(LY) - LXb(YL) - XLa(YL)\right) \quad (18)$$

The Lax equations (1) are hamiltonian with respect to both Poisson brackets

$$\partial_{t_k} l_X(L) = \{l_X, \mathcal{H}_k\}^{a,b}_{(2)}(L) \qquad (19)$$

and the hamiltonians are also in involution for both structures

$$\{\mathcal{H}_p, \mathcal{H}_q\}^{a,b}_{(2)} = 0. \qquad (20)$$

The linear bracket is a linearization of both quadratic brackets

$$\{l_X, l_Y\}^{a,b}_{(2)}(L + z\mathbf{1}) = \{l_X, l_Y\}^{a,b}_{(2)}(L) + z\{l_X, l_Y\}_{(1)}(L). \qquad (21)$$

The interpretation of the above conditions is as follows. i) ensures that both brackets (17,18) are skew-symmetric. Constraints (13,14) are nothing but the mYBe on a and d which are then skew-symmetric r-matrices. Constraints (15,16) mean that b and c are respectively homomorphisms from the Baxter algebras $(\mathcal{G}, [,]_a)$ and $(\mathcal{G}, [,]_d)$ to the Lie algebra $(\mathcal{G}, [,])$. iii) is needed for (19), (20) and (21) to hold. It is important to observe that this result is not constructive, that is to say a, b, c, d are not given in terms of R. In fact, neither the existence, nor the uniqueness of (a, b, c, d) are ensured.

2.4   Example: KP type hierarchies

Examples illustrating the previous results are provided by the integrable hierarchies constructed on the associative algebra $\mathcal{D}$ of one dimensional pseudo-differential operators, like the KP hierarchy. $\mathcal{D}$ is the set of formal series of the space derivative operator $\partial = \partial/\partial x$ of the type

$$A = \sum_{k \le M} a_k \partial^k \qquad (22)$$

where the coefficients a_k are functions of the space variable x. The associative product law on $\mathcal{D}$ is inferred from the commutation rule

$$\partial f = f\partial + f_x. \tag{23}$$

∂^{-1} is the formal inverse of ∂: $\partial^{-1}\partial = \partial\partial^{-1} = 1$. We define the residue of $A \in \mathcal{D}$ as the coefficient of ∂^{-1}: $\mathrm{res}(A) = a_{-1}(x)$. The property that the residue of a commutator is a total derivative enables us to equip $\mathcal{D}$ with a trace

$$\mathrm{tr}L = \int dx\, \mathrm{res}L, \quad \mathrm{tr}[L, L'] = 0 \tag{24}$$

and a non degenerate invariant bilinear form $\langle A, B \rangle = \mathrm{tr}(AB)$. With a ΨDO of order n

$$L = u_0^n \partial^n + \sum_{k \geq 1} u_k \partial^{n-k}, \tag{25}$$

for which u_0 is invertible, is associated a unique n^{th} root $L^{\frac{1}{n}}$ such that $\left(L^{\frac{1}{n}}\right)^n = L$ and

$$L = u_0\partial + \sum_{k \geq 0} \frac{Y_k}{u_0^{p_k}} \partial^{-k}, \tag{26}$$

where Y_k is a differential polynomial in terms of $u_0, u_1, \cdots, u_{k+1}$ and p_k is some positive integer. The nth root of L and then any power of it commute with L: $[L^{\frac{k}{n}}, L] = 0$.

We denote the space of monomials of order p by $\mathcal{D}_p = \{a_p \partial^p\}$ and define the spaces $\mathcal{D}_{\geq k} = \oplus_{p \geq k}\mathcal{D}_p$ and $\mathcal{D}_{<k} = \oplus_{p<k}\mathcal{D}_p$. There exists three fundamental decompositions of $\mathcal{D}$ into direct sum of subalgebras

$$\mathcal{D} = \mathcal{D}_{\geq k} \oplus \mathcal{D}_{<k} \tag{27}$$

for the values $k = 0, 1, 2$ [Reyman (1982)][Kupershmidt (1985)]. For $k = 0, 1$ these are splittings into associative subalgebras while for $k = 2$ it is a splitting into Lie subalgebras only. The projectors associated with such a decomposition are denoted by $P_{\geq k}$ and $P_{<k}$ and we denote by

$$R_k = \frac{1}{2}(P_{\geq k} - P_{<k}) \tag{28}$$

the corresponding r-matrix. Then with a ΨDO $L = u_0^n \partial^n + \sum_{j \leq n-1} u_{n-j} \partial^j$ of order n for which u_0 is invertible, we can associate the three hierarchies of consistent evolution equations on the infinite set of fields $\{u_j\}$

$$\partial_{t_p}^{(k)} L = [R_k(L^{\frac{p}{n}}), L]. \tag{29}$$

For k respectively equal to 0, 1, 2, these hierarchies are the KP, the non-standard KP and the Dym hierarchy. The hamiltonian properties of these hierarchies depend crucially on the algebraic nature of the splitting (27) as we shall expose now.

In the simplest case of the Dym hierarchy [Konopelchenko and Oevel (1993)], the decomposition into Lie subalgebras ensures that the equations

(29) are hamiltonian only with respect to the linear structure $\{,\}_{R_2}$ with the hamiltonians $\mathcal{H}_p = \frac{n}{p}\text{tr}(L^{\frac{p}{n}})$:

$$\partial_{t_p}^{(2)} l_X(L) = \{l_X, \mathcal{H}_{p+n}\}_{R_2}(L). \tag{30}$$

This hierarchy admits an hamiltonian reduction to a finite number of fields defined by the following Poisson subspace of the linear structure

$$L = u_0^n \partial^n + \sum_{j=0}^{n-1} u_{n-j}\partial^j + \partial^{-1}u_{n+1} + \partial^{-2}u_{n+2}. \tag{31}$$

It is worth noticing that although the restriction $L = P_{\geq 2}(L)$ is a consistent reduction for equations (29), it does not define a Poisson subspace of the linear structure. The corresponding hierarchy has then no local Poisson algebra.

In the case of non standard KP [Reyman (1982)][Kupershmidt (1985)], since we are faced to a splitting into associative subalgebras for which the r-matrix R_1 is not skew-symmetric, we expect this hierarchy to be hamiltonian with respect to a linear structure and, possibly, two quadratic $abcd$ structures. These two quadratic brackets $\{,\}_O^{a,b}$ have been found by Oevel [1994] and they are given by the choice of the quadruplet $a = R_1 + P_0 + r$, $b = -\partial^{-1}r\partial$, $c = P_0 + r$, $d = R_1 - \partial^{-1}r\partial$ where r is the non local map defined as $r(X) = D^{-1}.\text{res}(X)$. Although a, b, c and d are non local, the brackets $\{,\}_O^{a,b}$ are local since they can be written in a way such that r acts only on commutators of ΨDOs whose residues are total derivatives. The equations (29) may then be put in the hamiltonian form

$$\partial_{t_p}^{(1)} l_X(L) = \{l_X, \mathcal{H}_{p+n}\}_{R_1}(L) = \{l_X, \mathcal{H}_p\}_O^{a,b}(L). \tag{32}$$

This hierarchy admits, among others, two types of reductions depending on the Poisson structure under consideration. On one hand, the submanifold of $\mathcal{D}$ defined by

$$L = \partial^n + \sum_{j=1}^{n-1} u_{n-j}\partial^j \tag{33}$$

is a Poisson subspace of the quadratic bracket $\{,\}_O^a$ only, the corresponding Poisson algebra being the classical $\mathcal{W}_{n-1} \oplus \mathcal{U}(1)$ algebra. On the other hand, a Poisson subspace for $\{,\}_{R_1}$ and $\{,\}_O^b$ is defined by

$$L = \partial^n + \sum_{j=0}^{n-1} u_{n-j}\partial^j + \partial^{-1}u_{n+1}. \tag{34}$$

The corresponding hierarchy is then bi-hamiltonian. The Poisson algebra inherited from $\{,\}_O^b$ is $\mathcal{W}_{n+1} \oplus \mathcal{U}(1)$.

In the last case, the KP hierarchy [Dickey (1991)] is obtained from a splitting into associative subalgebras for which the r-matrix R_0 is skew-symmetric and we expect it to be bi-hamiltonian with respect to a linear structure (6) and the quadractic structure (9) of Adler-Gelfand-Dickey. The situation is somewhat peculiar since this hierarchy, correspondingly to a symmetry, admits a one parameter family of local quadratic structures [Di Francesco et al. (1991)]

$$\{l_X, l_Y\}_{GD}^\nu(L) = \{l_X, l_Y\}_{GD}(L) + \nu \int dx\, D^{-1}(\text{res}[L, X])\text{res}[L, Y]. \quad (35)$$

The flows (29) may then be written

$$\partial_{t_p}^{(0)} l_X(L) = \{l_X, \mathcal{H}_{p+n}\}_{R_0}(L) = \{l_X, \mathcal{H}_p\}_{GD}^\nu(L). \quad (36)$$

The bracket $\{,\}_{GD}^\nu$ has the *abcd* form with the choice $a_\nu = R_0 + \nu r$, $b_\nu = -\nu r$, $c_\nu = \nu r$ and $d_\nu = R_0 - \nu r$. This bracket, for the value $\nu = \frac{1}{n}$ only, and the linear one admit the Poisson subspace

$$L = \partial^n + \sum_{j=0}^{n-2} u_{n-j} \partial^j \quad (37)$$

defining the bi-hamiltonian n^{th} KdV hierarchy [Gelfand and Dickey (1976)]. The quadratic Poisson algebra is the $\mathcal{W}_n$ algebra in this case.

3 N=2 n^{th} KdV hierarchies

3.1 N=2 superspace

We shall consider N=2 supersymmetry with space coordinate x. The N=2 superspace contains x and two Grassmann coordinates θ, $\bar\theta$. We shall use the notation $\underline{x}$ for the triple of coordinates $(x, \theta, \bar\theta)$. The supersymmetric covariant derivatives are defined by

$$\partial = \frac{\partial}{\partial x}, \ D = \frac{\partial}{\partial \theta} + \frac{1}{2}\bar\theta\partial, \ \bar D = \frac{\partial}{\partial \bar\theta} + \frac{1}{2}\theta\partial \quad (38)$$

and satisfy the N=2 supersymmetry algebra

$$D^2 = \bar D^2 = 0, \ \{D, \bar D\} = \partial. \quad (39)$$

An ordinary superfield is a function of $\underline{x}$

$$H(\underline{x}) = j(x) + \theta\xi(x) + \bar\theta\bar\xi(x) + \theta\bar\theta h(x) \quad (40)$$

depending on two even and two odd components. One can define chiral superfields $\phi(\underline{x})$ satisfying $D\phi = 0$ and antichiral superfields $\bar\phi(\underline{x})$ satisfying $\bar D\bar\phi = 0$. We shall use the decomposition of an ordinary superfield in chiral and anti-chiral parts

$$H(\underline{x}) = \Phi(H) + \bar\Phi(H) \quad (41)$$

where $D\Phi(H) = \bar D\bar\Phi(H) = 0$. Notice that Φ and $\bar\Phi$ are non local operators on superfields. Finally, we define the integration over the N=2 superspace to be

$$\int d^3\underline{x}\, H(x, \theta, \bar\theta) = \int dx\, \bar D D H(x, \theta, \bar\theta)|_{\theta=\bar\theta=0}. \quad (42)$$

3.2 Definition of N=2 n^{th} KdV hierarchies

As we have seen in the previous section, the nth KdV or n^{th} Gelfand-Dickey hierarchy is an integrable system whose second hamiltonian structure is the classical $\mathcal{W}_n$ algebra. In other words, the hamiltonians of this hierarchy form a family of local quantities in involution with respect to the $\mathcal{W}_n$ Poisson algebra. Our aim is to generalize this property to the N=2 case. Hence we shall define a N=2 n^{th} KdV hierarchy to be an integrable system with the classical N=2 $\mathcal{W}_n$ algebra as a hamiltonian structure and whose evolution equations enjoy N=2 supersymmetry.

Such hierarchies have been intensively studied, in particular those with a low number of fields. Thus, in the simplest case of N=2 KdV ($n = 2$), it has been found not one but three different hierarchies satisfying the above criteria, and their N=2 Lax formulation [Laberge and Mathieu (1988)] [Labelle and Mathieu (1991)] [Popowicz (1993a)]. Accordingly, further computer calculations in the $n = 3, 4$ cases seem to indicate the existence of three N=2 n^{th} KdV hierarchies. As a matter of fact, the phase space of an N=2 n^{th} KdV hierarchy, the N=2 $\mathcal{W}_n$ algebra, contains $n-1$ bosonic superfields $U_2, \cdots, U_n$ of superconformal dimension $[U_k] = k-1$. The first non trivial flow of such a hierarchy is generated by this Poisson algebra together with the second order hamiltonian $\mathcal{H}_2 = \int d^3\underline{x}\, U_2^2 + \alpha U_3$ where α is a real parameter [3]. Computations for $n = 3$ [Bellucci et al. (1993)][Yung (1993)] and $n = 4$ [Yung and Warner (1993)] have shown that for three distinct values of the parameter α, this equation possesses non-trivial higher order conserved charges $\mathcal{H}_3$, $\mathcal{H}_4$... After detailed inspection of these results, one expects for all $n \geq 2$

(I) a hierarchy with $\mathcal{H}_{pn} = 0$

(II) a bi-hamiltonian [4] hierarchy with $\mathcal{H}_k \neq 0$ at each order

(III) a hierarchy with $\mathcal{H}_k \neq 0$ at each order.

In the search for a Lax formulation for these hierarchies, the most usual tool is the algebra of N=1 ΨDOs although these systems enjoy N=2 supersymmetry. In this framework, the hierarchies of type (I) were recognized [Inami and Kanno (1992)] as reductions of the N=1 non standard KP hierarchy [Brunelli and Das (1995a)][Ghosh and Paul (1995)].

The purpose of [Delduc and Gallot (1996)] was to develop a N=2 analog of the Gelfand-Dickey formalism. This formalism, which is an example of the AKS construction, is based on the use of the algebra $\mathcal{C}$ of chirality preserving ΨDOs rather than ordinary N=2 ΨDOs as was suggested in [Popowicz (1993b)]. An N=2 KP hierarchy is then defined using a splitting of $\mathcal{C}$ into two associative subalgebras. It turns out that the corresponding r-matrix is not skew-symmetric and that this KP hierarchy is hamiltonian with respect to the linear Poisson structure, denoted by $\{,\}_{(1)}$, and to two local quadratic

---

[3] This argument is valid for $n \geq 3$ only. The corresponding one for the KdV case ($n = 2$) is however similar since the KdV equation is generated by the third order hamiltonian, $\mathcal{H}_3 = \int d^3\underline{x}\, U_2^3 + \alpha U_2 D\bar{D}U_2$, instead of the second order one.

[4] We consider only local hamiltonian structures.

structures of the *abcd* type, denoted by $\{,\}^{a,b}_{(2)}$. A peculiarity of these quadratic brackets is that they are related to each other by a non trivial invertible map p of the phase space. This map does not preserve hamiltonians. The N=2 KP hierarchy admits distinct reductions corresponding to Poisson subspaces of the different Poisson structures. In particular, the first N=2 n^{th} KdV hierarchy is recovered as a reduction of the quadratic bracket $\{,\}^{a}_{(2)}$, while the second bihamiltonian N=2 n^{th} KdV is a reduction of both the linear and quadratic bracket $\{,\}^{b}_{(2)}$. These two hierarchies are related to each other by the Poisson map p and so they share the same Poisson algebra, the N=2 $\mathcal{W}_n$ algebra. However, they are not equivalent since p does not preserve the hamiltonians. The third N=2 n^{th} KdV hierarchy is not found in this construction which does not exhaust all possible cases.

In addition to these results about N=2 KdV hierarchies, we shall also present an other N=2 hierarchy, related to a different splitting of $\mathcal{C}$, which contains the Dym hierarchy (30) in its bosonic limit. The N=2 generalization of the reduction (31) is given.

3.3   Pseudo-differential operators preserving chirality

In order to generalize the Gelfand-Dickey construction, it is natural to consider the associative algebra of N=2 ΨDOs which are the operators

$$P = \sum_{i<M} (a_i + b_i[D, \bar{D}] + \alpha_i D + \beta_i \overline{D})\partial^i \qquad (43)$$

where a_i, b_i and α_i, β_i are respectively even and odd N=2 superfields. Such an operator will lead to a hierarchy with a bigger phase space than, say, a $\mathcal{W}_n$ algebra and hence one has to restrict suitably the form of the N=2 operators in such a way that to each power of ∂ corresponds only one bosonic N=2 superfield. It turns out that a possible restriction is to define the set $\mathcal{C}$ of pseudo-differential operators L preserving chirality of the form

$$L = D\mathcal{L}\bar{D}, \quad \mathcal{L} = \sum_{i<M} u_i \partial^i \qquad (44)$$

the coefficient functions u_i being bosonic N=2 superfields. The product rule in $\mathcal{C}$ is

$$LL' = D\left(\mathcal{L}\partial\mathcal{L}' + (D.\mathcal{L})(\bar{D}.\mathcal{L}')\right)\bar{D} \qquad (45)$$

where $(D.\mathcal{L}) = \sum_{i<M}(Du_i)\partial^i$. Notice that $I = D\partial^{-1}\bar{D}$ is the unit of the algebra $\mathcal{C}$. We could have used as well the algebra $\bar{\mathcal{C}}$ of ΨDOs of the form $\bar{L} = \bar{D}\mathcal{L}D$ but $\mathcal{C}$ and $\bar{\mathcal{C}}$ are related by the usual transposition, $\bar{L}^t = -D\mathcal{L}^t\bar{D} \in \mathcal{C}$, and hence the same hierarchies are obtained from the two algebras.

We define the residue of the pseudo-differential operator L by $\mathrm{res}L = u_{-1}$. The residue of a commutator is a total derivative, $\mathrm{res}[L, L'] = D\bar{\omega} + \bar{D}\omega$. The trace of L is the integral of the residue

$$\mathrm{Tr}L = \int \mathrm{d}^3\underline{x}\,\mathrm{res}L, \quad \mathrm{Tr}[L,L'] = 0. \tag{46}$$

An operator $L = D\mathcal{L}\bar{D}$ of the form $\mathcal{L} = v_0^n \partial^{n-1} + \sum_{i<n-1} v_{n-1-i}\partial^i$ has a unique n^{th} root in $\mathcal{C}$ of the form $L^{\frac{1}{n}} = D(v_0 + \sum_{i<0} w_{n-1-i}\partial^i)\bar{D}$. The nth root $L^{\frac{1}{n}}$ commute with L: $[L^{\frac{1}{n}}, L] = 0$. In the case where $v_0 = 1$, the coefficients functions w_i are N=2 differential polynomials in the v_j's.

3.4   N=2 KP and KdV hierarchies

The algebra $\mathcal{C}$ can be decomposed into the direct sum of two associative subalgebras $\mathcal{C} = \mathcal{C}_+ \oplus \mathcal{C}_-$ where $L = L_+$ is in $\mathcal{C}_+$ if $\mathcal{L}$ is a differential operator and $L = L_-$ is in $\mathcal{C}_-$ if $\mathcal{L}$ is a strictly pseudo-differential operator ($M = 0$ in (44)). The important fact here is that the corresponding classical r-matrix, defined by

$$\mathcal{R}(L) = \frac{1}{2}(L_+ - L_-) \tag{47}$$

is not skew-symmetric:

$$\mathrm{Tr}(\mathcal{R}(L)L' + L\mathcal{R}(L')) = -\int \mathrm{d}^3\underline{x}\,\mathrm{res}L\,\mathrm{res}L'. \tag{48}$$

This is due to the fact that while $\mathcal{C}_+$ is isotropic, $\mathcal{C}_-$ is not.

The commuting flows of a N=2 KP hierarchy, for an operator L such that

$$\mathcal{L} = \partial^n + \sum_{i=1}^{\infty} U_i \partial^{n-i-1} \tag{49}$$

are then defined by

$$\partial_{t_k} L = [\mathcal{R}(L^{\frac{k}{n}}), L]. \tag{50}$$

From the algebraic properties of the splitting of $\mathcal{C}$ here considered, we expect these equations to be hamiltonian with respect to the linear structure

$$\{l_X, l_Y\}_{(1)}(L) = \mathrm{Tr}\left(L[X,Y]_{\mathcal{R}}\right) \tag{51}$$

where $l_X(L) = \mathrm{Tr}(LX)$, X being some element of $\mathcal{C}$ with no dependence on the space phase fields $\{U_i\}$, and to two quadratic structures of the $abcd$ type that we now exhibit. From the relation (48) one obtains the admissible quadruplet of non local endomorphisms

$$a = R + c, \quad b(X) = D(\partial^{-1}\bar{\Phi}[\mathrm{res}X])\bar{D} \tag{52}$$
$$c(X) = D(\Phi[\mathrm{res}X]\partial^{-1})\bar{D}, \quad d = R + b. \tag{53}$$

Although a, b, c and d are non-local, the corresponding quadratic structures $\{,\}_{(2)}^{a,b}$ are local. For example, $\{,\}_{(2)}^a$ may be put in the form

$$\{l_X, l_Y\}^a_{(2)}(L) = \text{Tr}(LX(LY)_+ - XL(YL)_+)$$

$$+ \int d^3\underline{x} \, \text{res}(LX)\Phi(\text{res}[L,Y]) + \text{res}(XL)\bar{\Phi}(\text{res}[L,Y]) \quad (54)$$

where Φ and $\bar{\Phi}$ act on total derivatives. As usual, the first bracket is a linearization of the two quadratic ones, that is to say

$$\{l_X, l_Y\}^{a,b}_{(2)}(L + zD\partial^{-1}\bar{D}) = \{l_X, l_Y\}^{a,b}_{(2)}(L) + z\{l_X, l_Y\}_{(1)}(L) \quad (55)$$

and the linear bracket is compatible with each of the two quadratic brackets.

Introducing the hamiltonians $\mathcal{H}_k = \frac{n}{k}\text{Tr}(L^{\frac{k}{n}})$, the KP evolution equations (50) may be written in the hamiltonian form

$$\partial_{t_k} l_X(L) = \{l_X, \mathcal{H}_{k+n}\}_{(1)}(L) = \{l_X, \mathcal{H}_k\}^{a,b}_{(2)}(L). \quad (56)$$

The hamiltonians form a family of quantities in involution with respect to any of these three brackets. The Poisson algebra inherited from the quadratic brackets is the N=2 $\mathcal{W}_\infty$ algebra [Lu et al. (1991)].

The two quadratic brackets are related to each other by an invertible Poisson map p of $\mathcal{C}$ defined by

$$p(L) = D\partial^{-1}L^t\partial^{-1}\bar{D}. \quad (57)$$

One has the relation

$$\{l_X \circ p, l_Y \circ p\}^a_{(2)} = -\{l_X, l_Y\}^b_{(2)} \circ p \quad (58)$$

which gives an equivalence between the two quadratic structures. However, there is no relation between the hamiltonians $\text{Tr}(L^{\frac{k}{n}})$ and $\text{Tr}(p(L)^{\frac{k}{n-1}})$.

As in the bosonic case, we expect n^{th} KdV hierarchies to be defined as consistent reductions of the KP hierarchy, hence we shall find Poisson submanifolds of the KP phase space. From a technical point of view, the quadratic structure $\{,\}^a_{(2)}$ is simpler than the other. It admits a Poisson submanifold defined by $L = L_+$ that is to say

$$L = D(\partial^{n-1} + \sum_{i=1}^{n-1} U_{i+1}\partial^{n-i-1})\bar{D}. \quad (59)$$

The Poisson algebra on the fields U_i is the N=2 $\mathcal{W}_n$ algebra. It may be checked that this restriction does not define a Poisson subspace for the linear bracket and that hamiltonians of order multiple of n vanish. The hierarchies thus obtained form the series (I) of N=2 n^{th} KdV. The simplest example is provided by the Lax operator $L = D(\partial + U)\bar{D}$ and leads to the so-called N=2 $a = -2$ KdV hierarchy [Laberge and Mathieu (1988)].

The easiest way to find Poisson subspaces for the second quadratic bracket is to apply the map p on the Poisson subspaces of the first quadratic bracket. From (59), we are then led to the restriction

$$L = D(\partial^{n-2} + \sum_{i=1}^{n-2} V_{i+1}\partial^{n-i-2})\bar{D} + D\bar{D}\partial^{-1}V_n\partial^{-1}D\bar{D}. \tag{60}$$

From the equivalence relation (58), the Poisson algebra inherited from $\{,\}_{(2)}^b$ on the fields V_i is the N=2 $\mathcal{W}_n$ algebra. Moreover, adding the identity to L is equivalent to shift V_n by 1 so that $L + zD\partial^{-1}\bar{D}$ remains in the submanifold defined by (60). From the linearization property (55), it is then easily obtained that this restriction defines also a Poisson subspace for the linear bracket and so, this hierarchy is bi-hamiltonian. The presence in L of a pseudo-differential term implies the non-vanishing of hamiltonians at each order. Hence we have identified these hierarchies with the series (II) of N=2 n^{th} KdV. The simplest example in this case corresponds to the Lax operator

$$L = D\bar{D} + D\bar{D}\partial^{-1}W\partial^{-1}D\bar{D} \tag{61}$$

and leads to the so-called N=2 $a = 4$ KdV hierarchy [Laberge and Mathieu (1988)]. This Lax operator for N=2 $a = 4$ KdV is different from the one already known [Krivonos and Sorin (1995)].

It should be noticed, in order to make the link with different articles using N=1 superspace formulation, that the N=2 KP hierarchy is nothing but the so-called N=1 non-standard KP hierarchy [Brunelli and Das (1995a)][Ghosh and Paul (1995)]. The later is obtained using the algebra of ordinary N=1 ΨDOs with a non skew-symmetric r-matrix. The linear and two quadratic brackets thus obtained [Das and Panda (1996)] are the N=2 brackets obtained here written in N=1 formalism [Delduc and Gallot (1996)]. The reductions in N=1 formalism corresponding to (59) were obtained by Inami and Kanno [Inami and Kanno (1992)] while those corresponding to (60) are generalizations of the one studied by Brunelli and Das [1994]. In particular, their supersymmetric two bosons hierarchy is equivalent to the N=2 $a = 4$ KdV hierarchy.

The bosonic limit of the N=2 KP hierarchy has been studied in [Delduc and Gallot (1996)] where it was obtained that it is composed of two decoupled bosonic hierarchies, namely the standard and non-standard KP hierarchies. The bosonic limit of the N=2 n^{th} KdV (I) corresponding to the reduction (59) is composed of the usual n^{th} KdV (37) and of the reduction (33) of the non-standard KP hierarchy. The bosonic limit of the N=2 n^{th} KdV (II) corresponding to the reduction (60) is composed of the $(n-1)^{th}$ KdV and of the reduction (34) of the non-standard KP hierarchy with n replaced by $n-1$. The various linear and quadratic structures for these reductions discussed in the previous section can be obtained from the N=2 Poisson structures.

There exists many other properties of the N=2 KP hierarchy which we did not expose here. As an example, for both quadratic brackets, one can define some N=2 generalizations of constrained KP systems. In particular, the simplest cases correspond to the N=2 NLS and the N=4 KdV hierarchies.

3.5   A N=2 Dym hierarchy

Beside the N=2 KP hierarchy, there exists an other hierarchy linked to another splitting of the algebra $\mathcal{C}$. $\mathcal{C}$ can be decomposed into the direct sum of two subalgebras $\mathcal{C} = \mathcal{C}_{\geq} \oplus \mathcal{C}_{<}$ where $L = L_{\geq}$ is in $\mathcal{C}_{\geq}$ if $\mathcal{L} = \mathcal{L}_{\geq 1}$ is a differential operator with vanishing non-differential term and $L = L_{<}$ is in $\mathcal{C}_{<}$ if $\mathcal{L} = \mathcal{L}_{<1}$ is a pseudo-differential operator containing the zero order term ($M = 1$ in (44)). The important fact here is that, whereas $\mathcal{C}_{\geq}$ is an associative subalgebra, $\mathcal{C}_{<}$ is only a Lie subalgebra of $\mathcal{C}$. Hence, to the r-matrix defined by

$$S(L) = \frac{1}{2}(L_{\geq} - L_{<}) \tag{62}$$

will only correspond a linear Poisson structure

$$\{l_X, l_Y\}_S(L) = \mathrm{Tr}(L[X,Y]_S). \tag{63}$$

With the Lax operator

$$L = D(V_0^n \partial^{n-1} + \sum_{i<n-1} V_{n-1-i}\partial^i)\bar{D} \tag{64}$$

where v_0 is invertible, we associate the commuting consistent flows

$$\partial_{t_k} L = [S(L^{\frac{k}{n}}), L]. \tag{65}$$

They may be written, as usual, in the hamiltonian form

$$\partial_{t_k} l_X(L) = \{l_X, \mathcal{H}_{k+n}\}_S(L) \tag{66}$$

where the hamiltonians $\mathcal{H}_k = \frac{n}{k}\mathrm{Tr}(L^{\frac{k}{n}})$ are in involution.

We shall call this hierarchy an N=2 Dym hierarchy because, as we shall see now, it contains the usual Dym hierarchy (30) in its bosonic limit. We first introduce the linear maps μ_1, μ_2 from $\mathcal{C}$ to $\mathcal{D}$ [Delduc and Gallot (1996)] defined by

$$\mu_1(L) = \mathcal{L}|_0\partial \tag{67}$$
$$\mu_2(L) = \mathcal{L}|_0\partial + (\bar{D}D\mathcal{L})|_0 \tag{68}$$

where $|_0$ means that we take $\theta = \bar{\theta} = 0$ and that all fermionic components are set to zero. These maps are morphisms of associative algebras, *i.e.* $\mu_{1,2}(LL') = \mu_{1,2}(L)\mu_{1,2}(L')$, satisfying

$$\mu_1(S(L)) = R_2(\mu_1(L)) \tag{69}$$
$$\mu_2(S(L)) = R_1(\mu_2(L)) - P_1(\mu_1(L)) \tag{70}$$

where R_k is the r-matrix corresponding to the splitting (27) for $k = 1, 2$. The flows (65) are then mapped onto the coupled flows

$$\partial_{t_k} L_1 = [R_2(L_1^{\frac{k}{n}}), L_1] \tag{71}$$

$$\partial_{t_k} L_2 = [R_1(L_2^{\frac{k}{n}}) - P_1(L_1^{\frac{k}{n}}), L_2]. \tag{72}$$

We have used the short-hand notation $L_1 = \mu_1(L)$, $L_2 = \mu_2(L)$. The flows (71) are those of the Dym hierarchy while the flows (72) are a deformation of the non-standard KP hierarchy. However, there exists a transformation on L_2 mapping the flows (72) onto non standard KP flows decoupled from (71). Indeed, $P_1(L_1^{\frac{k}{n}}) = Q_k \partial$ acts as a space translation that we shall compensate as follows. Let us consider the change of coordinates

$$x \longrightarrow x'(x, t_k) \text{ such that } \frac{\partial x'}{\partial x} = \frac{1}{u_0} \tag{73}$$

$$t_k \longrightarrow t'_k = t_k \tag{74}$$

where $u_0^n \partial^n = L_1 + \mathcal{O}(\partial^{n-1}) = L_2 + \mathcal{O}(\partial^{n-1})$. Let $L'(x', t'_k)$, a formal series of $\partial' = \partial_{x'} = \frac{1}{u_0} \partial$, be linked to a given operator $L(x, t_k)$ by $L'(x', t'_k) = L(x, t_k)$. This transformation is such that $R_1(L) = R'_1(L')$ where the r-matrix R'_1 is defined with respect to ∂'. The equations of motion for u_0 can be deduced from (71) or (72) and read $\partial_{t_k} \left(\frac{1}{u_0} \right) = -\left(\frac{Q_k}{u_0} \right)_x$. Hence one obtains the identity between vector fields

$$\partial_{t'_k} = \partial_{t_k} + Q_k \partial = \partial_{t_k} + P_1(L_1^{\frac{k}{n}}) \tag{75}$$

which leads to the non standard KP flows

$$\partial_{t'_k} L'_2 = [R'_1((L'_2)^{\frac{k}{n}}), L'_2]. \tag{76}$$

It may be shown that the two linear Poisson structures for equations (71,76) emerge from the linear structure (63) of the N=2 Dym hierarchy and that they are decoupled [Gallot]. This shows that the bosonic limit of the N=2 Dym hierarchy is composed of two decoupled hierarchies, namely the Dym hierarchy and the non standard KP hierarchy.

The N=2 Dym hierarchy admits reductions to a finite number of fields. Considering the linear bracket (63), we rewrite it as $\{l_X, l_Y\}_S(L) = \mathrm{Tr} X \xi_Y$ where ξ_Y is the hamiltonian vector field associated with the functional l_Y. It reads

$$\xi_Y = [L, Y]_{\geq} - [L, Y_{\geq}] + \eta([L, Y]) \tag{77}$$

where

$$\eta(A) = D(A_0 + \partial^{-1} A_1)\bar{D} + D\bar{D}\partial^{-2}(A_{1x} + A_2)\partial^{-1}D\bar{D} \tag{78}$$

if $A = D(\cdots + A_0 + A_1\partial^{-1} + A_2\partial^{-2} + \cdots)\bar{D}$. Then one obtains that the submanifold defined by

$$L = D(V_0^n \partial^{n-1} + \sum_{i=1}^{n-1} V_i \partial^{n-i-1})\bar{D} + D\bar{D}\partial^{-1}(V_n + \partial^{-1}V_{n+1})\partial^{-1}D\bar{D} \tag{79}$$

is a Poisson subspace of the linear bracket. It is interesting, for consistency, to notice that this operator is mapped by μ_1 onto the bosonic operator (31) while the operator L_2' is of the type (34). As in the bosonic case, the restriction $L = L_{\geq}$ is a consistent reduction of the flows (65) but does not define a Poisson subspace of the Poisson bracket (63).

Acknowledgement I am grateful to F.Delduc for discussions and a careful reading of the manuscript.

References

M. Adler, Invent.Math. 50 (1979) 403.

S. Bellucci, E. Ivanov, S. Krivonos and A. Pichugin, Phys. Lett. B 312 (1993) 463.

J.C. Brunelli and A. Das , Phys. Lett. B 337 (1994) 303; Phys. Lett. B 354 (1995) 307; Int. J. Mod. Phys. A 10 (1995) 4563; preprint hep-th/9506096.

J.C. Brunelli and A. Das, Rev. Math. Phys. 7 (1995) 1181.

A. Das and S. Panda, Mod. Phys. Lett. A 11 (1996) 723.

F. Delduc and L. Gallot, preprint ENSLAPP-L-617-96, solv-int/9609008, to appear in Commun. Math. Phys.

L. A. Dickey, Soliton equations and hamiltonian systems, Advanced Series in Mathematical Physics, Vol.12, World Scientific (1991)

P. Di Francesco, C. Itzykson and J.B. Zuber, Commun. Math. Phys. **140** (1991) 543.

L. Freidel, J. M. Maillet, Phys. Lett. B 262 (1991) 278.

L. Gallot, Thesis (in french), work in progress

I.M. Gelfand and L.A. Dikii, Funct. Anal. Appl. 10 (1976) 259.

S. Ghosh and S. Paul, Phys. Lett. B 341 (1995) 293.

T. Inami and H. Kanno, Int. J. Mod. Phys. A 7, Suppl. 1A (1992) 419.

B.G. Konopelchenko and W. Oevel, Publ. RIMS, Kyoto Univ. **29**(1993) 581-666

S. Krivonos and A. Sorin, Phys. Lett. B 357 (1995) 94

B. A. Kupershmidt, Commun. Math. Phy. 99, (1985) 51.

P. Labelle and P. Mathieu, J. Math Phys. 32 (1991) 923.

C. A. Laberge, P. Mathieu, Phys. Lett. B 215 (1988) 718.

L.C. Li and S. Parmentier, Commun. Math. Phys. **125** (1989) 545.

H. Lu, C.N. Pope, L.J. Romans, X. Shen and X.J. Wang, Phys. Lett. B 264 (1991) 91.

W. Oevel, Phys. Lett. A 186 (1994) 79.

Z. Popowicz, Phys. Lett. A 174 (1993) 411.

Z. Popowicz, Phys. Lett. B 319 (1993) 478.

A.G. Reyman, J. Sov. Math. **19** (1982) 1507.

M.A. Semenov-Tian-Shansky, Funct. Anal. Appl. 17 (1983) 259.

Yu.B. Suris, Phys. Lett. **A 186** (1993) 419.

C.M. Yung, Phys. Lett. B 309 (1993) 175.

C.M. Yung and R.C. Warner, J. Math. Phys. 34 (1993) 4050.

Reduction of Self-Dual Yang-Mills Systems and Super Nonlinear Schrödinger Equations

M. Legaré

Department of Mathematical Sciences, University of Alberta, Edmonton, Alberta, Canada, T6G 2G1

Abstract. The symmetry reduction of the self-dual Yang-Mills equations and their linear systems with Lie superalgebra valued fields is reviewed. An example of reduction is given were supersymmetric generalized nonlinear Schrödinger equations are derived.

1 Introduction

The application of the method of symmetry reduction (cf for example Bluman and Kumei (1989), Olver (1986), Winternitz (1990)), with often the addition of algebraic constraints, to the self-dual Yang-Mills (hereafter : SDYM) equations has been quite fruitful in recovering known as well as providing new integrable systems in lower dimensions. Examples of such reductions abound. These reductions involves symmetry subgroups of the invariance group of the SDYM equations in four-dimensional Euclidean space ($\mathbb{E}^4$) and in space $\mathbb{R}^4$ with metric of signature $(2,2)$: $diag(+1,+1,-1,-1)$, ($\mathbb{E}^{(2,2)}$), also called Affine $(2,2)$ Minkowski space. Via translational symmetries, the Nahm (Chakravarty et al. (1990), Ivanova and Popov (1991), Ward (1985)), (modified, matrix) Korte–weg–de Vries (KdV) (Bakas and Depireux (1991a), Bakas and Depireux (1991b), Ivanova and Popov (1992), Ivanova and Popov (1995), Mason and Sparling (1989)), (generalized) nonlinear Schrödinger (NS) (Ivanova and Popov (1992), Ivanova and Popov (1995), Mason and Sparling (1989)), N–wave (Chakravarty and Ablowitz (1992)), Toda field (Ivanova and Popov (1995), Leznov and Saveliev (1980)), static non–Abelian Chern–Simons (Dunne et al. (1991), Dunne et al. (1992)), modified Chiral model in $\mathbb{R}^{(2,1)}$ (Ward (1988a), Ward (1988b)) and Davey-Stewartson (Chakravarty and Ablowitz (1992)) equations have been obtained, among others (Ablowitz and Clarkson (1991), Ward (1990)). With translations and rotations, the (modified or deformed) Nahm (Ivanova and Popov (1995), Kovalyov et al. (1993)) and Ernst (Fletcher and Woodhouse (1990), Woodhouse and Mason (1988)) equations can be retrieved. Moreover, tridimensional Abelian subgroups made of translations, and/or rotations, and/or dilations have been related to reductions to the Painlevé transcendents (Mason and Woodhouse (1993)).

Since the SDYM equations follow from the compatibility condition of linear systems, a reduction taking into account the space of the (spectral) parameter would lead to a corresponding reduced linear system whose involution is the set of reduced SDYM equations (Atiyah (1979), Ivanova and Popov (1995), Legaré and Popov (1995), Ward and Wells Jr (1990)), and therefore, as long as the reduction of the SDYM equations is known, a linear system (Lax pair) associated to the reduced equation(s) can be found by similarly reducing the SDYM linear system. The method of reduction by symmetry can also be used in self-dual gravity models to bring out the $sl(\infty)$–Toda (Boyer and Finley (1982), Gegenberg and Das (1984)) and Gibbons–Hawking (Gibbons and hawking (1979)) equations as well as reveal on a hierarchy with the SDYM equations as initial flow, hierarchies such as the KdV hierarchy (Ablowitz (1993)). "Hidden (infinitesimal) symmetries", which affinize the conformal symmetries, have also been uncovered for self–dual Yang–Mills and self–gravity systems (Popov and Preitschopf (1996), Popov et al. (1996)), and a twistor interpretation was given for the SDYM case (Ivanova (1997)).

With the addition of anticommuting variables (fields), these systems could still be integrable, or could even become supersymmetric. It is then natural to ask if an extension including anticommuting variables of the SDYM equations could lead to supersymmetric integrable systems through reductions. Indeed, there exists already a number of reductions of supersymmetric extensions of the SDYM equations to supersymmetric integrable models, such as the super Korteweg–de Vries and super Nonlinear Schrödinger equations (Gates Jr and Nishino (1993), Gilson et al. (1986)), where use of differential constraints was sometimes required.

A second approach to produce supersymmetric integrable systems relies on Lie superalgebras. For instance, integrable (super–) Toda theories can be constructed using the root system of Lie superalgebras or by carrying out a version of Hamiltonian reduction (Evans and Madsen (1996), Evans and Madsen (1997)). Another example is based on loop superalgebras, where a supersymmetric generalization of a hierarchy of the classical AKNS theory is obtained and, with the help of constraints on the variables, allows to derive known hierarchies of supersymmetric integrable systems with a zero curvature condition and a Lax pair setting (Morosi and Pizzocchero (1996)). Finally, a zero curvature condition in superspace and Lie superalgebras $(sl(n + 1, n))$ were involved in a Drinfeld–Sokolov version of a supersymmetric generalization of hierarchies of the AKNS type (Aratyn et al. (1997)).

From a combination of these two approaches to generate supersymmetric integrable systems : i.e. supersymmetric versions of SDYM systems and Lie superalgebras based models, emerges a simple way to introduce anticommuting field components through Lie superalgebra valued fields, and integrable systems with fermionic variables (Chaichian and Kulish (1978)), which are not necessarily supersymmetric, could then be found by reduction (under symmetry subgroups) of SDYM equations with Lie superalgebras as gauge algebras (Legaré M. (1997)). As above, if a reduction of the SDYM equations to an integrable sytem is determined, the corresponding Lax pair, or linear system, can be deduced from the original linear system of the SDYM equations.

The method of reduction (by symmetry) and Yang–Mills theories are not only useful in this context. Let us mention that Yang–Mills equations in higher dimensions with linear constraints (Corrigan et al. (1983), Ward (1984)), or closely, $D > 4$ dimensional extensions of SDYM equations, have been recently studied on $D > 4$ manifolds of reduced holonomy and have been shown to possess instanton–like solutions. Cohomological field theories in higher dimensions are helpful in the study of moduli spaces of these solutions and are related to (dimensional) reductions of 10 dimensional supersymmetric Yang–Mills models to Euclidean spaces or Riemannian manifolds (Acharya B.S. et al. (1997), Baulieu et al. (1997), Blau and Thompson (1997)). Futhermore, the quantum integrability of the SDYM theories could be investigated through its reduced systems, as well as in view of its relation to $N = 2$ (open,heterotic) strings (Ooguri and Vafa (1995)).

Below, a reminder of the SDYM equations and their linear systems in four dimensions with Lie superalgebras as gauge algebras is presented in section 2. Section 3 reviews the symmetry reduction of both the SDYM and linear systems on $\mathbb{E}^4$ and $\mathbb{E}^{(2,2)}$, and an example of reduction leading to a supersymmetric version of the $SO(3)/SO(2)$ generalized nonlinear Schrödinger equations is shown in section 4. Section 5 concludes with a summary and a discussion of future research avenues.

2   SDYM and Linear Systems in $\mathbb{E}^4$ and $\mathbb{E}^{(2,2)}$

Let $\mathcal{H}$ denote a Lie superalgebra of even dimensions m and odd dimensions n (Cornwell (1989), Frappat et al. (1989), Kac (1977)). On a gauge bundle, let $\omega = A_\mu \theta^\mu$, where $\mu = 1, ..., 4$ and $\{\theta^\mu\}$ is a 1-form basis of the co–tangent space, then the Lie superalgebra valued components (A_μ) of the gauge field can be decomposed as follows :

$$A_\mu = A_\mu^a M_a + \xi_\mu^\alpha \tilde{M}_\alpha, \tag{1}$$

where $\{M_a, a = 1, ..., m\}$ and $\{\tilde{M}_\alpha, \alpha = 1, ..., n\}$ are the corresponding even and odd part bases of $\mathcal{H}$, with A_μ^a and ξ_μ^α functions on $\mathbb{E}^4$(or $\mathbb{E}^{(2,2)}$) respectively with values in the even and odd parts of a Grassmann algebra (CB_L), where $m \geq 2^{L-1}$ and $n \geq 2^{L-1}$ (Cornwell (1989)).

The SDYM equations for the curvature 2-form $F = d\omega + \omega \wedge \omega$, with field strength $F_{\mu\nu} = \partial_\mu A_\nu - \partial_\nu A_\mu + [A_\mu, A_\nu]$, can be written as :

$$F = *F, \quad \text{or} \quad F_{\mu\nu} = \frac{1}{2}\epsilon_{\mu\nu\rho\sigma}F^{\rho\sigma}, \tag{2}$$

with $\mu, \nu, \rho, \sigma = 1, ..., 4$ and $\epsilon_{\mu\nu\rho\sigma}$, the completely antisymmetric tensor in four dimensions ($\epsilon_{1234} = 1$).

The involution of the following linear systems gives rise to the equations (2) (Atiyah (1979), Ward and Wells Jr (1990)) on the appropriate space. On $\mathbb{E}^4$:

$$[D_1 + iD_2 - \lambda(D_3 + iD_4)]\Psi = 0,$$
$$[D_3 - iD_4 + \lambda(D_1 - iD_2)]\Psi = 0, \tag{3}$$

and on $\mathbb{E}^{(2,2)}$:

$$[D_1 + iD_2 + \lambda(D_3 - iD_4)]\Psi = 0,$$
$$[D_3 + iD_4 + \lambda(D_1 - iD_2)]\Psi = 0, \tag{4}$$

with the covariant derivatives : $D_\mu = \partial_\mu + A_\mu$, and where the column vectors (multiplet) Ψ are holomorphic with respect to a parameter $\lambda \in \mathbb{CP}^1$ (or $\mathbb{H}^2$–sheet) (i.e. $\partial_{\bar{\lambda}}\Psi = 0$) and bear elements which are CB_L valued functions on $\mathbb{E}^4 \times \mathbb{CP}^1$ (or $\mathbb{E}^{(2,2)} \times (\mathbb{H}^2$–sheet)). The Cartesian coordinates on $\mathbb{E}^4$ (or $\mathbb{E}^{(2,2)}$) are denoted by $x^\mu, \mu = 1, ..., 4$. Since a linear Lie supergroup associated to a Lie superalgebra $(\mathcal{H})$ of even dimensions m and odd dimensions n can be described as a (linear) Lie group of $(m+n)2^{L-1}$ dimensions (Cornwell (1989)), this kind of gauge group could still allow a twistor characterization of the solutions (Ward and Wells Jr (1990)).

3   Reduction by Symmetry

Both the SDYM equations (2) and their respective linear system (3) for $\mathbb{E}^4$ and (4) for $\mathbb{E}^{(2,2)}$, are preserved under the conformal transformations $SO(5,1)$ in the case of $\mathbb{E}^4$ and $SO(3,3)$ in the case of $\mathbb{E}^{(2,2)}$. Since the vector parts of the linear system (3) (or (4)) constitutes a basis of antiholomorphic vectors, and therefore induce complex structures on $\mathbb{E}^4 \times \mathbb{CP}^1$ (or $\mathbb{E}^{(2,2)} \times (\mathbb{H}^2$–sheet)), the holomorphy of Ψ should be kept unchanged when a lift of these conformal actions to the 6-dimensional space $\mathbb{E}^4 \times \mathbb{CP}^1$ (or $\mathbb{E}^{(2,2)} \times (\mathbb{H}^2$–sheet)) is chosen.

The complex structures $(\mathcal{J})$ compatible with equations (3) are given by (Woodhouse (1985)) :

$$\mathcal{J} = (s_a (2Y_{a\,\rho\mu}), \, \epsilon_i{}^j), \tag{5}$$

where $s_a, a = 1, 2, 3$ $(s_a s_a = 1)$, are coordinates on $\mathbb{CP}^1$, $\epsilon_i{}^j$ is the 2d antisymmetric tensor $(\epsilon_1{}^2 = 1)$, and :

$$Y_{1\,\rho\mu} = -\frac{1}{2}(M_{23}-M_{14})_{\rho\mu}, Y_{2\,\rho\mu} = \frac{1}{2}(M_{13}+M_{24})_{\rho\mu}, Y_{3\,\rho\mu} = -\frac{1}{2}(M_{12}-M_{34})_{\rho\mu}, \tag{6}$$

are elements of a $so(3)$-subalgebra of $so(4)$ with :

$$[M_{\alpha\beta}]_{\mu\nu} = \delta_{\mu\alpha}\delta_{\nu\beta} - \delta_{\mu\beta}\delta_{\nu\alpha}, \tag{7}$$

$\alpha, \beta, \mu, \nu = 1, ..., 4$, where the $M_{\alpha\beta}$–matrices corresponds to $x^{\alpha}x^{\beta}$–plane rotation generators.

The complex structures $(\tilde{\mathcal{J}})$ related to the equations (4) have the form (Ivanova and Popov (1995)) :

$$\tilde{\mathcal{J}} = (-h_a(2\mathcal{B}_a{}^{\rho}{}_{\mu}), \epsilon_i{}^j), \tag{8}$$

where $h_a, a = 1, 2, 3$ $((h_1)^2 + (h_2)^2 - (h_3)^2 = -1)$, are coordinates on a sheet of $\mathbb{H}^2$, and :

$$\mathcal{B}_1{}^{\rho}{}_{\mu} = \frac{1}{2}(N_{23}+N_{14})^{\rho}{}_{\mu}, \mathcal{B}_2{}^{\rho}{}_{\mu} = -\frac{1}{2}(N_{13}-N_{24})^{\rho}{}_{\mu}, \mathcal{B}_3{}^{\rho}{}_{\mu} = \frac{1}{2}(N_{12}+N_{34})^{\rho}{}_{\mu}, \tag{9}$$

are elements of a $so(2,1)$ subalgebra of $so(2,2)$, with :

$$[N_{\alpha\beta}]^{\mu}{}_{\nu} = \delta_{\mu\alpha}\delta_{\nu\beta} + \delta_{\mu\beta}\delta_{\nu\alpha}, \quad \text{if} \quad \mu\nu \neq 12 \quad \text{and} \quad 34,$$
$$[N_{\alpha\beta}]^{\mu}{}_{\nu} = \delta_{\mu\alpha}\delta_{\nu\beta} - \delta_{\mu\beta}\delta_{\nu\alpha}, \quad \text{if} \quad \mu\nu \neq 12 \quad \text{or} \quad 34, \tag{10}$$

which are $x^{\alpha}x^{\beta}$–plane rotation and boosts generators.

A lift $(\tilde{X})$ of an element X of the conformal algebra $so(5,1)$ (or $so(3,3)$), which respects the holomorphy of the column vector Ψ on $\mathbb{E}^4 \times \mathbb{CP}^1$ (or $\mathbb{E}^{(2,2)} \times (\mathbb{H}^2\text{–sheet})$) endowed with the complex structures $\mathcal{J}$ (or $\tilde{\mathcal{J}}$) can be determined by solving the condition :

$$\mathcal{L}_{\tilde{X}}\mathcal{J} = 0, \tag{11}$$

where $\mathcal{L}_{\tilde{X}}$ denotes the Lie derivative along $\tilde{X}$.

For the isometry subalgebra $so(4) \triangleright t^4$ of $\mathbb{E}^4$ or $so(2,2) \triangleright t^4$ of $\mathbb{E}^{(2,2)}$, only the following elements are affected by a lift contribution :

$$\tilde{Y}_a = (Y_{a\,\mu\nu})x_{\mu}\partial_{\nu} - \epsilon_{abc}s^b\partial_{s^c},$$
$$\tilde{\mathcal{B}}_a = -(\mathcal{B}_a{}^{\mu}{}_{\nu})\, x^{\nu}\partial_{\mu} - f_{ab}^c h^b \partial_{h^c}, \tag{12}$$

where $a, b, c = 1, 2, 3$, ϵ_{abc} is the 3d antisymmetric tensor ($\epsilon_{123} = 1$) and f_{ab}^c stands for structure constants of $so(2,1)$. If not vanishing, $f_{12}^3 = 1$, $f_{23}^1 = -1$, and $f_{31}^2 = -1$.

The conformal transformations generators will also be modified, but the dilation will remain untouched, as the translations and other rotations (Ivanova and Popov (1995), Legaré and Popov (1995)). The equations (2),(3), and (4) are also invariant under gauge transformations derived from the gauge Lie superalgebra $\mathcal{H}$.

In order to carry out the symmetry reduction of the SDYM and linear systems under a given subgroup G, with Lie algebra $\mathcal{G}$, of the invariance group $SO(5,1)$ (or $SO(3,3)$), invariance conditions for the Yang-Mills fields (A_{μ}) and the $(m+n)$-column vector (Ψ), which transforms according to the fundamental representation of the gauge group, have to be specified.

For the (infinitesimal) G-invariance of the Yang-Mills fields, the following relation involving the Lie derivative $\mathcal{L}_X$ along the vector field associated to the element $X \in \mathcal{G}$ (Harnad J. et al. (1980), Hussin V. et al. (1994), Forgács and Manton (1980), Jackiw and Manton (1980)) :

$$\mathcal{L}_X A_\mu = \partial_\mu W + [A_\mu, W], \tag{13}$$

has to be satisfied for all $X \in \mathcal{G}$, where W is a function of $\mathcal{G} \times \mathbb{E}^4$ (or $\mathbb{E}^{(2,2)} \times (\mathbb{H}^2\text{-sheet})$) into the Lie superalgebra $\mathcal{H}$ (gauge algebra) classifying and inducing the lift of the G-action to the gauge bundle (Harnad J. et al. (1980), Hussin V. et al. (1994)).

As for the (infinitesimal) G-invariance of the $(m + n)$-column vector Ψ lying in the 6-dimensional with space coordinates x^μ and parameter variable λ, it obeys to the conditions :

$$\mathcal{L}_{\tilde{X}} \Psi = -W\Psi, \tag{14}$$

for all $X \in \mathcal{G}$ (as indicated above, $\tilde{X}$ is the lift of X). Let us mention that the function W in equations (13) and (14) allows an invariance up to an (infinitesimal) gauge transformation of A_μ and Ψ. Its vanishing implies the known (strict) invariance of a 1-form (A_μ) and 0-form(s) (Ψ). The global invariance conditions can be found in refs. Harnad J. et al. (1980), Hussin V. et al. (1994).

Since reductions by symmetry subalgebras equivalent under the adjoint action of the conformal (or isometry) group gives rise to equivalent reduced systems of equations (Winternitz (1990)), only representatives of conjugacy classes of the invariance group under its adjoint action would need to be considered in a systematic study of the reductions.

The method of reduction by symmetry under a N-dimensional subgroup G requires the knowledge of the orbit variables $\{\chi_m, m = 1, ..., d \leq N\}$ and invariant variables, here denoted by $\{\eta_i, i = 1, ..., 6 - d\}$, which are found using the lifted basis vectors of the symmetry algebra $\mathcal{G}$. Among the invariant variables, a parameter (ζ), which could be different from the original parameter λ, can be identified through the inequality : $\mathcal{L}_X \zeta \neq 0$, for some $X \in \mathcal{G}$.

Inserting the orbit coordinates (χ_m) and invariant coordinates (η_i) along with the G-invariant Yang–Mills fields (A_μ) and multiplet (Ψ) in the SDYM and linear systems will lead to the G-reduced SDYM equations and a G-reduced linear system.

4   Example: Super Nonlinear Schrödinger Equations

As indicated before, reductions with respect to translations do not involve any nontrivial lift contribution to the vector fields related to $\mathcal{G}$, contrary to those associated to pure conformal transformations and certain rotations in

$\mathbb{E}^4$ or $\mathbb{E}^{(2,2)}$. Reductions under the latter type of subgroups have been carried out in refs. Fletcher and Woodhouse (1990), Ivanova and Popov (1995), Kovalyov et al. (1993), Legaré M. (1997), Legaré and Popov (1995), Woodhouse and Mason (1988). In the following, an example of reduction of the SDYM equations in $\mathbb{E}^{(2,2)}$ under translations with Lie superalgebra $sl(2/1)$ as gauge algebra leading to a supersymmetric extension of the $SO(3)/SO(2)$ (class BD I) generalized nonlinear Schrödinger equations is shown. Thereby confirming that known supersymmetric integrable systems can also be retrieved via the reduction of the above SDYM systems with fermionic degrees of freedom.

On the superspace extension of the space variable with even coordinate x and odd coordinate θ, the supercovariant derivative D is defined as : $D = \partial_\theta + \theta\partial_x$, with $D^2 = \partial_x$. Let us introduce a bosonic (odd) superfield : $\Phi = \phi_1 + \theta\psi_1$, as well as a fermionic (even) superfield : $\xi = \psi_2 + \theta\phi_2$, the supersymmetric $SO(3)/SO(2)$ generalized Schrödinger equations can then be written as (Aratyn et al. (1997), Aratyn and Rasinariu (1997)) :

$$\partial_t \Phi = \Phi_{xx} - 2(\Phi\xi(D\Phi) - (D\xi)\Phi^2),$$
$$\partial_t \xi = -\xi_{xx} - 2(D\xi)\Phi\xi, \tag{15}$$

and are invariant under the supersymmetry transformations :

$$\delta_\varepsilon \Phi = \varepsilon D^\dagger \Phi,$$
$$\delta_\varepsilon \xi = \varepsilon D^\dagger \xi, \tag{16}$$

with the adjoint to the superderivative : $D^\dagger = \partial_\theta - \theta\partial_x$. A component form of these equations is provided in ref. Aratyn et al. (1997).

In order to recover the equations (15) as the reduction of the SDYM equations, $\mathcal{H}$ is set equal to $sl(2/1)$ and the space $\mathbb{E}^{(2,2)}$ is chosen. However, a different version of the linear system (4) is preferred. The $\mathbb{H}^2$–sheet parametrization, which originally is given in terms of λ, can also be realized via a linear fractional transformation to the unit disk with the complex parameter Λ. The new parameter is :

$$\Lambda = i\frac{(1-\lambda)}{(1+\lambda)}. \tag{17}$$

Then introducing the following new null variables (Ivanova and Popov (1992), Ivanova and Popov (1995)) :

$$X^1 = \frac{1}{2}(x^1 + x^3), \, X^2 = \frac{1}{2}(x^1 - x^3), \, X^3 = \frac{1}{2}(x^2 + x^4), \, t = \frac{1}{2}(x^2 - x^4), \tag{18}$$

with new components of the gauge fields :

$$A_{X^1} = (A_1 + A_3), \, A_{X^2} = (A_1 - A_3), \, A_{X^3} = (A_2 + A_4), \, A_t = (A_2 - A_4), \tag{19}$$

the linear system (4) becomes :

$$[D_t - \Lambda D_{X^2}]\Psi = 0, \quad [D_{X^1} + \Lambda D_{X^3}]\Psi = 0, \tag{20}$$

with the holomorphy condition on Ψ : $\partial_{\bar\Lambda}\Psi = 0$.

4.1   Example: $\mathcal{G} = \{P_{X^3}, P_{X^1-X^2}\}, \mathcal{H} = sl(2/1)$

One can now carry out a reduction by translations along the X^3 and $X^1 - X^2$ axes. Since the lift of translations does not have any parameter contribution to the vector fields, the orbit coordinates can be chosen as : $X^3, x^3 = X^1 - X^2$, and the invariant coordinates : $t, x \equiv x^1 = X^1 + X^2$.

Accordingly, the G-invariant gauge field components are equal to :

$$A_{X^1} = u_1(t, x), \ A_{X^2} = u_2(t, x), \ A_{X^3} = u_3(t, x), \ A_t = u_t(t, x), \tag{21}$$

and the G-invariant multiplet Ψ is simply :

$$\Psi = \psi(t, x, \Lambda). \tag{22}$$

The reduced linear system has the form (Ivanova and Popov (1992), Ivanova and Popov (1995)) :

$$[\partial_t - \Lambda\partial_x + u_t - \Lambda u_2]\psi = 0,$$
$$[\partial_x + u_1 + \Lambda u_3]\psi = 0, \tag{23}$$

or equivalently :

$$[\partial_t + u_t + \Lambda(u_1 - u_2) + \Lambda^2 u_3]\psi = 0,$$
$$[\partial_x + u_1 + \Lambda u_3]\psi = 0. \tag{24}$$

The linear system (24) is in fact identical to the Drinfeld-Sokolov formulation presented in ref. Aratyn and Rasinariu (1997) whose involution provides the supersymmetric version of the $SO(3)/SO(2)$ generalized nonlinear Schrödinger equations in refs. Aratyn et al. (1997), Aratyn and Rasinariu (1997) by identification (using their notation) of :

$$u_3 = -E = - \begin{bmatrix} 1 & 0 & 0 \\ 0 & 0 & 0 \\ 0 & 0 & 1 \end{bmatrix}, \tag{25}$$

an element of $sl(2/1)$ allowing a (canonical) decomposition through an involutive automorphism (ad_E),

$$u_1 = \mathcal{A} = \begin{bmatrix} 0 & \Phi & 0 \\ -D\xi + (\partial_x^{-1}\Phi\xi)\xi & 0 & \xi \\ 0 & D\Phi + (\partial_x^{-1}\Phi\xi)\Phi & 0 \end{bmatrix}, \tag{26}$$

$$u_2 = 0, \tag{27}$$

and

$$u_t = -\partial_x^{-1}([\mathcal{A}, \mathcal{R}(\mathcal{A})]) + \mathcal{R}(\mathcal{A}), \tag{28}$$

where (Aratyn et al. (1997)) :

$$\mathcal{R} = ad_E(\partial_x - ad_\mathcal{A} \partial_x^{-1} ad_\mathcal{A}). \tag{29}$$

The compatibility condition of the equations (24) with the expressions (25), (26), (27), (28) and (29) for the gauge field components produces the supersymmetric version of the $SO(3)/SO(2)$ generalized nonlinear Schrödinger equations (15).

5   Conclusion

In the above, the symmetry reduction of the SDYM equations and their linear systems (Lax pair) with Lie superalgebra valued gauge fields (A_μ) was summarized. The reduction of the linear system relies on a suitable lift of the group action to the parameter space which preserves the holomorphy of the $(m + n)$–column vector (Ψ). Its involution matches the similarly reduced SDYM equations. An advantage of this description resides in the fact that the knowledge of the reduction steps leading to the system under study from the SDYM equations allow the determination of a linear system, if unknown, by the same procedure applied to the SDYM linear system. Let us reiterate that the reduced systems are not necessarily supersymmetric, but reduced systems with supersymmetry transformations can be found as exemplified by the reduction of the $sl(2/1)$ valued SDYM equations to supersymmetric generalized nonlinear Schrödinger equations.

A natural future direction of interest of this work would be to find reductions to (extended) supersymmetric versions of the KdV equations starting from Lie superalgebra valued SDYM equations. May be this reduction coud be realized via an embedding of an AKNS formulation with quadratic polynomials of the parameter (λ) in the linear system of the KdV system in $sl(2/1)$ or in a larger Lie superalgebra. Other integrable systems (with anticommuting variables) could also be sought by performing reductions with respect to different subgroups of the conformal group as well as using different Lie superalgebras as gauge algebras, to generate systems such as supersymmetric extensions of generalized nonlinear Schrödinger equations based on higher dimensional Lie algebras and supersymmetric matrix KdV equations. Finally, it would be useful to find conditions leading to reduced integrable systems with supersymmetry, and investigate in a similar manner higher dimensional self-dual equations.

6   Acknowledgements

The author is grateful to the organizers for accepting his participation in the workshop, and acknowledges for the support of this work a grant from the National Sciences and Research Council (NSERC) of Canada.

References

Ablowitz M.J., Chakravarty S., and Takhtajan L.A. (1993), Commun. Math. Phys. 158, 289.

Ablowitz M.J. and Clarkson P.A. (1991) : *Solitons, Nonlinear Evolution Equations and Inverse Scattering*, (Cambrigde University Press, Cambridge), section 6.5 and references therein.

Acharya B.S., Figueroa-O'Farrill J.M., and Spence B. (1997) : *Euclidean D-branes and higher–dimensional gauge theory*, hep-th/9707118.

Aratyn H., Das A., and Rasinariu C. (1997) : *Zero curvature formalism for super-symmetric integrable hierarchies in superspace*, hep-th/9704119, and references therein.

Aratyn H. and Rasinariu C. (1997), Phys. Lett. B391, 99.

Atiyah M.F. (1979), : *Classical Geometry of Yang-Mills Fields* (Scuola Normale Superiore, Pisa)

Bakas I. and Depireux D.A. (1991a), Mod. Phys. Lett. A6, 399.

Bakas I. and Depireux D.A. (1991b), Mod. Phys. Lett. A6, 1561.

Baulieu L., Losev A., and Nekrasov N. (1997) : *Chern-Simons and twisted super-symmetry in various dimensions*, hep-th/9707174.

Blau M. and Thompson G. (1997) : *Euclidean SYM theories by time reduction and special holonomy manifolds*, hep-th/9706225.

Bluman G.W. and Kumei S. (1989) : *Symmetries and Differential Equations* (Springer-Verlag, N.Y.)

Boyer C. and Finley J. (1982), J. Math. Phys. 23, 1126.

Chaichian M. and Kulish P.P. (1978), Phys. Lett. 78B, 413.

Chakravarty S. and Ablowitz M.J. (1992), in *Painlevé transcendents, their asymptotics and physical applications*, eds : D. Levi and P. Winternitz, NATO ASI series B278, (Plenum Press, N.Y.), 331.

Chakravarty S., Ablowitz M.J. and Clarkson P.A. (1990), Phys. Rev. Lett. 65, 1085.

Cornwell J. (1989) : *Group Theory in Physics V. III* (Academic Press, N.Y.)

Corrigan E., Devchand C., Fairlie D.B., and Nuyts J. (1983), Nucl. Phys. B214, 452.

Dunne G.V., Jackiw R., Pi S.-Y., and Trugenberger C.A. (1992), Phys. Rev. D43, 1332.

Dunne G.V., Jackiw R., Pi S.-Y., and Trugenberger C.A. (1991), Phys. Rev. D45, 3012.

Evans J.M. and Madsen J.O. (1996), Phys. Lett. B384, 131.

Evans J.M. and Madsen J.O. (1997) : *Dynkin diagrams and integrable models based on Lie superalgebras*, hep-th/9703065, and references therein.

Fletcher J. and Woodhouse N.M.J. (1990), in *Twistors in Mathematics and Physics*, eds : T.N. Bailey and R.J. Baston, London Mathematical Society Note Ser. 156, (Cambridge University press, Cambridge), 260.

Forgács P. and Manton N.S. (1980), Commun. Math. Phys. 72, 15.

Frappat L., Sciarrino A. and Sorba P. (1989), Commun. Math. Phys. 121, 457.

Gates Jr S.J. and Nishino H. (1993), Phys. Lett. B299, 255.

Gegenberg J. and Das A. (1984), Gen. Rel. Grav. 16, 817.

Gibbons G.W. and Hawking S.W. (1979), Commun. Math. Phys. 66, 291.

Gilson C.R., Martin R., Restuccia A., and Taylor J.G. (1986), Commun. Math. Phys. 107, 377.

Harnad J., Shnider S. and Vinet L. (1980), J. Math. Phys. 21, 2719.

Hussin V., Negro J. and del Olmo M.A. (1994), Ann. Phys. 231, 211.

Ivanova T.A. (1997) : *On current algebra of symmetries of the self-dual Yang-Mills equations*, hep-th/9702144.

Ivanova T.A. and Popov A.D. (1991), Lett. Math. Phys. 23, 29.

Ivanova T.A. and Popov A.D. (1992), Phys. Lett. A170, 293.

Ivanova T.A. and Popov A.D. (1995), Theor. Math. Phys. 102, 280.
Jackiw R. and Manton N.S. (1980), Ann. Phys. 127, 257.
Kac V.G. (1977), Commun. Math. Phys. 53, 31.
Kovalyov M., Legaré M. and Gagnon L. (1993), J. Math. Phys. 34, 3425.
Legaré M. (1997), Int. J. Mod. Phys. A12, 219.
Legaré M. and Popov A.D. (1995), Phys. Lett. A198, 195.
Leznov A.N. and Saveliev M.V. (1980), Commun. Math. Phys. 74, 111.
Mason L.J. and Sparling G.A.J. (1989), Phys. Lett. A137, 29.
Mason L.J. and Woodhouse N.M.J. (1993), Nonlinearity 6, 569.
Morosi C. and Pizzochero L. (1996), Commun. Math. Phys. 176, 353.
Olver P.J. (1986): *Applications of Lie Groups to Differential Equations in Physics*
  (Springer-Verlag, N.Y.)
Ooguri H. and Vafa C. (1995), Nucl. Phys. B451, 121.
Popov A.D. and Preitschopf C.R. (1996), Phys. Lett. B374, 71.
Popov A.D., Bordemann M. and Römer H. (1996), Phys. Lett. B385, 63.
Ward R.S. (1984), Nucl. Phys. B236, 381.
Ward R.S. (1985), Phil. Trans. R. Soc. Lond. A315, 451.
Ward R.S. (1988a), J. Math. Phys. 29, 386.
Ward R.S. (1988b), Nonlinearity 1, 671.
Ward R.S. (1990) in : *Twistors in Mathematics and Physics* eds : T.N. Bailey
  and R.J. Baston, London Mathematical Society Lecture Note Ser., (Cambridge
  University Press, Cambridge), 156, 246 and references therein.
Ward R.S. and Wells Jr. R.O. (1990) : *Twistor Geometry and Field Theory* (Cam-
  bridge University press, Cambridge), and references therein.
Winternitz P. (1990) : in *Partially integrable Evolution equations in Physics*, eds
  : R. Conte and N. Boccara, NATO ASI, Ser C310 (Kluwer Academic Publ.,
  Dordrecht), 515 and references therein.
Woodhouse N.M.J. (1985), Class. Quantum Grav. 2, 257.
Woodhouse N.M.J. and Mason L.J. (1988), Nonlinearity 1, 73.

Darboux Transformations for SUSY Integrable Systems

Q. P. Liu* ** and Manuel Mañas***

Departamento de Física Teórica,
Universidad Complutense,
E28040-Madrid, Spain.

Abstract. Several types of Darboux transformations for supersymmetric integrable systems such as the Manin-Radul KdV, Mathieu KdV and SUSY sine-Gordon equations are considered. We also present solutions such as supersolitons and superkinks.

1 Introduction

Supersymmetric integrable systems constitute a subject of current interest, and as a consequence a number of well known integrable equations have been generalized into the supersymmetric (SUSY) context. We just mention the SUSY versions of sine-Gordon (Di Vecchia and Ferrara(1977), Chaichain and Kulish (1978), Ferrara et al (1978)), Nonlinear Schrödinger (Roelofs and Kersten(1992)), KP (Manin and Radul (1985), Rabin (1991), Mulase (1991)), KdV (Manin and Radul (1985), Mathieu (1988b)) and Boussinesq (Yung (1993), Bellucci et al (1993)). We also point out that there are two different generalizations, namely nonextended ($N = 1$, N being the number of fermionic independent variables) and extended ($N \geq 2$) generalizations. Here we are mainly interested in the former case. So far many of the tools used in the standard theory have been extended to this framework, such as Bäcklund transformations (Chaichain and Kulish (1978)), prologation theory (Roelofs and Kersten(1992)), Hamiltonian formalism (Oevel and Popowicz (1991), Morosi and Pizzocchero (1993)), Grassmanian description (Ueno et al (1989)), tau function (Ueno et al (1989), Martínez Alonso and Medina (1995)), the relationship with super W algebras (Mathieu (1988a), Bilal and Gervais (1988)), additional symmetries (Das et al (1992), Mañas et al (1994), Stanciu (1994)), etc..

However, there is one that only recently has been considered, we are talking about the so called *Darboux transformation*, that constitutes a very successful tool in the realm of integrable systems whenever one is interested

---

* On leave of absence from Beijing Graduate School, CUMT, Beijing 100083, China
** Supported by *Beca para estancias temporales de doctores y tecnólogos extranjeros en España: SB95-A01722297*
*** Partially supported by CICYT: proyecto PB95–0401

in the constructions of solutions. The roots of these techniques go back to geometrical studies of the last century. It was in Moutard (1878), see also Athorne and Nimmo (1990), where the two dimensional Schrödinger equation (as we called it today) was considered giving new wave functions and potentials from given ones. This was taken by Darboux (Darboux (1882)) and applied to the one-dimensional case. In fact these results are connected with the theory of conjugate nets (Darboux (1896), Eisenhart (1909)) and it was in Levy (1886) where a transformation of this type was applied for surfaces, and was iterated in Hammond (1920). New transformations for conjugate nets appeared in Jonas (1915) which were called fundamental in Eisenhart (1923) containing the Levy ones in appropriate limits.

In Crum (1955), independently of these geometrical studies, it was presented, for the Schrödinger equation, the iteration of the transformation found by Darboux, giving compact expressions in terms of Wroński determinants. Later on, in Wadati et al (1975), a *new* transformation for soliton equations was introduced. This tool was rapidly devoloped and it was in Matveev (1979) where they were named as Darboux transformations. This name is standard nowadays in the soliton community, however we have seen that is not completely appropriate. In Levi (1988) a further extension of the Darboux transformation was given and some people refer to it as binary or Darboux-Levi transformation, however this is just the fundamental type transformation mentioned above. Finally, we remark that recently (Guil and Mañas (1996), Mañas (1997)) a vectorial formulation of the binary Darboux transformation was given, allowing compact formulae for the iteration of the Darboux transformation, for the Nonlinear Schrödinger and Davey-Stewartson equations.

The paper Liu (1995) was the first one that considered Darboux transformations for the SUSY KdV system. Later on (Liu and Mañas (1997a), Liu and Mañas (1997b)) extensions of the binary Darboux transformation (fundamental transformation in Geometry) and Darboux transformation appeared. In this paper we want to present the SUSY version of the Darboux tranformations. In particular we will consider three important supersymmetric integrable systems, namely the Manin-Radul KdV and its reduction to the Mathieu KdV, and also the SUSY extension of the sine-Gordon equation. For the Manin-Radul KdV and vectorial binary Darboux transformations we improve the presentation of Liu and Mañas (1997a) giving the general transformation for wave functions and a permutability theorem. For the Mathieu KdV equation we present some technical improvements with respect to Liu (1995) and the part regarding the sine-Gordon equation is entirely new. Finally, let us remark that given the character of these proceedings we are not going to give any proof.

The layout of the paper is as follows. We start with the Manin-Radul KdV (MRKdV) by considering the vectorial binary Darboux transformations and the construction of solutions in terms of ordinary determinants of Grammian type, we also give the iteration of the Darboux transformation found in Liu (1995) to get Wroński superdeterminantal expressions for the solutions, here we present a genuine supersoliton. In §3 we study the application of the Darboux transformation for the Mathieu KdV to the SUSY sine-Gordon equation, presenting a superkink.

2 Darboux Transformations
for the Manin-Radul KdV Equation

The MRKdV system is defined in terms of three independent variables ϑ, x, t, where $\vartheta \in \mathbb{C}_a$ is an odd supernumber, and $x, t \in \mathbb{C}_c$ are even supernumbers, and two dependent variables $\alpha(\vartheta, x, t), u(\vartheta, x, t)$, where α is an odd function taking values in $\mathbb{C}_a$ and u is even function with values in $\mathbb{C}_c$. A basic ingredient is a superderivation defined by $D := \partial_\vartheta + \vartheta \partial_x$. The system is

$$
\begin{aligned}
\alpha_t &= \frac{1}{4}(\alpha_{xxx} + 3(\alpha D\alpha)_x + 6(\alpha u)_x), \\
u_t &= \frac{1}{4}(u_{xxx} + 6uu_x + 3\alpha_x Du + 3\alpha(Du_x)),
\end{aligned}
\tag{1}
$$

where we use the notation $f_x := \partial f/\partial x$ and $f_t := \partial f/\partial t$.

The following linear system for the wave function $\psi(\vartheta, x, t)$, that takes values in the Grassmann algebra $\Lambda = \mathbb{C}_c \oplus \mathbb{C}_a$,

$$
\begin{aligned}
L(\psi) &:= \psi_{xx} + \alpha D\psi + u\psi = \lambda\psi, \\
\psi_t &= M(\psi) := \frac{1}{2}\alpha(D\psi)_x + \lambda\psi_x + \frac{1}{2}u\psi_x - \frac{1}{4}\alpha_x D\psi - \frac{1}{4}u_x\psi,
\end{aligned}
\tag{2}
$$

where the spectral parameter $\lambda \in \mathbb{C}_c$ is an even supernumber, has as its compatibility condition Eqs. (1), and therefore it can be considered as a Lax pair for (1).

2.1 Vectorial Binary Darboux Transformations

The linear system (2) is of a scalar nature, $\lambda \in \mathbb{C}_c$, $\psi(\vartheta, x, t) \in \Lambda$. Nevertheless, it is possible to give a vector extension of these linear problem. Indeed, we may replace Λ by an arbitrary linear Grassmann space $\mathcal{E}$ over Λ and take b as an $\mathcal{E}$-valued eigenfunction, then the spectral parameter can be taken as $\ell \in L(\mathcal{E}_l) \oplus L(\mathcal{E}_\infty)$, an even operator.

Namely, the linear system

$$
\begin{aligned}
\psi_{xx} + \alpha D\psi + u\psi - \ell\psi &= 0, \\
\psi_t - \frac{1}{2}\alpha(D\psi_x) - \ell\psi_x - \frac{1}{2}u\psi_x + \frac{1}{4}\alpha_x D\psi + \frac{1}{4}u_x\psi &= 0,
\end{aligned}
\tag{3}
$$

has as its compatibility condition the MRKdV system (1).

Notice that Eqs. (1) are also the compatibility condition of adjoint linear system:

$$\phi_{xx} + D(\alpha\phi) + u\phi - \phi m = 0,$$

$$\phi_t + \frac{1}{2}\alpha D\phi_x - \phi_x m - \frac{1}{2}(u + D\alpha)\phi_x + \frac{1}{4}D(\alpha_x\phi) + \frac{1}{4}u_x\phi = 0, \tag{4}$$

where $\phi(\vartheta, x, t) \in \tilde{\mathcal{E}}^*$ is a linear function on the supervector space $\tilde{\mathcal{E}}$, and $m \in L(\tilde{\mathcal{E}_0}) \oplus L(\tilde{\mathcal{E}_1})$.

In order to construct Darboux transformations for these linear systems we need to introduce an operator, say $V[\psi, \phi] \in L(\mathcal{E}, \tilde{\mathcal{E}})$, bilinear in ψ and ϕ, defined by the compatible equations

$$DV[\psi, \phi] = \psi \otimes \phi,$$

$$\begin{aligned}
V[\psi, \phi]_t = {}&\ell V[\psi, \phi]_x + V[\psi, \phi]_x m \\
&- D(\psi_x \otimes \phi_x + \frac{1}{2}u DV[\psi, \phi]) - \frac{1}{4}\alpha_x DV[\psi, \phi] \\
&- \frac{1}{2}(D\psi) \otimes ((D\alpha)\phi - \alpha(D\psi)) + \frac{1}{2}\alpha(\psi \otimes \phi_x - \psi_x \otimes \phi)
\end{aligned} \tag{5}$$

such that

$$\ell V[\psi, \phi] - V[\psi, \phi]m = D(\psi_x \otimes \phi - \psi \otimes \phi_x) - \alpha\psi \otimes \phi. \tag{6}$$

Now we are ready to present the following:

Theorem 1 *Let $\psi_0(\vartheta, x, t) \in \mathcal{V}_l$ be an even vector satisfying Eq. (3) with spectral parameter ℓ_0, $\phi_0(\vartheta, x, t) \in \mathcal{V}_r^*$ an odd functional solving Eq. (4) with spectral parameter m_0 and $V[\psi_0, \phi_0] \in L(\mathcal{V}_l) \oplus L(\mathcal{V}_\infty)$ a non singular even operator, $\det V[\psi_0, \phi_0]_{body} \neq 0$, defined in terms of the compatible Eqs. (5) and (6). Then, the objects*

$$\hat{\psi} := \psi - V[\psi, \phi_0]V[\psi_0, \phi_0]^{-1}\psi_0,$$

$$\hat{\phi} := \phi - \phi_0 V[\psi_0, \phi_0]^{-1}V[\psi_0, \phi],$$

$$\hat{\alpha} = \alpha - 2D^3 \ln\det V[\psi_0, \phi_0],$$

$$\hat{u} = u + 2\hat{\alpha}D\ln\det V[\psi_0, \phi_0] + 2\left(\frac{\sum_j D(\psi_0)_j \ \det V[\psi_0, \phi_0]_j}{\det V[\psi_0, \phi_0]}\right)_x,$$

where $V[\psi_0, \phi_0]_j$ is an operator with associated supermatrix obtained from the corresponding one of $V[\psi_0, \phi_0]$ by replacing the j-th column by ψ_0, satisfy the Eqs. (3) and (4) whenever the unhatted variables do. Thus, $\hat{\alpha}$ and $\hat{u}$ are new solutions of the MRKdV (1). Moreover,

$$V[\hat{\psi}, \hat{\phi}] = V[\psi, \phi] - V[\psi, \phi_0]V[\psi_0, \phi_0]^{-1}V[\psi_0, \phi]. \tag{7}$$

Let us remark that this theorem extends Theorem 1 in our paper (Liu and Mañas (1997a)). In particular we stress the role of the general wave functions ψ and ϕ, and also the formula (7) that gives the path for iteration and it is also deeply connected, in the non SUSY case, with geometrical objects such as points of the transformed manifolds.

We shall call $(\mathcal{V}, \psi_0, \phi_0)$ as transformation data. The composition of two vectorial Darboux transformations yield a new Darboux transformation, and as it is shown in next proposition they commute as they can be expressed as a vectorial Darboux transformation:

Proposition 1 *The vectorial Darboux transformation with transformation data $\left(\mathcal{V}_1 \oplus \mathcal{V}_2, \left(\begin{smallmatrix} \psi_{0,(1)} \\ \psi_{0,(2)} \end{smallmatrix}\right), (\phi_{0,(1)}, \phi_{0,(2)})\right)$ coincides with the following composition of Darboux transformations:*

1. *First transform with data $(\mathcal{V}_2, \psi_{0,(2)}, \phi_{0,(2)})$, and denote the transformation by $'$.*
2. *On the result of this transformation apply a second one with data*

$$(\mathcal{V}_\infty, \psi'_{l,(\infty)}, \phi'_{l,(\infty)}).$$

A similar theorem in a completely different framework, namely discrete integrable systems: multidimensional quadrilateral lattices, appears in Doliwa et al (1997).

2.2 Wroński Superdeterminants Representation of Iterated Darboux Transformations

A Darboux transformation for the MRKdV equation is (Liu (1995))

Proposition 2 *Let ψ be a solution of (2) and θ_0 be a particular solution with $\lambda = \lambda_0$. Then, the quantities defined by*

$$\hat{\psi} := (D + \delta_0)\psi, \qquad \delta_0 := -\frac{D\theta_0}{\theta_0}, \quad (\theta_0 : even)$$

$$\hat{\alpha} := -\alpha - 2\partial\delta_0,$$

$$\hat{u} := u + (D\alpha) + 2\delta_0(\alpha + \partial\delta_0)$$

satisfy

$$\hat{L}\hat{\psi} = \lambda\hat{\psi},$$

$$\partial_t \hat{\psi} = \hat{M}\hat{\psi},$$

where $\hat{L}$ and $\hat{M}$ are obtained from L and M by replacing α and u with $\hat{\alpha}$ and $\hat{u}$, respectively.

As a consequence of this Proposition we conclude that $\hat{u}$ and $\hat{\alpha}$ are new solutions of the MRKdV system (1). We remark that, as usual, the Darboux transformation can be viewed as a gauge transformation:

$$\psi \to T_0 \psi,$$
$$L \to \hat{L} = T_0 L T_0^{-1},$$
$$M \to \hat{M} = \partial_t T_0 \cdot T_0^{-1} + T_0 M T_0^{-1},$$
$$T_0 := D + \delta_0.$$

To construct Crum type transformation, let us start with n solutions θ_i, $i = 0, ..., n-1$, of equation (2) with eigenvalues as $\lambda = k_i$, $i = 0, ..., n-1$. To make sense, we choose the θ_i in such way that its index indicates its parity: those with even indices are even and with odd indices are odd variables. We use θ_0 to do our first step transformation and then θ_i, $i = 1, ..., n-1$, are transformed to new solutions $\theta_i[1]$ of the transformed linear equation and θ_0 goes to zero. Next step can be effected by using $\theta_1[1]$ to form a Darboux operator and at this time $\theta_1[1]$ is lost. We can continue this iteration process until all the seeds are mapped to zero. In this way, we have

Proposition 3 *Let θ_i, $i = 0, ..., n-1$, be solutions of the linear system (2) with $\lambda = k_i$, $i = 0, ..., n-1$, and parities $p(\theta_i) = (-1)^i$, then after n iterations of the Darboux transformation of Proposition 1, one obtains a new Lax operator*

$$\hat{L} = T_n L T_n^{-1}, \quad T_n = D^n + \sum_{i=0}^{n-1} a_i D^i,$$

where the coefficients a_i of the gauge operator T_n are defined by

$$(D^n + \sum_{i=0}^{n-1} a_i D^i)\theta_j = 0, \qquad j = 0, ..., n-1. \tag{8}$$

The explicit form of the transformed field variables is given by $\hat{L} = T_n L T_n^{-1}$ from where it follows that the new fields $\hat{\alpha}$ and $\hat{u}$ can be written as

$$\hat{\alpha} = (-1)^n \alpha - 2\partial a_{n-1},$$

$$\hat{u} = u - 2\partial a_{n-2} - a_{n-1}((-1)^n \alpha + \hat{\alpha}) + \frac{1-(-1)^n}{2} D\alpha.$$

Now, we must recall the reader that the Berezinian or superdeterminant of an even matrix, say $\mathcal{M} = \left(\begin{smallmatrix} A & B \\ C & D \end{smallmatrix}\right)$, is

$$\text{sdet}\, \mathcal{M} = \frac{\det(\mathcal{A} - \mathcal{BD}^{-\infty}\mathcal{C})}{\det \mathcal{D}} = \frac{\det \mathcal{A}}{\det(\mathcal{D} - \mathcal{CA}^{-\infty}\mathcal{B})}.$$

To obtain the explicit expressions, we have to find out the a_{n-2} and a_{n-2} by solving the linear system (8). In the even case $n = 2k$ which is most interesting, we have

$$a_{2k-2} = -\frac{\text{sdet}\,\hat{\mathcal{W}}}{\text{sdet}\,\mathcal{W}}, \qquad a_{2k-1} = D\ln\text{sdet}\,\mathcal{W},$$

where

$$\boldsymbol{a}^{(0)} := (a_0, a_2, \ldots, a_{2k-2}), \qquad \boldsymbol{a}^{(1)} := (a_1, a_3, \ldots, a_{2k-1}),$$

$$\boldsymbol{\theta}^{(0)} := (\theta_0, \theta_2, \ldots, \theta_{2k-2}), \qquad \boldsymbol{\theta}^{(1)} := (\theta_1, \theta_3, \ldots, \theta_{2k-1}),$$

$$\boldsymbol{b}^{(i)} := \partial^k \boldsymbol{\theta}^{(i)}, \qquad W^{(i)} := \begin{pmatrix} \boldsymbol{\theta}^{(i)} \\ \partial\boldsymbol{\theta}^{(i)} \\ \vdots \\ \partial^{k-1}\boldsymbol{\theta}^{(i)} \end{pmatrix}, \qquad i = 0, 1,$$

$$\mathcal{W} := \begin{pmatrix} W^{(0)} & W^{(1)} \\ DW^{(0)} & DW^{(1)} \end{pmatrix}, \qquad \hat{\mathcal{W}} := \begin{pmatrix} \hat{W}^{(0)} & \hat{W}^{(1)} \\ DW^{(0)} & DW^{(1)} \end{pmatrix}$$

and $\hat{W}^{(0)}$ and $\hat{W}^{(1)}$ are obtained from the matrices $W^{(0)}$ and $W^{(1)}$ by replacing the last rows with $\boldsymbol{b}^{(0)}$ and $\boldsymbol{b}^{(1)}$, respectively. It should be noticed that the supermatrix $\mathcal{W}$ is even and has a Wronski type structure.

Summarizing the above results, we now have the following

Theorem 2 *Let α, u be a solution of (1) and $\{\theta_j\}_{j=0}^{n-1}$ be a set of $n(= 2k)$ solutions of the associated linear system (2), such that the parity is $p(\theta_j) = (-1)^j$. Then, we have new solutions $\hat{\alpha}$, $\hat{u}$ of (1) given by*

$$\hat{\alpha} = \alpha - 2D^3\ln\text{sdet}\,\mathcal{W},$$

$$\hat{u} = u + 2\partial\left(\frac{\text{sdet}\,\hat{\mathcal{W}}}{\text{sdet}\,\mathcal{W}}\right) + (\alpha + \hat{\alpha})D\ln\text{sdet}\,\mathcal{W}.$$

Let us remark that our Darboux transformations are useful even outside the MRKdV system. Indeed, the most obvious application is to the SUSY KP equation (Ueno et al (1989)) which is a closed system obtained from supersymmetric KP hierarchy. Since the Lax pair is essentially the one we had for MRKdV, our Darboux transformations can be used directly in this context.

Solutions of the MRKdV are found rarely, see for example Radul (1988), Ibort et al (1996). In Liu and Mañas (1997a) and Liu and Mañas (1997b) one can find examples of solutions, in particular in Liu and Mañas (1997b) we gave an interesting solution which behaves like a genuine supersoliton. Our solution is

$$\hat{\alpha} = -2\partial a_1, \qquad \hat{u} = -2\partial a_0.$$

with

$$a_0 = f - k\big(\gamma_-^{(1)}\gamma_+^{(1)} + \vartheta(\gamma_+^{(1)}\gamma_-^{(0)} - \gamma_-^{(1)}\gamma_+^{(0)})\big)g - \vartheta k\big(c_+^{(0)}\gamma_-^{(1)} - c_-^{(0)}\gamma_+^{(1)}\big)fg,$$

$$a_1 = \big(k(c_+^{(0)}\gamma_-^{(1)} - c_-^{(0)}\gamma_+^{(1)}) + \vartheta(c_+^{(0)}\gamma_-^{(0)} - c_-^{(0)}\gamma_+^{(0)})\big)g,$$

where

$$f := -k \left(\frac{c_+^{(0)} \exp(\eta) - c_-^{(0)} \exp(-\eta)}{c_+^{(0)} \exp(\eta) + c_-^{(0)} \exp(-\eta)} \right),$$

$$g := \frac{2}{\left(c_+^{(0)} \exp(\eta) + c_-^{(0)} \exp(-\eta) \right) \left(\gamma_+^{(0)} \exp(\eta) + \gamma_-^{(0)} \exp(-\eta) \right)}.$$

here $\eta = kx + k^3 t$, $k \in \mathbb{C}$ and $c_\pm^{(i)}$, $\gamma_\pm^{(i)}$ are supernumbers with parities indicated by the superfix.

Notice that our solution can be understood as a supersoliton which has the KdV soliton, $-2\partial f$, as its body, and that the choice $c_+^{(0)} = \gamma_+^{(0)}$ and $c_-^{(0)} = \gamma_-^{(0)}$ gives the solution found in Ibort et al (1996).

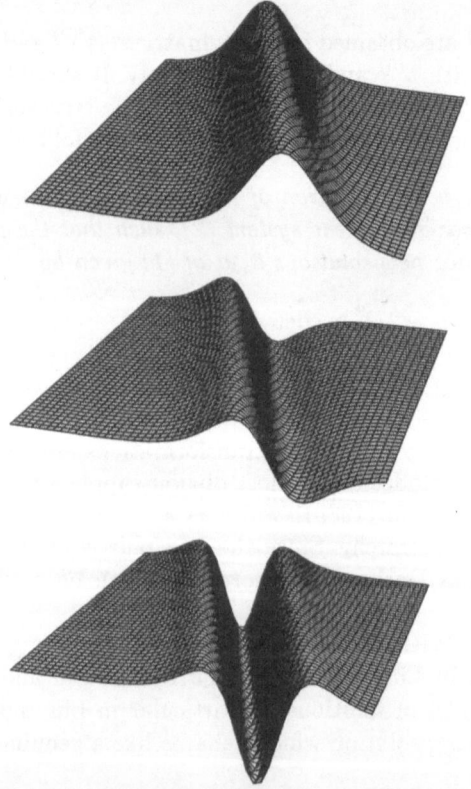

In the figures we plot, in the real x-t plane, the functions f_x, g_x and $(fg)_x$ that appear in the construction of the solution corresponding to the data: $k = 1.1$, $c_+^{(0)} = 1$, $c_-^{(0)} = 1.5$, $\gamma_+^{(0)} = 1.2$ and $\gamma_+^{(0)} = 2$.

3   Darboux Transformations for the Mathieu KdV and SUSY Sine-Gordon Equations

The Mathieu KdV equation reads (Manin and Radul (1985), Mathieu (1988b))

$$\alpha_t = \frac{1}{4}(\alpha_{xx} + 3\alpha D\alpha)_x,$$

which is obtained from (1) by setting $u = 0$, being the Lax operator

$$L = \partial^2 + \alpha D.$$

In Liu (1995) one of the authors presented a preliminary version of

Proposition 4 *If*

$$\psi_{xx} + \alpha D\psi = \lambda\psi, \tag{9}$$

and ψ_0 is an even solution of above equation with $\lambda = \lambda_0$ such that the constants

$$J(\psi_0) = \psi_{0,x}^2 + 2(D\psi_0)_x D\psi_0 - \lambda_0\psi_0^2,$$
$$I(\psi_0, \psi) = (\lambda - \lambda_0)\left(D^{-1}((D\psi_0)\psi)\right) + \lambda_0\ psi_0\psi - \psi_{0,x}\psi_x$$
$$- (D\psi_0)_x D\psi + (D\psi_0)D\psi_x$$

vanish, then

$$\hat{\psi} := \psi_0^{-1}D^{-1}(\psi_0 D\psi - (D\psi_0)\psi), \quad \hat{\alpha} := \alpha - 4D^3 \ln\psi_0$$

satisfy

$$\hat{\psi}_{xx} + \hat{\alpha}D\hat{\psi} = \lambda\hat{\psi},$$

Compared with the result of Liu (1995), the above Proposition is an improved version in the sense that the constant $I(\psi_0, \psi)$ has a much simpler structure. An application of this Darboux transformation is to the supersymmetric sine-Gordon system, which reads (Inami and Kanno (1991))

$$DD_t\Phi = 2\cosh(2\Phi) \tag{10}$$

where $D_t = \frac{\partial}{\partial\theta_t} + \theta_t\frac{\partial}{\partial t}$ and θ_t is another Grassmann odd variable.
    According to Inami and Kanno (1991), the linear problems are

$$D\psi_1 + \lambda\psi_3 + \psi_4 = 0,$$
$$D\psi_2 - \lambda\psi_3 + \psi_4 = 0,$$
$$D\psi_3 + 2(D\Phi)\psi_3 + \psi_1 + \psi_2 = 0,$$
$$D\psi_4 - 2(D\Phi)\psi_4 + \lambda(\psi_1 - \psi_2) = 0,$$

and

$$D_t \psi_1 + \exp(2\Phi)\psi_3 - \lambda^{-1} \exp(-2\Phi)\psi_4 = 0,$$
$$D_t \psi_2 + \exp(2\Phi)\psi_3 + \lambda^{-1} \exp(-2\Phi)\psi_4 = 0,$$
$$D_t \psi_3 + \lambda^{-1} \exp(-2\Phi)(\psi_1 - \psi_2) = 0,$$
$$D_t \psi_4 - \lambda^{-1} \exp(2\Phi)(\psi_1 + \psi_2) = 0.$$

Introducing $\tilde{\psi}_1 := \psi_1 - \psi_2$ and $\tilde{\psi}_2 := \psi_1 + \psi_2$, we easily see that the above linear system can be written as follows:

$$\begin{pmatrix} \tilde{\psi}_1 \\ \tilde{\psi}_2 \end{pmatrix}_x = 2 \begin{pmatrix} -(D\Phi)D & \lambda \\ \lambda & (D\Phi)D \end{pmatrix} \begin{pmatrix} \tilde{\psi}_1 \\ \tilde{\psi}_2 \end{pmatrix} \tag{11}$$

and

$$D_t \begin{pmatrix} \tilde{\psi}_1 \\ \tilde{\psi}_2 \end{pmatrix} = \begin{pmatrix} 0 & -\lambda^{-1} \exp(-2\Phi)D \\ \lambda^{-1} \exp(2\Phi)D & 0 \end{pmatrix} \begin{pmatrix} \tilde{\psi}_1 \\ \tilde{\psi}_2 \end{pmatrix}. \tag{12}$$

Differentiating (11) we get

$$\tilde{\psi}_{1,xx} = -\alpha_1 D\tilde{\psi}_1 + 4\lambda^2 \tilde{\psi}_1,$$
$$\tilde{\psi}_{2,xx} = \alpha_2 D\tilde{\psi}_2 + 4\lambda^2 \tilde{\psi}_2$$

where $\alpha_1 := \gamma_x - \gamma D\gamma$ and $\alpha_2 := -\gamma_x - \gamma D\gamma$ with $\gamma = 2D\Phi$.

Now a slight modification of the previous Darboux transformation yields

Theorem 3 *If Φ is a solution of the SUSY sine-Gordon equation (10) and $\tilde{\psi}_1$ and $\tilde{\psi}_2$ are particular solutions of (11) and (12) with spectral parameter $\lambda = \lambda_0$, such that $J(\tilde{\psi}_1) = J(\tilde{\psi}_2) = 0$, then*

$$\hat{\Phi} = \Phi + \ln \frac{\tilde{\psi}_1}{\tilde{\psi}_2},$$

is a new solution of (10).

Remarks

1. The method we used is a generalization of the one by Wadati et al (1975) for the classical sine-Gordon equation. We also notice that this idea was used in Nimmo (1993) for the two-dimensional sine-Gordon equation of Konopelchenko and Rogers.
2. We may also obtain Darboux type transformations for the SUSY modified KdV equation (Mathieu (1988b)).
3. We conjecture that Crum type iteration of the above Darboux transformation will be represented in terms of Pfaffians instead of Wronskians.

The application of this theorem to the most simple case, namely $\Phi = i\,\pi/4$ yields the following interesting SUSY extension of the kink solution. Namely, we have the following solution

$$\Phi = \frac{i\,\pi}{4} + \ln \frac{1 + \mathcal{A}\exp(-4\eta)}{1 - \mathcal{A}\exp(-4\eta)}$$

where

$$\mathcal{A}(\theta, \theta_t) := \frac{A_0(1 - 2\,i\,\theta\theta_t) + A_1(\theta + i\,\lambda^{-1}\theta_t)}{c^{-1}A_0(1 - 2\,i\,\theta\theta_t) + cA_1(\theta - i\,\lambda^{-1}\theta_t)},$$

$$\eta(x, t) := \lambda x + \lambda^{-1}t,$$

with A_0 and c even supernumbers with nonvanishing body, $\lambda \in \mathbb{C}$ and A_1 an odd supernumber.

This solution can be thought as a superkink, in fact its body is

$$\Phi_{\text{body}} = \frac{i\,\pi}{4} + \ln \frac{1 + c_{\text{body}}\exp(-4\eta)}{1 - c_{\text{body}}\exp(-4\eta)},$$

that after taking the appropriate Wick rotation goes into the standard kink solution of the sine-Gordon equation. Obviously, the soul of this solution is far from trivial.

Acknowledgement. The present paper is the extended version of the talk presented at the meeting by one of us (MM). A number of the results were obtained during or after the meeting. MM would like to thank the organizers for the hospitality and coverage of local expenses; and also *Bolsa de Viaje Complutense 1997.*

References

Athorne C., Nimmo J. J. C. (1990): On the Moutard transformation for integrable partial differential equations. Inv. Prob. **7**, 809–826.

Bellucci S., Ivanov E., Krivonos S., Pichugin A. (1993): $N = 2$ super Boussinesq hierarchy: Lax pairs and conservation laws. Phys. Lett. **B 312**, 463–470.

Bilal A., Gervais J.-L. (1988): Superconformal algebra and super-KdV equation. Phys. Lett. **B 211**, 95–100.

Chaichain M., Kulish P. P. (1978): On the method of inverse scattering problem and Bäcklund transformations for supersymmetric equations. Phys. Lett. **B 78**, 413–416.

Crum M. M. (1955): Associated Sturm-Liouville systems. Quart. J. Math. **6**, 121–127.

Darboux G. (1882): Sur une proposition relative aux équations linéaries, C. R. Acad. Sci. Paris **94**, 1456–1459.

Darboux G. (1896): Leçons sur la théorie générale des surfaces IV. (Gauthier-Villars, Paris). Reprinted in 1972 by (Chelsea Publishing Company, New York).

Das A., Sezgin E., Sin S. J. (1992): The super W_∞ symmetry of the Manin-Radul super KP hierarchy. Phys. Lett. **B 278**, 435–441.

Di Vecchia P., Ferrara S. (1977): Classical solutions in two-dimensional supersymmetric field theories. Nucl. Phys. **B 130**, 93–104.

Doliwa A., Santini P. M., Mañas M. (1997): Transformations of Quadrilateral Lattices (preprint).

Eisenhart L. P. (1909): A treatise on the differential geometry of curves and surfaces. (Ginn and Co., Boston).

Eisenhart L. P. (1923): Transformations of surfaces. (Princeton University Press, Princeton). Reprinted in 1962 by (Chelsea Publishing Company, New York).

Ferrara S., Girardello L., Sciuto S. (1978): An infinite set of conservation laws of the supersymmetric sine-Gordon theory. Phys. Lett. **B 76**, 303–306.

Guil F., Mañas M. (1996): Darboux transformations for the Davey-Stewartson equations. Phys. Lett. **A 217**, 1–6.

Hammond E. S. (1920): Periodic conjugate nets. Ann. Math. **22**, 238–261.

Ibort L. A., Martínez Alonso, L., Medina E. (1996): Explicit solutions of supersymmetric KP hierarchies: supersolitons and solitinos. J. Math. Phys. **37**, 6157–6172.

Inami T., Kanno H. (1991): Lie superalgebraic approach to super Toda lattice and generalized super KdV equations. Commun. Math. Phys. **136**, 519–542.

Jonas H. (1915): Über die Transformation der konjugierten Systeme and über den gemeinsamen Ursprung der Bianchischen Permutablitätstheoreme. Berl. Math. Ges. Ber. **14**, 96.

Levi D. (1988): On a new Darboux transformation for the construction of exact solutions of the Schrödinger equation. Inv. Prob. **4**, 165–172.

Levy L. (1886): Sur quelques équations linéares aux dérivées partieles. J. l'École Polytecnique **56**, 6.

Liu Q. P. (1995): Darboux transformations for supersymmetric Korteweg-de Vries equations. Lett. Math. Phys. **35**, 115–122.

Liu Q. P., Mañas M. (1997a): Darboux transformation for the Manin-Radul supersymmetric KdV equation. Phys. Lett. **B 394**, 337–342.

Liu Q. P., Mañas M. (1997b): Crum transformation and Wronskian type solutions for supersymmetric KdV equation, Phys. Lett. **B 396**, 133–140.

Mañas M. (1997): Darboux transformations for the nonlinear Schrödinger equations. J. Phys. A: Math. Gen. **29**, 7721–7737.

Mañas M., Martínez Alonso L., Medina E. (1994): Additional symmetries, Virasoro constraints and string equations for the super KP hierarchies. Phys. Lett. **B 336**, 178–182.

Manin Yu. I., Radul A. (1985): A supersymmetric extension of the Kadomtsev-Petviashvili hierarchy. Commun. Math. Phys. **98**, 65–77.

Martínez Alonso L., Medina E. (1995): Tau-function formalism for supersymmetric KP hierarchies. J. Math. Phys. **36** , 4898–4913..

Mathieu P. (1988a): Superconformal algebra and supersymmetric Korteweg-de Vries equation. Phys. Lett. **B 203**, 287–291.

Mathieu P. (1988b): Supersymmetric extension of the Korteweg-de Vries equation. J. Math. Phys. **29**, 2499–2506.

Matveev V. B. (1979): Darboux transformation and explicit solutions of the Kadomtsev-Petviashvili equation, depending on functional parameters. Lett. Math. Phys. **3**, 213–216.

Matveev V. B., Salle M. A. (1991): Darboux Transformations and Solitons. (Springer-Verlag, Berlin).

Morosi C., Pizzocchero L. (1993): On the biHamiltonian structure of the supersymmetric KdV hierarchies: a Lie superalgebraic approach. Commun. Math. Phys. **158**, 267–288.

Moutard Th. F. (1878): Sur la construction des équations de la forme $\frac{1}{z}\frac{\partial^2 z}{\partial x \partial y} = \lambda(x, y)$, qui admettent une intégrale générale explicite. J. l'École Polytecnique **45**, 1–11.

Mulase M. (1991): A new super KP system and a characterization of the Jacobians of arbitrary algebraic super curves. J. Diff. Geom. **34**, 651–680.

Nimmo J. J. C. (1993): Darboux transformations in (2+1)-dimensions. In "Applications of analytic and geometric methods to nonlinear differential equations", ed. Clarkson P. A., (Kluwer Academic Publisher).

Oevel W., Popowicz Z. (1991): The biHamiltonian structure of fully supersymmetric Korteweg-de Vries systems. Commun. Math. Phys. **139**, 441–460.

Rabin J. M. (1991): The geometry of the super KP flows. Commun. Math. Phys. **137**, 533–552.

Radul A. O. (1988): Algebro-geometric solutions to the super Kadomtsev-Petviashvili hierarchy. In "Seminar on supermanifolds" vol. 28, edited by Leites D. A., report Stockholm University.

Roelofs G. H. M., Kersten P. H. M. (1992): Supersymmetric extensions of nonlinear Schrödinger equation: symmetries and coverings. J. Math. Phys. **33**, 2185–2206.

Stanciu S. (1994): Additional symmetries of supersymmetric KP hierarchies. Commun. Math. Phys. **165**, 261–279.

Ueno K., Yamada H., Ikeda K. (1989): Algebraic study on the super-KP hierarchy and the ortho-sympletic super-KP hierarchy. Commun. Math. Phys. **124**, 57–78.

Wadati M., Sanuki H., Konno K. (1975): Relationships among inverse method, Bäcklund transformation and an infinite number of conservation laws. Prog. Theor. Phys. **53**, 419–436.

Yung C. M. (1993): The $N = 2$ supersymmetric Boussinesq hierarchies. Phys. Lett. **B 309**, 75–84 .

Susy Hierarchies and Affine Lie Algebras

Francesco Toppan

Shizuoka University, Department of Physics, Ohya 836, Shizuoka city, Japan

Abstract. We review some basic features of the Lie-algebraic classification of W-algebras and related integrable hierarchies in $1 + 1$ dimensions, pointing out the role of affine Lie algebras. We emphasize that the supersymmetric extensions of the above construction possibly lead, though some questions are still opened, to the classification of supersymmetric hierarchies based on "generic" supersymmetric affine Lie algebras. Here the word generic is used to make clear that well-known procedures, as those introduced by Inami and Kanno, are too restricted and do not lead to the full spectrum of supersymmetric integrable hierarchies one can construct. A particular attention is devoted to the large-N supersymmetric extensions (here $N = 4$). The attention paid by large-N theories being due to the fact that they arise as dimensional reduction of $N = 1$ models, and moreover that they realize an "unification" of known hierarchies.

1 Introduction

In the last few years a lot of attention has been devoted to inter-related topics which go under the name of W-algebras, integrable hierarchies (non-relativistic) in $1 + 1$ dimensions of KdV or NLS type, 2-dimensional reativistic equations like Liouville (more generally Toda field theories) or SG, 0-dimensional matrix models which describe discretized 2-dimensional gravity.

Some of the above topics have definitely a more mathematical flavour, like for instance the theory and classification of W-algebras; some others are definitely more physically grounded, KdV describes waves in shallow water, Liouville equation is an ubiquous one, but at least a lot of attention has been paid to it in connection with non-critical strings.

Despite the fact that the above models and theories seem all very different and can be constructed in apparently unrelated ways it turned out indeed that they are just manifestations of an underlying mathematical framework. Indeed W-algebras (i.e. non-linear Poisson-bracket algebras of 1-dimensional fields containing a Virasoro one, which satisfy the standard properties of antisymmetry, Jacobi identity and Leibniz rule in the classical case) turn out to be the Poisson bracket structures for both relativistic Toda field theories and non-relativistic integrable equations like KdV, Boussinesq and so on. Moreover the Ward identities of generalized matrix models generate the so-called W-constraints and their partition functions turn out to be related to the τ-functions of associated classical integrable hierarchies[1].

It deserves being mentioned that W-algebras themselves can be produced and classified via a truly algebraic approach, by putting restrictions to affine

Lie algebras; such restrictions can be realized either as hamiltonian reductions or coset constructions (by looking at some centralizer over some enveloping algebra [2]). While there is maybe no strict mathematical proof so far that all W-algebras can be obtained with the methods of [3], at least there is no need to believe that all W-algebras cannot be obtained that way.

The production of such a closed structure like a W-algebra is an interesting mathematical activity by itself, however there is much more than that. A very peculiar and absolutely non-trivial feature of W-algebras arises when they allow constructing towers of infinite hamiltonians in involution. In this way they turn out to be linked to a dynamical system of a special kind, an integrable one. The technical tool which allows to prove integrability consists in formulating the dynamical system (and its associated W-algebra) via a Lax operator which can be either of scalar (KP-like) or matrix type.

In the next section we will sketch the main features of the bosonic construction, postponing to the later section the introduction of supersymmetric integrable systems with the necessary modifications.

2   Bosonic Hierarchies

Let us first point out that $2D$ relativistic Toda models and $1 + 1$ non-relativistic integrable equations arise from constraining affine Lie algebras $\hat{\mathcal{G}}$ (and their associated enveloping algebras). The basic difference in the relativistic case is due to the fact that two copies of the affine algebra are considered, associated to the chiral and antichiral currents $J(z)$, $\overline{J}(\overline{z})$ respectively. The dynamical fields are group-valued $g(z, \overline{z})$ and possibly expressed through a Gauss decomposition. We have

$$J(z) = g^{-1}\partial_z g \qquad (1)$$

and a similar equation for $\overline{J}(\overline{z})$.

The simplest case is provided when $\mathcal{G} = sl(2)$. The three currents associated to $sl(2)$ are $J_\pm(x)$ and $J_0(x)$ ($J_0(x)$ generates the $\hat{U}(1)$ subalgebra).

In this simple case only two inequivalent constraints can be imposed on the (enveloping) affine algebra, either

A) constraining $J_+(x) = 1$ (hamiltonian constrain), or

B) selecting the $X(y)$ centralizer of the enveloping algebra, namely $\{J_0(x), X(y)\} = 0$ (coset).

Accordingly, we get in the relativistic (I) and in the non-relativistic (II) cases the following dynamical systems:

I A) The Liouville equation.

II A) The m-KdV (and KdV) equation.

I B) The $2D$ Witten's black hole.

II B) The Non-Linear Schrödinger Equation.

From now on we will concentrate only on the non-relativistic case, that is the system of integrable equations in $1 + 1$ dimension which can be solved through inverse scattering method. As mentioned above the integrability property is expressed by the fact that one can express the equations of motion through a Lax operator. We have two kinds of such operators, the scalar type

$$L = \partial + \sum_{i=1,...,\infty} u_i \partial^{-i} \tag{2}$$

associated to the KP hierarchy, and the matrix type

$$\mathcal{L} = \partial + \sum_i J_i(x)\tau^i + \Lambda \tag{3}$$

where the currents $J_i(x)$ are valued in some Lie algebra $\mathcal{G}$ generated by τ^i. Λ is a constant element, depending on a spectral parameter λ, such that the loop algebra $\tilde{G} = \mathcal{G} \otimes C(\lambda, \lambda^{-1})$ can be decomposed in the direct sum $\tilde{G} = K \oplus M$, with K, M respectively the Kernel and Image under the adjoint action of Λ (this technical property implies that $\mathcal{L}$ can be diagonalized under a similarity transformation).

In order to extract from scalar Lax operators integrable equations involving only a finite number of fields, we have to constrain the infinite fields $u_i(x)$ in a way consistent with the KP flows (constrained KP hierarchies). One possibility is requiring e.g. for a given n $L^n = L^n{}_+$ (that is to be a purely differential operator). This is indeed a consistent constraint (leading to the n-th KdV hierarchies), however it is known there exists many more inequivalent consistent constraints and a classification of them out of the scalar Lax operators alone appears rather impractical.

On the contrary it is well-known how to classify all possible hierarchies associated with affine algebras. They turn out to be related to the acceptable integral grading for any given loop algebra $\mathcal{G}$ and the choice of the regular element Λ (see e.g. [4] for details). Moreover it is possible to relate such solutions with the constrained KP hierarchies. [5]

In order to be explicit we recall that in the original Drinfeld-Sokolov paper the n-th KdV hierarchies were obtained by assuming the underlying algebra to be $sl(n)$ and the regular element Λ to be the sum over the $\tilde{sl(n)}$ simple roots.

The above scheme seems quite satisfactory from the point of view of bosonic hierarchies since it provides a well-defined construction for them and is quite plausible they can all be accomodated in it. Questions concerning the possible equivalence of hierarchies arising from different choices of algebras, integral grading and/or regular element seem more tehnical and less central.

So far for purely bosonic hierarchies, in the next section we will introduce the supersymmetric ones.

3   Supersymmetric Hierarchies

The first natural question when discussing supersymmetric integrable hierarchies is of course why should we worry about them. One can think e.g. to the fact that so far no supersymmetric matrix model providing a discretized $2D$ supergravity has been produced. Neverthless some achievement has been made like the introduction of a supereigenvalue model which is in a sense pull out of a hat but is related to a superintegrable hierarchy[6].

Morever the remarkable relation of KdV-type hierarchies with the conformal algebras (Virasoro and supersymmetric extensions) establishes a connection between such hierarchies and the (super-)string theories which has still to be fully appreciated.

From a purely mathematical point of view the role of supersymmetric integrable hierarchies and superalgebras is essential in at least two respects. Indeed, even when considering purely bosonic hierarchies, if we not allow for super-structures like super-algebras we cannot pretend to exhaust the full class of possible hierarchies, new integrable interacting purely bosonic hierarchies arise in fact from the bosonic sector (B-B and F-F submatrices) of supermatrix-valued superhierarchies.

Moreover investigating large N-extended superhierarchies corresponds to a sort of "unification or grandunification" program of known hierarchies. It happens in fact that unrelated bosonic hierarchies or lower supersymmetric $(N = 1, 2)$ hierarchies turn out to be different manifestations of a single "unifying" large N supersymmetric hierarchy. We will see later an example of this fact when discussing the $N = 4$ KdV hierarchy.

The point of view that we adopt here in discussing supersymmetric hierarchies is based on the (super-)Lie algebra framework, which one can reasonably hope will provide the key to classify all superhierarchies. The main reason of the difficulty involved in classifying superhierarchies w.r.t. the purely bosonic ones is due to the complications involving the presence of both even and odd generators.

We need to point out that (super)-Lie algebras appear in 3 different classes according whether they admit a presentation in terms of Dynkin diagrams with simple roots which are either:

i) purely fermionic,

ii) necessarily of mixed type, or

iii) purely bosonic (they are reduced to standard Lie algebras).

A simple argument made people believe for a long time that only the special class of super-Lie algebras admitting fermionic simple roots were relevant for the construction of superhierarchies. Inami and Kanno [7] gave it in the contest of super-KdV hierarchies. In order to extend the bosonic matrix Lax operator they were led to consider a supersymmetric Lax of the kind

$$\mathcal{L} = D + \Psi(X) + \Lambda \tag{4}$$

where now D is the $(N = 1)$ fermionic supersymmetric derivative. $\Psi(X)$ are superfields valued in some superalgebra and as such are fermionic. The regular element Λ should be given by the sum over the simple roots and in order to respect statistics it must be fermionic as well. Therefore it seemed that only class i) superalgebras had to be considered. A similar argument was given by Evans and Hollowood [8] in the case of superToda theories. This argument has paved the way to the standard supersymmetrization recipe of bosonic models which goes as follows: embed the given bosonic algebra which provides the bosonic system into a larger superLie algebra having "good properties" and perform the hamiltonian reduction on it. In this way $N = 1$ extension of KdV and Liouville equations were provided in terms of the $osp(1|2)$ superalgebra.

The Inami-Kanno scheme is a perfectly consistent one, leading to a classification of supersymmetric hierarchies much similar to the bosonic case. There would be no need to look for improving it if it would not turn out a too restricted one. Indeed it happens that well-known and interesting superhierarchies cannot be accomodated in it. To my knowledge Brunelli and Das were the first [9] who faced this problem when they realized that the superNLS equation admits a Lax operator based on the $sl(2)$ bosonic algebra (and not $osp(1|2)$ as one would have been expected). In [10] it was pointed out that the superNLS hierarchy arises as a coset construction (just as its bosonic counterpart) over a Poisson bracket structure based on the superaffinization (that is expressed in terms of superfields) of the bosonic $sl(2)$ algebra.

As a consequence there exists a much bigger class of supersymmetric integrable models than previously expected which need to be investigated. Despite the fact that we do not have yet a systematic way of constructing them in terms of matrix Lax operators just like the bosonic models or the Inami-Kanno superhierarchies, still we can develop some strategy to investigate them. This will be explained next.

4   Supersymmetric Hierarchies and Affine Algebras

Let us here discuss a possible strategy for constructing supersymmetric integrable hierarchies from generic superaffinizations of (super-)Lie algebras. But first let us point out that a superaffinization of a given (super-)Lie algebra $\mathcal{G}$ with generators τ^i and structure constants $f^{ij}{}_k$ is realized by $N = 1$ superfields $\Psi^i(X)$, with opposite statistics w.r.t. τ^i and such that

$$\{\Psi^i(X), \Psi^j(Y)\} = f^{ij}{}_k \Psi^k(Y)\delta(X,Y) + K^{ij} D_Y \delta(X,Y) \tag{5}$$

with $\delta(X, Y)$ the $N = 1$ delta-function and $K^{ij} = Str(\tau^i \tau^j)$ in some given (let's say the adjoint) representation of $\mathcal{G}$.

The following steps should be performed:

i) Take a superaffine (super-)Lie algebra which should be regarded as Poisson bracket structure.

ii) Make some Ansatz over the possible hamiltonians in involution; this would mean imposing symmetry requirements, cosets or hamiltonian reductions.

iii) Check the consistency of flows and if indeed at lower orders one gets hamiltonians in involution.

iv) Try to figure out the form of possible Lax operators (this is the most difficult task).

It can even happen that one finds more structures than expected. Indeed it is well-known that a relation exists between division algebras ad extended supersymmetries. Complex, Quaternionic ad Octonionic structures are associated to (global) $N = 2, 4, 8$ extensions respectively.

A complex structure for a (super-)algebra over the real fields is an operation J which satisfy $J^2 = -1$, while a quaternionic structure involves 3 complex structures J_i whose mutual algebra is that of the Pauli matrices.

If a theory admits a complex structure it necessarily has an extended supersymmetry. For instance the superNLS equation which arises from the $s\hat{l}(2)/u\hat{(}1)$ structure is automatically $N = 2$ since such a coset admits a complex structure (while $s\hat{l}(2)$ does not). An elegant (but equivalent) formulation can be realized through the coset $sl(2) \hat{\oplus} u(1)/u(1) \hat{\oplus} u(1)$ [11]. This construction allows a manifestly $N = 2$ superfield formulation since the extra $u(1)$ are added to give a complex structures for both numerator and denominator.

The $sl(2) \oplus u(1)$ algebra turns out to be vey interesting because it appears in the list given by [12] (actually these authors considered group-manifolds, out of which algebras can be immediately recovered) as the simplest example of non-abelian algebra (the even simpler abelian case being $u(1)^{\otimes 4}$) admitting a quaternionic structure.

A natural question therefore arises, namely if it is possible that the super-affine algebra $sl(2) \hat{\oplus} u(1)$, taken as a Poisson bracket algebra, would allow to play another game, not just the coset already mentioned, according to the above scheme. In particular we can ask ourselves if we can demand an $N = 4$ symmetry requirement which in turns imply an $N = 4$ hierarchy. In the next section we will show that this is indeed the case [13].

5   The $N = 4$ structure of $sl(2) \oplus u(1)$

The superaffine algebra $sl(2) \hat{\oplus} u(1)$ can be conveniently described in terms of $N = 2$ superfields. Let us introduce the $N = 2$ fermionic derivatives $D, \overline{D}$ whose algebra reads as follows

$$D^2 = \overline{D}^2 = 0$$
$$\{D, \overline{D}\} = -\partial_x \tag{6}$$

The spin $\frac{1}{2}$ $N = 2$ superfields are denoted as $H, \overline{H}, F, \overline{F}$.

H and $\overline{H}$ are associated to the $u(1) \oplus u(1)$ subalgebra.

They are constrained superfields, the constraints being non-linearly realized

$$DH = \overline{DH} = 0$$
$$(D + H)F = (\overline{D} - \overline{H})\overline{F} = 0 \tag{7}$$

The non-vanishing structure constants are given by

$$\{H(1), \overline{H}(2)\} = D\overline{D}\delta$$
$$\{H(1), F(2)\} = DF \cdot \delta$$
$$\{H(1), \overline{F}(2)\} = -D\overline{F}\delta$$
$$\{\overline{H}(1), F(2)\} = -\overline{D}F \cdot \delta$$
$$\{\overline{H}(1), \overline{F}(2)\} = \overline{DF} \cdot \delta$$
$$\{F(1), \overline{F}(2)\} = (D + H)(\overline{D} + \overline{H})\delta + F\overline{F}\delta \tag{8}$$

where $\delta \equiv \delta(1, 2)$ is the $N = 2$ delta-function and the derivatives in the r.h.s. are computed at $Z \equiv 1$.

In the last line a "fake" non-linear term appears. It is not present when the chiral constraints are solved in terms of $N = 1$ superfields or component fields.

There exists a second set of *global* $N = 2$ non-linear supersymmetries, expressed through the infinitesimal parameters $\epsilon, \overline{\epsilon}$, which results from the quaternionic structure associated with $sl(2) \oplus u(1)$. We have

$$\delta H = \epsilon D\overline{F} + \overline{\epsilon}HF$$
$$\delta \overline{H} = \overline{\epsilon}\overline{D}F - \epsilon\overline{HF}$$
$$\delta F = -\epsilon D\overline{H} - \epsilon(H\overline{H} + F\overline{F})$$
$$\delta \overline{F} = -\overline{\epsilon}\overline{D}H - \overline{\epsilon}(H\overline{H} + F\overline{F}) \tag{9}$$

It can be easily checked that the above transformations preserve the chirality constraints and that their commutators close to give, together with the original transformations, an $N = 4$ supersymmetry.

6   The $N = 4$ Hierarchy

We have seen that $sl(2) \hat{\oplus} u(1)$ carries an $N = 4$ structure. To prove the existence of globally invariant $N = 4$ dynamical systems we have to construct explicitly the $N = 4$ invariant hamiltonians. They indeed exist and moreover, at the lower dimensional integral spin dimension $d = 1, 2$, they are unique up to total derivatives (at least if a global chargeless condition is required, where $H, \overline{H}$ are chargeless, while F and $\overline{F}$ have charges $+1$ and -1 respectively).

We have indeed

$$H_1 = F\overline{F} + H\overline{H}$$
$$H_2 = F'\overline{F} - H'\overline{H} -$$
$$-(D\overline{H} + \overline{D}H)(H\overline{H} + F\overline{F}) - 2H\overline{H}F\overline{F} \qquad (10)$$

Higher dimensional $N = 4$ hamiltonians can be explicitly constructed and turn out to be in involution with the lower dimensional ones.

The resulting equations of motion (which is not needed to write here, see [13]) with respect to the second hamiltonian realize an $N = 4$ dynamical system which combines in a non-trivial way both the $N = 2$ mKdV equation and the $N = 2$ NLS equations. The latters are recovered by setting, consistently with the equations of motion, respectively $F = \overline{F} = 0$ and $H = \overline{H} = 0$. A third, more mysterious, $N = 2$ system can be obtained by performing a non-symmetrical reduction leading to $\overline{H} = F = 0$.

So far for what concerns the construction of the $N = 4$ system. An Ansatz has guided us towards its realization and we have seen that it is essentially unique. A point which has been left apart consists in explicitly proving that our system indeed corresponds to an integrable hierarchy admitting an infinite tower of hamiltonians in involution. In this particular case we have a very elegant procedure which proves that. Unfortunately, as already stated, we cannot rely so far on any systematic construction for the Lax operators valid for generic theories. The best we can do at present is based on a trial procedure.

However, for the $sl(2) \overset{\wedge}{\oplus} u(1)$ case the key property which allows to solve the problem is the existence of a (differential polynomial) $N = 4$ Sugawara construction [13]. This very remarkable transformation has at least four different consequences:

i) it provides a linearization of the $N = 4$ transformations,

ii) it furnishes a realization for the "minimal" $N = 4$ SuperConformal Algebra (SCA),

iii) it relates the "affine hierarchy" to the $N = 4$ KdV system[14] and

iv) it allows the construction of the Lax operator which proves the integrability.

The Sugawara transformation is a differential polynomial transformation which express the "superconformal fields" through the original affine superfields $H, \overline{H}, F, \overline{F}$. The transformed superfields are an $N = 2$ real superVirasoro superfield J (with component fields content $(1, \frac{3}{2}, \frac{3}{2}, 2)$) plus two chiral and antichiral spin 1 superfields (in components $(1, \frac{3}{2})$). We have explicitly

$$J = H\overline{H} + F\overline{F} + D\overline{H} + \overline{D}H$$
$$\Phi = D\overline{F}$$
$$\overline{\Phi} = \overline{D}F \qquad (11)$$

The presence of the Feigin-Fuchs terms in the r.h.s. for J is especially important. Without them J would be a nilpotent field ($J^3 = 0$ due to the fermionic character of $H, \overline{H}, F, \overline{F}$). Moreover they allow the second set of $N = 2$ transformations to close linearly on $J, \varPhi, \overline{\varPhi}$ as

$$\delta J = -\bar{\epsilon} D \overline{\varPhi} - \epsilon \overline{D} \varPhi$$
$$\delta \varPhi = \bar{\epsilon} D J$$
$$\delta \overline{\varPhi} = \epsilon \overline{D} \varPhi \tag{12}$$

The composite superfields $J, \varPhi, \overline{\varPhi}$ satisfy a closed algebra structure under the original $sl(2) \hat{\oplus} u(1)$ Poisson bracket structure and it coincides with the minimal version of the $N = 4$ SCA.

The hamiltonians in involution can be closely expressed through the superfields $J, \varPhi, \overline{\varPhi}$ alone. At the lowest order we have, for the hamiltonian densities

$$H_1 = J$$
$$H_2 = J^2 - 2\varPhi\overline{\varPhi}$$
$$H_3 = J[D, \overline{D}]J + 2\varPhi\overline{\varPhi}' + \frac{2}{3}J^3 - 4J\varPhi\overline{\varPhi} \tag{13}$$

As a consequence we have a closed system of dynamical equations for $J, \varPhi, \overline{\varPhi}$ which coincides with the $N = 4$ KdV hierarchy.

The Lax operator can be borrowed from the known Lax operator of KdV and is given by [15]

$$L = D\overline{D} + D\overline{D}\partial^{-1}(J + \overline{\varPhi}\partial^{-1}\varPhi)\partial^{-1}D\overline{D} \tag{14}$$

It should be noticed that in this particular case checking the integrability properties of the given hierarchy was immediate once the Sugawara construction has been taken into account since the above Lax operator for the $N = 4$ KdV was already known. However, even if this would have not been the case (as one could expect from constructions based on more general algebras), the Sugawara transformation itself would greatly simplify the task of finding the correct Lax pair, since it is much easier to deal with three spin 1 fields than with spin $\frac{1}{2}$ superfields. The dramatic simpification of the hamiltonians when expressed through $J, \varPhi, \overline{\varPhi}$ is also an example.

Some more comments are in order: the $N = 4$ KdV is the "unifying hierarchy" for two of the three inequivalent $N = 2$ KdV hierarchies labeled by $a = 1, -2, 4$. The $a = -2$ and $a = 4$ N=2 KdV are indeed obtained from different reductions of $N = 4$ KdV.

The construction based on the abelian $u(1)^{\otimes 4}$ algebra could lead to global $N = 4$ hierarchies realized through strictly chiral and antichiral superfields, but it can be easily checked that they are definitely not polynomial generalization of $N = 2$ NLS and are not $N = 4$ superconformal.

7 Conclusions

We have pointed out that supersymmetrical integrable hierarchies can be very naturally investigated (and hopefully classified) taking as a starting point the (super-)Lie algebras and their supersymmetric affinizations. Our approach is very much complementary with the point of view advocated by many authors in literature (like e.g. Z. Popowicz who is also author of a package for computing Lax operators by using Reduce). They rather use the converse attitude of actually producing integrable equations in terms of some consistent Lax operators, especially of scalar type. This approach has the advantage of furnishing indeed integrable systems, but leave aside questions concerning the algebraic interpretation of these results. The approach based on Lie algebras has just opposite merits and drawbacks. It furnishes from the very beginning the algebraic setting for defining dynamical systems and provides guidelines how to obtain them, while the burden is on proving the existence of a tower of hamiltonians in involution.

This situation is very specific to the supersymmetric case since, in contrast with bosonic hierarchies, we do not dispose of a hamiltonian reduction procedure which automatically leads to Lax operators. The examples where this is indeed the case, corresponding to the Inami-Kanno hierarchies, are of interest but they belong to a restricted class. Other interesting super-integrable systems like the $N = 4$ KdV equation previously discussed are left out of this scheme.

The approach based on (super-)Lie algebras is a very convenient one in the investigation of supersymmetric extended hierarchies. As discussed in this paper, one has to look for algebras admitting extra structures, complex, quaternionic and so far. Some partial results obtained in collaboratin with Ivanov and Krivonos show indeed that $sl(3)$, the next simplest quaternionic algebra, admits a global $N = 4$ structure which suggests the realization of at least global $N = 4$ hierarchies.

In conclusion it deserves being mentioned that investigating supersymmetric integrable hierarchies looks promising due to the presence of open problems.

References

1 P. Di Francesco, P. Ginsparg, J. Zinn Justin, Phys. Rept. **254**, 1 (1995)
2 F. Delduc, L. Frappat, P. Sorba, F. Toppan, E. Ragoucy, Phys. Lett. **B 318**, 457 (1993)
3 L. Frappat, E. Ragoucy, P. Sorba, Comm. Math. Phys.**157**, 93 (1993)
4 I. R. McIntosh, . *An Algebraic Study of Zero Curvature Equations* Ph.D. Thesis, Dept. Math., Imperial College, London (1988), unpublished; M.F. de Groot, T.J. Hollowood, J.L. Miramontes, Comm. Math. Phys. **145**, 57 (1992); L. Feher, J. Harnad, I. Marshall, Comm. Math. Phs. **154**, 181 (1993)

5 H. Aratyn, J.F. Gomes, A.H. Zimerman, J. Math. Phys. **36**, 3419 (1995): H. Aratyn, E. Nissimov, S. Pacheva, Int. Jou. Mod. Phys. A **12**, 1265 (1997)

6 L. Alvarez-Gaume', H. Itoyama, J.L. Manes, A. Zadra, Int. Jou. Mod. Phys. A **7**, 5337 (1992)

7 T. Inami, H. Kanno, Comm. Math. Phys. **136**, 519 (1991); Int. Jou. Mod. Phys. A **7** Suppl. 1A, 419 (1992); Jou. Phys. A **25**, 3729 (1992)

8 J. Evans, T. Hollowood, Nucl. Phys. **B 352**, 723 (1991)

9 J.C. Brunelli, A. Das, Phys. Lett. **B337**, 303 (1995); Jou. Math. Phys. **36**, 268 (1995)

10 F. Toppan, Int. Jou. Mod. Phys. A **10**, 895 (1995)

11 S. Krivonos, A. Sorin, F. Toppan, Phys. Lett. **A 206**, 146 (1995)

12 P.H. Spindel, A. Sevrin, W. Troost, A. Van Proeyen, Phys. Lett. **B 206**, 71 (1988)

13 E. Ivanov, S. Krivonos, F. Toppan, Preprint hep-th/9703224 (1997), to appear in Phys. Lett. B

14 F. Delduc, E. Ivanov, Phys. Lett. **B 309**, 312 (1993)

15 F. Delduc, L. Gallot, E. Ivanov, Preprint hep-th/9611033 (1996)

Part III

Supersymmetric Quantum Mechanics

A Semiclassical Approach to Level Crossing in Supersymmetric Quantum Mechanics

J. F. Beacom * and A. B. Balantekin **

Department of Physics, University of Wisconsin, Madison, WI 53706, USA

Abstract. Much use has been made of the techniques of supersymmetric quantum mechanics (SUSY QM) for studying bound-state problems characterized by a superpotential $\varphi(x)$. Under the analytic continuation $\varphi(x) \rightarrow i\varphi(x)$, a pair of superpartner bound-state problems is transformed into a two-state level-crossing problem in the continuum. The description of matter-enhanced neutrino flavor oscillations involves a level-crossing problem. We treat this with the techniques of supersymmetric quantum mechanics. For the benefit of those not familiar with neutrino oscillations and their description, enough details are given to make the rest of the paper understandable. Many other level-crossing problems in physics are of exactly the same form. Particular attention is given to the fact that different semiclassical techniques yield different results. The best result is obtained with a uniform approximation that explicitly recognizes the supersymmetric nature of the system.

1 SUSY QM and the Bound-State Problem

Starting with a superpotential $\varphi(x)$, one can generate two superpartner potentials

$$V_{\pm} = \varphi^2(x) \pm \frac{\hbar}{\sqrt{2m}} \varphi'(x). \tag{1}$$

For these two potentials the two corresponding Schrödinger equations are

$$\left[-\frac{\hbar^2}{2m} \frac{\partial^2}{\partial^2 x} + V_{\pm} \right] \Psi_{\pm}(x) = E\Psi_{\pm}(x). \tag{2}$$

It can be shown that the eigenspectrum of the "+" system can be obtained by shifting the quantum numbers n of the "−" system by $n \rightarrow n - 1$, with the ground state of the "−" system discarded. That is, the spectra of the two systems are identical except for a single state. In applications, this property is exploited in the following way. Given a potential $V(x)$, one attempts to find a superpotential $\varphi(x)$ that will generate $V(x)$ via Eq. (1), with one or the other sign. If this can be done, one can immediately generate the superpartner potential by choosing the opposite sign in Eq. (1). In some fortunate

---

* Speaker. Current address: Physics 161-33, Caltech, Pasadena, CA 91125, USA.
  beacom@citnp.caltech.edu
** baha@nucth.physics.wisc.edu

circumstances, the equations of motion for the second system are much easier to solve than the first. See Schwabl (1995) for more introductory material, and Cooper et al. (1995) and references therein for active areas of research.

The application of supersymmetric quantum mechanics to the solution of bound-state problems has been extensively developed. There has been particular interest in semiclassical techniques. A direct primitive semiclassical (WKB) approach to the Schrödinger equation yields the usual Bohr-Sommerfeld quantization condition:

$$\sqrt{2m} \int_{x_1}^{x_2} dx \sqrt{E - V(x)} = \left(n + \frac{1}{2} \right) \hbar \pi \,, \tag{3}$$

where x_1 and x_2 are the turning points (zeros of the integrand). Since the superpartner potentials depend explicitly on $\hbar$, the Schrödinger equations in Eq. (2) have a different dependence on $\hbar$ than the usual case. A primitive semiclassical (WKB) solution of Eq. (2), which however explicitly recognizes the supersymmetric nature of the system [Comtet et al. (1985)], therefore yields a modified quantization condition:

$$\sqrt{2m} \int_{x_1}^{x_2} dx \sqrt{E - \phi^2(x)} = n \hbar \pi \,, \tag{4}$$

where x_1 and x_2 are the turning points (zeros of the integrand). This modified quantization condition is exact for many systems (see Hruska et al. (1997) for a catalog of results), but not all [DeLaney and Nieto (1990)]. The WKB wave functions are singular at each turning point. It is possible to avoid this problem by using a uniform approximation – one that is valid for all x, including at the turning points. The approach is similar to WKB, but there is a special construction to cancel the turning-point singularity. A uniform semiclassical solution [Fricke et al. (1988)] of Eq. (2) recovers the modified quantization condition. However, it gives a much better wave function, which will be essential for the transition probability derived below.

2   Introduction to Neutrino Oscillations

2.1   Vacuum Neutrino Oscillations

It is possible that the flavor and mass eigenstates of neutrinos are not identical. Throughout this paper, we consider mixing between only two flavors, electron and muon. Then a general state can be written in the flavor basis:

$$|\nu(t)\rangle = \Psi_e(t)|\nu_e\rangle + \Psi_\mu(t)|\nu_\mu\rangle \tag{5}$$

or the mass basis:

$$|\nu(t)\rangle = \Psi_1(t)|\nu_1\rangle + \Psi_2(t)|\nu_2\rangle \,. \tag{6}$$

The amplitudes are defined as

Ψ_e (Ψ_μ) = amplitude for the neutrino to have flavor e (μ)

Ψ_1 (Ψ_2) = amplitude for the neutrino to have mass m_1 (m_2).

The amplitudes are taken to be time-dependent, and the kets to be time-independent. For two flavors, the flavor and mass bases must be related by a simple rotation. This rotation is taken to be between the amplitudes, with the kets held fixed, and is given by:

$$\begin{bmatrix} \Psi_1(t) \\ \Psi_2(t) \end{bmatrix} = \begin{bmatrix} \cos\theta_v & -\sin\theta_v \\ \sin\theta_v & \cos\theta_v \end{bmatrix} \begin{bmatrix} \Psi_e(t) \\ \Psi_\mu(t) \end{bmatrix}, \tag{7}$$

where θ_v is the vacuum mixing angle.

In the mass basis, the Schrödinger equation is

$$i\hbar \frac{\partial}{\partial t} \begin{bmatrix} \Psi_1(t) \\ \Psi_2(t) \end{bmatrix} = H_{\text{mass}} \begin{bmatrix} \Psi_1(t) \\ \Psi_2(t) \end{bmatrix}, \tag{8}$$

where

$$H_{\text{mass}} = \begin{bmatrix} E_1 & 0 \\ 0 & E_2 \end{bmatrix}, \tag{9}$$

and is diagonal by definition. Since the neutrino masses m_1 and m_2 are presumed small, we make an ultrarelativistic expansion (using $c = 1$ units here and below). Defining E to be the common energy, and defining the mass-squared difference as

$$\delta m^2 = m_2^2 - m_1^2, \tag{10}$$

one can show that H_{mass} is given by

$$H_{\text{mass}} = \left(E + \frac{m_1^2 + m_2^2}{4E} \right) \begin{bmatrix} 1 & 0 \\ 0 & 1 \end{bmatrix} + \frac{\delta m^2}{4E} \begin{bmatrix} -1 & 0 \\ 0 & 1 \end{bmatrix}. \tag{11}$$

Below, the term proportional to the identity matrix will be dropped.

Now we change to the flavor basis, using the rotation matrix above that relates the two bases. In the flavor basis, the Schrödinger equation is

$$i\hbar \frac{\partial}{\partial t} \begin{bmatrix} \Psi_e(t) \\ \Psi_\mu(t) \end{bmatrix} = H_{\text{flavor}} \begin{bmatrix} \Psi_e(t) \\ \Psi_\mu(t) \end{bmatrix}, \tag{12}$$

where

$$H_{\text{flavor}} = \frac{\delta m^2}{4E} \begin{bmatrix} -\cos 2\theta_v & \sin 2\theta_v \\ \sin 2\theta_v & \cos 2\theta_v \end{bmatrix}. \tag{13}$$

Using the rotation matrix, the amplitude to be of the electron type at a time t is

$$\Psi_e(t) = \cos\theta_v \Psi_1(t) + \sin\theta_v \Psi_2(t). \tag{14}$$

The time evolution of the mass amplitudes is trivial, so that just a phase relates the amplitudes at a point t to those at the point $t = 0$. The mass amplitudes at $t = 0$ can be expressed in terms of the flavor amplitudes at $t = 0$ by use of the rotation matrix. Taking as initial conditions $\Psi_e(0) = 1$, $\Psi_\mu(0) = 0$ (an electron-type neutrino produced at $t = 0$),

$$\Psi_e(t) = \cos^2\theta_v \exp\left(+i\frac{\delta m^2}{4E}\frac{t}{\hbar}\right) + \sin^2\theta_v \exp\left(-i\frac{\delta m^2}{4E}\frac{t}{\hbar}\right). \tag{15}$$

The probability for the neutrino to be of the electron type can then be shown to be

$$P(\nu_e \rightarrow \nu_e)(t) = |\Psi_e(t)|^2 = 1 - \sin^2 2\theta_v \sin^2(\pi t/L_{\text{osc}}), \tag{16}$$

where the oscillation length is $L_{\text{osc}} = 4\pi E\hbar/\delta m^2$, so called since this is the separation between extrema in the survival probability. Since the initial flavor state was not a stationary state of the Hamiltonian, the probability to be of either flavor oscillates. Note that if either the vacuum mixing angle or the mass-squared splitting is small, the effects of the oscillations are minimal. If either the source or detector has a finite size of order L_{osc} or greater, then after averaging over that region,

$$\langle P(\nu_e \rightarrow \nu_e)\rangle_{\text{avg}} = 1 - \frac{1}{2}\sin^2 2\theta_v, \tag{17}$$

which is independent of energy.

2.2   Matter-Enhanced Neutrino Oscillations

In this section we assume the vacuum oscillations discussed above. Here we consider that the neutrino is traveling through a medium with a varying density of electrons, e.g., the sun. We consider the scattering

$$\nu_x + e^- \rightarrow \nu_x + e^-, \tag{18}$$

where ν_x indicates either flavor. In such reactions, a neutrino of a given flavor enters and a neutrino of the same flavor leaves. Electron neutrinos undergo both charged- and neutral-current reactions with the electrons, but muon neutrinos undergo only neutral-current scattering. For elastic forward scattering, the coherence of the neutrino "beam" can be maintained. The effect of the medium is to modify the dispersion relation (refractive index) of the neutrino. Equivalently, one can say that the neutrino masses are modified in the medium, and that the modification is different for electron and muon neutrinos. Below, we neglect a term proportional to the identity, and consider only the difference between the electron and muon neutrinos. Then the Hamiltonian in Eq.(13) becomes

$$H_{\text{flavor}} = \frac{\delta m^2}{4E} \begin{bmatrix} \zeta(t) - \cos 2\theta_v & \sin 2\theta_v \\ \sin 2\theta_v & -\zeta(t) + \cos 2\theta_v \end{bmatrix}, \tag{19}$$

where $\zeta(t)$ is related to the electron number density $N_e(t)$ by

$$\zeta(t) = \frac{2\sqrt{2}G_F N_e(t)}{\delta m^2/E}, \tag{20}$$

and G_F is the Fermi coupling constant.

When $\zeta(t) \to 0$, the vacuum mixing case is recovered. When $\zeta(t) \to \infty$, the Hamiltonian is also considerably simplified. However, when $\zeta(t) = \cos 2\theta_v$, the effect of the off-diagonal coupling is maximal. This is known as the resonance point. In a medium with a varying density, such as the sun, it is possible for an electron neutrino to be produced at a density above the resonance density. Then during its passage out of the sun it will pass through the resonance density. That has profound consequences for the probability that it emerges in vacuum as an electron neutrino.

The flavor-basis Hamiltonian can be instantaneously diagonalized with a rotation matrix, the angle of which is called the matter angle $\theta(t)$. The bases are related as:

$$\begin{bmatrix} \Psi_1(t) \\ \Psi_2(t) \end{bmatrix} = \begin{bmatrix} \cos\theta(t) & -\sin\theta(t) \\ \sin\theta(t) & \cos\theta(t) \end{bmatrix} \begin{bmatrix} \Psi_e(t) \\ \Psi_\mu(t) \end{bmatrix}, \tag{21}$$

As the density varies, so does the matter angle. At high densities, the matter angle $\theta \to \pi/2$; at resonance, $\theta = \pi/4$; and at low densities, $\theta \to \theta_v$, the vacuum mixing angle (this is explained in Balantekin and Beacom (1996)).

The instantaneous eigenvalues of the flavor-basis Hamiltonian are (reintroducing the term proportional to the identity matrix):

$$E + \frac{m_1^2 + m_2^2 + \delta m^2 \zeta(t)}{4E} \mp \frac{\delta m^2}{4E} \sqrt{\sin^2 2\theta_v + (\zeta(t) - \cos 2\theta_v)^2}. \tag{22}$$

If the density changes slowly (adiabatically), then the instantaneous diagonalization will be almost exact and the mass eigenstates will be almost stationary states. (Note that the Hamiltonian is exactly diagonal in the mass basis for any constant density.) If the density changes quickly (nonadiabatically), then the neutrino can "hop" from one eigenstate to another in the neighborhood of the resonance (where the splitting between mass eigenvalues is at a minimum). The probability of this occurring is called the hopping probability P_{hop}. This occurs proportionally to the extent that the instantaneous diagonalization fails. The instantaneous eigenvalues of H_{flavor} as a function of density are shown schematically in Fig. 1.

Consider an electron neutrino created at a high density and a time $t = 0$. Since the matter angle will be near $\pi/2$, then $\Psi_e(0) \approx \Psi_2(0)$. If the density

Instantaneous eigenvalues of H_{flavor}

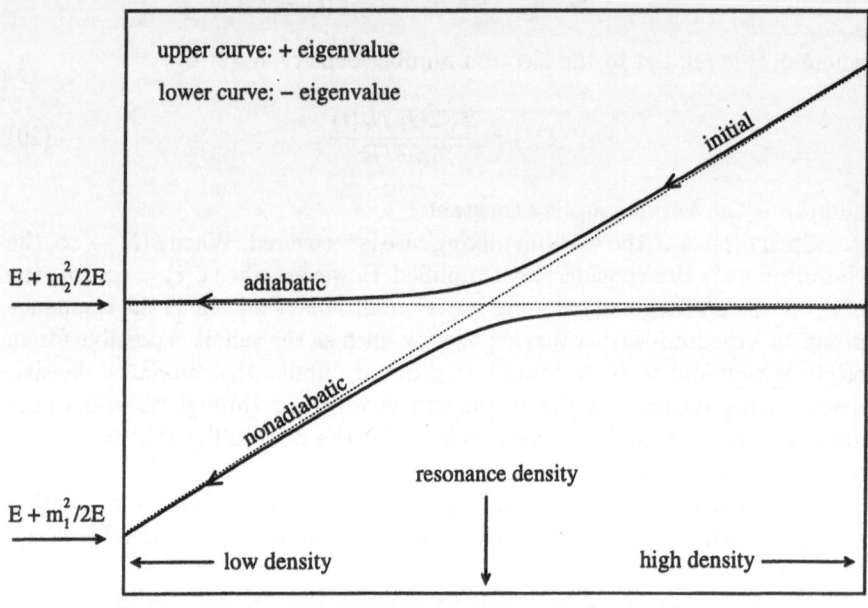

electron number density

Fig. 1. A schematic illustration of how the instantaneous energy eigenvalues of H_{flavor} vary as a function of electron number density. The details depend upon the mixing parameters, the energy, and the density profile. Beyond the left edge of the figure, the density is assumed to vanish; the eigenvalues are constant in vacuum and have the values indicated. In the figure, the initial state is assumed to be a mass $= m_2$ eigenstate. The final state may be the same eigenstate (adiabatic case) or the other eigenstate (nonadiabatic case). In the figure, the initial and final densities appear to be equally far from the resonance; that is not true in general.

changes slowly, the neutrino will remain in this eigenstate. Then, if the vacuum mixing angle is fairly small, then $\Psi_2(t) \approx \Psi_\mu(t)$ in vacuum. In this case, the electron neutrino created in the solar core will emerge as a muon neutrino, and it will be missed by experiments counting electron neutrinos from the sun. On the other hand, if the density changes rapidly the neutrino may emerge in the other instantaneous eigenstate. Then, if the vacuum mixing angle is fairly small, then $\Psi_1(t) \approx \Psi_e(t)$ in vacuum. In this case, the neutrino emerges as an electron neutrino after all.

In a later section, this problem is reformulated so that these extreme limits of the matter and vacuum angle are not necessary. In general then, whether the neutrino emerges from the sun as the electron or muon type is a complicated function of the mixing parameters, the energy, and the form of the density profile. The probability of it emerging as the original flavor is called the survival probability $P(\nu_e \rightarrow \nu_e)$; the general form is given in Eq. (31).

For further introductory reading on neutrino oscillations, see Boehm and Vogel (1992). For a recent review of neutrino astrophysics (including oscillations), see Balantekin (1997).

3   Supersymmetric Character of the Level-Crossing Problem

The Hamiltonian of Eq. (19) is a typical form for a level-crossing problem. Away from the resonance point, the off-diagonal elements may be neglected. The diagonal elements approximate the eigenvalues, and are allowed to vary. If the off-diagonal elements were exactly zero, the eigenvalue trajectories would cross (become equal) at the resonance point. This crossing will be avoided if the off-diagonal elements have any nonzero value, no matter how small. In the most general level-crossing problem, the off-diagonal elements would also be allowed to vary. Here they are taken to be constant. Nevertheless, that is not a large practical restriction. Approximating the off-diagonal elements as constant is reasonable as the resonance region (the only region in which they contribute significantly) is usually very narrow. The explicit representation of the supersymmetry in Eq. (19) is given in Balantekin et al. (1988).

Before proceeding further, we switch to working with dimensionless quantities. We define a length scale

$$L = \frac{\hbar \lambda}{\delta m^2 / 4E} \, , \tag{23}$$

and use this to define $x = t/L$. In Section 1, x was used to denote a generic coordinate (with dimensions of length). Here and below, x is dimensionless. Since we will be making a semiclassical expansion, we need to be able to keep track of formal powers of $\hbar$. For each $\hbar$ in the problem, we write λ and consider λ to be formally small; this is equivalent to saying that the length L is small. We will make expansions in powers of λ, truncating the higher orders. At the end of the calculation, we will set $\lambda = 1$. For notational convenience, we redefine the flavor-basis Hamiltonian as follows:

$$i\lambda \frac{\partial}{\partial x} \begin{bmatrix} \Psi_e(x) \\ \Psi_\mu(x) \end{bmatrix} = H_{\text{flavor}}(x) \begin{bmatrix} \Psi_e(x) \\ \Psi_\mu(x) \end{bmatrix} , \tag{24}$$

where

$$H_{\text{flavor}}(x) = \begin{bmatrix} \eta\varphi(x) & \sqrt{\Lambda} \\ \sqrt{\Lambda} & -\eta\varphi(x) \end{bmatrix} . \tag{25}$$

We have defined

$$\eta\varphi(x) = \zeta(x) - \cos 2\theta_v \tag{26}$$

and

$$\Lambda = \sin^2 2\theta_v . \tag{27}$$

(The definition of Λ was misprinted in Balantekin and Beacom (1996)). The scaled electron density is

$$\zeta(x) = \frac{2\sqrt{2}\, G_F E N_e(x)}{\delta m^2} . \tag{28}$$

Note that there are notation changes from previous related works [Fricke et al. (1988), Balantekin et al. (1988), Balantekin and Seger (1991)]; here we have made Λ and φ dimensionless. The factor η (taken to be ± 1), is introduced above to control the analytic behavior of the function $\varphi(x)$ in the complex plane, as explained in Balantekin and Beacom (1996). In the expressions with φ^2 below, we drop $\eta^2 = 1$.

The coupled first-order equations of Eq. (25) can be decoupled to yield

$$-\lambda^2 \frac{\partial^2 \Psi_e(x)}{\partial x^2} - \left[\Lambda + \varphi^2(x) + i\lambda\eta\varphi'(x) \right] \Psi_e(x) = 0 \tag{29}$$

and

$$-\lambda^2 \frac{\partial^2 \Psi_\mu(x)}{\partial x^2} - \left[\Lambda + \varphi^2(x) - i\lambda\eta\varphi'(x) \right] \Psi_\mu(x) = 0 , \tag{30}$$

where $\varphi(x)$ and Λ are defined in Eqs. (26) and (27). Such a simple decoupling is not possible in the mass basis. Eqs. (29) and (30) are explicitly of the supersymmetric form. This form follows directly from the (rather general) form of the level-crossing Hamiltonian above. The two levels are superpartners.

These Schrödinger-like equations are similar to those for non-relativistic particles in the presence of a complex barrier, and for convenience we use the language of wave mechanics to describe them. In particular, to the extent that we can ignore the imaginary terms in the potential, these correspond to particles above a barrier (since $\Lambda > 0$). There are two caveats regarding discussing this as a barrier penetration problem. First, that our boundary conditions do not correspond to the usual picture of incident, reflected, and transmitted waves; in general, there are waves moving in each direction on each side of the barrier. The boundary conditions make it a level-crossing problem instead of a barrier problem. (The condition of a pure electron neutrino at the initial point requires $\Psi_e = 1$, $\Psi_\mu = 0$ there.) Second, the pure imaginary terms in the potentials play an extremely important role here, even in the asymptotic regions. These terms are needed not only to represent nonadiabatic transitions, but also to allow the local matter angle to change. Since our problem is a level-crossing problem, the quantity of interest is not a reflection or transmission coefficient, but rather $P(\nu_e \to \nu_e) = |\Psi_e(x \to \infty)|^2$, the probability of the neutrino being of the electron type far from the source.

4   Semiclassical Solution of the Equations of Motion

4.1   General Form of the Solution

In general, the survival probability has terms which depend on the source and detector positions. In this paper, these interference terms are considered to be averaged away by the finite source and detector sizes (or by a varying distance between them, such as that due to the motion of the earth). The remaining term below must also be considered to have been appropriately averaged.

The general form of the survival probability is

$$P(\nu_e \to \nu_e) = \frac{1}{2}\left[1 + (1 - 2P_{\text{hop}})\cos 2\theta_i \cos 2\theta_v\right] . \tag{31}$$

The two matter angle terms account for the rotations between the mass and flavor bases at the initial and final (in vacuum) points. (The matter angle thus ranges from $\pi/2$ at infinite density to θ_v in vacuum. At the resonance, $\theta = \pi/4$.) P_{hop} is the probability of hopping from one mass eigenstate to the other during the transit through the resonance region. This result for the survival probability is completely general. The heart of the problem is to determine P_{hop}. Various semiclassical results for P_{hop} are discussed below. The variation of the averaged survival probability with $\delta m^2/E$ for two choices of the vacuum mixing angle is shown in Fig. 2.

The general shape of the survival probability vs $\delta m^2/E$ is that of a pit. High- and low-energy electron neutrinos have a relatively high probability to remain so, whereas medium-energy electron neutrinos have a relatively low probability to remain so (and hence a relatively high probability to transform into muon neutrinos). The solar neutrino detectors are primarily sensitive to electron neutrinos. The energy-dependent suppression of electron neutrinos as in Fig. 2 explains the solar neutrino data well.

4.2   Primitive Semiclassical Solution

A primitive semiclassical (WKB) solution can be made to Eqs. (29) and (30). It can be shown that this is exactly equivalent to the adiabatic solution. The adiabatic solution is obtained when the Hamiltonian changes so slowly that the instantaneous diagonalization always holds with high accuracy. That is, there are no transitions between mass eigenstates, so $P_{\text{hop}} = 0$ by definition. This result is not shown in Fig. 2. If it were, it would match the exact result in all but the left-hand rise of the pit. Where the other curves rise up, the WKB result would remain flat, at the level of the bottom of the pit.

In the bound-state problem, even the primitive semiclassical solution gave an excellent answer for the quantization condition. As noted, in the neutrino problem this approximation does not predict P_{hop}. The problem comes from the well-known singularity of the WKB wave functions near the turning

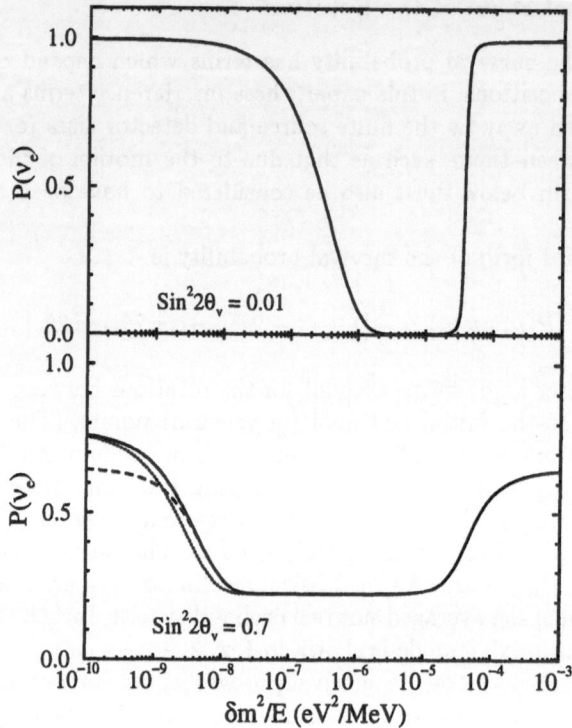

Average survival probability

Fig. 2. The electron neutrino survival probability vs. the mass-squared difference parameter for two different vacuum mixing angles. The solid line is given by the method of Balantekin and Beacom (1996). The dashed line is the exact (numerical) result. The dotted line is the linear Landau–Zener result. In the top figure, the lines are indistinguishable. An exponential density with parameters chosen to approximate the sun was used [Bahcall (1989)]. The region leftward of the lower left corner of the trough is the nonadiabatic region.

points. In this problem, the turning points are in the complex plane, near the resonance point (which is on the real axis). In the bound-state problem, the turning point singularity is not as crucial since the quantization condition only counts nodes between the turning points. In this problem, we need the wave function itself to be accurate, since its modulus squared determines the survival probability.

4.3 Landau-Zener Solution

Far from the resonance, the propagation is adiabatic, and can be trivially solved for any density profile. A reasonable approach to simplifying the problem is to approximate the density profile as linear in the resonance region. This is the basis of the Landau-Zener solution. In the flavor-basis Hamiltonian of Eq. (25), this gives the usual Landau-Zener setup: linear variation of the diagonal elements, and constant off-diagonal coupling. In order to make Fig. 1 more general, the variation is shown versus the density itself (instead of x).

The solution is obtained from the decoupled equations (29) and (30). With $\zeta(x) \sim x$, the differential equations can be reduced to the defining equation for the parabolic cylinder (Weber) functions. Given the initial conditions $\Psi_e = 1, \Psi_\mu = 0$, the solution for $\Psi_e(x)$ at any point is straightforward. (Actually, the boundary conditions require some care since the linear density eventually becomes negative, which is unphysical.) From the asymptotic form appropriate far after the resonance, one can extract the hopping probability P_{hop} from the the survival probability $P(\nu_e \to \nu_e) = |\Psi_e(x \to +\infty)|^2$. The result is

$$P_{hop} = \exp\left(-\pi \frac{\delta m^2}{4E\hbar} \frac{\sin^2 2\theta_v}{\cos 2\theta_v} \left| \frac{\dot{\zeta}(t)}{\zeta(t)} \right|^{-1}_{res} \right). \tag{32}$$

Because the Landau-Zener solution is based on the exact solution for the linear density, there are no turning-point singularities as with the WKB solution. Thus the approximate wave function and hence P_{hop} are fairly reasonable. However, there is some inaccuracy due to the fact that a general density does vary more than linearly through the resonance region. The fact that the Landau-Zener result has the right general behavior but is not very accurate can be seen in Fig. 2.

4.4 Uniform Semiclassical Solution

In this section we outline a uniform semiclassical solution to the problem [Balantekin and Beacom (1996)]. An arbitrary monotonic density profile is allowed, as are nearly arbitrary mixing parameters. We start with Eq. (29):

$$-\lambda^2 \frac{\partial^2 \Psi_e(x)}{\partial x^2} - \left[\Lambda + \varphi^2(x) + i\lambda\eta\varphi'(x) \right] \Psi_e(x) = 0. \tag{33}$$

There is no known exact solution of this for arbitrary $\zeta(x)$. Recall that λ is a formally small perturbation parameter (we set $\lambda = 1$ later), and that $\eta = \pm 1$. Compare this to

$$-\lambda^2 \frac{\partial^2 U(S)}{\partial S^2} - \left[\Omega + S^2 \pm i\lambda\eta \right] U(S) = 0, \tag{34}$$

which is solvable in terms of parabolic cylinder functions. Identification of the turning points of each of these will be crucial. At lowest order in λ, they are the points for which $\Lambda + \varphi^2(x) = 0$ and $\Omega + S^2 = 0$, respectively. The turning points are complex conjugate pairs.

Suppose we could find a change of variables $S = S(x)$. Using this, we could deform the simple parabolic "barrier" of the second equation into the more complicated "barrier" shape of the first. Then the solutions of the general case could be expressed in terms of the solutions of the simple case. Such a formal solution can be written as

$$\Psi_e(x) = \frac{1}{\sqrt{S'(x)}} U(S(x)).\tag{35}$$

Starting with this formal solution, and the two Schrödinger equations above for $\Psi_e(x)$ and $U(S)$, one can determine a differential equation for $S = S(x)$. Unsurprisingly, that is a nonlinear equation, and at least as difficult as the original Schrödinger equation. However, this equation can be profitably subjected to a semiclassical solution. We expand as

$$S(x) = S_0(x) + \lambda S_1(x)\ldots,\tag{36}$$

and truncate at second order. Separating the various orders in λ determines a series of equations. The solution of the $\mathcal{O}(\lambda)$ equation determines $S_0(x)$:

$$\int_{i\sqrt{\Omega}}^{S_0(x)} dS_0 \sqrt{\Omega + S_0^2} = \int_{x_0}^{x} dt \sqrt{\Lambda + \varphi^2(x)}\tag{37}$$

and also Ω:

$$\Omega = \frac{2i}{\pi} \int_{x_0}^{x_0^{\bullet}} dx \sqrt{\Lambda + \varphi^2(x)},\tag{38}$$

where x_0 and x_0^* are the turning points (zeros of the integrand). The $\mathcal{O}(\lambda^1)$ equation determines $S_1(x)$:

$$S_1(x) = \frac{i\eta}{2\sqrt{\Omega + S_0^2(x)}}$$
$$\times \left\{ \ln \left[\frac{\varphi + \sqrt{\Lambda + \varphi^2(x)}}{\sqrt{\Lambda}} \right] + \ln \left[\frac{\sqrt{\Omega}}{S_0(x) + \sqrt{\Omega + S_0^2(x)}} \right] \right\}\tag{39}$$

The change of variables $S = S(x)$ is actually a mapping of the complex x-plane to the complex S-plane. In order to avoid spurious branch cut discrepancies, the mapping must be chosen to not fold or flip the plane. This is accomplished with some appropriate choices of signs. Further, the turning points in the x-plane must be mapped onto the turning points in the S-plane. That demand fixes the definition of Ω to be the one given above.

The approximate (but uniformly valid) solution is

$$\Psi_e(x) \approx \left[\frac{\Omega + S_0^2(x)}{\Lambda + \varphi^2(x)}\right]^{1/4} U(S_0(x) + \lambda S_1(x)) \tag{40}$$

At the turning points, the denominator vanishes, just as in the WKB solution. However, because of the matching of turning points in the mapping, the numerator vanishes at the same points. That cancels the singularity and gives an excellent approximation to the wave function at all points, including near the turning points (and hence near the resonance, which is on the real axis near the turning points).

As noted above, the functions $U(S)$ are parabolic cylinder functions. Using the defining equations for $S_0(x)$ and $S_1(x)$ given above, one can solve for $\Psi_e(x)$ for any x. The implicit definition of $S_0(x)$ makes it analytically solvable only for large $|x|$ and large $|S_0|$, which holds far from the resonance. However, both the initial and final point can be taken to be far from the resonance. At each point, the general solution can be written in terms of two independent parabolic cylinder functions with arbitrary coefficients. After taking the asymptotic forms (meaning well away from the resonance), one has:

$$\Psi_e(x \to -\infty) \tag{41}$$
$$= C_- \cos\theta(x) \exp\left(+iI_p(x, x_i)/\lambda\right) + D_- \sin\theta(x) \exp\left(-iI_p(x, x_i)/\lambda\right),$$

$$\Psi_e(x \to +\infty) \tag{42}$$
$$= C_+ \sin\theta(x) \exp\left(-iI_p(x, x_i)/\lambda\right) + D_+ \cos\theta(x) \exp\left(+iI_p(x, x_i)/\lambda\right)$$
$$= \left[c_1 \sin\theta_i \exp\left(+i\mathrm{Re}I_p(x_i, x_0)/\lambda\right) + c_2 \cos\theta_i \exp\left(-i\mathrm{Re}I_p(x_i, x_0)/\lambda\right)\right]$$
$$\times \sin\theta(x) \exp\left(-i\mathrm{Re}I_p(x, x_0)/\lambda\right)$$
$$+ \left[c_1^* \cos\theta_i \exp\left(-i\mathrm{Re}I_p(x_i, x_0)/\lambda\right) - c_2^* \sin\theta_i \exp\left(+i\mathrm{Re}I_p(x_i, x_0)/\lambda\right)\right]$$
$$\times \cos\theta(x) \exp\left(+i\mathrm{Re}I_p(x, x_0)/\lambda\right),$$

where

$$I_p(x, x_i) = \int_{x_i}^{x} dx \sqrt{\Lambda + \varphi^2(x)}. \tag{43}$$

Above, x is a general point, x_i is the initial point, and x_0 is a turning point. Now one must solve the problem of connecting the coefficients C_- and D_- (known from the initial conditions) to the coefficients C_+ and D_+ (or equivalently c_1 and c_2). Tedious algebra reveals the solution to the connection problem to be:

$$c_1 = -\frac{\Gamma(-\nu)}{\sqrt{2\pi}} \Omega^{-i\Omega/2 + \eta/2} \left(\frac{e^{-i\pi/2}}{2}\right)^{\nu} \frac{e^{-3i\pi/4}}{\sqrt{2}} \exp\left(+\frac{i\Omega}{2}\right) 2i\sin(\nu\pi) \tag{44}$$

$$c_2 = e^{-i\nu\pi} \tag{45}$$

Comparison of $|\Psi_e(x \to +\infty)|^2$ above to the general form

$$P(\nu_e \to \nu_e) = \frac{1}{2}\left[1 + (1 - 2P_{\mathrm{hop}})\cos 2\theta_i \cos 2\theta_\nu\right] \tag{46}$$

reveals that:

$$1 - P_{\text{hop}} = |c_1|^2 = 1 - e^{-\Omega \pi}, \tag{47}$$

$$P_{\text{hop}} = |c_2|^2 = e^{-\Omega \pi}, \tag{48}$$

where

$$\Omega = \frac{2i}{\pi} \left(\frac{\delta m^2}{4E\hbar} \right) \int_{t_0}^{t_0^*} dt \sqrt{\sin^2 2\theta_v + (\zeta(t) - \cos 2\theta_v)^2}. \tag{49}$$

The dimensions have been restored and Λ and φ replaced by their definitions; recall that $\zeta(t) \sim$ density and t_0, t_0^* are the turning points (zeros of the integrand). For a linear density, the Landau-Zener result is recovered, i.e.,

$$\text{linear density} : \Omega = \frac{\delta m^2}{4E\hbar} \frac{\sin^2 2\theta_v}{\cos 2\theta_v} \left| \frac{\dot{\zeta}(t)}{\zeta(t)} \right|_{\text{res}}^{-1}. \tag{50}$$

However, any arbitrary monotonic density may be used. For example, for an exponential density,

$$\text{exponential density} : \Omega = \frac{\delta m^2}{4E\hbar} (1 - \cos 2\theta_v) \left| \frac{\dot{\zeta}(t)}{\zeta(t)} \right|^{-1}, \tag{51}$$

which is the leading term in the exact result. That the approximation is rather good over a wide variety of parameters can be seen in Fig. 2.

5   Conclusions

In a semiclassical approximation, one always finds that the wave functions have terms like

$$\Psi \sim \exp\left((\ \ )/\hbar\right), \tag{52}$$

where (.) indicates a phase integral and various numerical factors. There is an essential singularity in Ψ in the formal limit $\hbar \to 0$. That is, Ψ does not have a well-defined value in this limit. Different approximations (i.e., different ways of limiting $\hbar \to 0$) therefore give different results. Not all semiclassical approximations are equivalent; other criteria must be used to decide which approach to use. In this level-crossing problem, the best results come from a uniform approximation (valid for all x, including the turning points) that explicitly recognizes the supersymmetric nature of the potential. Given the success of similar approximations in the bound-state problem, this is perhaps to be expected.

Even the uniform supersymmetric approximation eventually breaks down. In the original problem (in the figure above, based on the exponential density), we considered only the two turning points closest to the real axis. In general, there are more, further out in the complex plane. The distance of the the turning points from the real axis scales like $|\zeta/\dot{\zeta}|$ (and for the exponential density, integer multiples of this). The analog potential (based on the linear density), only has two turning points. This mismatch in turning-point topologies eventually makes the mapping function multivalued, at which point the approximation breaks down. The problem is worst when the length scale $E\hbar/\delta m^2$ becomes large and "sees" further into the complex plane.

This is an interesting practical problem, in part because there is good evidence for matter-enhanced neutrino oscillations. The technique given here is applicable for any monotonic density profile (and not just in the sun), and for a large range of mixing parameters [Balantekin and Beacom (1996)]. In addition, the techniques here should also be useful in a variety of the continuum level-crossing problems that are ubiquitous in quantum mechanics. As noted, certain kinds of two-level systems are automatically of the supersymmetric form. The techniques of semiclassical SUSY QM for the bound-state problem can be carried over, as shown in our example.

Acknowledgments

This work was supported in part by the U.S. National Science Foundation Grant No. PHY-9605140 at the University of Wisconsin, and in part by the University of Wisconsin Research Committee with funds granted by the Wisconsin Alumni Research Foundation.

References

J.N. Bahcall (1989): *Neutrino Astrophysics* (Cambridge, New York)

A.B. Balantekin, S.H. Fricke, P.J. Hatchell (1988): "Analytical and Semiclassical Aspects of Matter-Enhanced Neutrino Oscillations," Phys. Rev. D **38**, 935

A.B. Balantekin, J.E. Seger (1991): "Dynamical Effects in Pair Production by Electric Fields," Int. J. Mod. Phys. A **6**, 695

A.B. Balantekin, J.F. Beacom (1996): "Semiclassical Treatment of Matter–Enhanced Neutrino Oscillations for an Arbitrary Density Profile," Phys. Rev. D **54**, 6323

A.B. Balantekin (1997): "Selected Topics in Neutrino Astrophysics," astro-ph/9706256, to be published in the *Proceedings of the 1997 Jorge Andre Swieca Summer School, Brazil*

F. Boehm, P. Vogel (1992): *Physics of Massive Neutrinos, 2nd ed.* (Cambridge, New York)

A. Comtet, A.D. Bandrauk, D.K. Campbell (1985): "Exactness of Semiclassical Bound State Energies for Supersymmetric Quantum Mechanics," Phys. Lett. B **150**, 159

F. Cooper, A. Khare, U. Sukhatme (1995): "Supersymmetry and Quantum Mechanics," Phys. Rep. **251**, 267

D. DeLaney, M.M. Nieto (1990): "SUSY-WKB is Neither Exact Nor Never Worse than WKB for all Solvable Potentials," Phys. Lett. B **247**, 301

S.H. Fricke, A.B. Balantekin, P.J. Hatchell, T. Uzer (1988): "Uniform Semiclassical Approximation to Supersymmetric Quantum Mechanics," Phys. Rev. A **37**, 2797

M. Hruska, W.-Y. Keung, U. Sukhatme (1997): "Accuracy of Semiclassical Methods for Shape-Invariant Potentials," Phys. Rev. A **55**, 3345

F. Schwabl (1995): *Quantum Mechanics, 2nd ed.*, Ch. 19 (Springer, Berlin)

A more extensive list of references is given in Balantekin and Beacom (1996).

One-Dimensional Disordered Supersymmetric Quantum Mechanics: A Brief Survey

Alain Comtet[1,2] and Christophe Texier[1]

[1] Division de Physique Théorique‡, IPN Bât. 100, 91406 Orsay Cédex, France.
[2] L.P.T.P.E, Université Paris 6, 4 place Jussieu, 75252 Paris Cédex 05, France.

E-mail: comtet@ipno.in2p3.fr
E-mail: texier@ipno.in2p3.fr
‡ Unité de recherche des Universités Paris 11 et Paris 6 associée au CNRS.

Abstract. We consider a one-dimensional model of localization based on the Witten Hamiltonian of supersymmetric quantum mechanics. The low energy spectral properties are reviewed and compared with those of other models with off-diagonal disorder. Using recent results on exponential functionals of a Brownian motion we discuss the statistical properties of the ground state wave function and their multifractal behaviour.

IPNO/TH 97-21

1 Introduction

Many physical systems can be described using the concept of a random Hamiltonian. Such a formulation is useful when the Hamiltonian depends on a set of quenched variables. In most cases it is convenient to describe them by a set of random variables distributed according to some probability law. Consider, for instance, the quantum dynamics of a particle interacting with randomly distributed scatterers. If the potential is a sum of two-body potentials the Hamiltonian may be taken in the form

$$ H = \frac{\mathbf{p}^2}{2m} + \sum_{k=1}^{N} V(\mathbf{r} - \mathbf{r}_k). $$

Here the quenched variables are the positions of the impurities $\mathbf{r}_k$ and the number N of scatterers. They can, for instance, be modelled by independently random variables distributed according to a Poisson distribution.

In this context one of the most elementary quantities of physical interest is the average density of states $\rho(E)$. If the potential is repulsive and short range then $\rho(E)$ vanishes exponentially at the bottom of the spectrum

$$ \rho(E) \underset{E \to 0}{\sim} e^{-C\,E^{-d/2}} $$

where d is the dimension of the space. This non-analytic behaviour was first discussed by Lifshits (Lifshits 1965) and then studied by a number of authors (see for example Luttinger and Waxler 1987). The physical mechanism which leads to this behaviour is the occurence of large regions of space that are free from impurities and where the particle can move freely. Although these are exponentially rare events, they nevertheless contribute in the thermodynamic limit. This singular behaviour may also be derived using instanton techniques (Neuberger 1982).

However there do exist some systems for which the density of states has a very different behaviour:

1. The vibrations of a chain consisting of harmonic strings and random masses gives a spectral density with an accumulation of states at low frequency. This model, first introduced by Dyson (Dyson 1953), is in fact equivalent to the one-dimensional Anderson model with off-diagonal disorder.

2. In particle physics, the investigation of the random Dirac operator has stimulated interesting conjectures related to chiral symmetry breaking (Floratos et al. 1980). A particular model in dimension $2+1$ is to take as a Hamiltonian the square of the Euclidean Dirac operator coupled to a random magnetic field $H = -\not{D}^2 = -(\partial_\mu + iA_\mu)^2 + \frac{1}{2}\sigma_{\mu\nu}F^{\mu\nu}$. In this case the low energy density of states must fullfil the inequality (Casher et al. 1984) $\rho(E) > \rho_o(E)$, where $\rho_o(E)$ is the free density of states. This means that there is an accumulation of low energy states which contributes to the chiral condensate.

3. The one-dimensional Schrödinger Hamiltonian $H = -\frac{d^2}{dx^2} + \phi^2(x) + \sigma_3\phi'(x)$ which was first introduced by Witten (Witten 1981) as a toy model of supersymmetric quantum mechanics (for a review see for example Cooper et al. 1995) provides a localization model with very unusual low energy spectral properties. In certain cases it gives rise to an accumulation of levels at low energy. The density of states displays either a power law behaviour $\rho(E) \underset{E\to 0}{\sim} E^{\mu-1}$ or a logarithmic singularity $\rho(E) \underset{E\to 0}{\sim} \frac{1}{E\ln^3 E}$ of the same form as in the Dyson model (Dyson 1953).

A common feature of these three models is the fact that the zero energy wave function is exactly known for any realization of the disorder. Using this property we have presented, for the supersymmetric model (Comtet et al. 1995), a physical picture that accounts for the different behaviours of the density of states at the bottom of the spectrum. We believe that this model is probably generic, by which we mean representative of a whole class of systems in which the disorder is encoded in the ground state. Since this model is easier to handle, because one can use a wealth of techniques specific to one-dimensional systems, we will concentrate on this case. The recent literature shows a revival of interest for these problems, mainly in the context of condensed matter physics. We will briefly comment on this work and also draw attention to earlier work which is scattered in the literature and has so far remained unnoticed.

In part 2 we will review the basic mechanism which leads to these singularities and underline the differences with the usual Lifshits singularities. We will also mention some recent applications of the supersymmetric model to quantum spin chains. Applications to classical diffusion in a random medium (Bouchaud et al. 1990, Bouchaud and Georges 1990, Oshanin et al. 1993b) will not be discussed here.

In part 3, following (Broderix and Kree 1995) and (Shelton and Tsvelik 1997) we compute the correlation function of the zero energy states.

In part 4 we discuss the fluctuation properties of the ground state wave function using two different approaches. When $\phi(x)$ is white noise the wave function is an exponential functional of the Brownian motion. Such functionals were studied extensively both in the mathematical (Yor 1992) and physical literature (Monthus and Comtet 1994, Comtet et al. 1996). We will use our previous work to compute their statistical properties.

2   Spectral properties

The one-dimensional Schrödinger Hamiltonians

$$H_{\pm} = -\frac{\mathrm{d}^2}{\mathrm{d}x^2} + \phi^2(x) \pm \phi'(x) \tag{1}$$

may be rewritten in the factorized forms $H_+ = Q^\dagger Q$ and $H_- = QQ^\dagger$, where $Q \equiv -\frac{\mathrm{d}}{\mathrm{d}x} + \phi(x)$. This implies that H_+ and H_- have the same spectrum for $E > 0$. When $\phi(x)$ is random, they are characterized by the same localization length and density of states. These quantities have been computed exactly in two cases for which we now recall the main results.

2.1   White noise potential

$$\begin{cases} \langle \phi(x) \rangle = F_0 \\ \langle \phi(x)\phi(x') \rangle - \langle \phi(x) \rangle^2 = \sigma\, \delta(x - x') \end{cases}$$

The integrated density of states $N(E)$ and the localization length $\lambda(E)$ are respectively

$$N(E) = \frac{2\sigma}{\pi^2} \frac{1}{J_\mu^2(z) + N_\mu^2(z)} \tag{2}$$

$$\lambda^{-1}(E) = -\frac{\sigma z}{2} \frac{\mathrm{d}}{\mathrm{d}z} \ln\left(J_\mu^2(z) + N_\mu^2(z)\right)$$

where $z \equiv \frac{\sqrt{E}}{\sigma}$ and $\mu \equiv \frac{F_0}{\sigma}$. $J_\mu(z)$ and $N_\mu(z)$ are Bessel functions.

Equation (2) was first obtained by Ovchinnikov and Erikmann (Ovchinnikov and Erikmann 1977) and then rediscovered independently in (Bouchaud et al. 1987). These results can be derived either by the node counting method (Luck 1992) or by the replica trick (Bouchaud et al. 1990). By this latter approach one can also compute the Green's function at non-coinciding points.

The low energy behaviour of the density of states and the localization length are given for $\mu = 0$ by

$$N(E) \underset{E \to 0}{\sim} \frac{1}{\ln^2 E}$$
$$\lambda(E) \underset{E \to 0}{\sim} -\ln E$$

and by

$$N(E) \underset{E \to 0}{\sim} E^\mu$$
$$\lambda(E) \underset{E \to 0}{\sim} E^{\mu-1}$$

when $\mu \neq 0$.

2.2 Random telegraph process

The function $\phi(x)$ is described by an ensemble of rectangular barriers (Benderskiĭ and Pastur 1974) with alternating heights ϕ_0 and ϕ_1 of random length l distributed according to an exponential law $p_{0,1}(l) = n_{0,1}e^{-n_{0,1}l}$.

In the case $\phi_0 = -\phi_1$ this model yields the same low energy behaviour as above (Comtet et al. 1995). The parameter μ is now given by $\mu = \frac{n_1 - n_0}{2\phi_0}$. The main interest of this model is to provide a physical picture of the low energy behaviour.

For $\mu \neq 0$ the potential $V(x) = \phi^2 + \phi'$ is constant everywhere except at the positions where $\phi(x)$ has a discontinuity. One thus obtains a sequence of δ functions with alternating signs. The attractive δ potentials can support bound states which would have exactly zero energy if one would ignore the couplings to the other peaks. By taking carefully into account these couplings, one can recover the low energy power law behaviour (Comtet et al. 1995). The physical picture that emerges from this analysis is that the *low energy states are localized at the positions of the impurities*. Therefore this is just the opposite mechanism from that in the Lifshits case. It would be interesting to generalize this approach to higher dimensions where similar behaviour can also occur.

For $\mu = 0$, since the positive and negative δ functions play a symmetric role this argument doesn't apply anymore. A study of the low energy states on a finite interval shows that the existence of quasi zero-modes can account for the logarithmic behaviour of the density of states.

Imposing Dirichlet boundary conditions on a finite interval $[-R, R]$ one finds that the ground state energy is (Comtet et al. 1995, Monthus et al. 1996)

$$E_0(R, \{\phi\}) \simeq \frac{1}{\int_{-R}^{R} dx' \, \psi_0^2(x')} \left(\frac{1}{\int_{-R}^{0} \frac{dx}{\psi_0^2(x)}} + \frac{1}{\int_{0}^{R} \frac{dx}{\psi_0^2(x)}} \right).$$

This result is obtained by approximating the true ground state wave function near the boundaries by a suitable linear combination of the two linearly independent solutions of $H_+ \psi = 0$:

$$\psi_0(x) = e^{\int^x dx' \, \phi(x')} \tag{3}$$

and

$$\psi_1(x) = \psi_0(x) \int^x \frac{dx'}{\psi_0^2(x')}. \tag{4}$$

If $\phi(x)$ is a white noise or a random telegraph process with Poissonian lengths, the typical behaviour of $\psi_0(x)$, given by the central limit theorem, is $\psi_0(x) \sim e^{\pm\sqrt{x}}$. Replacing $\psi_0(x)$ by its typical behaviour one finds that the energy E_0 is exponentially small in the length of the system $E_0 \sim e^{-\sqrt{R}}$. Therefore a quasi zero mode of energy E has a typical spatial extension $2R$ such that $R \sim \ln^2 E$. Coming back to the whole line one finds that the number of such states per unit length is

$$N(E) = \frac{1}{2R} \sim \frac{1}{\ln^2 E}.$$

Obviously this argument may be generalized to any one-dimensional disordered system for which the zero energy wave functions can be expressed in terms of the potential. This is in particular the case of the Anderson model with off-diagonal disorder. The discrete Schrödinger equation may be written in the form

$$\beta_{n+1}\varphi_{n+1} + \beta_n\varphi_{n-1} = E\varphi_n \tag{5}$$

where β_n are random variables. The model of Dyson of an harmonic chain with random masses belongs to this class. For any configuration of the disorder one can write down two independent zero energy solutions. One of them is obtained by solving the recurrence relation (5) with $E = 0$. A zero energy state satisfying the boundary conditions $\varphi_0 = 1$ and $\varphi_1 = 0$ is

$$\varphi_{2n} = \prod_{k=1}^{n} \frac{\beta_{2k-1}}{\beta_{2k}}.$$

If the β_n are independent identically distributed random variables, the typical behaviour of φ_n, given by the central limit theorem, is again of the form $|\varphi_n| \sim e^{\pm\sqrt{n}}$. One will therefore get the same low energy behaviour as before. For earlier references see (Theodorou and Cohen 1976, Markos 1988, Bovier 1989); another derivation of the logarithmic singularity is given in (Eggarter and Riedinger 1978).

2.3   Remarks

1. All these arguments are based on typical realizations of disorder. There exist however certain quantities whose behaviour cannot be obtained by this type of reasoning. This is in particular the case of the average ground state energy. Its dependence on the size of the sample has been obtained in (Monthus et al. 1996). It is given by the stretched-exponential function $\langle E_0 \rangle = \exp\left(-R^\xi\right)$ where the exponent ξ depends only on the nature of the correlations in the potential (the Gaussian white noise corresponds to $\xi = 1/3$). In (Monthus et al. 1996) it is shown that this behaviour is indeed supported by atypical realizations of the random potential.

2. The existence of a singular behaviour in the density of states implies, by the Thouless formula, that there will be a corresponding singularity in the localization length $\lambda(E)$. This quantity indeed diverges as $\ln E$ which reflects the appearance of a critical state at $E = 0$. It was recently pointed out (Steiner et al. 1997, Balents and Fisher 1997) that there exists another length scale in the system - the correlation length which controls the decay of the two-point Green's function. It behaves like $\ln^2 E$ and thus diverges faster than the localization length. These results are in agreement with those presented in (Bouchaud et al. 1990); although the full correlations were not computed exactly, it is shown in this paper that the two-point Green's function is indeed given by

$$\overline{\langle x | \frac{1}{H-E} | y \rangle} \underset{E \to 0}{\simeq} \sum_n c_n e^{-\frac{\sigma \pi^2 (2m+1)^2}{2 \ln^2 E} |x-y|}$$

It would be interesting to compare this method with Berezinskiĭ's diagrammatic technique (Berezinskiĭ 1974) recently used in (Steiner et al. 1997).

3. The existence of a critical state at $E = 0$ is best understood when this model is reinterpreted in the context of classical diffusion in a random medium. A diffusive behaviour at large time requires the existence of an extended state at $E = 0$ (Tossatti 1990).

4. The thermodynamic properties of some one-dimensional spin systems with random exchange couplings can be reinterpreted by using a mapping of the spin system onto a model of free fermions. This approach, which can be traced back to the pioneering work of Lieb, Schultz and Mattis (Lieb et al. 1961), was used by Smith (Smith 1970) in the context of the X-Y model. Exact result for quantum phase transition in random X-Y spin chains with a comparison with the renormalization group approach (Fisher 1994) were obtained by McKenzie (McKenzie 1996). Quite recently, a similar approach with a different type of disorder was developed by Fabrizio and Mélin (Fabrizio and Mélin 1997). It is also worth mentioning the nice paper of Steiner, Fabrizio and Gogolin (Steiner et al. 1997) extending this analysis to the case of correlations and boundary effects.

3 Correlation Functions
of the Ground State Wave Function

The localization properties of the wave function can be characterized by the
density-density correlation function. Various techniques have been developed,
mainly by the Russian school (Lifshits et al. 1988), to compute such quantities in the weak disorder limit. In the supersymmetric model, one may take
advantage of the fact that the ground state wave function is known exactly
as a functional of the disorder. If the disordered potential $\phi(x)$ is white noise,
this allows one to compute the corresponding n-point function by using a
mapping with Liouville quantum mechanics. Such a calculation was recently
carried out by Shelton and Tsvelik (Shelton and Tsvelik 1997) for the case
$n = 2$ and $n = 3$ in the context of spin Peierls systems. An extension of
this result to arbitrary n is given below. We first consider periodic boundary
conditions. For completness we also give the corresponding formula due to
Broderix and Kree in the case of free boundary conditions (Broderix and
Kree 1995).

3.1 Periodic Boundary Conditions

We consider the supersymmetric Hamiltonian (1) in which we set $\frac{\beta U(x)}{2} \equiv$
$\int_0^x dx'\, \phi(x')$ for consistency with the notation of previous work (Monthus
and Comtet 1994, Comtet et al. 1996). We are interested in the following two
sections in the statistical properties of the zero mode wave function

$$\psi_0(x) = \frac{e^{\frac{\beta U(x)}{2}}}{\left[\int_0^L dx'\, e^{\beta U(x')}\right]^{1/2}} \tag{6}$$

when the disordered potential $\phi(x)$ is white noise. We consider a system of
length L and impose periodic boundary conditions. This is achieved if the
disordered potential is a Brownian bridge ($U(0) = U(L)$). The average over
the disorder is performed through the Wiener measure

$$\langle \cdots \rangle = \mathcal{N} \int_{U(0)=0}^{U(L)=0} \mathcal{D}U(x) \cdots e^{-\frac{1}{2\sigma}\int_0^L dx \left(\frac{dU(x)}{dx}\right)^2} \tag{7}$$

where $\mathcal{N}$ is a normalization to be determined. Our aim is to compute the
correlation functions of the square wave function $\psi_0^2(x)$:

$$C_n(x_1,\cdots,x_n) \equiv \left\langle |\psi_0(x_1)|^2 \cdots |\psi_0(x_n)|^2 \right\rangle. \tag{8}$$

In order to perform the average over $U(x)$ it is convenient to exponentiate the
denominator coming from the normalization of the wave function. By using
the integral representation of the Γ function, one gets

$$C_n(x_1,\cdots,x_n) = \frac{\mathcal{N}}{\Gamma(n)}\int_0^\infty dp\, p^{n-1}\int_{U(0)=0}^{U(L)=0}\mathcal{D}U(x)$$

$$\times\, e^{-\int_0^L dx\left[\frac{1}{2\sigma}\left(\frac{dU(x)}{dx}\right)^2+pe^{\beta U(x)}\right]+\beta U(x_1)+\cdots+\beta U(x_n)}$$

This expression establishes a link with one-dimensional Liouville quantum mechanics. For a detailed discussion of this model we refer to (Monthus and Comtet 1994) and (Kolokolov 1993, Kolokolov 1994). Considerable simplification occurs if one makes use of the fact that, in this theory, a change in the coupling constant can be interpreted as a translation in U space. This suggests the change of variable $p = \alpha e^{\beta U}$, where $\alpha \equiv \frac{\sigma\beta^2}{2}$, which leads to

$$C_n(x_1,\cdots,x_n) = \frac{2\sqrt{\pi L}\,\alpha^{n+\frac{1}{2}}}{\Gamma(n)}\int_{-\infty}^{+\infty}dU\int_{U(0)=U}^{U(L)=U}\mathcal{D}U(x)\,e^{-S_L}e^{\beta\sum_{i=1}^n U(x_i)}$$

$$(9)$$

where the action

$$S_L = \int_0^L dx\left[\frac{1}{2\sigma}\left(\frac{dU(x)}{dx}\right)^2+\alpha e^{\beta U(x)}\right] \qquad (10)$$

is associated with the Liouville Hamiltonian

$$H_L = -\frac{\sigma}{2}\frac{d^2}{dU^2}+\alpha e^{\beta U}. \qquad (11)$$

The role of the normalization constant $\mathcal{N}$ is to insure the relation $\int dx_1\cdots dx_n\, C_n(x_1,\cdots,x_n) = 1$. Since $\mathcal{N}$ is independent of n this relation may be used in the case $n = 1$ to find $\mathcal{N}$.

One may express the correlation function in terms of the Liouville propagator $G_x(U,U') \equiv \langle U|e^{-xH_L}|U'\rangle$. Choosing $L \geq x_1 \geq x_2 \geq \cdots \geq x_n \geq 0$ where $x_{ij} \equiv x_i - x_j$ one gets

$$C_n(x_1,\cdots,x_n) = \frac{2\sqrt{\pi L}\,\alpha^{n+\frac{1}{2}}}{\Gamma(n)}\int_{-\infty}^{+\infty}dU_1 dU_2\cdots dU_n\, G_{L-x_{1n}}(U_n,U_1)e^{\beta U_1}$$

$$G_{x_{12}}(U_1,U_2)e^{\beta U_2}\cdots G_{x_{n-1n}}(U_{n-1},U_n)e^{\beta U_n}.$$

Using the eigenstates $|k\rangle$ of the Liouville Hamiltonian[1] gives

$$C_n(x_1,\cdots,x_n) = \frac{2\sqrt{\pi L}\,\alpha^{n+\frac{1}{2}}}{\Gamma(n)}\int_0^{+\infty}dk_1\cdots dk_n\,\langle k_n|e^{\beta U}|k_1\rangle\cdots\langle k_{n-1}|e^{\beta U}|k_n\rangle$$

$$e^{-\frac{\alpha(L-x_{1n})}{4}k_1^2-\cdots-\frac{\alpha x_{n-1n}}{4}k_n^2}. \qquad (12)$$

Knowledge of the wave functions gives the matrix elements

---

[1] The Liouville Hamiltonian has a continuous spectrum $H_L\psi_k(U) = \frac{\alpha k^2}{4}\psi_k(U)$ where the wave function is $\psi_k(U) = \frac{\sqrt{\beta k\sinh\pi k}}{\pi}K_{ik}\left(2e^{\frac{\beta U}{2}}\right)$.

$$\langle k \,|e^{\beta U}|\, k' \rangle = \frac{1}{8}\sqrt{kk' \sinh \pi k \sinh \pi k'}\,\frac{k^2 - k'^2}{\cosh \pi k - \cosh \pi k'}. \tag{13}$$

Expression (12) allows one to get the long range behaviour (when all the distances involved are large compared to $\frac{1}{\alpha}$). Using the Laplace method one eventually finds

$$C_n(x_1,\cdots,x_n) \simeq \frac{1}{(4\pi\alpha)^{\frac{n-1}{2}}\,\Gamma(n)}\frac{\sqrt{L}}{[(L-x_1+x_n)(x_1-x_2)\cdots(x_{n-1}-x_n)]^{3/2}}. \tag{14}$$

Despite the coefficients being different to those found by Shelton and Tsvelik (Shelton and Tsvelik 1997) when $n = 2$ and $n = 3$, we get the same behaviour as a function of the distances. This expression shows the existence of long range correlations. A nice interpretation of the algebraic tail was recently given in (Laloux and Le Doussal 1997). It is suggested that the exponent $3/2$ is associated with configurations of the disorder where $U(x)$ returns to its starting point. It is interesting to point out that the same exponent appears in conventional localization theory. In this case wave functions with the same (Berezinskiǐ 1974, Gogolin 1975) or nearly equal (Gor'kov et al. 1983) energy are weakly correlated at large distances. The correlation function decays exponentially with the power law preexponential factor $\left(\frac{1}{x}\right)^{3/2}$.

3.2   Free boundary conditions

Computation of correlation functions of the type in (8) was first performed by Broderix and Kree (Broderix and Kree 1995) in a different context: $|\psi_0(x)|^2$ is interpreted as the equilibrium Gibbs measure for the potential $U(x)$. Setting $U(0) = 0$ and letting $U(L)$ be free they get

$$C_n^{\mathrm{BK}}(x_1,\cdots,x_n) \simeq \frac{1}{\pi(4\pi\alpha)^{\frac{n-1}{2}}\,\Gamma(n)}\frac{1}{\sqrt{x_n(L-x_1)}}\frac{1}{[(x_1-x_2)\cdots(x_{n-1}-x_n)]^{3/2}} \tag{15}$$

which is very similar to (14). The different boundary prescription therefore induces a slight change in the x dependance near the boundary.

4   Multifractality

In a recent series of publications, de Chamon et al. (de Chamon et al. 1996) have considered a two-dimensional localization model which exhibits a localization transition at zero energy. They consider a two-dimensional Dirac Hamiltonian in a random magnetic field $B(\mathbf{r}) = \Delta\phi(\mathbf{r})$. In this case the zero energy solution (Aharonov and Casher 1979) can be constructed explicitly for any realization of $\phi(\mathbf{r})$. They consider the particular "ground state" solution $\psi(\mathbf{r}) \propto (\mathrm{e}^{-\phi(\mathbf{r})}, 0)$. For a Gaussian disorder $P[\phi] = \exp -\frac{1}{2g}\int d\mathbf{r}\,(\boldsymbol{\nabla}\phi)^2$

the successive moments of $\psi_(\mathbf{r})$ are encoded in the partition function $\mathcal{Z}(q) \equiv \int d\mathbf{r} \, e^{-2q\phi(\mathbf{r})}$ since $\int d\mathbf{r} \, \psi^{2q}(\mathbf{r}) = \frac{\mathcal{Z}(q)}{\mathcal{Z}(1)^q}$. The multifractal exponents of the wave function can be obtained by using a formal equivalence with the problem of a directed polymer on the Cayley tree (Derrida and Spohn 1988).

In the following we consider a one-dimensional version of this model with, however, a different type of disorder and take into account the wave function normalization as in (Kogan et al. 1996). The starting point is the two-dimensional Euclidean Dirac operator

$$i\slashed{D} = i\sigma_1 \left(\partial_y + iA_y\right) + i\sigma_2 \left(\partial_x + iA_x\right)$$

σ_i are the Pauli matrices and the gauge field is given by

$$\begin{cases} A_y = f(x) \\ A_x = 0. \end{cases}$$

We may take eigenstates of the form $\psi(x,y) = \left(\chi(x)e^{i\omega y}, 0\right)$. The eigenvalue equation $-\slashed{D}^2\chi = E^2\chi$ then becomes

$$\left[-\partial_x^2 + (\omega + f(x))^2 - \sigma_3 f'(x)\right]\chi(x) = E^2\chi(x).$$

We are thus led to a one-dimensional Schrödinger equation with a supersymmetric potential $V(x) = \varphi^2(x) - \sigma_3\varphi'(x)$ where $\varphi(x) = \omega + f(x)$.

In an earlier work by one of us (Comtet et al. 1988) this approach was used to study the density of states of the two-dimensional Dirac operator. If $f(x)$ is white noise we may use the density of states of the one-dimensional problem and integrate over the free motion on the y axis. The resulting expression displays an enhancement at low energy as compared to the free case.

Here our purpose is to characterize the fluctuation properties of the ground state wave function $\chi(x) = e^{-\int dy\varphi(y)}$. By using two different approaches for two boundary prescription we compute exactly the successive moments of the normalized ground state

$$\psi_0(x) = \frac{\chi(x)}{\left[\int_0^L dx' \chi^2(x')\right]^{1/2}}$$

In order to keep unified conventions we will parametrize $\psi_0(x)$ as in equation (6).

4.1   Moments of $\psi_0(x)$ for periodic boundary conditions

If the random potential obeys periodic boundary conditions, the situation is the one described in section 3. The moment of order $2n$ is then related to the n-point correlation function at coinciding points

$$\left\langle |\psi_0(x)|^{2n} \right\rangle = C_n(x, \cdots, x).$$

In terms of the Liouville propagator this is expressed as

$$\langle |\psi_0(x)|^{2n} \rangle = \frac{2\sqrt{\pi L}\,\alpha^{n+\frac{1}{2}}}{\Gamma(n)} \int_{-\infty}^{+\infty} dU_1\, G_L(U_1, U_1) e^{n\beta U_1}$$

$$= \frac{2\sqrt{\pi L}\,\alpha^{n+\frac{1}{2}}}{\Gamma(n)} \int_0^\infty dk\, \langle k\,|e^{n\beta U}|\,k \rangle e^{-\frac{\alpha L}{4}k^2}$$

where the matrix element is

$$\langle k\,|e^{n\beta U}|\,k \rangle = \frac{\Gamma(n)^2}{4\pi\Gamma(2n)} \prod_{m=0}^{n-1} (m^2 + k^2)\,.$$

The moments are eventually given by

$$\langle |\psi_0(x)|^{2n} \rangle = \frac{\alpha^n}{2^n(2n-1)!!} \sum_{m=1}^n a_m^n 2^m (2m-1)!! \frac{1}{(\alpha L)^m} \tag{16}$$

where the coefficients a_m^n are defined by the equation $\prod_{m=0}^{n-1}(m^2 + X) = \sum_{m=1}^n a_m^n X^m$. For example $a_n^n = 1$ and $a_1^n = \frac{n(n-1)(2n-1)}{6}$.

One may extract from this expression the asymptotic dependance of the moments (when L is large compared to $\frac{1}{\alpha}$)

$$\langle |\psi_0(x)|^{2n} \rangle \simeq \frac{\alpha^{n-1} n(n-1)}{2^n 3(2n-3)!!} \frac{1}{L}\,. \tag{17}$$

4.2   Moments of $\psi_0(x)$ for free boundary conditions

If one leaves $U(L)$ free instead of imposing periodic boundary, then the Brownian that enters in the wave function (6) is no longer a Brownian bridge but a free Brownian motion starting from $U(0) = 0$.

In the average over the disordered potential one must now take into account all the Brownian paths starting from 0 without any restriction on the final point. Except for this slight modification the formalism is the same as in previous sections. One is led to

$$\langle \psi_0^{2n}(x_1) \rangle = \frac{\beta\alpha^n}{\Gamma(n)} \int_{-\infty}^{+\infty} dU \int_{U(0)=U} \mathcal{D}U(x)\, e^{-S_L} e^{n\beta U(x_1)}$$

$$= \frac{\beta\alpha^n}{\Gamma(n)} \int_{-\infty}^{+\infty} dU\, dU_1\, dU_2\, G_{L-x}(U_2, U_1) e^{n\beta U_1} G_x(U_1, U)\,.$$

It is interesting to present another derivation of the moments using the language of previous works (Oshanin et al. 1993a, Monthus and Comtet 1994, Comtet et al. 1996) devoted to the study of the statistical properties of the exponential functional $Z_L^{(\mu)} \equiv \int_0^L dx\, e^{-(\mu\alpha x + \sqrt{2\alpha}W(x))}$. Here μ is the drift for the Brownian motion $\beta U(x) = -(\alpha \mu x + \sqrt{2\alpha} W(x))$ and $W(x)$ is a Brownian motion of average 0 and variance 1 defined on $[0, L]$.

We will prove that the moments of $\psi_0(x)$ can be expressed in terms of the characteristic function $\phi^{(\mu)}(p, L) \equiv \langle e^{-pZ_L^{(\mu)}} \rangle$ (Laplace transform of the distribution law of $Z_L^{(\mu)}$). For this purpose we may note that $\psi_0(x)$ involves two exponential functionals. We may write

$$\psi_0^2(x) = \frac{1}{\int_0^L dy\, e^{\beta(U(y) - U(x))}}$$

and separate the denominator into two parts

$$\int_0^L dy\, e^{\beta(U(y) - U(x))} = \int_0^x dy\, e^{-[-\mu\alpha y + \sqrt{2\alpha} B(y)]} + \int_0^{L-x} dy\, e^{-[\mu\alpha y + \sqrt{2\alpha} \tilde{B}(y)]}$$

where $B(y) \equiv W(x - y) - W(x)$ and $\tilde{B}(y) \equiv W(x + y) - W(x)$, respectively defined on $y \in [0, x]$ and $y \in [0, L-x]$, are two independent Brownian motions starting from zero $B(0) = \tilde{B}(0) = 0$. The denominator is then a sum of two statistically independent exponential functionals $Z_x^{(-\mu)}$ and $\tilde{Z}_{L-x}^{(\mu)}$.

The moments may then be rewritten as

$$\langle \psi_0^{2n}(x) \rangle = \left\langle \frac{1}{\left(Z_x^{(-\mu)} + \tilde{Z}_{L-x}^{(\mu)}\right)^n} \right\rangle$$

$$= \frac{1}{\Gamma(n)} \int_0^\infty dp\, p^{n-1} \left\langle e^{-pZ_x^{(-\mu)}} \right\rangle \left\langle e^{-p\tilde{Z}_{L-x}^{(\mu)}} \right\rangle$$

$$= \frac{1}{\Gamma(n)} \int_0^\infty dp\, p^{n-1} \phi^{(-\mu)}(p, x)\, \phi^{(\mu)}(p, L - x).$$

Where the characteristic functions are given in (Monthus and Comtet 1994, Comtet et al. 1996). In the case $\mu = 0$ which is of interest for us

$$\phi^{(0)}(p, L) = \frac{2}{\pi} \int_0^\infty ds\, \cosh \frac{\pi s}{2} K_{is}\left(2\sqrt{\frac{p}{\alpha}}\right) e^{-\frac{\alpha L}{4} s^2} \tag{18}$$

One may extract from this expression the dominant behaviour when $\alpha x \gg 1$ and $\alpha(L - x) \gg 1$

$$\langle \psi_0^{2n}(x) \rangle \simeq \left(\frac{\alpha}{2}\right)^{n-1} \frac{\Gamma(n)^2}{\pi(2n-1)!!} \frac{1}{\sqrt{x(L-x)}}. \tag{19}$$

One may check that this result agrees with the one of Broderix and Kree (15) in the case $n = 1$. Whenever x belongs to the interval $[0, L]$, but is not on the edges, the moments still behave as $\langle \psi_0^{2n}(x) \rangle \sim \frac{1}{L}$.

We may also explore the behaviour of the moments of the wave function on the edges $\langle \psi_0^{2n}(0) \rangle = \langle \psi_0^{2n}(L) \rangle$. This quantity is equal to the moments of the partition function $Z_L^{(0)}$ for negative orders

$$\langle\psi_0^{2n}(0)\rangle = \left\langle\frac{1}{(Z_L^{(0)})^n}\right\rangle = \frac{1}{\Gamma(n)}\int_0^\infty dp\,p^{n-1}\phi^{(0)}(p,L).$$

We may easily extract the dominant behaviour of this quantity with the help of (18) and eventually find

$$\langle\psi_0^{2n}(0)\rangle \simeq \frac{\alpha^n\,\Gamma(n)}{\sqrt{\pi\alpha L}}. \tag{20}$$

This shows that the wave function fluctuates more on the edges of the interval (when $x \ll \frac{1}{\alpha}$ or $L-x \ll \frac{1}{\alpha}$) than in the bulk (when $x \gg \frac{1}{\alpha}$ and $L-x \gg \frac{1}{\alpha}$).

The square of the wave function on the edge is interpreted in the problem of classical diffusion in the random potential $U(x)$, as the steady current density when a constant density of particle is imposed at $x = 0$ in the presence of a trap at $x = L$. The distribution of the steady current was found in (Oshanin et al. 1993b, Monthus and Comtet 1994). Setting $J \equiv \frac{1}{\alpha}\psi_0^2(0)$ the distribution of J is a log-normal law for small values of J

$$P(J) \underset{J\to 0}{\simeq} \frac{1}{2\sqrt{\pi\alpha L}}e^{-\frac{1}{4\alpha L}\ln^2 J}$$

and possesses an exponential tail for large values of J

$$P(J) \underset{J\to\infty}{\simeq} \frac{1}{\sqrt{\pi\alpha L}}\frac{1}{J}e^{\frac{\pi^2}{4\alpha L}-J}$$

which is responsible of the behaviour of the moments given in (20).

4.3   Discussion of the results

We have just seen that the one-dimensional model is closely related to the model studied in (de Chamon et al. 1996). Since the one-dimensional model is exactly solvable, it is interesting to study its multifractal properties, although one cannot expect it to give the same behaviour. In fact, in both models the potential is long-range correlated; though in $d = 2$ the correlations are logarithmic whereas in $d = 1$ they are linear.

We define the scaling exponent $\tilde{\tau}(q)$ which characterizes the behaviour of the critical wave function at $E = 0$ as

$$\langle|\psi(\mathbf{r})|^{2q}\rangle \underset{L\to\infty}{\sim} L^{-d-\tilde{\tau}(q)}.$$

Using equation (17) we get $\tilde{\tau}(q) = 0$. This exponent is the scaling exponent of the average moments of the wave function. We may introduce a slightly different scaling exponent

$$\tau(q) = \lim_{L\to\infty}\frac{\langle\ln\int d\mathbf{r}\,|\psi_0(\mathbf{r})|^{2q}\rangle}{\ln(1/L)},$$

which is the mean exponent of the critical wave function. In terms of the partition function $Z_L^{(0)}(\beta) = \int_0^L dx\, e^{\beta U(x)}$ introduced before, it may be rewritten

$$\tau(q) = \lim_{L \to \infty} \frac{\left\langle \ln Z_L^{(0)}(q\beta) \right\rangle - q \left\langle \ln Z_L^{(0)}(\beta) \right\rangle}{\ln(1/L)}.$$

The mean free energy that appears has been calculated in (Comtet et al. 1996) to be

$$\left\langle \ln Z_L^{(0)}(\beta) \right\rangle = 2\sqrt{\frac{\alpha L}{\pi}} + C - \ln \alpha - \frac{\pi}{3\sqrt{\alpha L}} + O\left(\frac{1}{(\alpha L)^{3/2}}\right),$$

where $C = -\Gamma'(1)$ is the Euler-Mascheroni constant. Again, one eventually gets a scaling exponent $\tau(q) = 0$ which agrees with the behaviour found in (de Chamon et al. 1996) in the strong disorder regime. By using the mapping with the random directed polymer model (Derrida and Spohn 1988), de Chamon et al. have shown that this regime corresponds in fact to the low temperature phase of this model. The existence of such a link also appears in the one-dimensional case. A comparison of the moments of $Z_L^{(0)}(\beta)$ with those of the random energy model (REM) (Derrida 1981) reveals some striking similarities. For the one-dimensional case the expression of the moments given in (Monthus and Comtet 1994) read

$$\langle (Z_L^0)^n \rangle = \frac{1}{\alpha^n} \left(\frac{\Gamma(n)}{\Gamma(2n)} \sum_{k=1}^{n} (-1)^{n-k} e^{\alpha L k^2} C_{2n}^{k+n} + \frac{(-1)^n}{n!} \right).$$

In particular, for large L one obtains

$$\langle (Z_L^0)^n \rangle \underset{\alpha L \gg 1}{\simeq} \frac{1}{\alpha^n} \frac{\Gamma(n)}{\Gamma(2n)} e^{\alpha L n^2}.$$

These moments grow in the same manner as in the REM. The main difference is that, unlike in the REM, there is here no transition above which the behaviour would change from $e^{\alpha L n^2}$ to $e^{\alpha L n}$. In some sense one can consider that the transition temperature is sent to infinity. This explains why, in this one-dimensional case, one only probes the low temperature phase. It would be extremely interesting to explore intermediate cases with weaker correlations, as recently suggested in (Bouchaud and Mézard 1997).

Acknowledgments

Many of these results are based on earlier works done in collaboration with Jean-Philippe Bouchaud, Jean Desbois, Pierre Le Doussal, Antoine Georges, Gleb Oshanin, Cécile Monthus and Marc Yor. A great thanks to all of them. We especially thank Cécile Monthus for drawing our attention to the recent literature and for interesting remarks. We thank Gleb Oshanin and Serguei Nechaev for interesting discussion.

References

Aharonov, Y., Casher, A. (1979): Ground state of a spin-1/2 charged particle in a two-dimensional magnetic field, Phys. Rev. A $19(6)$, 2461–2462.

Balents, L., Fisher, M. P. (1997): Delocalization transition via supersymmetry in one dimension, preprint , cond-mat/9706069.

Berezinskiĭ, V. (1974): Kinetics of a quantum particle in a one-dimensional random potential, Sov. Phys. JETP $38(3)$, 620.

Benderskiĭ, M., Pastur, L. (1974): Sov. Phys. JETP 40, 241.

Bouchaud, J.-P., Comtet, A., Georges, A., Le Doussal, P. (1987): Europhys. Lett. 3, 653.

Bouchaud, J.-P., Comtet, A., Georges, A., Le Doussal, P. (1990): Classical diffusion of a particle in a one-dimensional random force field, Ann. Phys. (N.Y.) 201, 285–341.

Bouchaud, J.-P., Georges, A. (1990): Anomalous diffusion in disordered media: Statistical mechanisms, models and physical application, Phys. Rep. 195, 267.

Bouchaud, J.-P., Mézard, M. (1997): Universality classes for extreme value statistic, preprint , cond-mat/9707047.

Bovier, A. (1989): J. Stat. Phys. 56, 645.

Broderix, K., Kree, R. (1995): Thermal equilibrium with the Wiener potential: testing the replica variational approximation, Europhys. Lett. 32, 343–348.

Casher, A., Neuberger, H. (1984): Phys. Rev. B 139, 67.

de C. Chamon, C., Mudry, C., Wen, X.-G. (1996): Localization in two dimensions, Gaussian field theory and Multifractality, Phys. Rev. Lett. 77, 4194–4197.

Comtet, A., Georges, A., Le Doussal, P. (1988): Exact density of states of a two-dimensional Dirac operator in a random magnetic field, Phys. Lett. B $208(3/4)$, 487.

Comtet, A., Desbois, J., Monthus, C. (1995): Localization properties in one-dimensional disordered supersymmetric quantum mechanics, Ann. Phys. (N.Y.) 239, 312–350.

Comtet, A., Monthus, C., Yor, M. (1996): Exponential functionals of Brownian motion and disordered systems, to appear in J. of Appl. Prob. , preprint cond-mat/9601014.

Cooper, F., Khare, A., Sukhatme, U. (1995): Supersymmetry and quantum mechanics, Phys. Rep. 251, 267–385.

Derrida, B. (1981): Random-energy model: an exactly solvable model of disordered systems, Phys. Rev. B $24(5)$, 2613–2626.

Derrida, B., Spohn, H. (1988): Polymers on disordered trees, spin glasses and traveling waves, J. Stat. Phys. 51, 817.

Dyson, F. J. (1953): Phys. Rev. 92, 1331.

Eggarter, T., Riedinger, R. (1978): Phys. Rev. 18, 569.

Fabrizio, M., Mélin, R. (1997): Coexistence of antiferromagnetism and dimerization in a disordered spin-Peierls model: exact results, preprint , cond-mat/9701149.

Fisher, D. (1994): Phys. Rev. B $50(6)$, 3799.

Floratos, E., Stern, J. (1980): Phys. Lett. B 119, 419.

Gogolin, A., Melnikov, V., Rashba, E. (1975): Zh. Eksp. Teor. Fiz. 69, 328.

Gor'kov, L., Dorokhov, O., Prigara, F. (1983): Sov. Phys. JETP $58(4)$, 852.

Kogan, I. I., Mudry, C., Tsvelik, A. M. (1996): Liouville Theory as a model for prelocalized states in disordered conductors, Phys. Rev. Lett. $77(4)$, 707.

Kolokolov, I. (1993): The method of functional integration for a one-dimensional localization, higher correlators and the average current flowing in a mesoscopic ring in an arbitrary magnetic field, Sov. Phys. JETP **76**, 1099.

Kolokolov, I. (1994): Europhys. Lett. **28**, 193.

Laloux, L., Le Doussal, P. (1997): Aging and diffusion in low dimensional environments, preprint , cond-mat/9705249.

Lieb, E., Schultz, T., Mattis, D. (1961): Ann. Phys. (N.Y.) **16**, 407.

Lifshits, I. M. (1965): Energy spectrum structure and quantum states of disordered condensed systems, Sov. Phys. Usp. **18**(4), 549.

Lifshits, I. M., Gredeskul, S. A., Pastur, L. A. (1988): *Introduction to the theory of disordered systems*, John Wiley & Sons.

Luck, J.-M. (1992): *Systèmes désordonnés unidimensionnels*, Aléa Saclay.

Luttinger, J., Waxler, R. (1987): Ann. Phys. (N.Y.) **175**, 319.

McKenzie, R. H. (1996): Phys. Rev. Lett. **77**(23), 4804.

Markos, P. (1988): Z. Phys. B - Cond. Matt. **73**, 17.

Monthus, C., Comtet, A. (1994): On the flux distribution in a one-dimensional disordered system, J. Phys. I (France) **4**, 635–653.

Monthus, C., Oshanin, G., Comtet, A., Burlatsky, S. (1996): Sample-size dependence of the ground-state energy in a one-dimensional localization problem, Phys. Rev. E **54**, 231–242.

Oshanin, G., Mogutov, A., Moreau, M. (1993a): Steady flux in a continuous-space Sinai chain, J. Stat. Phys. **73**, 379.

Oshanin, G., Burlatsky, S., Moreau, M., Gaveau, B. (1993b): Behaviour of transport characteristics in several one-dimensional disordered systems, Chem. Phys. **177**, 803.

Ovchinnikov, A., Erikmann, N. (1977): Zh. Eksp. Teor. Fiz. **73**, 650.

Neuberger, H. (1982): Phys. Lett. B **112**(4/5), 341.

Shelton, D. G., Tsvelik, A. M. (1997): An effective theory for midgap states in doped spin ladder and spin-Peierls systems: Liouville quantum mechanics, preprint , cond-mat/9704115.

Smith, E. (1970): J. Phys. C **3**, 1419.

Steiner, M., Fabrizio, M., Gogolin, A. O. (1997): Random mass Dirac fermions in doped spin-Peierls and spin-ladder systems: one-particle properties and boundary effects, preprint , cond-mat/9706096.

Theodorou, G., Cohen, M. (1976): Phys. Rev. B **13**, 4597.

Tossatti, E., Vulpiani, A., Zannetti, M. (1990): Physica A **164**, 705.

Yor, M. (1992): Adv. Appl. Prob. **24**, 509.

Witten, E. (1981): Nucl. Phys. B [FS] **188**, 513.

Zero Modes and Self-Isospectral Potentials in Periodic Supersymmetric Quantum Mechanics

Gerald Dunne[1] and Joshua Feinberg[2]

[1] Physics Department, University of Connecticut, Storrs, CT 06269
[2] Institute for Theoretical Physics, University of California, Santa Barbara, CA 93106

Abstract. In standard discussions of nonperiodic supersymmetric quantum mechanics either one or neither (but not both) of the isospectral pair of potentials has a zero mode. In contrast, in supersymmetric quantum mechanical models with periodic potentials it is possible for *both* isospectral potentials to support zero modes. Thus it is possible to have supersymmetry unbroken and yet also have a vanishing Witten index. We present some explicit exactly soluble examples for which the isospectral potentials have identical band spectra, and which are "self-isospectral" in the sense that the potentials have identical shape, but are translated by one half period relative to one another.

Supersymmetry and supersymmetry breaking are fundamental issues in theoretical particle physics, and supersymmetric (SUSY) quantum mechanics provides an important testing ground for both physical and computational aspects of SUSY theories (Witten (1981),Witten (1982),Cooper *et al* (1995)). There are also many applications to the theory of solitons (Rajaraman (1982),Rebbi and Soliani (1984)). Of particular interest for particle physics are possible mechanisms for breaking SUSY dynamically. Typically, one considers models with discrete spectra, and then the Witten index, which characterizes the difference between the number of bosonic and fermionic zero modes, may be used to indicate whether or not SUSY is broken (Witten (1981),Witten (1982)). Interesting subtleties arise for potentials with continuum states (Akhoury and Comtet (1984)) or with singularities (Jevicki and Rodrigues (1984),Casahorran and Nam (1991),Das and Pernice (1996)).

In this paper we consider SUSY quantum mechanics for *periodic* potentials (which therefore have band spectra). The main new feature is that it is possible for the periodic isospectral bosonic and fermionic potentials to have *exactly* the same spectrum, *including zero modes*. This is in contrast to the usual (nonperiodic and fast decaying) case for which at most one potential of an isospectral pair can have a zero mode.

Consider one dimensional SUSY quantum mechanical models on the real line. The bosonic and fermionic Hamiltonians $H_\pm$ correspond to an isospectral pair of potentials $V_\pm(x)$ defined in terms of the "superpotential" $W(x)$ as

$$V_\pm(x) = W^2(x) \pm W'(x) \tag{1}$$

The Hamiltonians may be factorized into products of hermitean conjugate operators as

$$H_+ = [\frac{d}{dx} + W(x)][-\frac{d}{dx} + W(x)], \quad H_- = [-\frac{d}{dx} + W(x)][\frac{d}{dx} + W(x)] \tag{2}$$

which indicates that $H_\pm$ are formally positive operators. Thus, their energy spectrum cannot go below zero. [We exclude singular potentials from the present discussion.] The factorization (2) also implies that $V_\pm$ have (almost) the same spectrum because there is a one-to-one mapping between the energy eigenstates $\psi_E^{(\pm)}$:

$$\psi_E^{(+)} = \frac{1}{\sqrt{E}} \left(\frac{d}{dx} + W(x) \right) \psi_E^{(-)}; \quad \psi_E^{(-)} = \frac{1}{\sqrt{E}} \left(-\frac{d}{dx} + W(x) \right) \psi_E^{(+)} \tag{3}$$

The caveat 'almost' is needed above because this mapping between states does not apply to the "zero modes" (eigenstates with $E = 0$), which due to the positivity of $H_\pm$, are the lowest possible states in the spectrum. From (2) it is easy to see that the Schrödinger equation $[-\partial_x^2 + V_\pm(x)]\psi_E^{(\pm)} = E\psi_E^{(\pm)}$ has zero modes

$$\psi_0^{(\pm)}(x) = e^{\pm \int^x W} \tag{4}$$

provided these functions $\psi_0^{(\pm)}$ belong to the Hilbert space. SUSY is said to be unbroken if at least one of the $\psi_0^{(\pm)}$ is a true zero mode. Otherwise, SUSY is said to be broken dynamically. In the broken SUSY case there are no zero modes and so the spectra of $V_\pm$ are identical [due to the mapping (3)].

In the "standard cases" (Witten (1981),Witten (1982)), in which $V_\pm(x)$ tend to positive asymptotic values as $x \to \pm\infty$, this means that $\psi_0^{(\pm)}$ must be normalizable in order to be true zero modes. But it is clear that in these cases, at most only one of the functions $\psi_0^{(\pm)}$ may be normalizable. Thus the spectra of the potentials $V_\pm$ coincide except possibly for the zero energy ground state level. For example, if the (well-behaved) superpotential tends to asymptotic values with opposite signs as $x \to \pm\infty$, then one of the zero modes in (4) is normalizable, and SUSY is unbroken; superpotentials that are odd functions, $W(-x) = -W(x)$, belong to this class. Dynamically broken SUSY occurs when the superpotential tends to asymptotic values with equal signs as $x \to \pm\infty$; superpotentials that are even in x are of this type.

Two simple representative examples of unbroken SUSY are: (i) $W(x) = x$, which gives the harmonic oscillator. Of the two functions $\psi_0^{(\pm)} = e^{\pm x^2/2}$, clearly only $\psi_0^{(-)}$ is normalizable and hence a zero mode. (ii) $W(x) = j \tanh x$ (with j a positive integer), which gives the Pöschl-Teller potentials

$$V_\pm = j^2 - j(j \mp 1)sech^2 x \tag{5}$$

Once again, of the two possibilities $\psi_0^{(\pm)} = [\cosh x]^{\pm j}$, only $\psi_0^{(-)}$ is normalizable. V_- has j discrete bound states [with energies $E_n = n(2j - n)$ for $n = 0, 1, \ldots, j - 1$] and a continuum beginning at $E = j^2$; on the other hand, V_+ has $j - 1$ discrete bound states [with energies $E_n = n(2j - n)$ for $n = 1, \ldots, j - 1$] and a continuum beginning at $E = j^2$. The two spectra coincide manifestly, except for the zero mode.

Now consider the superpotential $W(x)$ to be periodic, with period L: $W(x + L) = W(x)$. The potentials $V_{\pm}(x)$ in (1) are therefore also periodic with period L. From the Bloch-Floquet theory (Kittel (1976)), the Hilbert space consists of quasi-periodic functions: functions that satisfy $\psi_k(x + L) = (\exp ikL)\psi_k(x)$, where the real quantity k is the crystal momentum.

From (4) we have $\psi_0^{(\pm)}(x + L) = e^{\pm\phi_L}\psi_0^{(\pm)}(x)$ where the real constant ϕ_L is given by

$$\phi_L = \int_x^{x+L} W(y)dy \tag{6}$$

For either one of the functions $\psi_0^{(\pm)}$ to belong to the Hilbert space, we must identify $\pm\phi_L = ikL$. But ϕ_L is real, which means that $\phi_L = kL = 0$. Thus, the two functions $\psi_0^{(\pm)}$ either *both* belong to the Hilbert space, in which case they are strictly periodic with period L: $\psi_0^{(\pm)}(x + L) = \psi_0^{(\pm)}(x)$, or (when $\phi_L \neq 0$) *neither of them* belongs to the Hilbert space. Note that this is the exact opposite of the situation for nonperiodic potentials where if $\psi_0^{(\pm)}$ is a zero mode of $V_{\pm}$, then $\psi_0^{(\mp)}$ is *not* a zero mode of $V_{\mp}$.) Thus, in the periodic case the spectra of V_+ and V_- match *completely*.

This conclusion is valid provided the $\psi_0^{(\pm)}$ do not have nodes (i.e. zeros, which means that the other function has poles). This condition is violated for example in the case of the Scarf potential (Scarf (1958)), $V(x) \sim cosec^2(x)$, for which $W(x) = j\cot x$. We shall exclude such singular potentials from the present discussion, but note that these deserve further study - even in the nonperiodic case, singular superpotentials naturally exhibit interesting properties (Jevicki and Rodrigues (1984),Casahorran and Nam (1991)).

To summarize, we see that

$$\phi_L = \int_0^L W(y)dy = 0 \tag{7}$$

is a necessary condition for unbroken SUSY, and when this condition is satisfied, the bosonic and fermionic sectors have identical spectra, including zero modes. The Witten index then vanishes.

It is instructive to consider some simple special classes of periodic superpotentials which satisfy (7). First, suppose the superpotential is antisymmetric on a half-period:

$$W(x + \frac{L}{2}) = -W(x), \tag{8}$$

Then, from (1) and (8) we obtain

$$V_\pm(x + \frac{L}{2}) = V_\mp(x) \tag{9}$$

The potentials $V_\pm$ are simply translations of one another by half a period, and thus are essentially identical in shape. Therefore, they must support exactly the same spectrum, as SUSY indeed tells us they do. We refer to such a pair of isospectral $V_\pm$ that are identical in shape as "self-isospectral". A simple example of a superpotential of this type is $W(x) = \sin x$, with $V_+(x) = \sin^2(x) + \cos x = V_-(x + \pi)$.

Second, consider periodic superpotentials that are even functions of x:

$$W(-x) = W(x) \tag{10}$$

but which also satisfy the condition $\phi_L = 0$ (by subtracting an appropriate constant, any even $W(x)$ can be brought into this class). The function $dW(x)/dx$ is odd and so (1) implies

$$V_\pm(-x) = V_\mp(x) \tag{11}$$

The two potentials are then simply reflections of one another. They have the same shape and therefore give rise to exactly the same spectrum, as we know from SUSY. Such potentials are also "self-isospectral". A simple example of a superpotential of this type is $W(x) = \cos x$, with $V_+(x) = \cos^2(x) - \sin x = V_-(-x)$.

Third and last, consider periodic superpotentials that are odd functions of x:

$$W(-x) = -W(x) \tag{12}$$

Then $\phi_L = 0$ is satisfied trivially. The function $dW(x)/dx$ is even and thus $V_\pm(x)$ are also even. In this case, $V_\pm(x)$ are not necessarily related by simple translations or reflections. They are isospectral, but may not be "self-isospectral". As an example, the superpotential $W(x) = \sin x + \sin 2x$ gives rise to an isospectral pair which is not self-isospectral, while $W(x) = \sin x + \sin 3x$ [which also belongs to the first special class mentioned above: $W(x + \pi) = -W(x)$] gives rise to a self-isospectral pair.

To make these general ideas more explicit, we now present a class of exactly soluble models. We illustrate this class beginning with the simplest case. Consider the superpotential

$$W(x) = m\frac{sn(x|m)cn(x|m)}{dn(x|m)} \tag{13}$$

Here $sn(x|m)$, $cn(x|m)$ and $dn(x|m)$ are the Jacobi elliptic functions (Abramowitz and Stegun (1990), Whittaker and Watson (1980)), and the (real) elliptic modulus parameter m can be chosen $0 < m \le 1$. Given this superpotential, the isospectral pair (1) of potentials is

$$V_\pm = \begin{cases} 2 - m + 2(m-1)/dn^2(x|m) \\ 2 - m - 2dn^2(x|m) \end{cases} \tag{14}$$

Some relevant properties of the Jacobi elliptic functions are listed here.

1. Periodicity properties:

$$sn(x + 2K(m)|m) = -sn(x|m)$$
$$cn(x + 2K(m)|m) = -cn(x|m)$$
$$dn(x + 2K(m)|m) = dn(x|m) \tag{15}$$

Here $K(m) \equiv \int_0^{\pi/2} d\theta/\sqrt{1 - m\sin^2\theta}$, is the "real elliptic quarter period".

2. Differentiation properties:

$$\frac{d}{dx}sn(x|m) = cn(x|m)\, dn(x|m)$$
$$\frac{d}{dx}cn(x|m) = -sn(x|m)\, dn(x|m)$$
$$\frac{d}{dx}dn(x|m) = -m\, sn(x|m)\, cn(x|m) \tag{16}$$

3. Quadratic relations:

$$-dn^2(x|m) + 1 - m = -m\, cn^2(x|m) = m\, sn^2(x|m) - m \tag{17}$$

Finally, we note that when $m = 1$ these relations all reduce to those for the familiar hyperbolic functions since

$$sn(x|1) = tanh x; \quad cn(x|1) = sech x; \quad dn(x|1) = sech x \tag{18}$$

Thus, when $m = 1$ the superpotential in (13) reduces to $tanh x$ and the isospectral potentials (14) reduce to the $j = 1$ case of the example in (5). [The reader is urged to consider the $m \to 1$ limit at all stages of the subsequent discussion.]

From the periodicity properties (15), the superpotential $W(x)$ in (13) and the potentials $V_\pm$ in (14) have period $2K(m)$:

$$V_\pm(x + 2K(m)) = V_\pm(x) \tag{19}$$

The zero modes (4) are [note that the function $dn(x|m)$ has no nodes or poles on the real axis]

$$\psi_0^{(\pm)}(x) = e^{\mp \log dn(x|m)} = [dn(x|m)]^{\mp 1} \tag{20}$$

Both $\psi_0^{(+)}$ and $\psi_0^{(-)}$ have period $2K(m)$, and are *both* good zero modes. Thus the spectra of $V_\pm$ should be *identical*. This can be checked explicitly because the spectrum can be computed exactly, since the Schrödinger equation for the potentials (14) is an example of the Lamé equation, whose explicit solution is known in terms of elliptic functions (Whittaker and Watson (1980)). Each spectrum has a single bound band and a continuum, as shown in Fig. 1.

The upper edge of the bound band has energy $E_1 = 1 - m$, with Bloch wavefunctions

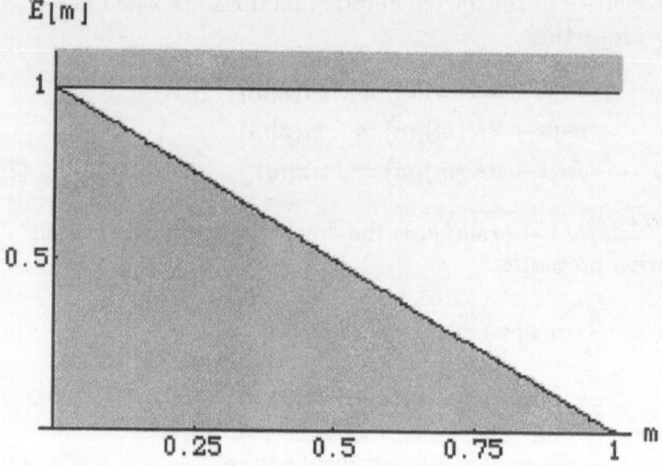

Fig. 1. The spectrum of the isospectral pair of potentials in (14), as a function of the elliptic parameter m. There is a single bound band, bounded below by energy $E_0 = 0$, and above by energy $E_1 = 1 - m$. There is also a continuum band beginning at $E_2 = 1$. Note that when $m = 1$ the bound band smoothly degenerates into a single discrete bound level of energy $E_0 = 0$; this is just the zero mode of the $j = 1$ Pöschl-Teller system in (5).

$$\psi_1^{(\pm)} = \begin{cases} sn(x|m)/dn(x|m) \\ cn(x|m) \end{cases} \tag{21}$$

while the the lower edge of the continuum band has energy $E_2 = 1$, and Bloch wavefunctions

$$\psi_2^{(\pm)} = \begin{cases} cn(x|m)/dn(x|m) \\ sn(x|m) \end{cases} \tag{22}$$

These band-edge properties may be verified directly using the various properties listed in (16,17).

Concentrating on the potential V_- [an analogous analysis holds for V_+], the Schrödinger equation can be written [using (17)] in Lamé form

$$\psi'' = [2m\, sn^2(x|m) - m - E]\psi \tag{23}$$

This equation has two independent solutions

$$\psi(x) = \frac{H(x \pm \alpha)}{\Theta(x)} e^{\mp x Z(\alpha)} \tag{24}$$

where the parameter α is related to the energy eigenvalue by $E = dn^2(\alpha|m)$; $H(x)$ is the Jacobi eta function, $\Theta(x)$ the Jacobi theta function, and $Z(\alpha)$ the Jacobi zeta function (Whittaker and Watson (1980)). It is an instructive exercise to verify that at the band edges these solutions reduce to the wave-functions in (20,21,22), and furthermore that when $m = 1$ they reduce to the well-known bound state and continuum states for the Pöschl-Teller potential $V_- = 1 - 2\tanh^2 x$.

Given the exact solution (24) we can use Bloch's theorem to find the exact dispersion relation between the energy E and the crystal momentum k:

$$\psi(x + 2K(m)) = e^{ik2K(m)}\psi(x)$$

$$k = \mp \frac{\pi}{2K(m)} \pm iZ(dn^{-1}(\sqrt{E}|m)) \qquad (25)$$

We plotted this dispersion relation in Fig. 2, which clearly shows the band-gap. Note that it is rare to have an exact solution for these band features.

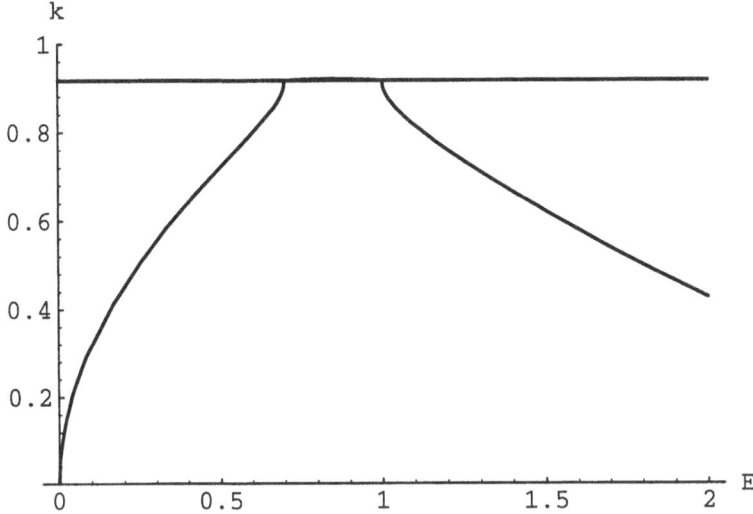

Fig. 2. The *exact* dispersion relation (25) between energy E and crystal momentum k for the isospectral system (14). This plot is for $m = 0.3$, and we clearly see the band gap between $E = 0.7$ and $E = 1$. The horizontal line marks the edge of the Brillouin zone, at which $k = \pi/(2K(m))$.

The isospectral potentials $V_\pm$ in (14) are also self-isospectral. Indeed, using the properties

$$sn(x + K(m)|m) = cn(x|m)/dn(x|m)$$
$$cn(x + K(m)|m) = -\sqrt{1-m}\, sn(x|m)/dn(x|m)$$
$$dn(x + K(m)|m) = \sqrt{1-m}/dn(x|m) \qquad (26)$$

we see that the superpotential (13) satisfies the condition (8), and the two potentials are identical up to a displacement by half a period (see Fig. 3):

$$V_+(x + K(m)) = V_-(x) \tag{27}$$

Since each potential extends indefinitely and periodically, they are indistinguishable as fas as their spectrum is concerned.

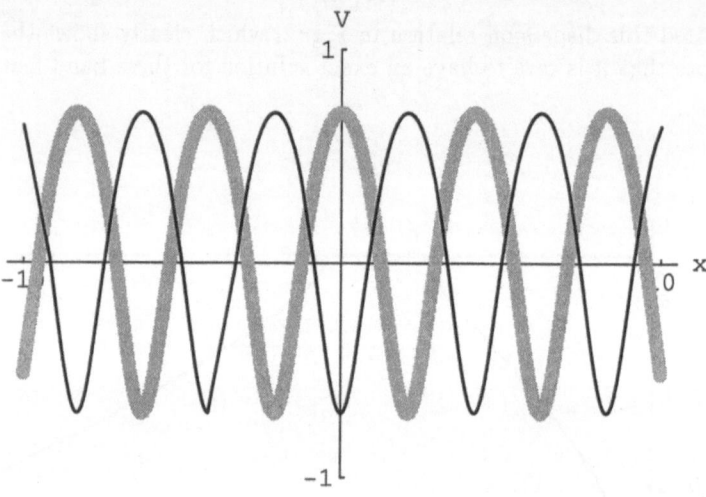

Fig. 3. The self-isospectral potentials (14): V_+ (thick line) and V_- (thin line). Note that they are identical, except for being displaced by half a period. These plots are for elliptic parameter $m = 0.7$.

This raises the question of what happens in the $m \to 1$ limit, because we know that when $m = 1$ the potentials $V_\pm$ are genuinely different and only V_- has a zero mode. The situation is best illustrated by Fig. 4. Consider a single period $-K(m) \le x \le K(m)$. As $m \to 1$, $K(m) \to \infty$, and this single period becomes our real line. On this domain, the potential V_- becomes $1 - 2tanh^2x$, and its bound band collapses smoothly into a single discrete bound level (see Fig. 1). Moreover, the Bloch wavefunctions (20) and (21) at the lower and upper edges of the bound band each tend smoothly to the normalizable wavefunction $\psi_0^{(-)} = sechx$ of this single bound state. On the other hand, on this domain V_+ flattens out and becomes 1, which has no bound states (only a continuum $E > 1$). Correspondingly, the Bloch wavefunctions (20) and (21) at the lower and upper edges of the bound band tend smoothly to $coshx$ and $sinhx$ (respectively), which are *not* normalizable. Displacing this picture by $K(m)$ (i.e. by half a period), the roles of V_- and V_+ are interchanged.

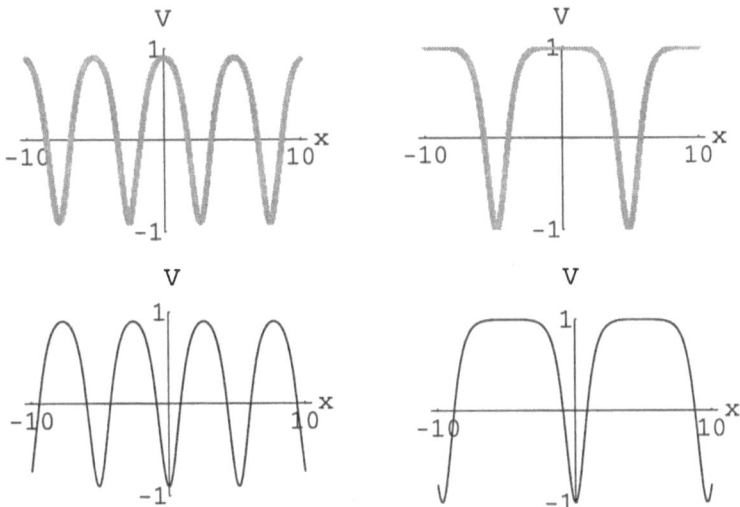

Fig. 4. The potentials V_+ (thick lines) and V_- (thin lines) for two values of the elliptic modulus parameter $m = 0.9$ and $m = 0.999$, going from left to right. Notice that V_+ is identical to V_-, but shifted by half a period. Also note that as m approaches 1, within the central period, $-K(m) \leq x \leq K(m)$, V_- approaches a Pöschl-Teller potential with a single binding well, while V_+ flattens out to a constant. The situation is interchanged if we displace the picture by $K(m)$, which is half a period of the potentials.

This is just the simplest example of a general class of exactly solvable periodic potentials with bound bands. Indeed, it is a classic result that the spectrum of the Lamé equation

$$\psi'' = [j(j+1)m\,sn^2(x|m) - E]\psi \qquad (28)$$

has j bound bands and a continuum band (Whittaker and Watson (1980)). Moreover, the exact solution [analogous to (24)] can be written in terms of elliptic functions (although for $j \geq 2$ the relation between the energy and the crystal momentum becomes more difficult to specify explicitly). In order to make the connection between these Lamé equations and SUSY quantum mechanics we must shift the Lamé potential by a constant to ensure that the lower edge of the lowest band has energy $E = 0$. For example, for $j = 2$ the self-isospectral pair of potentials is

$$V_\pm = \begin{cases} -2m + 4 + 2\sqrt{m^2 - m + 1} + 6(m-1)/dn^2(x|m) \\ -2m + 4 + 2\sqrt{m^2 - m + 1} - 6\,dn^2(x|m) \end{cases} \qquad (29)$$

The energy spectrum is shown in Fig. 5, and the band-edge Bloch wavefunctions for V_- are

$$E_0 = 0 : \quad \psi_0^{(-)} = m + 1 + \sqrt{m^2 - m + 1} - 3msn^2(x|m)$$

$$E_1 = -1 - m + 2\sqrt{m^2 - m + 1} : \quad \psi_1^{(-)} = cn(x|m)dn(x|m)$$

$$E_2 = -1 + 2m + 2\sqrt{m^2 - m + 1} : \quad \psi_2^{(-)} = sn(x|m)dn(x|m)$$

$$E_3 = 2 - m + 2\sqrt{m^2 - m + 1} : \quad \psi_3^{(-)} = sn(x|m)cn(x|m)$$

$$E_4 = 4\sqrt{m^2 - m + 1} : \psi_4^{(-)} = m + 1 - \sqrt{m^2 - m + 1} - 3msn^2(x|m) \quad (30)$$

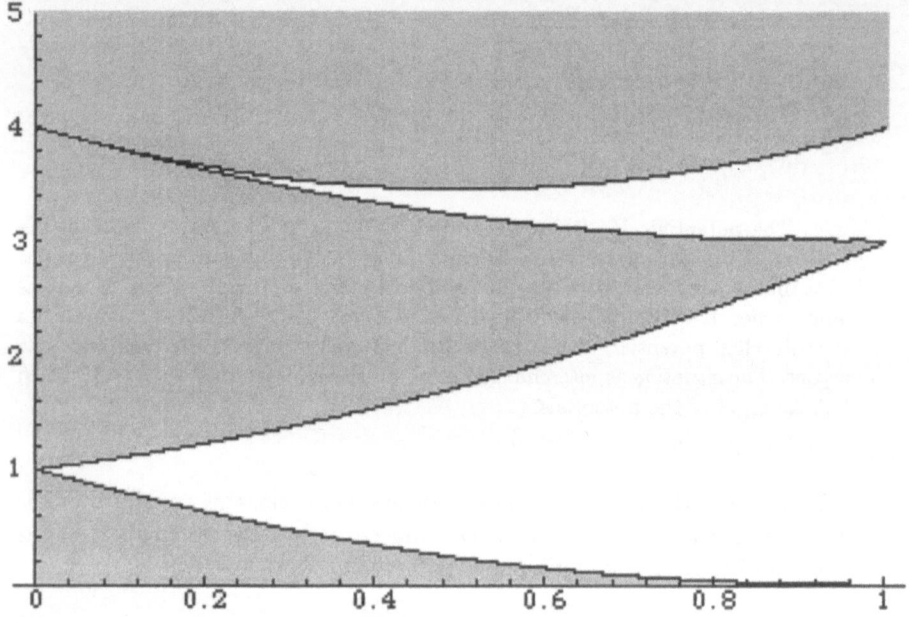

Fig. 5. The spectrum of the isospectral pair of potentials in (29), as a function of the elliptic parameter m. There are two bound bands, and a continuum band beginning at $E = 4\sqrt{m^2 - m + 1}$. Note that when $m = 1$ the bound bands smoothly degenerate into two discrete bound levels of energy $E = 0$ and $E = 3$ and a continuum threshold at $E = 4$; these are the two bound states and continuum threshold of the $j = 2$ Pöschl-Teller system in (5).

The band-edge Bloch wavefunctions $\psi_n^{(+)}$ for V_+ are obtained simply by shifting the $\psi_n^{(-)}$ in (30) by half a period. The superpotential for the isospectral pair (29) is determined by the zero-mode $\psi_0^{(-)}$:

$$W = -\frac{d}{dx}\log\psi_0^{(-)} = \frac{6msn(x|m)cn(x|m)dn(x|m)}{m + 1 + \sqrt{m^2 - m + 1} - 3msn^2(x|m)} \quad (31)$$

When $m = 1$ this reduces to $W = 2\tanh x$, which is the $j = 2$ Pöschl-Teller model in (5).

This procedure may be repeated for higher integer values of j in the Lamé equation (28), leading to a general class of self-isospectral periodic potentials. The band-edge Bloch wavefunctions (there are $2j + 1$ of them since there are j bound bands) are always polynomials of order j in the Jacobi elliptic functions (known as Lamé functions (Whittaker and Watson (1980))). Thus, it is a straightforward algebraic problem to determine the band-edge wavefunctions and energies. Indeed, in a beautiful paper (Alhassid et al (1983)), Alhassid et al showed that the band-edge energies are simply the eigenvalues of the $su(2)$ operator $J_x^2 + mJ_y^2$, with J_x and J_y being the standard $su(2)$ generators in a $(2j+1)$ dimensional matrix reprentation (see also (Ward (1987))).

We conclude by noting that it would be interesting to extend some of these ideas to field theoretic examples.

Note Added : After completing this work we learned that the closely related problem of elliptic supersymmetric models on a circle, of circumference L, was studied in (Braden and Macfarlane (1985)), where it was noted that both supersymmetric sectors carried zero modes with periodic boundary conditions. These models have discrete spectra, and the precise form of the discrete spectrum is determined by the boundary conditions, which in general may be parametrized as $\psi(x + L) = e^{i\alpha x/L}\psi(x)$; thus, in particular, $\alpha = 0$ corresponds to periodic boundary conditions while $\alpha = \pi$ corresponds to anti-periodic boundary conditions. In fact, the closure of all the discrete spectra as α changes from $\alpha = 0$ to $\alpha = 2\pi$ fills out the band spectrum of the corresponding periodic potential problem (Titchmarsh (1958)).

Acknowledgements : This work has been supported by the DOE grant DE-FG02-92ER40716.00 (GD) and NSF grant PHY89-04035 (JF).

References

Witten E. (1981): "Dynamical Breaking of Supersymmetry", Nucl. Phys. B **185**, 513.

Witten E. (1982): "Constraints on Supersymmetry Breaking", Nucl. Phys. B **202**, 253.

Cooper F., Khare A. and Sukhatme U. (1995): "Supersymmetry and Quantum Mechanics", Phys. Rep. **251**, 267.

Rajaraman R. (1982): *Solitons and Instantons* (North-Holland, Amsterdam).

Rebbi C. and Soliani G. (1984): *Solitons and Particles* (World Scientific, Singapore).

Akhoury R. and Comtet A. (1984): "Anomalous Behavior of the Witten Index - Exactly Soluble Models", Nucl. Phys. B **246**, 253.

Jevicki A. and Rodrigues J. (1984): "Singular Potentials and Supersymmetry Breaking", Phys. Lett. B **146**, 55.

Casahorran J. and Nam S. (1991): "Singular Superpotentials and Explicit Breaking of Supersymmetry", Int. J. Mod. Phys. A **6**, 2729.

Das A. and Pernice S. (1996): "'Induced' Super-Symmetry Breaking with a Vanishing Vacuum Energy", hep-th/9612122.

Kittel C. (1976): *Introduction to Solid State Physics* (5th Edition) (Wiley, New York).

Scarf F. (1958): "New Soluble Energy Band Problem", Phys. Rev. **112**, 1137.

Abramowitz M. and Stegun I. (1990): *Handbook of Mathematical Functions*, (Dover).

Whittaker E. and Watson G. (1980): *A Course of Modern Analysis*, (Cambridge U. Press).

Alhassid Y., Gürsey F. and Iachello F. (1983): "Potential Scattering, Transfer Matrix, and Group Theory", Phys. Rev. Lett. **50**, 873.

Ward R. (1987): "The Nahm Equations, Finite-Gap Potentials and Lamé Functions", J. Phys. A: Math. Gen. **20**, 2679.

Braden H. W. and Macfarlane A. J. (1985): "Supersymmetric Quantum Mechanical Models With Continuous Spectrum and the Witten Index", J. Phys. A **18**, 3151.

Titchmarsh E. C. (1958): *Eigenfunction Expansions Associated with Second-order Differential Equations, Volume II*, (Oxford U. Press). See Section 21.10.

Shape Invariance and Its Connection to Potential Algebra

Asim Gangopadhyaya[1], Jeffry V. Mallow[1], and Uday P. Sukhatme[2]

[1] Department of Physics, Loyola University Chicago, Chicago, IL 60626, U.S.
[2] Department of Physics, University of Illinois at Chicago, (m/c 273), 845 W. Taylor Street, Chicago, IL 60607-7059

Abstract. Exactly solvable potentials of nonrelativistic quantum mechanics are known to be shape invariant. For these potentials, eigenvalues and eigenvectors can be derived using well known methods of supersymmetric quantum mechanics. The majority of these potentials have also been shown to possess a potential algebra, and hence are also solvable by group theoretical techniques. In this paper, for a subset of solvable problems, we establish a connection between the two methods and show that they are indeed equivalent.

1 Introduction

It is well known that most of the exactly solvable potentials of nonrelativistic quantum mechanics fall under the Natanzon class (Natanzon (1979)) where the Schrödinger equation reduces either to the hypergeometric or the confluent hypergeometric differential equations. A few exceptions are known (Spiridinov (1992), Barclay et al. (1993)), where solvable potentials are given as a series, and can not be written in closed form in general. With the exception of Ginnochio potential, all exactly solvable potentials are known to be shape invariant (Gendenshtein (1983), Gendenshtein and Krive (1985)); i.e. their supersymmetric partners are of the same shape, and their spectra can be determined entirely by an algebraic procedure, akin to that of the one dimensional harmonic oscillator, without ever referring to the underlying differential equations (Cooper et al. (1995)).

Several of these exactly solvable systems are also known to possess what is generally referred to as a potential algebra (Alhassid et al. (1983), Barut et al. (1987), Englefield and Quesne (1987), Englefield (1987), Wu et al. (1990), Tangerman (1993)). The Hamiltonian of these systems can be written as the Casimir of an underlying SO(2,1) algebra, and all the quantum states of these systems can be determined by group theoretical methods.

Thus, there appear to be two seemingly independent algebraic methods for obtaining the complete spectrum of these Hamiltonians. In this paper, we analyze this ostensible coincidence. For a category of solvable potentials, we find that these two approaches are indeed related.

In the next section, we briefly describe supersymmetric quantum mechanics (SUSY-QM), and discuss how the constraint of shape invariance suffices to determine the spectrum of a shape invariant potential (SIP). In sec. 3, we judiciously construct some algebraic operators and show that the shape invariance constraint can be expressed as an algebraic condition. For a set of shape invariant potentials, we find that the shape invariance condition leads to the presence of a SO(2,1) potential algebra, and we thus establish a connection between the two algebraic methods. In sec. 4, for completeness, we provide a brief review of SO(2,1) representation theory. In sec. 5, we derive the spectrum of a class of potentials and explicitly show that both methods indeed give identical spectrum.

2   SUSY-QM and Shape Invariance

A quantum mechanical system specified by a potential $V_-(x)$ can alternatively be described by its ground state wavefunction $\psi_0^{(-)}$. Apart from a constant (chosen suitably to make the ground state energy zero), it follows from the Schrödinger equation that the potential can be written as $V_-(x) = \left(\frac{\psi_0''}{\psi_0} \right)$, where prime denotes differentiation with respect to x. In SUSY-QM, it is customary to express the system in terms of the superpotential $W(x) = -\left(\frac{\psi_0'}{\psi_0} \right)$ rather than the potential, and the ground state wavefunction is then given by $\psi_0 \sim \exp\left(-\int_{x_0}^{x} W(x)dx \right)$, where x_0 is an arbitrarily chosen reference point. We are using units with $\hbar$ and $2m = 1$. The Hamiltonian H_- can now be written as

$$H_- = \left(-\frac{d^2}{dx^2} + V_-(x) \right) = \left(-\frac{d^2}{dx^2} + W^2(x) - \frac{dW(x)}{dx} \right) . \qquad (1)$$

However, as we shall see, there is another Hamiltonian H_+ with potential $V_+(x) = \left(W^2(x) + \frac{dW(x)}{dx} \right)$, that is almost iso-spectral with the original potential $V_-(x)$. In particular, the eigenvalues E_n^+ of $H_+(x)$ satisfy $E_n^+ = E_{n+1}^-$, where E_n^- are eigenvalues of $H_-(x)$ and $n = 0, 1, 2, \cdots$, i.e. except the ground state all other states of H_- are in one-to-one correspondence with states of H_+. The potentials $V_-(x)$ and $V_+(x)$ are known as supersymmetric partners.

In analogy with the harmonic oscillator, we now define two operators: $A \equiv \left(\frac{d}{dx} + W(x) \right)$, and and its Hermitian conjugate $A^+ \equiv \left(-\frac{d}{dx} + W(x) \right)$. Hamiltonians H_- and its superpartner H_+ are given by operators A^+A and AA^+ respectively.

Now we shall explicitly establish the iso-spectral relationship between states of H_+ and H_-. Let us denote the eigenfunctions of $H_\pm$ that correspond to eigenvalues $E_n^\pm$, by $\psi_n^{(\pm)}$. For $n = 1, 2, \cdots$,

$$H_+\left(A\psi_n^{(-)}\right) = AA^+\left(A\psi_n^{(-)}\right) = A\left(A^+A\psi_n^{(-)}\right) = AH_-\left(\psi_n^{(-)}\right)$$
$$= E_n^-\left(A\psi_n^{(-)}\right) . \tag{2}$$

Hence, excepting the ground state which obeys $A\psi_0^{(-)} = 0$, for any state $\psi_n^{(-)}$ of H_- there exists a state $A\psi_n^{(-)}$ of H_+ with exactly the same energy, i.e. $E_{n-1}^+ = E_n^-$, where $n = 1, 2, \cdots$, i.e. $A\psi_n^{(-)} \propto \psi_{n-1}^{(+)}$. Conversely, one also has $A^+\psi_n^{(+)} \propto \psi_{n+1}^{(-)}$. Thus, if the eigenvalues and the eigenfunctions of H_- were known, one would automatically obtain the eigenvalues and the eigenfunctions of H_+, which is in general a completely different Hamiltonian.

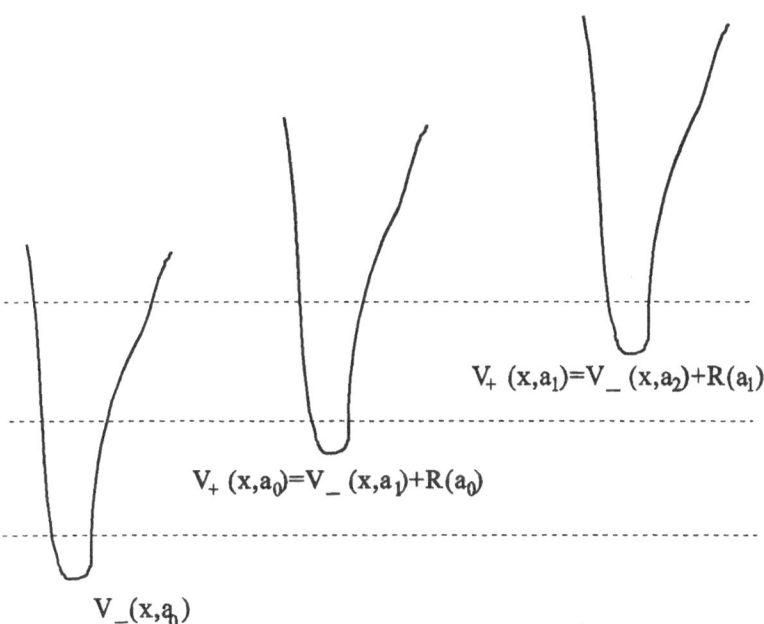

$V_+(x,a_1) = V_-(x,a_2) + R(a_1)$

$V_+(x,a_0) = V_-(x,a_1) + R(a_0)$

$V_-(x,a_0)$

Fig. 1. A series of potentials related by shape invariance. Dashed lines represent eigenvalues of the sytem with potential $V_-(x, a_0)$.

Now, let us consider the special case where $V_-(x)$ is a SIP. This implies that $V_-(x)$ and $V_+(x)$ have the same functional form; they only differ in values of other discrete parameters and possibly an additive constant. To be explicit, let us assume that in addition to the continuous variable x, the potential $V_-(x)$ also depends upon a constant parameter a_0; i.e., $V_- \equiv V_-(x, a_0)$. The ground state of the system of H_- is given by

$\psi_0(x, a_0) \sim \exp\left(-\int_{x_0}^{x} W(x, a_0)dx\right)$. Now, for a shape invariant $V_-(x, a_0)$, one has, $V_+(x, a_0) = V_-(x, a_1) + R(a_0)$, where $R(a_0)$ is the additive constant mentioned above. Since potentials $V_+(x, a_0)$ and $V_-(x, a_1)$ differ only by $R(a_0)$, their common ground state is $\psi_0(x, a_1) \sim \exp\left(-\int_{x_0}^{x} W(x, a_1)dx\right)$. Now using SUSY-QM algebra, the first excited state of $H_-(x, a_0)$ is given by $A^+(x, a_0)\psi_0^{(-)}(x, a_1)$. Its energy is $E_1^{(-)}$, which is equal to $E_0^{(+)}$. But since $E_0^{(-)} = 0$, $E_0^{(+)}$ must be $R(a_0)$. Continuing up the ladder of series of potentials $V_-(x, a_i)$, we can obtain the entire spectrum of H_- by algebraic methods of SUSY-QM. The eigenvalues are given by

$$E_0^{(-)} = 0; \quad \text{and} \quad E_n^{(-)} = \sum_{k=0}^{n-1} R(a_k) \quad \text{for } n > 0,$$

and the n-th eigenstate is given by

$$\psi_{n+1}^{(-)}(x, a_0) \sim A^+(a_0)\, A^+(a_1) \cdots A^+(a_{n-1})\, \psi_0^{(-)}(x, a_{n-1})\,.$$

(To avoid notational complexity, we have suppressed the x-dependence of operators $A(x, a_0)$ and $A^+(x, a_0)$.)

3   Shape Invariance and Potential Algebra

Let us consider the special case of a potential $V_-(x, a_0)$ with an additive shape invariance; i.e. $V_+(x, a_0) = V_-(x, a_1) + R(a_0)$, where $a_n = a_{n-1} + \delta = a_0 + n\delta$, where δ is a constant. Most SIP's fall into this category. For the superpotential $W(x, a_m) \equiv W(x, m)$, the shape invariance condition implies

$$W^2(x, m) + W'(x, m) = W^2(x, m+1) - W'(x, m+1) + R(m) \qquad (3)$$

As described in the last section, this constraint suffices to determine the entire spectrum of the potential $V_-(x, m)$. In this section, we shall explore the possible connection of this method with the potential algebra discussed by several authors (Alhassid et al. (1983), Barut et al. (1987), Englefield and Quesne (1987), Englefield (1987), Wu et al. (1990), Tangerman (1993)).

Since for a SIP, the parameter m is changed by a constant amount each time as one goes from the potential $V_-(x, m)$ to its superpartner, it is natural to ask whether such a task can be formally accomplished by the action of a ladder-type operator.

With that in mind, we first define an operator $J_3 = -i\frac{\partial}{\partial \phi}$, analogous to the z-component of the angular momentum operator. It acts upon functions in the space described by two coordinates x and ϕ, and its eigenvalues m play the role of the parameter of the potential. We also define two more operators, J^- and its Hermitian conjugate J^+ by

$$J^{\pm} = e^{\pm i \phi} \left[\pm \frac{\partial}{\partial x} - W \left(x, -i \frac{\partial}{\partial \phi} \pm \frac{1}{2} \right) \right] \quad . \tag{4}$$

The factors $e^{\pm i \phi}$ in $J^{\pm}$ ensure that they indeed operate as ladder operators for the quantum number m. Operators $J^{\pm}$ are basically of the same form as the $A^{\pm}$ operators described earlier in sec. 2, except that the parameter m of the superpotential is replaced by operators $\left(J_3 \pm \frac{1}{2} \right)$. With explicit computation we find

$$[J_3, J^{\pm}] = \pm J^{\pm} \quad , \tag{5}$$

and hence operators $J^{\pm}$ change the eigenvalues of the J_3 operator by unity, similar to the ladder operators of angular momentum $(SU(2))$. Now let us determine the remaining commutator $[J^+, J^-]$. The product $J^+ J^-$ is given by

$$
\begin{aligned}
J^+ J^- &= e^{i \phi} \left[\frac{\partial}{\partial x} - W \left(x, J_3 + \frac{1}{2} \right) \right] e^{-i \phi} \left[-\frac{\partial}{\partial x} - W \left(x, J_3 - \frac{1}{2} \right) \right] \\
&= \left[-\frac{\partial^2}{\partial x^2} + W^2 \left(x, J_3 - \frac{1}{2} \right) - W' \left(x, J_3 - \frac{1}{2} \right) \right]
\end{aligned} \tag{6}
$$

Similarly,

$$J^- J^+ = \left[-\frac{\partial^2}{\partial x^2} + W^2 \left(x, J_3 + \frac{1}{2} \right) + W' \left(x, J_3 + \frac{1}{2} \right) \right] . \tag{7}$$

Hence the commutator of operators J_+ and J_- is given by

$$
\begin{aligned}
[J^+, J^-] &= \left[-\frac{\partial^2}{\partial x^2} + W^2 \left(x, J_3 - \frac{1}{2} \right) - W' \left(x, J_3 - \frac{1}{2} \right) \right] \\
&\quad - \left[-\frac{\partial^2}{\partial x^2} + W^2 \left(x, J_3 + \frac{1}{2} \right) + W' \left(x, J_3 + \frac{1}{2} \right) \right] \\
&= -R \left(J_3 + \frac{1}{2} \right) ,
\end{aligned} \tag{8}
$$

where we have used the constraint of shape invariance, i.e. $V_-(x, J_3 - \frac{1}{2}) - V_+(x, J_3 + \frac{1}{2}) = -R(J_3 + \frac{1}{2})$. Thus, we see that Shape Invariance enables us to close the algebra of J_3 and $J^{\pm}$ to

$$[J_3, J^{\pm}] = \pm J^{\pm} \quad , \quad [J^+, J^-] = -R \left(J_3 + \frac{1}{2} \right) \quad . \tag{9}$$

Now, if the function $R(J_3)$ were linear in J_3, the algebra of eq.(9) would reduce to that of a SO(3) or SO(2,1). Several SIP's are of this type, among them are the Morse, the Rosen-Morse and the Pöschl-Teller I and II potentials. For these potentials, $R \left(J_3 + \frac{1}{2} \right) = 2 J_3$, and eq.(9) reduces to an SO(2,1) algebra and thus establishes the connection between shape invariance and potential algebra. Even though there is much similarity between SO(2,1) and SO(3) algebras, there are some important differences between their representations. Hence, for completeness, we will briefly describe the unitary representations of SO(2,1) and refer the reader to Adams et al. (1987) for a more detailed presentation.

4   Unitary Representations of SO(2,1) Algebra

In this section, we shall briefly review the SO(2,1) algebra and its unitary representations (unireps). This description is primarily based upon a review article by B.G. Adams, J. Cizeka and J. Paldus (1987). The generators of the SO(2,1) algebra satisfy

$$[J_3, J^\pm] = \pm J^\pm \;\; ; [J_+, J_-] = -2J_3 \;\; , \tag{10}$$

where $J^\pm$ are related to their Cartessian counterparts by $J^\pm = J_1 \pm J_2$. (For the familiar SO(3) case, one has $[J_+, J_-] = +2J_3$). The Casimir of the SO(2,1) algebra is

$$J^2 = -J^+ J^- + J_3^2 - J_3 \;\; = -J^- J^+ + J_3^2 + J_3 \;\; . \tag{11}$$

In analogy to the representation of angular momentum algebra, one can choose J^2 and one of the J_i's as two commuting observables. However, unlike the SO(3) case, each such choice of a pair generates a different set of inequivalent representations. For bound states, we choose the familiar representation space of states $|j, m\rangle$ on which the operators $\{J^2, J_3\}$ are diagonal: $J^2|j, m\rangle = j(j + 1)|j, m\rangle$, $J_3|j, m\rangle = m|j, m\rangle$. Operators $J^\pm$ act upon $|j, m\rangle$ states as ladder operators: $J^\pm|j, m\rangle = [-(j \mp m)(j \pm m + 1)]^{\frac{1}{2}} |j, m + 1\rangle$. Since the quantum number m increases in unit steps for a given j, the general value for m is of the form $m_0 + n$, where n is an integer and m_0 is a real number. There is also another constraint on the quantum numbers m and j. In unitary representations, J^+ and J^- are Hermitian conjugates of each other, and J^+J^- and J^-J^+ are therefore positive operators. This implies $[-(j \mp m)(j \pm m + 1)] = -\left[\left(j + \frac{1}{2}\right)^2 - \left(m + \frac{1}{2}\right)^2\right] \geq 0$. These constraints can be illustrated on a two dimensional planar diagram [Fig.2] depicting the allowed values of m and j. Only the open triangular areas DFB, HEG and the square AEFC are the allowed regions. The values of $|m|$ are no longer bounded by j, and depending on the m_0 (the starting value of m), representations multiplets are either semi-infinite (bounded from below or above) or completely unbounded. Thus there is no finite (nontrivial) unitary representation of SO(2,1). In general, there are four classes of unireps.

$D^+(j)$	Bounded from below (j, m_0) lie along the segment AB	$\begin{cases} m = -j + n; & n = 0, 1, 2, \cdots, \\ j < 0, \end{cases}$
$D^-(j)$	Bounded from above (j, m_0) lie along the segment AG	$\begin{cases} m = j + n; & n = 0, -1, -2, \cdots, \\ j < 0, \end{cases}$
$D_s(j, m_0)$	(j, m_0) lie in the square area	$\begin{cases} m = m_0 + n; & n = 0, \pm 1, \pm 2, \cdots, \\ j(j+1) < (|m_0| - 1)|m_0|; \\ -\frac{1}{2} < m_0 < -\frac{1}{2}, \end{cases}$
$D_p(j, m_0)$	Unbounded and complex j	$\begin{cases} m = m_0 + n; & n = 0, \pm 1, \pm 2, \cdots, \\ -\frac{1}{2} < m_0 < -\frac{1}{2}, \\ j = -\frac{1}{2} + i\beta. \end{cases}$

Here we will be interested in representations that are bounded from either below or above. Such representations fall in triangular areas DFB and HEG. For the D^+ representation, the starting value of m can be anywhere on the darkened part of the line AB; other allowed values of m are then obtained by the action of the ladder operator J^+. Owing to the equivalence of $D^+(j)$ and $D^+(-j-1)$, they correspond to the same value of $j(j+1)$. One could have equivalently started anywhere on the segment CD as well and used $D^+(-j-1)$. Both are equivalent and each is unique. Similarly, for complete $D^-(j)$ ($D^-(-j-1)$) representation, one starts from AG (GH) and generates all other states by the action of the J^- operator.

5   Example

As a concrete example, we will examine the Scarf potential which can be related to the Pöschl-Teller II potential by a redefinition of the independent variable. We will show that the shape invariance of the Scarf potential automatically leads to its potential algebra: SO(2,1). (Exactly similar analysis can be carried out for the Morse, the Rosen-Morse, and the Pöschl-Teller potentials.) The Scarf potential is described by its superpotential $W(x, a_0, B) = a_0 \tanh x + B \operatorname{sech} x$. The potential $V_-(x, a_0, B) = W^2(x, a_0, B) - W'(x, a_0, B)$ is then given by

$$V_-(x, a_0, B) = \left[B^2 - a_0(a_0 + 1) \right] \operatorname{sech}^2 x + B(2a_0 + 1)\operatorname{sech} x \, \tanh x + a_0^2 . \tag{12}$$

The eigenvalues of this system are given by (Cooper et al. (1995))

$$E_n = a_0^2 - (a_0 - n)^2 . \tag{13}$$

The partner potential $V_+(x, a_0, B) = W^2(x, a_0, B) + W'(x, a_0, B)$ is given by

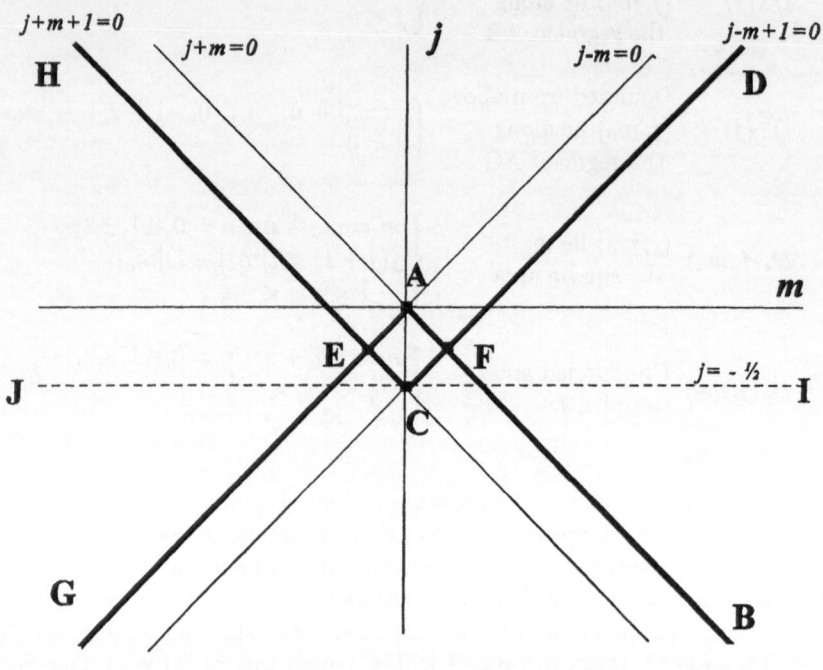

Fig. 2. Two dimension plot showing the allowed region for m and j.

$$V_+(x, a_0, B) = \left[B^2 - a_0(a_0 - 1)\right] \operatorname{sech}^2 x + B(2a_0 - 1)\operatorname{sech} x \tanh x + a_0^2 \ .$$
$$= V_-(x, a_1, B) + a_0^2 - a_1^2 \ , \tag{14}$$

where $a_1 = a_0 - 1$. Thus, $R(a_0)$ for this case is $a_0^2 - a_1^2 = 2a_0 - 1$, linear in a_0.

Now, following the mechanism of the sec. 2, consider a set of operators $J^\pm$ which is given by

$$J^\pm = e^{\pm i\,\phi}\left[\pm\frac{\partial}{\partial x} - \left\{\left(-i\frac{\partial}{\partial\phi} \pm \frac{1}{2}\right) \tanh x + B \operatorname{sech} x\right\}\right] \ . \tag{15}$$

Note the similarity between the operators $J^\pm$ and operators $A^\pm$ defined in sec. 2. Since only the parameter a_0 changes in the shape invariance condition, it is replaced by $J_3 \pm \frac{1}{2}$. Commutators of these operators with $J_3 = -i\frac{\partial}{\partial\phi}$ can be shown to close on $J^\pm$, as discussed in general in Sec. 2. Now, from eq.(9) and (14), the commutator of $J^\pm$ operators is given by $-2J_3$, thus forming a closed SO(2,1) algebra. Moreover, the operator J^+J^-, acting on the basis $|j, m\rangle$ gives:

$$J^+ J^- \equiv \left[B^2 - \left(m^2 - \frac{1}{4} \right) \right] \operatorname{sech}^2 x$$

$$+ B \left(2 \left(m - \frac{1}{2} \right) + 1 \right) \operatorname{sech} x \tanh x + \left(m - \frac{1}{2} \right)^2 . \quad (16)$$

which is just the $H_{\text{scarf}} \left(x, m - \frac{1}{2}, B \right)$, i.e. the Scarf Hamiltonian with a_0 replaced by $m - \frac{1}{2}$. Thus the energy eigenvalues of the Hamiltonian will be the same as that of the operator $J^+ J^- = J_3^2 - J_3 - J^2$. Hence, the energy is given by $E = m^2 - m - j(j+1)$. Substituting $j = n - m$, one gets

$$E_n = m^2 - m - (n - m)^2$$

$$= (m - \frac{1}{2})^2 - \left[n - (m - \frac{1}{2}) \right]^2 . \quad (17)$$

which is the same as eq.(13), with a_0 replaced by $\left(m - \frac{1}{2} \right)$. Thus for this potential, as well as for the other three potentials mentioned above, there are actually an infinite number of potentials characterised by all allowed values of the parameter m that correspond to the same value of j and hence to the same energy E. Hence the name "potential algebra" (Alhassid et al. (1983), Wu et al. (1990)).

Conclusion: The algebra of Shape Invariance plays an important role in the solvability of most exactly solvable problems in quantum mechanics. Their spectrum can be easily generated simply by algebraic means. Many of these systems also have been shown to possess a potential algebra, which provides an alternate algebraic method to determine the eigenvalues and eigenfunctions. An obvious question is whether these are two unrelated algebraic methods or there is a link between them. For a subset of exactly solvable potentials, those with $R(a_0)$ linear in parameter a_0, we have shown the equivalence of their shape invariance property to an SO(2,1) potential algebra. As a concrete example, we started with the Scarf potential and showed explicitly how shape invariance translates into the SO(2,1) potential algebra. We determined the spectra using the algebra of SO(2,1) and showed them to be the same as that obtained from shape invariance.

However, we only worked with solvable models for which $R(J_3)$ is a linear function of J_3. There are many systems for which the above is not true. Also there were new Shape Invariant problems discovered in 1992 (Barclay et al. (1993)) for which it is not possible to write the potential in closed form. It will be interesting to know whether there are potential algebras that describe these system, and whether they are connected to their Shape Invariance. These are open problems and are currently under investigation.

One of us (AG) would like to thank the Physics Department of the University of Illinois for warm hospitality. We would also like to thank Dr. P. Panigrahi for many related discussions.

References

Adams, B.G, Cizeka, J., Paldus, J. (1987): *Lie Algebraic Methods And Their Applications to Simple Quantum Systems*. (From the Advances in Quantum Chemistry, Vol. 19, Edited by Per-Olov Löwdin. Academic Press).

Alhassid, Y., Gürsey, F. and Iachello, F. (1983): Ann. Phys. **148**, 346.

Barclay, D., Dutt, R., Gangopadhyaya, A., Khare, A., Pagnamenta, A. and Sukhatme, U. (1993): Phys. Rev. **A48** 2786.

Barut, A.O., Inomata, A., Wilson, R. (1987): J. Phys. A: Math. Gen. **20** 4075; J. Phys. A: Math. Gen. **20** 4083.

Cooper, F., Khare, A., and Sukhatme, U. (1995): Phys. Rep. **251**, 268 and references therein.

Englefield, M.J., Quesne, C. (1987): J. Phys. A: Math. Gen. **24** 827(1987).

Englefield, M.J. (1987): J. Phys. A: Math. Gen. **28** 3557.

Gendenshtein, L. E. (1983): JETP Lett. **38**, 356;

Gendenshtein, L. E., Krive, I. V. (1985): Sov. Phys. Usp. **28**, 645.

Natanzon, G. A. (1979): Teor. Mat. Fiz. **38**, 219.

Spiridonov, V. P. (1992): Phys. Rev. Lett. **69**, 298.

Tangerman R.D., Tjon, J.A. (1993): Phys. Rev. A **48** 1089.

Wu J., Alhassid, Y. (1990): Phys. Rev. A **31** 557.

Supersymmetry in Quantum Mechanical Models: a Quantum Hamilton-Jacobi Approach

R.S. Bhalla[2], A.K. Kapoor[1] and P.K.Panigrahi[1]

[1] School of Physics, University of Hyderabad, Hyderabad-500 046, India
[2] Department of Mathematical and Computer Sciences, Loyola University, Chicago, IL 60626.

Abstract. Using quantum Hamilton-Jacobi formalism, we analyze various aspects of supersymmetric quantum mechanical models, *viz.*, origin of solvability, breaking of supersymmetry and the exactness of supersymmetric WKB approximation scheme.

1 Introduction

In recent times, supersymmetry (SUSY) has played a significant role in the understanding of solvability of conventional quantum mechanical problems [1]. It has also been used to develop better semi-classical approximation schemes [2]. SUSY inspired WKB scheme gives exact eigenvalues for a class of potentials for which WKB fails to yield the spectra. On the other hand, although SUSY was first discovered in relativistic field theories, the quantum mechanical models have been extremely useful in throwing light on subtle issues like breaking of SUSY. Keeping in mind the importance of SUSY in quantum mechanics, it is worthwhile to study the supersymmetric quantum mechanical (SUSY-QM) models using other approaches for obtaining a different perspective as to the workings of SUSY. In particular, the solvability of certain potentials as compared to others, the exactness of SUSY WKB and the breaking of SUSY needs deeper understanding.

In what follows, we apply the quantum Hamilton-Jacobi (QHJ) formalism as developed by Leacock and Padgett [3] to SUSY-QM problems. As will be clear from the text, the quantization condition in QHJ method bears a striking resemblance to SUSY WKB and hence this formalism is ideal for illuminating the SUSY inspired semi-classical methods. The plan of the text is as follows: In the first Sect., we give a brief introduction to SUSY-QM and proceed to elucidate the workings of the QHJ formalism in the second Sect. The third Sect., deals with the Hamilton-Jacobi approach to both unbroken and spontaneously broken SUSY-QM models. In the fourth Sect., we explain the exactness of SUSY WKB scheme for certain potentials from the Hamilton-Jacobi point of view and then conclude with some speculations about future directions of work.

2   Supersymmetry in Non-Relativistic Quantum Mechanics

In non-relativistic quantum mechanical problems, the $N = 2$ SUSY algebra involving the supercharge $Q = (Q_1 + iQ_2)/\sqrt{2}$, its adjoint and the Hamiltonian H can be written as,

$$H = \{Q, Q^\dagger\} , \quad Q^2 = (Q^\dagger)^2 = 0 , \tag{1}$$

and

$$[Q, H] = [Q^\dagger, H] = 0 . \tag{2}$$

A matrix realization of the supercharges can be given by,

$$Q = \begin{pmatrix} 0 & 0 \\ A & 0 \end{pmatrix} , \quad Q^\dagger = \begin{pmatrix} 0 & A^\dagger \\ 0 & 0 \end{pmatrix} . \tag{3}$$

Here,

$$A = \hbar \frac{d}{dx} + \omega(x) , \tag{4}$$

and $\omega(x)$ is called the superpotential. A straightforward calculation yields,

$$H = \begin{pmatrix} H_- & 0 \\ 0 & H_+ \end{pmatrix} . \tag{5}$$

The partner Hamiltonians are explicitly given by $(2m = 1)$,

$$H_\pm = -\hbar^2 \frac{\partial^2}{\partial x^2} + V_\pm(x) ,$$

where,

$$V_\pm = \omega^2(x) \pm \hbar \frac{\partial \omega(x)}{\partial x} . \tag{6}$$

The eigenstates of H, apart from the ground-state for the unbroken case, are doublets, given by

$$\psi_n(x) = \begin{pmatrix} \psi_{n+1}^{(-)} \\ \psi_n^{(+)} \end{pmatrix} , \tag{7}$$

where n represents the number of nodes of the wave function. It can be easily shown that $E_{n+1}^{(-)} = E_n^{(+)}$ (apart from the ground-state, which is unpaired). Here $E_{n+1}^{(-)}$ and $E_n^{(+)}$ are the energy eigenvalues for H_- and H_+ respectively.

For the case where SUSY is unbroken, the square-integrable ground-state of H_- having zero energy satisfies,

$$A\psi_0^{(-)} = 0 , \tag{8}$$

implying

$$\omega(x) = -\hbar \frac{1}{\psi_0^{(-)}} \frac{\partial \psi_0^{(-)}}{\partial x} \quad . \tag{9}$$

In the spontaneously broken case, the ground state, having non-zero energy, is no longer annihilated by either of the supercharges and for this eigenvalues of H_- and H_+ coincide. To differentiate between these two phases of SUSY, given $\psi_n^{(-)}$ and $\psi_m^{(+)}$ are the wave functions of H^- and H^+, Gozzi gave an index [4], in terms of the following quantities,

$$p^{\pm}(E) \equiv \frac{\hbar}{i} \frac{\partial \psi_E^{\pm}}{\partial x} \frac{1}{\psi_E^{\pm}} \quad , \tag{10}$$

as,

$$\delta \equiv \oint p^{(-)}(E) dx - \oint p^{(+)}(E) dx \quad . \tag{11}$$

The integral is taken around a contour in the complex x plane enclosing part of the real axis, between the turning points. A contour integral enclosing $p(E)$ between the turning points counts the number of nodes of $\psi_E^{(\pm)}(E)$ and thus gives,

$$\oint P^{(\pm)}(E) = n\hbar \quad .$$

Therefore depending upon phases of spontaneously broken and unbroken SUSY (setting $\hbar = 1$), δ can take values 0 and ± 1 respectively. The relationship between the QHJ formalism and the Gozzi index will be further elucidated in the later part of the text.

3   Quantum Hamilton-Jacobi Formalism

A quantum Hamilton-Jacobi formalism, in parallel to the classical case, has been developed recently [3,5], which provides the energy eigenvalues of a bound state problem, without explicitly solving the Schrödinger equation. In the QHJ formalism, the QHJ equation,

$$\frac{\hbar}{i} \frac{\partial p(x, E)}{\partial x} + p^2(x, E) = (E - V(x)) \equiv p_c^2 \quad , \tag{12}$$

replaces the Schrödinger equation as the dynamical equation. The connection of these two equations can be seen by substituting

$$p(x, E) = \frac{\hbar}{i} \frac{1}{\psi} \frac{\partial \psi(x, E)}{\partial x} \quad , \tag{13}$$

in the QHJ equation, under which it goes over to the Schrödinger equation,

$$\left(-\hbar^2 \frac{\partial^2}{\partial x^2} + V(x) \right) \psi = E\psi \quad . \tag{14}$$

The quantum action variable can be defined as,

$$J(E) \equiv \frac{1}{2\pi} \oint_C dx \, p(x, E) \ . \tag{15}$$

Here, C is a counter clockwise contour in the complex x plane, enclosing the real line between the classical turning points obtained from $p_c^2(x, E) = 0$. The nodes of the wave function correspond to poles of the quantum momentum function (QMF) $p(x, E)$. By expanding $p(x, E)$ near a zero of the wave function located at x_0, it can be seen that $p(x, E)$ has a first order pole at x_0 with residue $-i\hbar$. These set of first order poles are of quantum mechanical origin and their positions are energy dependent. These poles will be referred to as the moving poles. The QHJ equation shows that, $p(x, E)$ can also have singularities at the singular points of the potential term $V(x)$; these energy independent poles will be called fixed poles. For a given energy level the quantum number 'n' equals the number of nodes of the wave function and hence it counts the number of moving poles of $p(x, E)$ inside the contour C. Hence, we have

$$J(E) = n\hbar \ , \tag{16}$$

as an *exact quantization condition*.

In order to unambiguously determine a physically acceptable solution for $p(x)$ the following boundary condition must be imposed.

$$p(x, E) \overset{\hbar \to 0}{\to} p_c(x, E) \ . \tag{17}$$

One notices from QHJ equation that, the function $p_c(x)$ is double valued in the complex x plane, having branch points where $p_c(x)$ vanishes. After defining $x = x_R - i\epsilon$, where $\epsilon > 0$ and x_R lies on the real axis between the physical turning points, we take $p_c(x)$ to be positive when $\epsilon \to 0$.

A priori, the location and the number of the moving poles are not known; however, a suitable deformation of the contour and change of variables allows one to compute $J(E)$, for many potentials, in terms of the fixed poles whose locations and residues are known. In what follows, the QHJ formalism is applied to study different phases of SUSY in the context of the generalized Pöschl-Teller potential. Interestingly, for appropriate ranges of the parameters appearing in the above potential, the Hamiltonian can exhibit either broken or unbroken phases of SUSY. It is found that, the boundary condition in the QHJ method plays a crucial role in naturally differentiating the two phases of SUSY, for different ranges of the parameters appearing in the potentials.

4   Generalized Pöschl-Teller Potential

For the generalized Pöschl-Teller potential [6,7], the superpotential is given by

$$\omega(x) = A\coth\alpha x - B\,\text{cosech}\,\alpha x \qquad\qquad (x \geq 0) \ . \qquad (18)$$

The partner potentials are

$$V_{\pm}(x) = A^2 + (A^2 + B^2 \mp A\alpha\hbar)\text{cosech}^2\alpha x - B(2A \mp \alpha\hbar)\coth\alpha x\,\text{cosech}\,\alpha x \ . \qquad (19)$$

The calculations for only one of the partner Hamiltonians (H_-) will be presented as the calculational details for the other is similar. For the convenience of calculation, we introduce the variable y:

$$\coth\alpha x + \text{cosech}\,\alpha x = y \ , \qquad (20)$$

$$\coth\alpha x - \text{cosech}\,\alpha x = \frac{1}{y} \ , \qquad (21)$$

and the corresponding QHJ equation is

$$\tilde{p}^2(y, E) + \frac{i\hbar\alpha}{2}\frac{\tilde{p}(y, E)}{\partial y}(1 - y^2)$$
$$= E - A^2 - (A^2 + B^2 + A\alpha)\frac{(y^2 - 1)^2}{4y^2} + B(2A + \alpha\hbar)\frac{(y^4 - 1)}{4y^2} \quad (22)$$

The quantization condition is given by,

$$J(E) \equiv \frac{-1}{\pi}\oint_{C_1}\frac{\tilde{p}(y, E)}{\alpha(y^2 - 1)} = n\hbar \ , \qquad (23)$$

where $\tilde{p}(y)$ is the QMF in the y variable. The quantum action integral around contour C_1 will be denoted by I_{C_1}. The mapping given earlier has introduced singularities in the above integrand at $y = \pm 1$. QMF has a fixed pole at $y = 0$ and in addition $\tilde{p}(y)$ has extra energy dependent poles in the region $y < -1$.

In order to calculate the quantum action integral I_{C_1}, we deform the contour C_1, so as to enclose all the singular points of the integrand and the extra energy dependent poles in the non-classical region. I_{C_1} is calculated, by considering a contour integral I_{Γ_R} for a circle Γ_R of radius R, which is large enough so as to enclose the above mentioned singularities in $\tilde{p}(y)$ and is such that there is no other singularity outside Γ_R. We thus have

$$I_{\Gamma_R} = I_{C_1} + I_{C_2} + I_{\gamma_1} + I_{\gamma_2} + I_{\gamma_3} \ , \qquad (24)$$

where I_{C_2} represents the integral around the contour C_2 enclosing the moving poles between the turning points in the non-classical region of the complex y plane; γ_1, γ_2, and γ_3 represent the contours around the poles at $y = 1, -1$ and 0 respectively.

One notices that the symmetry $y \to -y$ in $\tilde{p}_c(y)$ interchanges the turning points in the classical region of the complex y plane with those in the non-classical region. Therefore I_{C_1} and I_{C_2} give equal contribution. The contour integral I_{Γ_R} is calculated by one more change of variable: $z = 1/y$. So in terms of the new variable z, the integral I_{Γ_R} becomes,

$$I_{\Gamma_R} = -\frac{1}{\pi} \oint_{\gamma_0} \frac{\tilde{\tilde{p}}(z,E)}{\alpha(1-z^2)} \tag{25}$$

$$\equiv I_{\gamma_0} , \tag{26}$$

where γ_0 is a small circle in the complex z plane that encloses the singular point at $z = 0$ and $\tilde{\tilde{p}}(z,E) = \tilde{p}(1/z,E)$.

Therefore the quantization condition can now be written as,

$$n\hbar = (I_{\gamma_0} - I_{\gamma_1} - I_{\gamma_2} - I_{\gamma_3})/2 . \tag{27}$$

The above integrals can be evaluated in terms of four constants a_0, a_0', b_0, and b_0' which are, respectively, the values of $\tilde{p}(y)$ at $y = +1, -1$, the residue of $\tilde{p}(y)$ at $y = 0$ and the residue of $\tilde{\tilde{p}}$ at $z = 0$. The quantization condition then takes the form

$$(a_0' - a_0) + 2(b_1 + b_1') = 2in\hbar\alpha . \tag{28}$$

Each of the four constants, introduced above, is determined by substituting the relevant Taylor or Laurent series expansions, around the appropriate point, into QHJ and comparing the coefficients of different powers of y or z on both sides. Because the QHJ is quadratic, this process leads, in general, to two answers for each of the four constants. The correct choice for each of the four constants a_0, a_0', b_0, and b_0' depends on the values of the parameters A and B appearing in the superpotential. Interestingly, the correct choices for the above parameters lead to either unbroken or spontaneously broken phases of SUSY [7]. The values of QMF and its residues listed in Table 1 can be utilized to compute the energy eigenvalues. A similar treatment can be given for the partner potential H_+. The final energy eigenvalues are listed in Table 2. This table also gives the Gozzi index, which is the difference between the quantum action integral for the partner Hamiltonians H_+ and H_- . A close look at these energy eigenvalues reveals that SUSY is intact for a set of ranges of A and B, where as for another range of values of these parameters SUSY is broken. It should also be noted that for yet another set of values of parameters A and B, the roles of $H^{(+)}$ and $H^{(-)}$ are interchanged in the sense that the $E = 0$ is the ground state of $H^{(+)}$, instead of $H^{(-)}$. It can be seen from the Table 2 that for the broken phases, as expected, Gozzi index is 0 and for the unbroken case it is ± 1. It is amusing to note that, in the QHJ approach the boundary condition unambiguously leads to the two possible phases of SUSY. Unlike the SUSY-QM approach, there is no complication for determining the eigenvalues in the broken phase.

5   SUSY WKB Scheme and the QHJ Method

In this Sect., we elaborate on the usefulness of QHJ in explaining the exactness of SUSY inspired WKB schemes. In the context of SUSY-QM, it has been observed [2] that a semi-classical WKB-type approximation,

$$\frac{1}{\pi} \int_{x_1}^{x_2} \sqrt{E - \omega^2(x)}\,dx = n\hbar \quad , \tag{29}$$

gave the exact energy eigenvalues for many well-known potentials. For these potentials, SUSY was unbroken. However, for spontaneously broken SUSY, although the above formula gave better answer than the standard WKB approximations, it was not exact. Interestingly, it was noticed that

$$\frac{1}{\pi} \int_{x_1}^{x_2} \sqrt{E - \omega^2(x)}\,dx = (n + \frac{1}{2})\hbar \quad , \tag{30}$$

reproduces the energy spectra for known cases of broken SUSY [8]. Using the techniques of complex integration, these integrals can be written in the form

$$J_{SWKB} \equiv \frac{1}{2\pi} \oint_C \sqrt{E - \omega^2(x)}\,dx \quad . \tag{31}$$

One finds an intriguing similarity between J_{SWKB} and the quantum action variable given earlier. The above integral can be evaluated in a manner analogous to the previous calculation of the quantum action variable $J(E)$. For the unbroken SUSY potentials, where the SUSY WKB gave exact answers, the location of the poles and the corresponding residues of the above integral matches identically with those of $J(E)$ [9]. Since these poles completely determine the values of the integrals, J_{SWKB} and $J(E)$, turn out to be equal. The exactness of quantization condition $J(E) = n\hbar$, in the QHJ formalism, then explains the success of SUSY WKB approximation scheme. Amusingly, explicit calculation for the broken SUSY case, in the Pöschl-Teller example yields, $J_{SWKB} = J(E) + \frac{\hbar}{2}$. Using $J(E) = n\hbar$, one immediately concludes that $J_{SWKB} = (n + \frac{1}{2})\hbar$ should yield the exact eigenspectrum for this case.

In conclusion, we have used the QHJ formalism to analyze the generalized Pöschl-Teller potential and showed that, the proper implementation of the boundary condition can pick out both the phases of SUSY. An appropriate generalization of this approach to field theoretical models can be of immense help in shedding light on the puzzling question of supersymmetry breaking. It is worth pointing out that, unlike the SUSY-QM analyses, the finding of the eigenvalues for the broken phase posed no special difficulty for the QHJ formalism; this is because of the fact that the locations of the singularities of the QMF remains unchanged in both the phases. The only difference between the two phases is the presence of additional $\hbar$ dependent terms in the residue of the poles in the broken phase. The study of the SUSY WKB scheme revealed that, it is the matching of the location and the residues of the quantum action variable and the SUSY WKB integral, that lies at the origin of the exactness of these semi-classical methods. This approach can be of potential use for constructing better approximation schemes. Although we studied only SUSY related models here, the QHJ approach can be applied to other potentials and may be useful for obtaining new solvable potentials. The powerful technique of complex variables may be of help in this regard.

Acknowledgements: We acknowledge useful discussions with Profs. V. Srinivasan, U. Sukhatme, A. Khare and Pankaj Sharan. P.K.P would like to acknowledge the kind hospitality of the Department of Physics, University of Illinois at Chicago, during this workshop.

References

1. E. Witten, *Nucl. Phys.* **B 188**, 513 (1981); for a recent review of the application of SUSY to quantum mechanics and references see F. Cooper, A. Khare and U. Sukhatme, *Phys. Rep.* **251**, 267 (1995).

2. A. Comtet, A. Bandrauk and D. Campbell. *Phys. Lett.* **B 150**, 159 (1985); A. Khare, *Phys. Lett.* **B 161**, 131 (1985).

3. R.A. Leacock and M.J. Padgett, *Phys. Rev. Lett.* **50**, 3 (1983); R.A. Leacock and M.J. Padgett, *Phys. Rev.* **D28**, 2491 (1983).

4. E. Gozzi, *Phys. Rev.* **D33**, 3665 (1986).

5. R.S. Bhalla, A.K. Kapoor and P.K. Panigrahi, preprint, quant-ph/9512018, accepted for publication in Am. Jour. Phys.

6. C.X. Chuan, *Jour. of Phys.* **A23**, L659 (1992); Y. Ralchenko and V. Semenov, *Jour. of Phys.* **A24**, L1305 (1992); R. Dutt, A. Gangopadhyaya, A. Khare, A. Pagnamenta and U.P. Sukhatme, *Phys. Rev.* **A48**, 1845 (1993).

7. R.S. Bhalla, A.K. Kapoor and P.K. Panigrahi, *Int. Jour. Mod. Phys.* **A 12**, 1875 (1997).

8. A. Inomata and G. Junker, in *Proc. Adriatic Research Conf. on path-integration and its applications*, (ICTP, Italy, 1991); R. Dutt, A. Gangopadhyaya, A. Khare, A. Pagnamenta and U.P. Sukhatme, *Phys. Lett.* **A174**, 363 (1993).

9. R.S. Bhalla, A.K. Kapoor and P.K. Panigrahi, *Phys. Rev.* **A 54**,951 (1996).

Table 1 : Correct values and residues of the QMF for different sets of values of A and B for generalized Pöschl-Teller potential ($V_-(y)$). The four constants a_0, a'_0, b_1, and b'_1, are pure imaginary.

S.No.	Parameter ranges	$\mathrm{Im}\,a_0$	$\mathrm{Im}\,a'_0$	$\mathrm{Im}\,b_1$	$\mathrm{Im}\,b'_1$
1.	$A+B>0$ and $A-B<0$	$\sqrt{\lvert E-A^2 \rvert}$	$-\sqrt{\lvert E-A^2 \rvert}$	$\frac{1}{2}(A+B)$	$\frac{1}{2}(A-B)$
2.	$A+B>0$ and $A-B>0$	$\sqrt{\lvert E-A^2 \rvert}$	$-\sqrt{\lvert E-A^2 \rvert}$	$\frac{1}{2}(A+B)$	$\frac{1}{2}(B-A-\alpha\hbar)$
3.	$A+B<0$ and $A-B>0$	$\sqrt{\lvert E-A^2 \rvert}$	$-\sqrt{\lvert E-A^2 \rvert}$	$-\frac{1}{2}(A+B+\alpha\hbar)$	$\frac{1}{2}(B-A-\alpha\hbar)$
4.	$A-B<0$ and $A+B<0$	$\sqrt{\lvert E-A^2 \rvert}$	$-\sqrt{\lvert E-A^2 \rvert}$	$-\frac{1}{2}(A+B+\alpha\hbar)$	$\frac{1}{2}(A-B)$

Table 2: Ranges of A and B for different phases of SUSY for generalized Pöschl-Teller potential $(V_-(y))$.

Parameter ranges	$E_n^{(-)}$	$E_n^{(+)}$	Gozzi index δ	Remarks
Set I	$A^2 - (A - n\hbar\alpha)^2$	$A^2 - (A - (n+1)\hbar\alpha)^2$	1	SUSY is unbroken
Set II	$A^2 - (B - (n+\frac{1}{2})\hbar\alpha)^2$	$A^2 - (B - (n+\frac{1}{2})\hbar\alpha)^2$	0	SUSY is broken
Set III	$A^2 - (A + (n+1)\hbar\alpha)^2$	$A^2 - (A + n\hbar\alpha)^2$	-1	Roles of $H^{(-)}$ & $H^{(+)}$ interchanged
Set IV	$A^2 - (B + (n+\frac{1}{2})\hbar\alpha)^2$	$A^2 - (B + (n+\frac{1}{2})\hbar\alpha)^2$	0	SUSY is broken

For all the cases listed above, n takes integer values ≥ 0.

Quantum-Mechanical Supersymmetry in Traps

V. Alan Kostelecký and Neil Russell

Physics Department, Indiana University, Bloomington, IN 47405, U.S.A.

Abstract. We discuss the application of quantum-mechanical supersymmetry to particle traps. The supersymmetric-partner wave functions may be used to describe a valence fermion in a trap system with an isotropic harmonic-oscillator potential. Interactions with the core are incorporated analytically. The close similarity of this approach to the application of supersymmetry in atomic systems is made explicit by means of a radial mapping between the two systems.

Supersymmetry has been an active research area for well over two decades. Despite this, very few physical supersymmetries are experimentally known. One is the appearance of an effective radial potential in the context of atomic systems (Kostelecký and Nieto 1984, 1985a). We discuss a possible further application in the context of particle traps. More details are given in the two references listed under our names.

The isotropic harmonic oscillator has a radial equation admitting a supersymmetric partner. The physical implications of this mathematical fact may be investigated using trap systems for which an isotropic potential can be established. In the Ioffe-Pritchard trap (Gott et al. 1962, Pritchard 1983) and the time-averaged orbiting-potential (TOP) trap (Petrich et al. 1995), this condition may be satisfied. Both are neutral-particle traps that use the interaction of the magnetic dipole moment μ of the particle with a confining magnetic field. The traps select all dipoles aligned opposite to the direction of the magnetic field and draw them into the region of weakest field at the center. The Ioffe-Pritchard trap is purely magnetostatic, with the field provided by two coils and four linear conductors. The TOP trap comprises six coils, four of which have alternating currents, creating a high-frequency rotating magnetic field. An averaging procedure removes the time dependence, yielding an effective magnetostatic potential. With suitable choices of currents in the conductors of these systems (Bergeman et al. 1987, Kostelecký and Russell 1997), it is possible to ensure isotropy of the potential energy near the center of the trap,

$$U(r) = \mu B_0 \left(1 + r^2/r_0^2\right) \ . \tag{1}$$

Here, r_0 is a length characteristic of the trap system, and $B_0 \neq 0$ is the magnitude of the magnetic field at the center of the trap. The radial wave functions describing a single trapped dipole may be expressed in terms of the generalized Laguerre polynomials $L_N^{(\alpha)}(z)$. Ignoring a factor of r that removes the first-order derivative in the differential equation, the functions are

$$W_{N,L}(r) = C_{N,L}\,(r/r_0)^{L+1}\exp\left(-r^2/2r_0^2\right)L_{N/2-L/2}^{(L+1/2)}\left(r^2/r_0^2\right)\ . \qquad (2)$$

Here, $L = 0, 1, 2, \ldots$ is the angular momentum, and $N = L,\ L+2,\ L+4,\ \ldots$ is the principal quantum number. Normalization is ensured via the constant $C_{N,L}$. The full solutions of the harmonic oscillator $|N, L, M\rangle$ behave as

$$W_{N,L}(r)\,Y_{L,M}(\theta, \phi)\ , \qquad (3)$$

where the azimuthal quantum number M takes the usual values.

Details of supersymmetric quantum mechanics may be found elsewhere (Nicolai 1976, Witten 1981, Kostelecký 1994, Cooper et al. 1995). We use the term *bosonic sector* to refer to the given radial system with fixed angular momentum L and spectrum shifted to have zero lowest-state energy. The term *fermionic sector* is used for the partner system.

One of our objectives is to regard the radial-equation fermionic sector as providing an effective potential for an excited valence particle in a trap. It experiences not only the trapping potential, but also interactions with a core of other particles. Before motivating this application of supersymmetry further, let us consider the case of a valence particle with angular momentum $L = 0$ that is excluded from occupying levels below $N = 2$ by a filled core. This exclusion can only hold for fermions, to which we restrict ourselves here. If the interactions between the trapped dipoles are small compared with the natural spacing of the energies in the oscillator, the number of particles in the core may be found by counting the levels in the single-particle bosonic system lying below that of the valence fermion. There are four in this case. One is the ground state, $|N = 0, L = 0, M = 0\rangle$, and the other three can be labelled as $|N = 1, L = 1, M = 0, \pm 1\rangle$. Similarly, the core would have 20 fermions if the $L = 0$ valence fermion was restricted to $N \geq 4$.

The physical interpretation of the fermionic sector is motivated by several observations. It is well known that the fermionic sector is degenerate with the bosonic sector except for the lowest bosonic state, which has no corresponding fermionic state. One may visualize a situation in which *physically* such a corresponding state exists, but is inaccessible to the valence fermion because it is occupied by a core fermion. The absence of a zero-energy state in the fermionic sector of the *mathematical* formalism reflects this physical picture. The core need not be occupied by only one fermion, since there might be others of different angular momenta. Adopting this interpretation, the Pauli principle is seen to underlie radial supersymmetry, and it becomes natural to interpret the fermionic sector as describing a valence fermion in a multifermion system. For $L = 0$, the effective radial potential for the fermionic sector differs from the corresponding one for the bosonic sector by an expression that includes the term $\hbar^2/mr^2$ where m is the valence fermion mass.

This additional repulsion is thus consistent with the meaning of the Pauli principle.

So, the effect of the supersymmetry is to fill an inner core with fermions. This procedure must leave the angular momentum of the valence particle unchanged. We therefore construct the full three-dimensional wave functions from the product of the fermionic radial wave functions $W_{N_s-1,L+1}(r)$ and the same spherical harmonics as in the bosonic sector (3):

$$W_{N_s-1,L+1}(r)\, Y_{L,M}(\theta,\phi)\ , \tag{4}$$

where $N_s = L + 2,\ L + 4,\ L + 6,\ \dots$ is the principal quantum number. Even though the constant L in the radial function appears to have been shifted, the angular momentum of the system is defined by the spherical harmonics and is unchanged. These full wave functions (4) for the fermionic sector differ from the full wave functions (3) for the three-dimensional isotropic harmonic oscillator.

An alternative description for multifermion traps could account for the filled core by using the standard harmonic-oscillator solutions (3) but requiring the values of the principal quantum number for the valence fermion to exclude numbers corresponding to the filled core. However, the valence fermion would then be described by an incomplete set of states and the lowest valence state would have too many nodes. Both of these drawbacks are absent for the fermionic functions obtained via supersymmetry, which form a complete orthonormalizable set of states and for which the lowest state has no nodes. They therefore resemble solutions for other conventional bound systems in quantum mechanics.

The number of fermions in the core of a particular trap described by the functions (4) depends on the angular momentum of the valence fermion. For $L = 0$, the principal quantum number takes values $N_s = 2,\ 4,\ 6,\ \dots$ and by the reasoning considered above, this trap has four core fermions. The filling of another shell in the core can be accomplished by shifting the fermionic-sector spectrum to have zero lowest-energy state, thereby treating it as a new bosonic sector. A new fermionic sector is then obtained via the usual supersymmetry procedure. This describes a trap with 20 core fermions and a valence fermion with zero angular momentum. Further iterations of this procedure fills further shells, giving cores with $56,\ 120,\ 220,\ \dots$ fermions. For valence fermions with $L = 1$, the concept is the same, and there are $1, 10, 35, 84, \dots$ core particles. General formulae for these sequences can be obtained (Kostelecký and Russell 1997). They assume only one spin orientation since the dipoles in the Ioffe-Pritchard and TOP traps are oriented against the magnetic field. The formulae differ for other trap systems.

Although it accounts for the Pauli principle, radial supersymmetry ignores interactions between the valence fermion and the core. We discuss one method of incorporating interactions developed in analogy with analytical supersymmetry-based quantum-defect theory for atomic systems (Kostelecký

and Nieto 1985b). The modifications in the oscillator energy spectrum due
to interactions plausibly generate the form

$$E_{N^*} = \mu B_0 + \hbar\omega_0\left(N^* + 3/2\right) ,\tag{5}$$

for $\omega_0 = (2\mu B_0/mr_0^2)^{1/2}$. The modified eigenvalues are thus incorporated via
a shifted principal quantum number N^*. Denoting the shift by $\Delta = \Delta(N, L)$
and including also an integral shift $I = I(L)$, we define $N^* = N + I - \Delta$.
Equivalently, with $N_s = N + 2I$, we write $N^* = N_s - I - \Delta$. If in addition we
shift the angular momentum, $L^* = L + I - \Delta$, and add the effective potential

$$V_{\text{EFF}}(r) = \frac{\hbar^2}{2m}\frac{L^*(L^* + 1) - L(L + 1)}{r^2} + \hbar\omega_0(N - N^*)\tag{6}$$

to the differential operator in the Schrödinger equation for (2), analytical
radial wave functions that correspond to the modified eigenspectrum (5) are
obtained. They may be expressed in terms of the functional form (2) as
$W_{N^*,L^*}(r)$. This analytical defect theory extends the radial supersymmetry
quite naturally, and if the defects are switched off appropriately the exact
bosonic and noninteracting fermionic sectors are recovered.

The application of supersymmetry in traps closely follows the application
of supersymmetry in multi-electron atoms and in ions (Kostelecký and Nieto
1984, 1985a). The similarity in the two applications is more than a mathematical parallel and can be made explicit in the form of a mapping. The
existence of a natural correspondence between the radial three-dimensional
Coulomb problem and the radial harmonic oscillator in two or four dimensions
was first noted more than fifty years ago (Schrödinger 1941), and since then
has received much attention (Bergmann and Frishman 1965, Čížek and Pál-
dus 1977, Kostelecký, Nieto and Truax 1985, Kostelecký and Russell 1996).
We consider Coulomb dimensions $d > 1$, to avoid normalization issues associated with the one-dimensional case, and oscillator dimensions $D \geq 1$. For
these arbitrary-dimensional cases, the radial equations may still be separated
(Louck 1960), and we write the radial solutions as $W_{D,N,L}(r)$ for the oscillator and $w_{d,n,l}(r)$ for the Coulomb case. Details of these functions may be
found elsewhere (Kostelecký and Russell 1996). As an example, $W_{D=3,L,N}(r)$
is identical to (2). We adopt lower-case symbols for the Coulomb system and
upper-case symbols for the oscillator systems, with an exception made for
the oscillator radial variable r. In the Coulomb system, this convention gives
angular momentum l and principal quantum number n.

A natural mapping between these two radial systems exists subject to
certain conditions on the dimensions, the angular momenta, and the principal
quantum numbers. The relationship between the radial wave functions is
(Kostelecký, Nieto and Truax 1985)

$$W_{D,N,L}(r) = K_{d,n,\lambda}\, r^{-1/2}\, w_{d,n,l}\left((n + \gamma)r^2\right) ,\tag{7}$$

where $\gamma = (d - 3)/2$ is a dimension parameter for the Coulomb system that
vanishes in the three-dimensional case. The constant $K_{d,n,\lambda}$ is selected to

preserve the normalization. The restrictions on this correspondence may be expressed as

$$D = 2d - 2 - 2\lambda \ , \tag{8}$$

$$N = 2n - 2 + \lambda \ , \tag{9}$$

$$L = 2l + \lambda \ , \tag{10}$$

where it can be seen from (10) that λ, which gives an extra degree of freedom in the mapping, has to be integer valued. For $d = 3$, λ may equal zero or one, yielding oscillator dimensions of $D = 4, 2$. This is Schrödinger's original result. From (8), there is no such correspondence between the physically interesting cases of the $D = 3$ oscillator system and the $d = 3$ Coulomb system. The oscillator is limited to even dimensions D only.

To circumvent these dimensional restrictions on the correspondence between the exact systems, we broaden the class of systems considered to include ones with analytical modifications of the type introduced above for interactions between particles in a trap. So, whereas there is no natural mapping between the single-particle Coulomb and oscillator systems, a map may exist between a trap with *several* fermions and the exact Coulomb system.

We allow for an integral shift J in the oscillator dimension via the definition and requirement $D^* = D + J \geq 1$. If we also define a dimension parameter $\Gamma^* = (D^* - 3)/2$, which vanishes for $D^* = 3$, then with the choice of effective potential

$$V_{\text{EFF}}(r) = \frac{\hbar^2}{2m} \frac{(L^* + \Gamma^*)(L^* + \Gamma^* + 1) - (L + \Gamma)(L + \Gamma + 1)}{r^2} + \hbar\omega_0(N - N^* + \Gamma - \Gamma^*) \tag{11}$$

we obtain a differential equation with analytical solutions $W_{D^*, N^*, L^*}(r)$. The values of the parameters Δ, I, and J are restricted if normalizability and orthogonality is desired (Kostelecký and Russell 1996). With this broader class of oscillator radial systems, the case of a mapping from the $D^* = 3$ oscillator to the $d = 3$ Coulomb system becomes possible. It is obtained by setting

$$\Delta - I = \lambda - 1/2 \ , \tag{12}$$

and since λ can only take values zero or one, the mapping requires a nonzero defect Δ in the oscillator. Explicitly, the relationship and its constraints are

$$W_{3,N^*,L^*}(r) = K_{3,n,1/2} \, r^{-1/2} w_{3,n,l}\left(n \, r^2\right) \ , \tag{13}$$

$$N^* = 2n - 3/2 \ , \tag{14}$$

$$L^* = 2l + 1/2 \ . \tag{15}$$

Condition (14) ensures that the entire stack of states of the one system maps across to the entire stack for the other, with the lowest states in the stacks identified with each other, the second lowest states with each other, and so on.

This mapping is not the only one possible between the three-dimensional systems. An alternative method involves allowing for an analytical modification in the Coulomb system instead of in the oscillator and then following methods similar to those leading to (13). Such an analytical quantum defect may be introduced by shifting the principal quantum number to give eigenenergies according to the well-known Rydberg formula $E_0/(n^*)^2$ (Rydberg 1890), where E_0 is the ground state energy of the Coulomb system and $n^* = n - \delta$, with $\delta = \delta(n, l)$. Analytical solutions exist for this system, and their applications include the study of highly excited valence electrons (Bluhm and Kostelecký 1994a, 1994b, 1995, Bluhm, Kostelecký and Tudose 1995, 1996). Since the Rydberg formula models the spectra of multi-electron atoms, this option establishes a mapping between a single-particle trap and a multiparticle atom such as an alkali-metal atom. A more general third option involves analytical defects in *both* systems and provides a mapping from a multiparticle trap to a multi-electron atom. This most general form of the mapping incorporates in special cases the bosonic sectors, the fermionic sectors, and the quantum-defect sectors for various dimensions. For example, in a mapping between the three-dimensional cases, the generalization of (12) can take the form

$$\Delta - I = 2(\delta - i) + \lambda - 1/2 \ , \tag{16}$$

where i is the analogue for the Coulomb problem of I for the oscillator.

Unlike the defects Δ for traps, the defects δ have been measured for many atoms (Kuhn 1969). They depend on l but are asymptotically independent of n for large n. This feature is attractive as it allows approximate orthogonality of the solutions. It would be equally attractive if it could be established for the oscillator system. Indeed, if the oscillator defects Δ could be measured, it would be amusing to know if the constraint (16) on the radial mappings between the three-dimensional systems is consistent with the known atomic defects δ.

References

T. Bergeman, G. Erez and H.J. Metcalf (1987): Phys. Rev. A **35**, 1535 .

D. Bergmann and Y. Frishman (1965): J. Math. Phys. **6**, 1855 .

R. Bluhm and V.A. Kostelecký (1994a): Phys. Rev. A **49**, 4628 (quant-ph/9508020).

R. Bluhm and V.A. Kostelecký (1994b): ibid., **50**, R4445 (hep-ph/9410325).

R. Bluhm and V.A. Kostelecký (1995): ibid., **51**, 4767 (quant-ph/9506009).

R. Bluhm, V.A. Kostelecký and B. Tudose (1995): Phys. Rev. A **52**, 2234 (quant-ph/9509010).

R. Bluhm, V.A. Kostelecký and B. Tudose (1996): ibid., **53**, 937 (quant-ph/9510023).

E. Chacón, D. Levi and M. Moshinsky (1976): J. Math. Phys. **17**, 1919 .

J. Čížek and J. Paldus (1977): Internat. J. Quantum Chem. **12**, 875 .

F. Cooper, A. Khare and U. Sukhatme (1995): Phys. Rep. **251**, 267 .

Y.V. Gott, M.S. Ioffe and V.G. Tel'kovskii (1962): Nucl. Fusion, Suppl. Pt. 3, 1045.

D.E. Pritchard (1983): Phys. Rev. Lett. **51**, 1336 .

V.A. Kostelecký (1994): *Symmetries in Science VII*, eds. B. Gruber and T. Otsuka (Plenum, New York), 295 (quant-ph/9508015).

V.A. Kostelecký and M.M. Nieto (1984): Phys. Rev. Lett. **53**, 2285 .

V.A. Kostelecký and M.M. Nieto (1985a): Phys. Rev. A **32**, 1293 .

V.A. Kostelecký and M.M. Nieto (1985b): ibid., **32**, 3243 .

V.A. Kostelecký, M.M. Nieto and D.R.Truax (1985): Phys. Rev. D **32**, 2627 .

V.A. Kostelecký and N.E. Russell (1996): J. Math. Phys. **37**, 2166 (quant-ph/9602007).

V.A. Kostelecký and N.E. Russell (1997): Phys. Lett. A. , in press (quant-ph/9708020).

H.G. Kuhn (1969): *Atomic Spectra* (Academic, New York).

J.D. Louck (1960): J. Mol. Spectr. **4**, 298 .

H. Nicolai (1976): J. Phys. A **9**, 1497 .

W. Petrich, M.H. Anderson, J.R. Ensher and E.A. Cornell (1995): Phys. Rev. Lett. **74**, 3352 .

J.R. Rydberg (1890): Kongl. Sven. vetensk.-akad. hand. **23**, no. 11 .

E. Schrödinger (1941): Proc. Roy. Irish Acad. Sect. A **46**, 183 .

E. Witten (1981): Nucl. Phys. B**185**, 513 .

Cyclic Shape Invariant Potentials

U. Sukhatme[1], C. Rasinariu[1], A. Khare[2], A. Gangopadhyaya[3]

[1] Department of Physics (m/c 273), University of Illinois at Chicago, 845 W. Taylor Street, Chicago, Illinois 60607, USA
[2] Institute of Physics, Sachivalaya Marg, Bhubaneswar 751005, India
[3] Department of Physics, Loyola University Chicago, Chicago, Illinois 60626, USA

Abstract. Using the concept of shape invariance in supersymmetric quantum mechanics, we derive new types of potentials which have an infinite number of energy levels with a systematic pattern of periodic spacing. These cyclic superpotentials are obtained as solutions of a set of coupled nonlinear differential equations. They are found to have a harmonic oscillator behaviour with superposed oscillations consisting of several systematically varying frequencies. As an application, we identify the energy level spacings with the periodic points generated by the logistic map $z_{k+1} = r z_k (1 - z_k)$, and study the progression of cyclic potentials as the value of r in increased and one follows the bifurcation route to chaos.

1 Introduction

Cyclic shape invariant potentials [Gangopadhyaya and Sukhatme (1996), Sukhatme et al. (1997)] are new types of quantum mechanical potentials with an infinite number of energy levels having a systematic pattern of periodic spacing. The simplest type of periodic spacing is of course equal spacing. If the ground state is taken to be at $E = 0$, the energy level spacing is ω_0, and if one considers symmetric potentials over the real line $-\infty < x < \infty$, this spectrum comes from the simple harmonic oscillator potential $\frac{1}{4}\omega_0^2 x^2 - \frac{1}{2}\omega_0$. Now consider a more complicated pattern of period 2. The ground state is at $E = 0$, and the subsequent energy level spacings are $\omega_0, \omega_1, \omega_0, \omega_1, \ldots$. Similarly, the case of period 3 corresponds to energy level spacings $\omega_0, \omega_1, \omega_2, \omega_0, \omega_1, \omega_2, \ldots$. More generally, for period p, the level spacings are given by $\omega_0, \omega_1, \ldots, \omega_{p-1}, \omega_0, \omega_1, \ldots, \omega_{p-1}, \omega_0, \omega_1, \ldots$. The ground state is at zero energy; the next $(p-1)$ eigenvalues are $E_l = \sum_{k=0}^{l} \omega_k$, $l = 0, 1, \ldots, (p-2)$, and all other eigenvalues are obtained by adding arbitrary multiples of the quantity $\Omega_p \equiv \omega_0 + \omega_1 + \cdots + \omega_{p-1}$. The general formula which gives all the excited energy levels is

$$n\Omega_p + \sum_{k=0}^{l} \omega_k \quad ; \quad \{n = 0, 1, 2, \ldots, \infty \ ; \ l = 0, 1, \ldots, (p-1)\} \ . \tag{1}$$

Effectively, the spectrum is obtained by starting with p infinite sets of energy levels with equal spacing Ω_p and shifting these sets by successive arbitrary amounts $\omega_0, \omega_1, \ldots, \omega_{p-1}$.

In this paper we want to discuss how one generalizes the harmonic oscillator to produce new potentials with the more elaborate periodic level spacings of the type described above. The answer will be provided in the framework of supersymmetric quantum mechanics [Witten (1981), Cooper et al. (1995)] using the powerful concept of shape invariance [Gendenshtein (1983)].

Our plan is to first review some basic ideas of supersymmetric quantum mechanics and the concept of shape invariance for a change of parameters $a_1 = f(a_0)$. Next, we formulate the general set of coupled nonlinear differential equations governing cyclic shape invariant potentials of arbitrary period p. We study and discuss the solutions of the cyclic shape invariance conditions. The special cases of periods $p = 3, 4$ are treated in some detail. Many properties generalize to arbitrary values of p. Numerical solutions for $W^{(p)}(x)$ and the corresponding potentials $V_-^{(p)}(x)$ illustrate the general results.

The superpotentials $W^{(p)}(x)$ corresponding to cyclic shape invariant potentials turn out to have rather interesting behavior at large x. One finds a linear dependence $\Omega_p x/2p$ corresponding to a harmonic oscillator of angular frequency Ω_p/p with superposed sinusoidal oscillations which become more closely spaced with increasing x. We find that these oscillations at large x are given by the functional form $b_{pj} \sin(q_{pj} x^2)$. For any choice of the period p, there are several oscillation "frequencies" q_{pj}, with $j = 1, 2, \ldots, [p]/2$, where $[p]$ is the largest even integer less than p. These "frequencies" are given by $q_{pj} = \Omega_p \tan(\pi j/p)/2p$. The amplitudes of the oscillations b_{pj} depend on the energy spacings $\omega_0, \omega_1, \ldots$. The superpotentials also have unusual behavior at small x. For even values of p, one finds that $W^{(p)}(x)$ diverges like $1/x$, whereas for odd values of p, $W^{(p)}(0) = 0$.

As an application, we examine cyclic shape invariant potentials corresponding to energy level spacings ω_k which are the periodic points generated by the logistic map $z_{k+1} = r z_k (1 - z_k)$ [Peitgen (1992)]. As the parameter r increases from 1 to 3.569946, the period p also increases and the energy level spacings become increasingly more chaotic. This is reflected in the fact that the potentials $V_-(x)$ show increasingly more oscillation frequencies as one follows the period doubling route toward chaos.

2   General Formalism for Cyclic Shape Invariant Potentials

Recall that the superpotential $W(x, a_0)$ generates the supersymmetric partner potentials

$$V_\pm(x, a_0) = W^2(x, a_0) \pm W'(x, a_0) \ , \tag{2}$$

where a_0 is a set of parameters. These partner potentials are shape invariant if they both have the same x-dependence upto a change of parameters $a_1 = f(a_0)$ and an additive constant $R(a_0)$. The shape invariance condition is [Gendenshtein (1983)] $V_+(x, a_0) = V_-(x, a_1) + R(a_0)$, or equivalently

$$W^2(x, a_0) + W'(x, a_0) = W^2(x, a_1) - W'(x, a_1) + R(a_0) \ . \tag{3}$$

The property of shape invariance permits an immediate analytic determination of energy eigenvalues [Gendenshtein (1983)], eigenfunctions [Dutt et al. (1986)] and scattering matrices [Khare and Sukhatme (1988)]. For unbroken supersymmetry, the eigenstates of the potential $V_-(x)$ are:

$$E_0^{(-)} = 0 \ , \ E_n^{(-)} = \sum_{k=0}^{n-1} R(a_k) \ , \psi_0^{(-)} \propto e^{-\int_{x_0}^{x} W(y, a_0) dy} \ ,$$

$$\psi_n^{(-)}(x, a_0) = \left[-\frac{d}{dx} + W(x, a_0) \right] \psi_{n-1}^{(-)}(x, a_1) \ (n = 1, 2, 3, \ldots) \ . \tag{4}$$

The lowest lying eigenstate is at zero energy, characteristic of unbroken supersymmetry.

Solutions of the shape invariance condition (3] corresponding to a translational change of parameters $a_1 = f(a_0) = a_0 + \beta$, $a_n = f^n(a_0) = a_0 + n\beta$ are well-known [Cooper et al. (1995)]. Also, the scaling change of parameters $a_1 = f(a_0) = qa_0$, $a_n = f^n(a_0) = q^n a_0$ has been investigated and shown to produce new classes of shape invariant potentials [Khare and Sukhatme (1993), Barclay et al. (1993)], which included self-similar potentials [Shabat and Spiridonov (1992)] as a special case.

Very recently, another type of change of parameters defined by $f^2(a_0) = a_0$ was studied in detail [Gangopadhyaya and Sukhatme (1996)]. Here, the parameters repeat after a cycle of two iterations and the spectrum consists of two shifted sets of equally spaced eigenvalues. In this paper, we consider a much broader change of parameters defined by

$$f^p(a_0) = a_0, \tag{5}$$

where the parameters will repeat after a cycle of p iterations. For this case, one has $a_0 = a_p$, $f(a_0) = a_1 = a_{p+1}$, etc. The problem of finding explicit real transformation functions $f(a_0)$ which satisfy eq. (5) is discussed in ref. [Sukhatme et al. (1997)]. One solution is to use projective (Moebius) transformations with appropriate constraints on the parameters.

For a cyclic set of parameters, we now formulate the corresponding shape invariance conditions (3) for any period p. Defining $R(a_k) \equiv \omega_k$ for $k = 0, 1, \ldots, (p-1)$, and using $a_p = a_0$, one obtains a set of p coupled nonlinear differential equations

$$W^2(x, a_k) + W'(x, a_k) = W^2(x, a_{k+1}) - W'(x, a_{k+1}) + \omega_k \ , \ k = 0, \ldots, (p-1) \ . \tag{6}$$

There are p unknown superpotentials $W(x, a_k)$ which are related to each other by cyclic permutations of the indices $0, 1, 2, \ldots, (p-1)$. We are seeking superpotentials with odd parity, which will yield symmetric potentials $V_-(x)$. It is convenient to denote $W(x, a_k) \equiv W_k$. The set of shape invariance conditions now has a deceptively succinct form:

$$W_k^2 + W_k' = W_{k+1}^2 - W_{k+1}' + \omega_k \ , \ k = 0, 1, \ldots, (p-1) \ . \tag{7}$$

3   Solutions of the Cyclic Shape Invariance Conditions

We now turn our attention to studying the solutions of the cyclic set of equations (7). A very general result comes from adding all the equations. It gives $\sum_{k=0}^{p-1} W_k' = \Omega_p/2$, which on integration yields

$$\sum_{k=0}^{p-1} W_k(x) = \frac{1}{2}\Omega_p x \ . \tag{8}$$

This condition is valid for all x. There is no constant of integration in eq. (8) since we want the superpotentials W_k to be odd functions of x. In particular, at large x, constraints (7) and (8) give all W_k the linear asymptotic behaviour $\Omega_p x/2p$, corresponding to a simple harmonic oscillator potential of angular frequency Ω_p/p. Also, at small x, in order to satisfy eqs. (7), all superpotentials can at most have a $1/x$ singularity.

The general analytic solution of equations (7) is not easy to obtain. However, a numerical solution can be obtained via the Runge-Kutta method and we will also obtain several analytic results at small and large values of x. For ease of presentation, we systematically consider small values of the cycle period p before coming to general results.

$p = 1$:   Here $a_1 = f(a_0) = a_0$ and all energy levels are equally spaced. As expected, the shape invariance condition (7) gives $W_0^{(1)} = \omega_0 x/2$ which is a simple harmonic oscillator superpotential.

$p = 2$:   This is the case of two shifted sets of equally spaced eigenvalues considered in detail in ref. [Gangopadhyaya and Sukhatme (1996)]. The shape invariance conditions (7) are:

$$W_0^2 + W_0' = W_1^2 - W_1' + \omega_0 \quad , \quad W_1^2 + W_1' = W_0^2 - W_0' + \omega_1 \ . \tag{9}$$

The solution is obtained by straightforward manipulations. The superpotential $W_0^{(2)} \equiv W^{(2)}(x, a_0)$ and the corresponding potential $V_-^{(2)}(x)$ obtained using eq. (2) are given by

$$W_0^{(2)}(x) = \frac{\Omega_2 x}{4} + \frac{(\omega_0 - \omega_1)}{2\Omega_2 x} \ , \ V_-^{(2)}(x) = \frac{(\omega_0 - \omega_1)(3\omega_0 + \omega_1)}{4\Omega_2^2 x^2} - \frac{\omega_1}{2} + \frac{\Omega_2^2 x^2}{16} \ . \tag{10}$$

Note that for period $p = 2$, the superpotential $W_0^{(2)}$ diverges like α_0/x near the origin, which produces a singular inverse square potential in the transition region [Landau and Lifshitz (1977), Frank et al. (1971), Gangopadhyaya et al. (1994)]. A detailed discussion of the potential as well as the eigenfunctions $\psi_n^{(-)}$ obtained from eq. (4) is given in ref. [Gangopadhyaya and Sukhatme (1996)].

$p = 3$:   There are now three shape invariance conditions (7):

$$W_0^2 + W_0' = W_1^2 - W_1' + \omega_0, W_1^2 + W_1' = W_2^2 - W_2' + \omega_1, W_2^2 + W_2' = W_0^2 - W_0' + \omega_2. \tag{11}$$

These equations are not easy to solve analytically. In order to obtain a numerical solution, it is best to first solve eqs. (11) for the derivatives W_0', W_1', W_2'. One gets

$$W_0' = W_1^2 - W_2^2 + \frac{1}{2}(\omega_0 - \omega_1 + \omega_2) \ ,$$

$$W_1' = W_2^2 - W_0^2 + \frac{1}{2}(\omega_1 - \omega_2 + \omega_0) \ ,$$

$$W_2' = W_0^2 - W_1^2 + \frac{1}{2}(\omega_2 - \omega_0 + \omega_1) \ . \tag{12}$$

In this form the Runge-Kutta method is immediately applicable, and the superpotentials can be determined for any choice of the energy spacings $\omega_0, \omega_1, \omega_2$. An illustration of the results is shown in Fig. 1. [The superpotential $W_0^{(3)}(x)$ for period $p = 3$ and a choice of energy level spacings $\omega_0 = 1.0$, $\omega_1 = 0.5$, $\omega_2 = 2.3$. The superpotential was obtained by a numerical solution of eqs. (12) by the Runge-Kutta method. The oscillation frequency given by eq. (22) is $q_{31} = 1.097$] The overall average behavior of

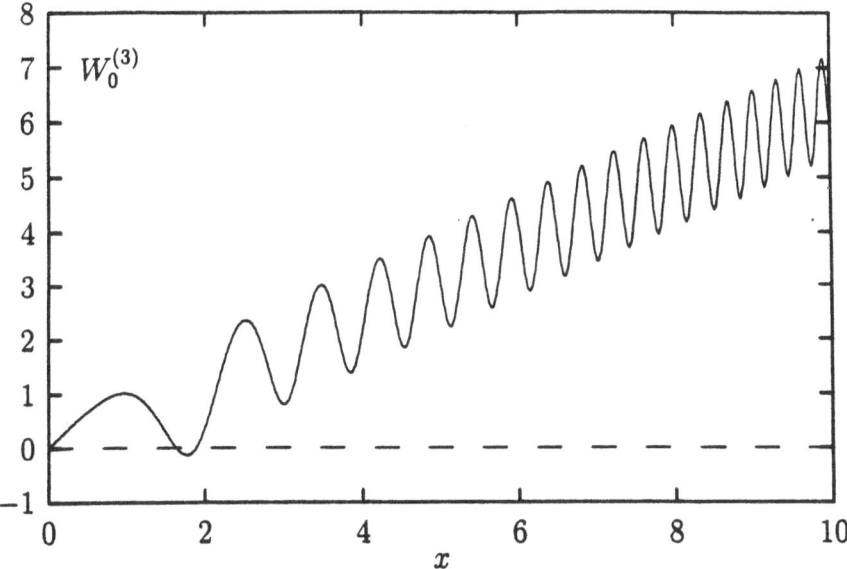

Fig. 1. Superpotential $W_0^{(3)}(x)$ for period $p = 3$.

$W_0^{(3)}$ at large x is indeed seen to be $\Omega_3 x/6$ as expected. On this linear dependence are clear oscillations which get more closely spaced as x increases. They correspond to a behavior $b \sin(qx^2)$ where the amplitude b and the frequency

q both depend on the choice of energy spacings $\omega_0, \omega_1, \omega_2$. In particular, when one chooses $\omega_0 = \omega_1 = \omega_2$, the numerical solution shows no oscillations. This is reasonable since for this case one expects a simple harmonic superpotential $\Omega_3 x/6$.

One can get a number of analytic results for eqs. (12) at small x as well as large x. At small x, if one feeds the series expansions

$$W_k = \alpha_k/x + \beta_k x + \gamma_k x^3 + \delta_k x^5 + \cdots \tag{13}$$

for $k = 0, 1, 2$ into eq. (12) and equates coefficients of x, one gets

$$W_0^{(3)} = \frac{x}{2}(\omega_0 - \omega_1 + \omega_2) + \frac{x^3}{3}\omega_1(\omega_0 - \omega_2) + \frac{x^5}{15}\omega_1(\omega_0\omega_1 + \omega_1\omega_2 - \omega_0^2 - \omega_2^2) + \cdots \tag{14}$$

with $W_1^{(3)}, W_2^{(3)}$ obtained by cyclic permutations of the indices $0, 1, 2$. $W_0^{(3)}$ vanishes at $x = 0$. Note that $\sum_{k=0}^{2} \gamma_k = \sum_{k=0}^{2} \delta_k = 0$ in agreement with eq. (8). At large x, one finds oscillations with one frequency $q_{31} = \Omega_3\sqrt{3}/6$. This is a special case of the result for arbitrary p which will be derived below.

$p = 4$:   It is worth discussing period 4 briefly, since several results are different from period 3. There are four cyclic shape invariance conditions:

$$W_0^2 + W_0' = W_1^2 - W_1' + \omega_0 , \quad W_1^2 + W_1' = W_2^2 - W_2' + \omega_1 ,$$
$$W_2^2 + W_2' = W_3^2 - W_3' + \omega_2 , \quad W_3^2 + W_3' = W_0^2 - W_0' + \omega_3 . \tag{15}$$

Unlike the period 3 case, one cannot directly solve for W_0', W_1', W_2', W_3', since their coefficients in eqs. (15) have a zero determinant. Indeed, this situation occurs for all even periods p. For the case of period 4, one can proceed by noting that eqs. (15) give the constraint $W_0^2 - W_1^2 + W_2^2 - W_3^2 = (\omega_0 - \omega_1 + \omega_2 - \omega_3)/2$, which on differentiation yields the result $W_0 W_0' - W_1 W_1' + W_2 W_2' - W_3 W_3' = 0$. This equation, along with any three of eqs. (15), permits one to solve for W_0', W_1', W_2', W_3', and proceed with a numerical determination of the superpotentials by the Runge-Kutta method.

Analytically, at small x, substitution of expansions of the form (13) into eqs. (15) and collecting powers of x gives

$$W_0^{(4)} = \frac{(\omega_0 - \omega_1 + \omega_2 - \omega_3)}{2\Omega_4 x} + \frac{x\Omega_4(\omega_0\omega_3 + \omega_0\omega_1 + \omega_2\omega_3 - \omega_2\omega_1)}{4(\omega_0 + \omega_2)(\omega_1 + \omega_3)} + \cdots \tag{16}$$

with the other superpotentials $W_1^{(4)}, W_2^{(4)}, W_3^{(4)}$ obtained by cyclic permutations of the indices $0, 1, 2, 3$. Note the singular $1/x$ dependence, similar to period 2. This is a feature of all even periods. At large x, period 4 superpotentials show oscillations with a single frequency $q_{41} = \Omega_4/8$. We will derive the general result for arbitrary p below.

Results for Arbitrary p: The experience gained from small values of the period p can be used to derive a number of general analytic statements valid for any arbitrary period p. At small x, the superpotential $W_0^{(p)}(x)$ has the form

$$W_0^{(p)} = \frac{x}{2} \left(\omega_0 - \omega_1 + \cdots + \omega_{p-1} \right)$$

$$- \frac{x^3}{3} \Big[\omega_1 \left(\omega_2 - \omega_3 + \cdots + \omega_{p-1} - \omega_0 \right) + \omega_3 \left(\omega_4 - \omega_5 + \cdots + \omega_1 - \omega_2 \right)$$

$$+ \cdots + \omega_{p-2} \left(\omega_{p-1} - \omega_0 + \omega_1 - \cdots - \omega_{p-3} \right) \Big] + O(x^5) \; ; \qquad (p \;\; \text{odd})$$

$$W_0^{(p)} = \frac{1}{2\Omega_p x} \left(\omega_0 - \omega_1 + \omega_2 - \cdots - \omega_{p-1} \right) + \frac{x\Omega_p}{4} \Big[1 - 2 \times$$

$$\frac{\omega_1 \left(\omega_2 + \omega_4 + \cdots + \omega_{p-2} \right) + \omega_3 \left(\omega_4 + \cdots + \omega_{p-2} \right) + \cdots + \omega_{p-3}\omega_{p-2}}{\left(\omega_0 + \omega_2 + \cdots + \omega_{p-2} \right) \left(\omega_1 + \omega_3 + \cdots + \omega_{p-1} \right)} \Big]$$

$$+ O(x^3) \; ; \qquad (p \;\; \text{even}) \tag{17}$$

The behaviour of the other superpotentials $W_k^{(p)}(x)$ can be seen by cyclically permuting the indices $0, 1, 2, \ldots, (p-1)$. Note the $1/x$ singularity at the origin for even periods. As for the case of periods $p = 2, 4$, it can be readily verified that the coefficient of the $1/x^2$ singularity in the potential $V_k^{(p)}$ lies in the transition region $-1/4$ to $3/4$. In this region, both solutions of the Schrödinger equation are square integrable at the origin [Landau and Lifshitz (1977), Frank et al. (1971), Gangopadhyaya et al. (1994)]. The infinite discontinuity in $W^{(p)}(x)$ at the origin for even p produces a $\delta(x)$ term in the potential $V_-^{(p)}(x)$ as described in ref. [Gangopadhyaya and Sukhatme (1996)].

At large x, one removes the asymptotic linear behaviour by defining

$$Q_k(x) \equiv W_k(x) - \Omega_p x/2p \; , \quad k = 0, 1, \ldots, p-1 \; . \tag{18}$$

Since numerical solutions suggest sinusoidal oscillations in x^2, we make a change of variables to $u \equiv x^2$. In terms of u, the equations satisfied by Q_k are obtained by substituting eq. (18) into (7). Retaining only the leading powers of u yields

$$Q_0' + Q_1' = \frac{\Omega_p}{2p} \left(-Q_0 + Q_1 \right) \tag{19}$$

and cyclic permutations. This set of equations can be written in a more compact form by putting all the superpotentials into a column vector Q and making use of the matrix C_p whose only non-zero elements are $(C_p)_{12} = (C_p)_{23} = \cdots = (C_p)_{p-1,p} = (C_p)_{p1} = 1$. The equations now read

$$(C_p + 1)Q' - \frac{\Omega_p}{2p}(C_p - 1)Q = 0 \; . \tag{20}$$

The most general solution is a linear combination of eigenstates of the form $Q \propto \exp(\lambda u)$. The eigenvalues λ are determined from

$$\det[\lambda(C_p + 1) - (\Omega_p/2p)(C_p - 1)] = 0 \; . \tag{21}$$

Defining $\tilde{\lambda} = 2p\lambda/\Omega_p$, and expanding the determinant, eq. (21) can be re-written as

$$(\tilde{\lambda} + 1)^p - (-1)^p(\tilde{\lambda} - 1)^p = 0 \ ,$$

which is a polynomial equation of degree p for p odd and of degree $p - 1$ for p even. This situation occurs because eq. (21) is more general than the standard eigenvalue equation. The solution is

$$\frac{1 - \tilde{\lambda}}{1 + \tilde{\lambda}} = \exp\left(\frac{2\pi j}{p} i\right) \ , \quad j = 0, \pm 1, \pm 2, \cdots, \pm[p]/2 \ .$$

$[p]$ is the largest even integer less than p. This gives rise to distinct oscillation frequencies

$$q_{pj} = (\Omega_p/2p) \tan(\pi j/p) \ , \quad j = 1, 2, \ldots, [p]/2 \ . \tag{22}$$

At large x, the superpotential $W_0^{(p)}(x)$ has the form

$$W_k^{(p)}(x) = \frac{\Omega_p x}{2p} + \sum_j b_{pj} \sin(q_{pj} x^2 + \phi_j^{(k)}) + O(1/x) \tag{23}$$

where the phases are

$$\phi_j^{(k)} = \phi_j + 2\pi k/p \ . \tag{24}$$

Note that the requirement of having odd superpotentials can be satisfied by inserting an asymptotically constant odd function such as $\tanh(\alpha x)$ or $\tan^{-1}(\alpha x)$ with the coefficient b_{pj}. This does not affect the above discussion of oscillation frequencies. Also, if one uses eq. (4) to get the asymptotic behaviour of wave functions, the linear term in $W^{(p)}$ gives the familiar Gaussian dependence of a harmonic oscillator and the sinusoidal correction terms in $W^{(p)}$ give Fresnel integrals producing oscillations on the Gaussian.

As an example, for period 5 , taking $\Omega_5 = 6.5$, we predict two oscillation frequencies $q_{51} = 0.472$ and $q_{52} = 2.001$ from eq. (22), and we have checked that these values are in agreement with the numerical solutions shown in Fig. 2. [Fig. 2 is a plot of the superpotential $W_0^{(5)} - \frac{\Omega_5 x}{10}$ for period $p = 5$ and energy level spacings $\omega_0 = 0.8$, $\omega_1 = 1.5$, $\omega_2 = 1.0$, $\omega_3 = 3.0$, $\omega_4 = 0.2$. The asymptotic behavior $\frac{\Omega_5 x}{10}$ has been subtracted in order to clearly reveal the two oscillation frequencies $q_{51} = 0.4722$ and $q_{52} = 2.001$, predicted by eq. (22). The maxima are spaced in agreement with the smaller frequency q_{51} whereas their "envelope" corresponds to the larger frequency q_{52}.]

4   Approach to Quantum Chaos Using the Logistic Map

We have now established the method of generating cyclic shape invariant potentials. This method can be applied using any transformation $f(y)$ which generates parameters which repeat with period p. For the special projective

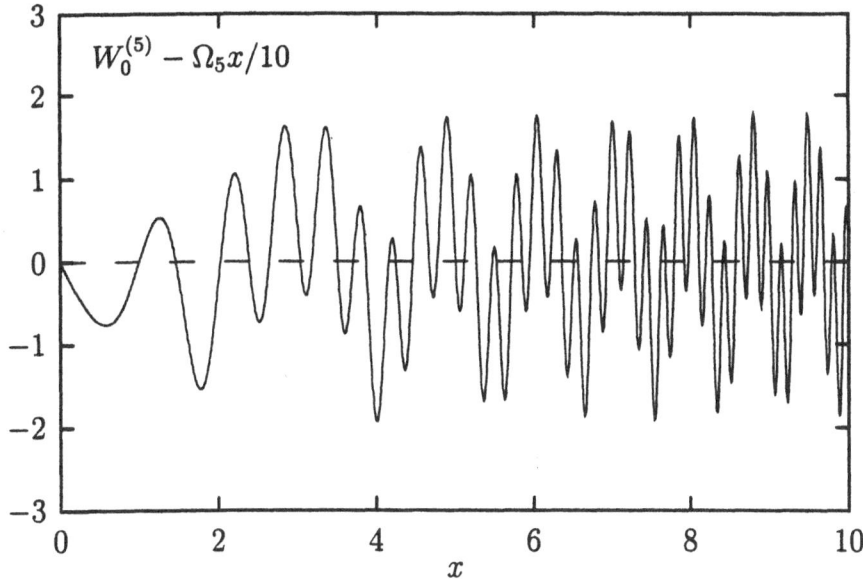

Fig. 2. Superpotential $W_0^{(5)} - \frac{\Omega_5 x}{10}$ for period $p = 5$.

transformations described previously, the initial parameter a_0 can be chosen arbitrarily. There are also other possible transformations in which cyclic parameters are obtained if one starts from specific choices for a_0. One such interesting possibility, which we discuss in detail here, is to use a set of cyclic parameters given by the periodic points of the logistic map

$$z_{k+1} = r z_k (1 - z_k) \ . \tag{25}$$

This famous transformation has been extensively studied [Peitgen (1992)]. For $1 < r < 3$, one has a single non-zero fixed point at $z = (r-1)/r$; for $3 < r < 3.449499$, there is a stable 2-cycle with periodic points at $z = [(r+1) \pm \sqrt{(r-3)(r+1)}]/2r$; for $3.449499 < r < 3.544090$, there is a stable 4-cycle, etc. As r is increased, the period keeps doubling, giving finally the period 2^∞ at 3.569946. For $r > 3.569946$, one obtains all odd periods and their harmonics, and at $r > 4$, there is complete chaos.

In our application, we will identify the energy spacings ω_k with the periodic points of the logistic map. For example, if we choose $r = 3.20$, the periodic points are 0.799 and 0.513, and we determine the potentials which have energy spacings $\omega_0 = 0.799$ and $\omega_1 = 0.513$. The corresponding superpotential is shown in Fig. 3(a). Other examples with larger values of r and correspondingly larger periods $p = 4, 8$ are also shown in Figs. 3(b) and 3(c). [Figures 3 show how the superpotential $W_0^{(p)}(x) - \frac{\Omega_p x}{2p}$ gets progressively more complicated and aquires more oscillation frequencies as the

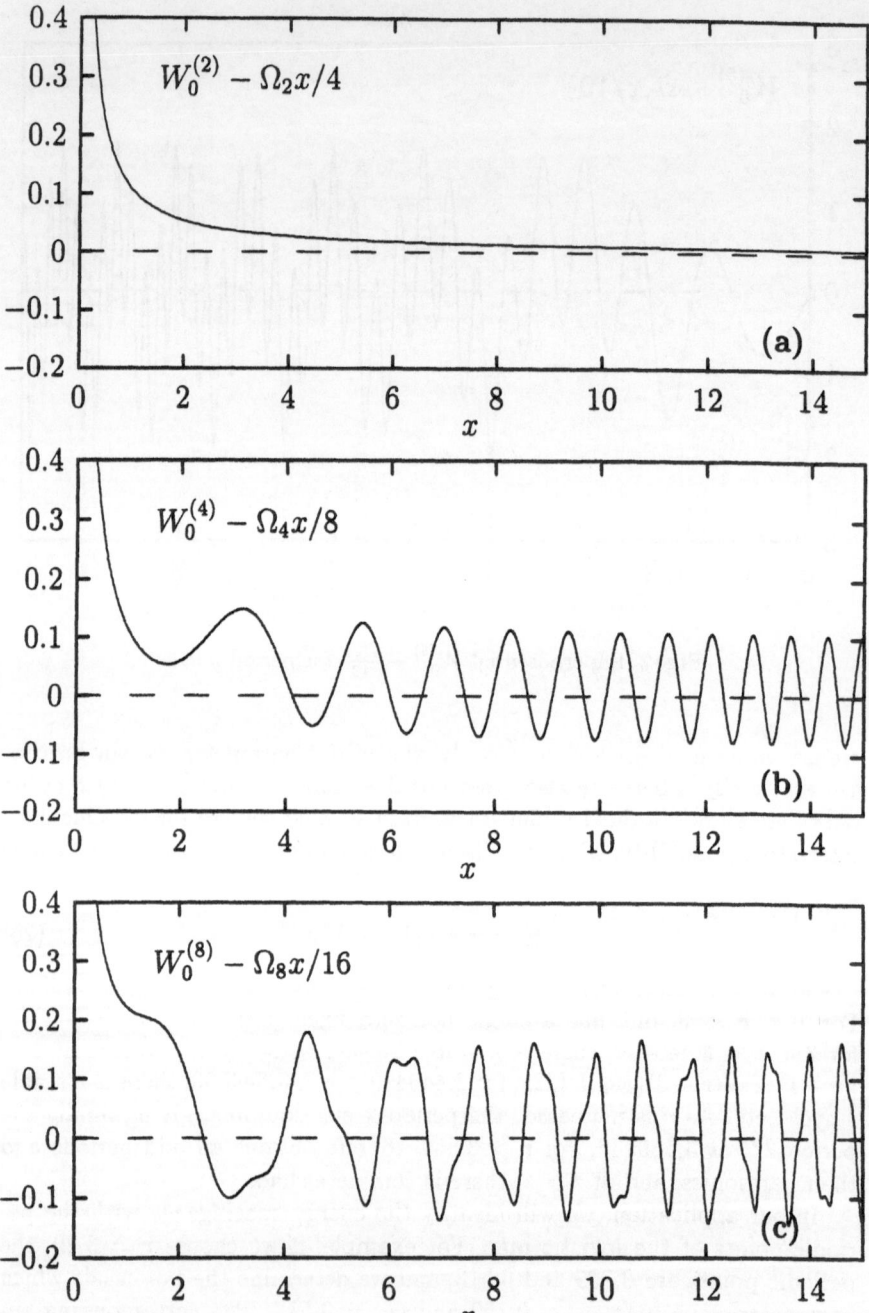

Fig. 3. Variation of superpotentials with increasing r in the logistic map.

period p is successively doubled along the bifurcation route to chaos via the logistic map (25). The parameter values chosen for the figures are: (a) $r = 3.200$ which gives period $p = 2$ and a stable 2-cycle with energy spacings $\omega_0 = 0.799$, $\omega_1 = 0.513$; (b) $r = 3.500$ which gives period $p = 4$ and a stable 4-cycle with $\omega_0 = 0.827$, $\omega_1 = 0.501$, $\omega_2 = 0.875$, $\omega_3 = 0.383$. The oscillation frequency from eq. (22) is $q_{41} = 0.323$; (c) $r = 3.561$ which gives period $p = 8$ and a stable 8-cycle with $\omega_0 = 0.890$, $\omega_1 = 0.348$, $\omega_2 = 0.808$, $\omega_3 = 0.552$, $\omega_4 = 0.881$, $\omega_5 = 0.374$, $\omega_6 = 0.834$, $\omega_7 = 0.494$. The three oscillation frequencies given by eq. (22) are $q_{81} = 0.1341$, $q_{82} = 0.3238$, $q_{83} = 0.7818$.]

Clearly, as one advances along the period doubling route to chaos, more and more oscillation frequencies come into play - which is certainly a plausible result. Specifically, for period 2^n, there are $2^{n-1} - 1$ frequencies given by eq. (22). It is quite possible that these increasingly chaotically spaced energy levels correspond to potentials with fractal dimensions greater than one [Wu and Sprung (1993)], but the general connection to quantum chaos needs further investigation.

U.S. acknowledges the hospitality of the Institute of Physics, Bhubaneswar where part of this work was done. This research was supported in part by the U.S. Department of Energy.

References

Barclay, D., Dutt, R., Gangopadhyaya, A., Khare, A., Pagnamenta, A. and Sukhatme, U., Phys. Rev. **A48** (1993) 2786.

Cooper, F., Khare, A. and Sukhatme, U., Phys. Rep. **251** (1995) 267 give a recent review of supersymmetric quantum mechanics and additional references.

Dutt, R., Khare, A. and Sukhatme, U., Phys. Lett. **181B** (1986) 295; Dabrowska, J., Khare, A. and Sukhatme, U., Jour. Phys. **A21** (1988) L195.

Frank, W. Land, D. and Spector, R., Rev. Mod. Phys. **43** (1971) 36.

Gangopadhyaya, A., Panigrahi, P. and Sukhatme, U., J. Phys. **A27** (1994) 4295.

Gangopadhyaya, A. and Sukhatme, U., Phys. Lett. **A224** (1996) 5.

Gendenshtein, L., JETP Letters **38** (1983) 356.

Khare, A. and Sukhatme, U., Jour. Phys. **A21** (1988) L501.

Khare, A. and Sukhatme, U., Jour. Phys. **A26** (1993) L901.

Landau, L. and Lifshitz, E., *Quantum Mechanics*, Pergamon Press (1977).

Peitgen, H., Jürgens, H. and Saupe, D., *Fractals*, Springer-Verlag (1992) gives a simple description of the logistic map.

Shabat, A., Inverse Prob. **8** (1992) 303; Spiridonov, V., Phys. Rev. Lett. **69** (1992) 398.

Sukhatme, U., Rasinariu, C. and Khare, A., UIC preprint UICHEP-TH/97-6; to appear in Phys. Lett. A (1997).

Witten, E., Nucl. Phys. **B188** (1981) 513.

Wu, H. and Sprung, D., Phys. Rev. **E48** (1993) 2595.